Molecular Pathogenesis of Periodontal Disease

Molecular Pathogenesis of Periodontal Disease

Edited by

Robert Genco
Department of Oral Biology, State University of New York at Buffalo
Buffalo, New York

Shigeyuki Hamada
Department of Oral Microbiology, Faculty of Dentistry, Osaka University
Osaka, Japan

Thomas Lehner
Department of Immunology, Guy's Hospital
London, United Kingdom

Jerry McGhee
Department of Microbiology, University of Alabama
Birmingham, Alabama

Stephan Mergenhagen
Laboratory of Immunology, National Institute of Dental Research
Bethesda, Maryland

ASM Press • *Washington DC*

Copyright © 1994 American Society for Microbiology
1325 Massachusetts Ave., N.W.
Washington, DC 20005

Library of Congress Cataloging-in-Publication Data

Molecular pathogenesis of periodontal disease / edited by Robert Genco . . . [et al.].
   p.   cm.
   Includes index.
   ISBN 1-55581-075-6
   1. Periodontal disease—Molecular aspects—Congresses.   2. Periodontal disease—Pathogenesis—Congresses.   3.Periodontal disease—Microbiology—Congresses.   I. Genco, Robert J.
   [DNLM: 1.   Periodontal   Diseases—immunology—congresses.   2.   Periodontal   Diseases—microbiology—congresses.   3.   Porphyromonas   gingivalis—pathogenicity—congresses.   4.   Bacteriodaceae   Infections—immunology—congresses.   5.   Wound   Healing—physiology—congressees.   WU 240 M718 1994]
RK361.A2M65   1994
617.6'3207—dc20
DNLM/DLC
for Library of Congress   94-4855
                                                                                                                                        CIP

Cover photo: Actinobacillus actinomycetemcomitans SUNY 465, a bacterium which was cultured aerobically on agar. The fibrillar membranous extensions emanating from the cell surface are membrane vesicles. Extracellular vesicles which have budded off from the membrane vesicles are also evident. A portion of the micrograph was published previously (**Meyer and Fives-Taylor,** *Infect. Immun.* **62:**928–935). The micrograph was produced by Gregory M. Hendricks, Electon Microscopy Research Facility Coordinator, University of Vermont College of Medicine.

Contents

SECTION I. MICROBIAL VIRULENCE FACTORS

SECTION II. HOST FACTORS: CYTOKINES AND OTHER EFFECTOR MOLECULES

SECTION III. HOST FACTORS: NEUTROPHILS, MAPPING T- AND B-CELL EPITOPES, AND GENETIC FACTORS

SECTION IV. PERIODONTAL WOUND HEALING

Contributors

Kenneth W. Beagley · Department of Microbiology, University of Alabama at Birmingham, UAB Station, Birmingham, Alabama 35294

Gurrinder Bedi · Maginin Pharmaceuticals, Inc., 5110 Campus Drive, Plymouth Meeting, Pennsylvania 19462

Henning Birkedal-Hansen · National Institute of Dental Research, National Institutes of Health, Building 30, Room 132, Bethesda, Maryland 20892

Amy L. Bodeau · Section of Oral Biology and Dental Research Institute, University of California School of Dentistry, Los Angeles, California 90024-1668

Marc Charette · Creative BioMolecules, 45 South Street, Hopkinton, Massachusetts 01748

A. Childerstone · Department of Immunology, United Medical and Dental Schools, Guy's and St. Thomas' Hospitals, Guy's Tower, Floor 28, London Bridge, London SE1 9RT, United Kingdom

Moon-Il Cho · Department of Oral Biology, School of Dental Medicine, State University of New York at Buffalo, Foster Hall, 3435 Main Street, Buffalo, New York 14214-3092

Sebastian G. Ciancio · Department of Periodontology, State University of New York at Buffalo, 250 Squire Hall, Buffalo, New York 14214

Robert E. Cohen · Department of Oral Biology and Periodontology, School of Dental Medicine, State University of New York at Buffalo, Foster Hall, 3435 Main Street, Buffalo, New York 14214-3092

John G. Collins · Dental Research Center, School of Dentistry, University of North Carolina at Chapel Hill, Chapel Hill, North Carolina 27599-7455

Ernesto De Nardin · Department of Oral Biology, School of Dental Medicine, State University of New York at Buffalo, Foster Hall, 3435 Main Street, Buffalo, New York 14214-3092

Robert Dunford · Department of Oral Biology, School of Dental Medicine, State University of New York at Buffalo, Foster Hall, 3435 Main Street, Buffalo, New York 14214-3092

David W. Dyer · Department of Microbiology and Immunology, University of Oklahoma Health Sciences Center, Oklahoma City, Oklahoma 73190

J. W. Eastcott · Department of Immunology, Forsyth Dental Center, 140 The Fenway, Boston, Massachusetts 02115

Jeffrey Ebersole · Department of Periodontics, University of Texas Health Science Center at San Antonio, San Antonio, Texas 78284

Richard Ellen · Faculty of Dentistry, University of Toronto, 124 Edward Street, Toronto, Ontario M5G 1G6, Canada

David Engel · Department of Periodontics, SM-44, University of Washington, Seattle, Washington 98195

Pamela R. Erickson · School of Dentistry, University of Minnesota, 515 Delaware Street, S.E., Minneapolis, Minnesota 55455

M. J. Evans · Department of Oral Biology, State University of New York at Buffalo, Foster Hall, 3435 Main Street, Buffalo, New York 14214-3092, and Roswell Park Cancer Institute, Buffalo, New York 14263

R. T. Evans · Department of Oral Biology, State University of New York at Buffalo, Foster Hall, 3435 Main Street, Buffalo, New York 14214-3092

Paula M. Fives-Taylor · Department of Microbiology and Molecular Genetics, University of Vermont, Stafford Hall, Burlington, Vermont 05405

Kohtaro Fujihashi · Department of Oral Biology, University of Alabama at Birmingham, Birmingham, Alabama 35294

Taku Fujiwara · Department of Oral Microbiology, Osaka University Faculty of Dentistry, Yamadaoka, Suita-Osaka 565, Japan

J. Gegner · Department of Immunology, The Scripps Research Institute, La Jolla, California 92037

Robert J. Genco · Department of Oral Biology, School of Dental Medicine, State University of New York at Buffalo, Foster Hall, 3435 Main Street, Buffalo, New York 14214-3092

William V. Giannobile · School of Dental Medicine, Harvard University, Boston, Massachusetts 02115

M. Jane Gillespie · Departments of Oral Biology and Periodontology, School of Dental Medicine, State University of New York at Buffalo, Foster Hall, 3435 Main Street, Buffalo, New York 14214-3092

Jan Gottlow · Guidor Research Center, Gruvgatan 8, S-421 30 V. Frolunda, Sweden

Sara Grossi · Department of Oral Biology, School of Dental Medicine, State University of New York at Buffalo, Foster Hall, 3435 Main Street, Buffalo, New York 14214-3092

Shigeyuki Hamada · Department of Oral Microbiology, Osaka University Faculty of Dentistry, Yamadaoka, Suita-Osaka 565, Japan

Majed M. Hamawy · Laboratory of Immunology, National Institute of Dental Research, National Institutes of Health, Bethesda, Maryland 20892

J. Han · Department of Immunology, The Scripps Research Institute, La Jolla, California 92037

Geoffrey Haradon · Dental Research Center, School of Dentistry, University of North Carolina at Chapel Hill, Chapel Hill, North Carolina 27599-7455

Koichi Harano · Department of Microbiology, Tokyo Dental College, 1-2-2 Masago, Chiba 261, Japan

Violet I. Haraszthy · Department of Oral Biology, School of Dental Medicine, State University of New York at Buffalo, Foster Hall, 3435 Main Street, Buffalo, New York 14214-3092

Mark C. Herzberg · School of Dentistry, University of Minnesota, 17-164 Moos Tower, 515 Delaware Street, S.E., Minneapolis, Minnesota 55455

Keith L. Hines · Laboratory of Immunology, National Institute of Dental Research, National Institutes of Health, Bethesda, Maryland 20892-0001

Kaname Hirai · Department of Microbiology, Tokyo Dental College, 1-2-2 Masago, Chiba 261, Japan

Takachika Hiroi · Mucosal Immunization Research Group, University of Alabama at Birmingham, Birmingham, Alabama 35294

Stanley Holt · Department of Periodontology, University of Texas Health Science Center at San Antonio, San Antonio, Texas 78284

Tomozumi Imamichi · Laboratory of Immunology, National Institute of Dental Research, National Institutes of Health, Bethesda, Maryland 20892-0001

John R. Kalmar · Division of Oral and Maxillofacial Pathology, Eastman Dental Center, 625 Elmwood Avenue, Rochester, New York 14620

Tetsuo Kato · Department of Microbiology, Tokyo Dental College, 1-2-2 Masago, Chiba 261, Japan

Kyoko Katsuragi · Department of Periodontology and Endodontology, Okayama University Dental School, 2-5-1 Shikata-cho, Okayama 700, Japan

C. G. Kelly · Department of Immunology, United Medical and Dental Schools, Guy's and St. Thomas' Hospitals, Guy's Tower, Floor 28, London Bridge, London SE1 9RT, United Kingdom

H. Kendal · Department of Immunology, United Medical and Dental Schools, Guy's and St. Thomas' Hospitals, Guy's Tower, Floor 28, London Bridge, London SE1 9RT, United Kingdom

Irene R. Kieba · Department of Pathology, School of Dental Medicine, University of Pennsylvania, 4010 Locust Street, Philadelphia, Pennsylvania 19104-6002

Caroline King · Department of Pathology, School of Dental Medicine, University of Pennsylvania, 4010 Locust Street, Philadelphia, Pennsylvania 19104-6002

T. Kirkland · Department of Pathology and Medicine, Veterans Administration Hospital Medical Center, and University of California, San Diego, San Diego, California 92161

Hiroshi Kiyono · Department of Oral Biology, University of Alabama at Birmingham, UAB Station, Birmingham, Alabama 35294

B. Klausen · University of Copenhagen, Copenhagen, Denmark

V. Kravchenko · Department of Immunology, The Scripps Research Institute, La Jolla, California 92037

Howard K. Kuramitsu · Department of Oral Biology, School of Dental Medicine, State University of New York at Buffalo, Foster Hall, 3435 Main Street, Buffalo, New York 14214-3092

Hidemi Kurihara · Department of Periodontology and Endodontology, Okayama University Dental School, 2-5-1 Shikata-cho, Okayama 700, Japan

Edward T. Lally · Department of Pathology, School of Dental Medicine, University of Pennsylvania, 4010 Locust Street, Philadelphia, Pennsylvania 19104-6002

J.-D. Lee · Department of Immunology, The Scripps Research Institute, La Jolla, California 92037

Jin-Yong Lee · Department of Oral Biology, School of Dental Medicine, State University of New York at Buffalo, Foster Hall, 3435 Main Street, Buffalo, New York 14214-3092

T. Lehner · Department of Immunology, United Medical and Dental Schools, Guy's and St. Thomas' Hospitals, Guy's Tower, Floor 28, London Bridge, London SE1 9RT, United Kingdom

Robert I. Lehrer · Department of Medicine, University of California School of Medicine, Los Angeles, California 90024, and Wadsworth Division, Department of Veterans Affairs, Los Angeles, California 90073

D. Leturcq · The Robert Wood Johnson Pharmaceutical Research Institute, La Jolla, California 92121

Yi-Ping Li · Department of Cytokine Biology, Forsyth Dental Center, 140 The Fenway, Boston, Massachusetts 02115

W.-L. Lin · Department of Oral Biology and Periodontal Disease Research Center, State University of New York at Buffalo, Buffalo, New York 14214

Peixin Liu · School of Dentistry, University of Minnesota, 515 Delaware Street, S.E., Minneapolis, Minnesota 55455

Harald Löe · National Institute of Dental Research, Building 31, Room 2C-39, Bethesda, Maryland 20892

Samuel E. Lynch · Institute of Molecular Biology, Inc., One Innovation Drive, Worcester, Massachusetts 01605-4308

J. K.-C. Ma · Department of Immunology, United Medical and Dental Schools, Guy's and St. Thomas' Hospitals, Guy's Tower, Floor 28, London Bridge, London SE1 9RT, United Kingdom

Gordon D. MacFarlane · School of Dentistry, University of Minnesota, 515 Delaware Street, S.E., Minneapolis, Minnesota 55455

J. C. Mathison · Department of Immunology, The Scripps Research Institute, La Jolla, California 92037

N. Matsuda · Sunstar, Inc., 3-1, Asahi-Machi, Takatsuki, Osaka 569, Japan

Nancy L. McCartney-Francis · Laboratory of Immunology, National Institute of Dental Research, National Institutes of Health, Bethesda, Maryland 20892-0001

Jerry R. McGhee · Department of Microbiology, University of Alabama, UAB Station, Birmingham, Alabama 35294

Junichi Mega · Immunobiology Vaccine Center, University of Alabama at Birmingham, Birmingham, Alabama 35294

Christian Ménard · Groupe de Recherche en Ecologie Buccale, Faculté de Médecine Dentaire, Université Laval, Québec, Québec G1K 7P4, Canada

Stephan E. Mergenhagen · Laboratory of Immunology, National Institute of Dental Research, National Institutes of Health, Building 30, Room 332, Bethesda, Maryland 20892

Diane H. Meyer · Department of Microbiology and Molecular Genetics, University of Vermont, Stafford Hall, Burlington, Vermont 05405

Joji Mihara · Department of Oral Microbiology, Osaka University Faculty of Dentistry, Yamadaoka, Suita-Osaka 565, Japan

Keith P. Mintz · Department of Microbiology and Molecular Genetics, University of Vermont, Stafford Hall, Burlington, Vermont 05405

Kenneth T. Miyasaki · Section of Oral Biology, 63-050 CHS, and Dental Research Institute, University of California School of Dentistry, Los Angeles, California 90024-1668

A. Moriarty · The Robert Wood Johnson Pharmaceutical Research Institute, La Jolla, California 92121

Seiji Morishima · Department of Oral Microbiology, Osaka University, Osaka, Japan

Christian Mouton · Groupe de Recherche en Ecologie Buccale, Faculté de Médecine Dentaire, Université Laval, Québec, Québec G1K 7P4, Canada

G. Munro · Department of Immunology, United Medical and Dental Schools, Guy's and St. Thomas' Hospitals, Guy's Tower, Floor 28, London Bridge, London SE1 9RT, United Kingdom

Yoji Murayama · Department of Periodontology and Endodontology, Okayama University Dental School, 2-5-1 Shikata-cho, Okayama 700, Japan

Patricia A. Murray · Department of Periodontology, New Jersey Dental School, University of Medicine and Dentistry, 110 Bergen Street, Newark, New Jersey 07103

A. Rekha K. Murthy · Department of Medicine, University of California School of Medicine, Los Angeles, California 90024

Lien Nguyen · Department of Cytokine Biology, Forsyth Dental Center, 140 The Fenway, Boston, Massachusetts 02115

R. J. Nisengard · Department of Periodontology, School of Dental Medicine, and Department of Microbiology, School of Medicine, State University of New York at Buffalo, 250 Squire Hall, Buffalo, New York 14214

Steven Offenbacher · Dental Research Center, School of Dentistry, University of North Carolina at Chapel Hill, CB# 7455, Chapel Hill, North Carolina 27599-7455

Katsuji Okuda · Department of Microbiology, Tokyo Dental College, 1-2-2 Masago, Chiba 261, Japan

Roy C. Page · Center for Research in Oral Biology, University of Washington, Seattle, Washington 98195

Jan Pohl · Microchemical Facility, Winship Cancer Center, School of Medicine, Emory University, Atlanta, Georgia 30322

Hans Preus · Department of Periodontology, University of Oslo, Box 1109 Blindern, N-0317 Oslo, Norway

J. Pugin · Department of Immunology, The Scripps Research Institute, La Jolla, California 92037

P. R. Ramakrishnan · Department of Oral Biology and Periodontal Disease Research Center, State University of New York at Buffalo, Buffalo, New York 14214

N. S. Ramamurthy · State University of New York at Stony Brook, Stony Brook, New York 11790

A. H. Reddi · Laboratory of Musculoskeletal Cell Biology, Department of Orthopaedic Surgery, Johns Hopkins University School of Medicine, Baltimore, Maryland 21205

David Rueger · Creative BioMolecules, 35 South Street, Hopkinton, Massachusetts 01748

Bruce Rutherford · Department of Cariology and General Dentistry, School of Dentistry, University of Michigan, 1101 North University, Ann Arbor, Michigan 48109-1078

Atsushi Saito · Department of Periodontics, Tokyo Dental College, 1-2-2 Masago, Chiba 261, Japan

Harvey A. Schenkein · Clinical Research Center for Periodontal Diseases, School of Dentistry, Virginia Commonwealth University, MCV Station Box 980566, Richmond, Virginia 23298-0566

Robert E. Schifferle · Department of Oral Biology, School of Dental Medicine, State University of New York at Buffalo, Foster Hall, 3435 Main Street, Buffalo, New York 14214-3092

William M. Shafer · Department of Microbiology and Immunology, School of Medicine, Emory University, Atlanta, Georgia 30322, and Laboratories of Microbial Pathogenesis and Research Sciences, Veterans Affairs Medical Center, Decatur, Georgia 30033

Ashu Sharma · Department of Oral Biology, School of Dental Medicine, State University of New York at Buffalo, Foster Hall, 3435 Main Street, Buffalo, New York 14214-3092

Bruce J. Shenker · Department of Pathology, School of Dental Medicine, University of Pennsylvania, 4010 Locust Street, Philadelphia, Pennsylvania 19104-6002

Sharnn Shepheard · Laboratory of Immunology, National Institute of Dental Research, National Institutes of Health, Bethesda, Maryland 20892-0001

H. Shimauchi · Department of Immunology, Forsyth Dental Center, 140 The Fenway, Boston, Massachusetts 02115

Reuben P. Siraganian · Laboratory of Immunology, National Institute of Dental Research, National Institutes of Health, Bethesda, Maryland 20892

D. J. Smith · Department of Immunology, Forsyth Dental Center, 140 The Fenway, Boston, Massachusetts 02115

Hakimuddin T. Sojar · Department of Oral Biology, School of Dental Medicine, State University of New York at Buffalo, Foster Hall, 3435 Main Street, Buffalo, New York 14214-3092

Prem K. Sreenivasan · Laboratory of Molecular Infectious Diseases, The Rockefeller University, 1230 York Avenue, New York, New York 10021

Philip Stashenko · Department of Cytokine Biology, Forsyth Dental Center, 140 The Fenway, Boston, Massachusetts 02115

S. Steinemann · Department of Immunology, The Scripps Research Institute, La Jolla, California 92037

Ichiro Takahashi · Research Center in Oral Biology, University of Alabama at Birmingham, Birmingham, Alabama 35294

Shogo Takashiba · Department of Periodontology and Endodontology, Okayama University Dental School, 2-5-1 Shikata-cho, Okayama 700, Japan

O. Takeichi · Department of Immunology, Forsyth Dental Center, 140 The Fenway, Boston, Massachusetts 02115

M. A. Taubman · Department of Immunology, Forsyth Dental Center, 140 The Fenway, Boston, Massachusetts 02115

Hongsheng Tian · Laboratory of Immunology, National Institute of Dental Research, National Institutes of Health, Bethesda, Maryland 20892-0001

P. S. Tobias · Department of Immunology, The Scripps Research Institute, La Jolla, California 92037

S. Todryk · Department of Immunology, United Medical and Dental Schools, Guy's and St. Thomas' Hospitals, Guy's Tower, Floor 28, London Bridge, London SE1 9RT, United Kingdom

R. J. Ulevitch · Department of Immunology, The Scripps Research Institute, La Jolla, California 92037

Sharon Wahl · Laboratory of Immunology, National Institute of Dental Research, National Institutes of Health, Building 30, Room 331, Bethesda, Maryland 20892-0001

P. Walker · Department of Immunology, United Medical and Dental Schools, Guy's and St. Thomas' Hospitals, Guy's Tower, Floor 28, London Bridge, London SE1 9RT, United Kingdom

Sharon Wannberg · Department of Pathology, School of Dental Medicine, University of Pennsylvania, 4010 Locust Street, Philadelphia, Pennsylvania 19104-6002

Mark E. Wilson · Department of Oral Biology, School of Dental Medicine, State University of New York at Buffalo, Foster Hall, 3435 Main Street, Buffalo, New York 14214-3092

J. Leslie Winston · Department of Oral Biology, School of Dental Medicine, State University of New York at Buffalo, Foster Hall, 3435 Main Street, Buffalo, New York 14214-3092

Behnaz Yalda · Dental Research Center, School of Dentistry, University of North Carolina at Chapel Hill, Chapel Hill, North Carolina 27599-7455

Masafumi Yamamoto · Department of Oral Biology, University of Alabama at Birmingham, UAB Station, Birmingham, Alabama 35294

Joseph J. Zambon · Departments of Oral Biology and Periodontology, School of Dental Medicine, State University of New York at Buffalo, Foster Hall, 3435 Main Street, Buffalo, New York 14214-3092

Preface

Advances in understanding the molecular basis of biologic phenomena have been rapid in the last 2 decades, giving rise to what is known as "The Golden Era of Biology." A symposium addressing the current status of knowledge regarding the molecular basis of pathogenesis and molecular targeting in periodontal diseases was held in Buffalo, New York, and this book represents the edited proceedings of that symposium.

The overall goal of this book is to provide a current evaluation of the role of periodontal pathogens, their virulence factors, and host factors contributing to the pathogenesis of periodontal diseases. In addition, the molecular and cellular bases of growth and development and the application of molecular technology to periodontal regeneration are discussed. Hopefully, this book will lead to future therapies and preventive regimens which will benefit the patient in a cost-effective manner. Specifically, the organizers invited participants who are active in investigating questions of virulence, tissue destruction, and wound healing to prepare chapters for this volume. The emphasis is on the presentation of data leading to new hypotheses which challenge old concepts. It is hoped that this publication will lead to increased research activity in the molecular basis of infectious disease and will catalyze opportunities for worldwide research collaborations. The organizers, sponsors, and participants are confident that these research efforts, when properly applied to clinical practice, will ultimately result in better management of chronic infections including periodontal diseases.

Robert J. Genco

Acknowledgments

The symposium was kindly supported by Sunstar Inc., the J.O. Butler Company, and the State University of New York at Buffalo School of Dental Medicine. We express our appreciation to Hiroo Kaneda, President of Sunstar Inc., and Charles Brearley, President of the J.O. Butler Company, for their generous financial support of the symposium. We thank Dean William Feagans, Dean Designate Louis Goldberg, Provost Aaron Bloch, and President William Greiner of the University at Buffalo for their support; we also congratulate them on the School of Dental Medicine Centennial, which the symposium helped celebrate. We are especially thankful to Rose Parkhill for her untiring efforts and attention to detail in many aspects of this meeting and in the preparation of this volume. The authors and organizers also thank Pamela Wilks, Patrick Fitzgerald, and the staff of the American Society for Microbiology for their conscientious efforts in publication of these proceedings.

Introduction

In June 1993, scientists gathered for a symposium which celebrated the hundredth anniversary of the State University of New York at Buffalo School of Dental Medicine in a manner that reflected the excellence of this premier institution of learning. It was a symposium that brought together world scientists to discuss the very frontiers of knowledge in periodontal disease: the molecular biology of pathogenesis and molecular targeting.

A hundred years ago, molecules were beyond the scope of biologists and even atomic theory was questioned. Radioactivity was yet to be discovered, so, for all the physicists knew, the earth was only a few hundred million years old. However, biology was on the move. Cell theory had been firmly established by mid-century and Virchow's classic paper on cellular pathology appeared in 1858, one year before Darwin's *Origin of Species*. The next few decades would see the revolution take fire with the seminal papers on germ theory and infectious disease by Pasteur and Koch and with the studies of inflammation and immunity by Cohnheim and Ehrlich. The oral health sciences were also on the move. W. D. Miller developed the bacterio-chemical theory of dental caries in Berlin, working next door to Robert Koch himself.

The past hundred years also mark the transition from the cell as the basic unit of study in biology to the vastly more complex world of subcellular organelles, genes, regulatory elements, and the growing number of growth factors, cytokines, receptors, and ligands that govern the behavior of cells and their extracellular matrices. The term "molecular biology" relates not only to the microcosmic entities under study but also to the laws of biophysics, structural biology, and biochemistry that they obey. We owe the recent developments in this field to the advances in tools and techniques that launched the biotechnology revolution in the 1980s and that began with the discovery of the structure of DNA by Watson and Crick, published in 1953, over 40 years ago.

By the 1960s the genetic code had been worked out by Nirenberg, Khorana, and others, and Francis Crick's central dogma of information flow from DNA to RNA to protein was firmly established. For a time, many geneticists grew disheartened; it seemed that all the great discoveries had been made, that the answers were all in, and that the rest was mere details and mop up.

Needless to say, they were wrong. Transcription and translation were not so easy. Turning genes on and off depended on operons, promoters, repressors, and stop signals. Genes were not pure informational molecules either. They contained intervening sequences, jumping genes, multiple repeats, and "junk" DNA. The central dogma was then amended with the discovery of reverse transcriptase and restriction enzymes. Today we are seeing the clinical and commercial fallout of those all-important findings. Routinely, scientists cut and reassemble DNA, transfect cells, exploit bacteria as gene and protein factories, make transgenic animals, and use restriction fragment length polymorphisms to study genetic disease in humans. Severe combined immunodeficient (SCID) mice for the study of the human immune system and knockout mice to tell us which genes are indispensable and why are just two of the latest developments. Gene therapy for incurable diseases is well and truly on its way, as is the effort to map the entire human genome.

Against such a background it is not surprising that we are seeing an explosion of knowledge with regard to the molecular biology of cell-cell, cell-bacteria, and cell-matrix interactions. We at the National Institute of Dental Research have been leaders in the discoveries of the family of collagens and of fibronectin, osteonectin, laminin, the amelogenins and enamelins, the bone morphogenetic proteins, and other growth factors and cytokines. We are significantly advancing the state of the science with regard to the cell adhesion molecules: the superfamily of integrins that bridge the gap across the plasma membrane to the cytoskeleton.

Studies of immunity and inflammation have revealed an equally dynamic world of messengers and mediators, the interleukins and transforming growth factors and, lately, nitric oxide, a versatile gas that functions as a neurotransmitter as well as an inflammatory mediator. These are the molecules that are going to lead to a more profound understanding of periodontal diseases, their pathogenesis, diagnosis, and clinical management. When we learn what these molecules do and how they do it and when we apply the tools of molecular biology to manipulate and regulate them, we may well be on our way to the control and the ultimate cure of not only periodontal diseases but other chronic degenerative inflammatory diseases that plague mankind as well.

We have already seen the value of exploiting molecular biology in the context of periodontal microbiology. Back in the 1950s and 1960s, when the important discovery that periodontal diseases were infections associated with bacteria in dental plaque was made, microbiologists labored to isolate and cultivate dozens of species in their search for the prime suspects. As we now know, more than one species and probably combinations of bacteria are involved in each of the various subcategories of the periodontal diseases. Later, investigators began to study adhesion as well as to look at interactions among bacteria and at plaque development and maturity. The field of microbial ecology came into being.

Today we can use polymerase chain reaction techniques to identify strains of bacteria in the periodontal pocket, and commercial test kits using genetic probes or monoclonal antibodies are available as diagnostic aids in the clinic. Meanwhile, investigators are using genetic engineering techniques to determine bacterial virulence factors. The search for mechanisms of attachment, aggregation, and coaggregation and the identification of families of adhesins and receptors are proceeding apace. Genes and gene products are being used to fabricate nonpathogenic bacterial forms, to detoxify bacteria, and to further our understanding of pathogenesis in other ways. Perhaps the next step will be to take these laboratory successes to clinical studies and trials using replacement therapy, antiadhesin molecules, pathogen-specific antibiotics, and the like to prevent or treat periodontitis.

The new immunology is also rapidly advancing our understanding of the host response. We have known for some time that both the cellular and humoral immune systems are activated during periodonal disease. The evidence indicates that in a general way they are both involved in immunopathologic processes that can contribute to tissue destruction, as well as play a major role in protection of the host. Much of the excitement in the field today lies in the deeper understanding of the cascade of events that follow tissue injury. It is now possible to track the mobilization and migration of immune cells in far greater detail, to watch their action at the site of the lesion, and to assess the relative roles of the major players. The challenge will be to come up with agents to counter host-mediated destruction of connective tissue and bone, choosing the ones that will be the most effective without introducing other risks or unwanted effects.

Toward this end, research on the tetracycline family of drugs has been proceeding for some time. It has long been known that the tetracyclines can inhibit bone resorption. What is new and important in relation to periodontal pathology is that they also inhibit collagenase production by tissues. Chemically modified tetracyclines, with the antibiotic properties removed, are now being used to experimentally treat a variety of chronic inflammatory diseases, including periodontitis.

The identification of virulence factors of periodontal disease bacteria is providing candidates for immunogens for use in genetically engineered oral vaccines. The potential for immunization is especially important in relation to early-onset periodontitis, in which a single bacterial organism is assumed to play the prominent role. The generation of monoclonal antibodies against periodontal pathogens in adult periodontitis is another possibility, especially if they can be applied to oral surfaces in a form of passive immune protection.

The issue of host response can also be framed in terms of questions that go beyond immunology to the intriguing subject of individual differences. We know that in the case of early-onset periodontitis, there is considerable evidence for genetic susceptibility. We might expect that genetic linkage studies in large families of affected individuals or a front-line attack on the suspected immune cell defects may yield the gene or genes involved. The challenge will then be to develop diagnostic screening tests and possibly gene therapy for prevention.

The real issue in the understanding of the pathogenesis of periodontal disease and the mechanism of the progression of the periodontal lesion is what turns it on and what turns it off. For approximately 20 years longitudinal studies of populations in Sri Lanka have been being performed, the aims of which were to describe the natural history of periodontal disease, its initiation and subsequent progression, and the resultant tooth loss. In brief, this study has shown that in a group homogeneous with regard to ethnicity, environment, education, and nutrition and in the absence of oral hygiene and oral health care, all teeth in all participants exhibited visible plaque, as well as overt gingivitis and supra- and subgingival calculus. With time and without intervention, periodontal destruction progressed beyond gingivitis in 90% of this population. But despite the remarkable homogeneity of the group, the severity of periodontal disease varied greatly. On the basis of the rate of periodontal attachment loss, this population could be divided into three groups: (i) a relatively small group (approximately 8%) who showed rapidly progressing periodontitis, (ii) a large group (approximately 80%) who showed moderately progressive periodontitis, and (iii) another small, but significant, group (approximately

10%) who exhibited essentially no progression of periodontal disease beyond chronic gingivitis.

Elsewhere I have discussed the implications of this study in terms of the rate and pattern of progression of the periodontal lesion (1). At this time, I am more intrigued by the 10% of the study group who did not progress beyond gingivitis. What protects them? Is it something in the genes? Is it a factor in the saliva or something in the diet? Is it a more-powerful immune response or a non-immune defense mechanism? Could it be that these individuals possess a more-efficient repair and regenerative capacity? There are no answers to such questions.

Whatever the explanations are, they shall require some very basic investigations into the molecular mechanisms of bacterial behavior and host responses. There is much to be done and a great need for the whole spectrum of scientific methodologies. Despite the relative ignorance on the subject, these are interesting times, and we have witnessed exciting developments in the history of periodontal disease research, indeed of all the oral health sciences. Enormous progress in understanding the transition from health to disease and from disease to health has been made as we have passed from the level of the organism to the tissue, the cell, and now the molecule. Let me remind you, however, that the reverse journey is equally important. Once the cell and the molecular derangements in disease processes are understood, we must go up the scale to understand the whole organism and to educate society itself and the community of societies that constitutes the world today, a world in which people can live longer and richer lives with their natural teeth for a lifetime.

Harald Löe
National Institute of Dental Research
Bethesda, Maryland

REFERENCE

1. **Löe, H.** 1986. Progression of natural untreated periodontal disease in man, p. 11–29. *In* T. Lehner and G. Gimasoni (ed.), *The Borderland between Caries and Periodontal Disease III.* Editions Medicine et Hygiene, Geneva.

SECTION I

MICROBIAL VIRULENCE FACTORS

Molecular Pathogenesis of Periodontal Disease
Edited by Robert Genco et al.
© 1994 American Society for Microbiology, Washington, DC 20005

Chapter 1

Epidemiology of Subgingival Bacterial Pathogens in Periodontal Disease

Joseph J. Zambon, Sara Grossi, Robert Dunford, Violet I. Haraszthy, Hans Preus, and Robert J. Genco

This chapter presents the current paradigm for the pathogenesis of the most common forms of human periodontal disease as well as the prevalence of subgingival periodontal pathogens in a large cross-sectional study, the Erie County Study. Bacterial, demographic, and other risk indicators of periodontal attachment loss identified in the Erie County Study are described, and evidence for the familial transmission of periodontal pathogens is given. This evidence is based on a novel method of discriminating among bacterial strains, the arbitrarily primed PCR.

ETIOLOGY OF PERIODONTAL DISEASE

The current paradigm for the pathogenesis of human periodontal diseases proposes that these afflictions occur as the result of subgingival plaque infection with specific bacteria. As shown in Fig. 1, the first event is the transmission of periodontal pathogens such as *Actinobacillus actinomycetemcomitans* (21), *Bacteroides forsythus*, *Campylobacter rectus* (formerly *Wolinella recta*), *Porphyromonas gingivalis* (formerly *Bacteroides gingivalis*), and *Prevotella intermedia* (formerly *Bacteroides intermedius*) (18) from other family members. Bacterial colonization of oral mucosa and, more specifically, gingival crevicular epithelium is then facilitated both by bacterial factors such as cell surface fimbriae and by host factors such as salivary molecules. Further, certain species of periodontal pathogens (*A. actinomycetemcomitans*, for example) not only can adhere to gingival crevicular epithelium but also can invade gingiva (5). Periodontal bacterial pathogens persist in colonized sites by evading host defense mechanisms through the production of factors such as leukotoxins or lymphocyte-suppressing factors. They also destroy

Joseph J. Zambon, Sara Grossi, Robert Dunford, Violet I. Haraszthy, and Robert J. Genco • Departments of Oral Biology and Periodontology, School of Dental Medicine, State University of New York at Buffalo, Foster Hall, 3435 Main Street, Buffalo, New York 14214-3092. *Hans Preus* • Department of Periodontology, University of Oslo, Box 1109 Blindern, N-0317 Oslo, Norway.

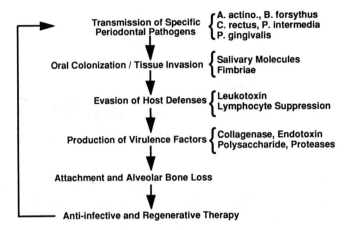

Figure 1. Current paradigm for the pathogenesis of periodontal diseases.

periodontal tissues through virulence factors such as collagenase and other proteolytic enzymes and by triggering destructive host responses such as bone resorption. Following pathogen-mediated tissue destruction, the clinical signs of periodontal disease, including gingival inflammation, loss of connective tissue attachment, periodontal pocket formation, and alveolar bone loss, become apparent. At this point, periodontal therapy can be instituted. This therapy is empirically or specifically directed toward eliminating pathogens from periodontal pockets and adjacent infected gingiva and to regenerating connective tissue and alveolar bone. However, a hallmark of periodontal disease is that, like other chronic infections, it has a propensity to recur. This recurrence is likely facilitated by the retransmission of pathogenic species to patients in whom periodontal therapy has eradicated the pathogens from the oral cavity. The patient may thus enter a cycle of infection, anti-infective therapy, reinfection, and retreatment.

Several clinically relevant questions arise from consideration of this paradigm. The first questions relate to the identity of the bacterial etiology. Which species are important and/or etiologic in periodontal disease? How does oral and/or subgingival infection with these species alter the risk for developing periodontal disease? Once the relative risks of the key periodontal pathogens have been determined, patients infected with these species can be targeted for more intensive monitoring and treatment. A second group of questions resulting from consideration of the paradigm concerns the route and mode of transmission of periodontal pathogens. From whom are these species acquired, and how are they transmitted? Such information could be important in preventing initial or recurrent infection with periodontal pathogens.

THE ERIE COUNTY STUDY

Over the past 5 years, the Periodontal Disease Research Center at Buffalo, N.Y., has examined a large cohort of subjects in order to determine risk factors for the development of periodontal disease (12). This Erie County Study (Fig. 2) examined 1,426 subjects aged 25 to 74 years. Clinical examinations and laboratory studies of each subject were carried out by a team of trained and calibrated examiners. Demographic and socioeconomic data, complete medical and dental histories, a history of occupational exposure to potential hazards in the workplace, and psychosocial information derived from a number of assessment instruments were obtained from each subject. This data collection was followed by complete oral and periodontal examinations employing standard measures of supragingival plaque, gingival inflammation, calculus, probing pocket depth, and clinical attachment level. Blood samples were obtained for laboratory analyses, including determination of serum antibody levels to periodontal pathogens. For assessment of the periodontal microflora, subgingival plaque samples were taken from the mesiobuccal surfaces of six maxillary teeth (teeth numbers 3, 5, 7, 9, 12, and 14) and six mandibular teeth (teeth numbers 19, 21, 23, 25, 28, and 30). Samples from the maxilla and the mandible were pooled separately, and the two resulting samples were analyzed by immunofluorescence microscopy for the presence and relative levels of subgingival bacterial species (2, 24) including *A. actinomycetemcomitans*, *B. forsythus*, *C. rectus*, *Capnocytophaga* species, *Eubacterium saburreum*, *Fusobacterium nucleatum*, *P. gingivalis*, and *P. intermedia*. A subject was considered positive for a target species if it made up at least 1% of the total cell count in either pooled sample.

The resulting microbiological, demographic, and clinical data were analyzed to evaluate the relationship of these factors to periodontal disease. The prevalence of periodontal pathogens was thus determined in this large adult population, as was the extent to which oral colonization by periodontal pathogens constitutes a risk factor for periodontitis. For these analyses, age was stratified into five decades; gender into "male" or "female"; race into "African Americans," "white," or "others," which included Native Americans, Asians, and Pacific Islanders; education into "high school" for subjects completing or not completing high school and "college" for subjects completing at least 1 year of college; and income into "moderate" for subjects with an annual income of <$30,000 or "high" for subjects with an annual income of ≥$30,000.

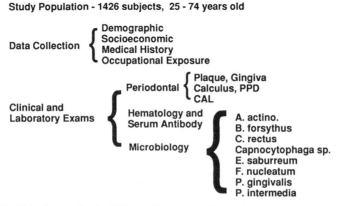

Figure 2. The Erie County Study. PPD, probing pocket depth; CAL, clinical attachment level.

RELATIONSHIPS OF PERIODONTAL PATHOGENS
TO GENDER, RACE, AND AGE

Table 1 shows the prevalence of the target periodontal pathogens in this cohort according to gender. Males were more often infected with the target species than females. *B. forsythus* occurred in highest prevalence in males, 42.88% of whom carried this species. *P. intermedia* was present in 42.31% of males, and *Capnocytophaga* species were present in 18.3%. Females were most often infected with *P. intermedia* (38.73%) and then by *B. forsythus* (34.95%) and *Capnocytophaga* species (15.52%). There was a significant difference between genders for *B. forsythus* and *P. gingivalis*, with both species having a significantly higher prevalence in males. *A. actinomycetemcomitans* was the target pathogen found least often among both males and females at prevalences of 3.81 and 3.64%, respectively.

Table 2 shows the distribution of the target periodontal pathogens in this cohort according to race. In white subjects, *P. intermedia* was the most prevalent species (39.66%). *B. forsythus* and *Capnocytophaga* species were found at intermediate levels, and *A. actinomycetemcomitans* was found least frequently (3.34%). As in white subjects, *P. intermedia* and *B. forsythus* were the most prevalent species in African-American subjects (51.69 and 46.07%, respectively). *P. gingivalis* was present in 25.84% of African-Americans, while *C. rectus* was found least often, at a prevalence rate of 3.37%. In the category of "other," *B. forsythus* was most prevalent (47.22%). It was followed by *P. intermedia* (36.11%) and *P. gingivalis* (30.56%).

P. gingivalis was the only target species for which the prevalence was significantly different between racial groups. *P. gingivalis* was higher in African-Americans and in "others" than in white subjects. This finding of a higher prevalence of *P. gingivalis* in African-American subjects than in white subjects is consistent with previous studies of the Piedmont Cohort, in which subgingival plaque samples from 366 African-Americans and 297 whites 65 years or older were analyzed in the same laboratory (J.J.Z.) as was used in the current study (1, 8). A study by Schenkein et al. (17), who used bacterial culture to analyze the subgingival flora, found that *P. gingivalis* was significantly associated with African-American subjects with periodontitis but not with white subjects. In contrast to the present study, Schenkein

Table 1. Epidemiology of periodontal organisms: gender relationships

Species	% of females infected		% of males infected	
	All subjects	Females only	All subjects	Males only
A. actinomycetemcomitans	1.90	3.64	1.82	3.81
B. forsythus[a]	18.21	34.95	20.54	42.88
C. rectus	2.18	4.18	2.10	4.39
Capnocytophaga spp.	8.07	15.52	8.78	18.30
E. saburreum	4.35	8.37	5.06	10.54
F. nucleatum	5.55	10.66	6.32	13.18
P. gingivalis[a]	6.95	13.36	8.71	18.16
P. intermedia	20.15	38.73	20.30	42.31

[a] Occurrence is significantly different between groups ($P < 0.01$).

et al. (17) did not find significant differences between the composition of the subgingival flora in males and females.

Table 3 examines the prevalence of the target periodontal pathogens in relation to age. *P. intermedia* and *B. forsythus* were the species most prevalent overall at rates of 40.45 and 38.75%, respectively, and they were the most prevalent target species in each age group. *A. actinomycetemcomitans* was least prevalent overall at 3.72%. The prevalence of all of the target species except *A. actinomycetemcomitans* and *C. rectus* was significantly different between age groups.

Three of the target species (*Capnocytophaga* species, *B. forsythus*, and *P. gingivalis*) were significantly associated ($P < 0.05$) with clinical attachment levels. *Capnocytophaga* spp. were negatively associated with attachment loss, while *B. forsythus* and *P. gingivalis* were both positively associated with attachment loss in this group. This relationship is shown in Fig. 3. Subjects were divided into healthy, low, moderate, high, and severe groups based on clinical attachment loss. As shown, there was a linear, positive relationship between the presence of *B. forsythus* or *P. gingivalis* in a subject and the severity of clinical attachment loss. *Capnocytophaga* spp., on the other hand, were inversely related to the degree of attachment loss.

TARGET PATHOGENS AS RISK INDICATORS FOR PERIODONTITIS

Figure 4 shows the results of a logistic regression model of the risk for periodontal disease (defined by clinical attachment levels) that examines a number of variables, including the nine target periodontal pathogens. Three variables (*Capnocytophaga* spp., education, and allergies) were associated with reduced risk for clinical attachment loss. That is, subgingival infection with *Capnocytophaga* spp., higher education, and a medical history of allergies all decreased the risk for clinical attachment loss. Subgingival infection with *P. gingivalis* or *B. forsythus* increased a subject's risk for attachment loss 1.59- and 2.54-fold, respectively. Age, however, was the factor associated with the greatest risk for attachment loss. Subjects 65 to

Table 2. Epidemiology of periodontal organisms: race relationships

Species	% of Afr-Am[a] infected		% of whites infected		% of other groups infected	
	All subjects	Afr-Am only	All subjects	Whites only	All subjects	Other groups only
A. actinomycetemcomitans	0.49	7.87	3.02	3.34	0.07	2.78
B. forsythus	2.88	46.07	34.32	38.01	1.20	47.22
C. rectus	0.21	3.37	3.93	4.35	0.14	5.56
Capnocytophaga spp.	0.77	12.36	15.31	16.95	0.56	22.22
E. saburreum	0.63	10.11	8.29	9.18	0.28	11.11
F. nucleatum	0.49	7.87	10.96	12.13	0.42	16.67
P. gingivalis[b]	1.62	25.84	13.20	14.62	0.77	30.56
P. intermedia	3.23	51.69	35.82	39.66	0.91	36.11

[a] Afr-Am, African-Americans.
[b] Occurrence is significantly different between groups ($P < 0.001$).

Table 3. Epidemiology of periodontal organisms: age relationships

Species	Total % infected	25–34 yr		35–44 yr		% Infected in age group: 45–54 yr		55–64 yr		65–74 yr	
		All subjects	This group only	All subjects	This group only	All subjects	This group only	All subjects	This group only	All subjects	This group only
A. actinomycetemcomitans	3.72	1.05	5.19	0.77	3.58	0.56	2.76	0.64	3.27	0.70	3.80
B. forsythus[a]	38.75	6.26	30.80	8.02	37.25	9.14	44.83	8.30	43.07	7.03	38.02
C. rectus	4.28	0.70	3.46	0.84	3.91	1.48	7.24	0.70	3.64	0.56	3.04
Capnocytophaga spp.[a]	16.85	2.95	14.53	2.11	9.77	3.72	18.28	3.86	20.00	4.21	22.81
E. saburreum[a]	9.41	1.19	5.88	1.83	8.47	2.46	12.07	1.68	8.73	2.25	12.17
F. nucleatum[a]	11.87	0.98	4.84	2.04	9.45	3.23	15.86	3.09	16.00	2.53	13.69
P. gingivalis[a]	15.66	2.11	10.38	2.88	13.36	3.86	18.97	3.72	19.27	3.09	16.73
P. intermedia[a]	40.45	7.59	37.37	6.95	32.25	10.32	50.69	8.15	42.18	7.44	40.30

[a] Occurrence significantly different between groups ($P < 0.001$).

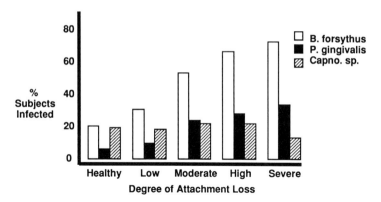

Figure 3. Relationship between infection with target periodontal pathogens and degree of attachment loss.

74 years of age had a 9.01-fold greater risk of attachment loss than subjects 25 to 34 years of age. Subjects in younger age categories were at lower risk for clinical attachment loss. Subjects 55 to 64 years of age had 4.01-fold greater risk, subjects 45 to 54 years of age had 3.54-fold greater risk, and subjects 35 to 44 years of age had 1.72-fold greater risk than subjects 25 to 34 years of age. Other variables significantly associated with an increased risk of attachment loss were (i) gender, with males at a 1.34-fold greater risk for attachment loss than females; (ii) smoking, which increased risk in a dose-dependent manner by 2.05-, 2.77-, or 4.75-fold depending on whether the subjects were judged to be light, moderate, or heavy smokers, respectively; and (iii) diabetes mellitus, which increased the risk for attachment loss by 2.32-fold.

In summary, analysis of data from this cross-sectional examination of a large subject cohort shows that subgingival infection with certain periodontal bacteria significantly increases the risk for periodontitis. This, however, does not exclude

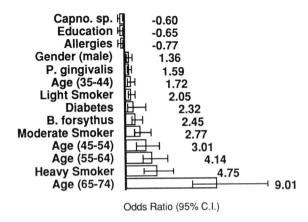

Figure 4. Risk indicators for attachment loss. C.I., confidence interval.

the other target pathogens as being significant in certain patients or groups of patients. *A. actinomycetemcomitans*, for example, does not constitute a significant risk factor for attachment loss and, as in similar studies (1, 3), was found in very low prevalence in this cohort. This, however, does not mean that *A. actinomycetemcomitans* is not a periodontal pathogen. On the contrary, a great deal of evidence strongly implicates this species as the cause of juvenile periodontitis, rapidly progressing periodontitis, and refractory periodontitis, diseases that occur at a rather low prevalence in the overall population (9, 20) and that are not likely to be well represented in the Erie County Study. The presence of this microorganism in post-surgical sites also adversely affects the course of guided tissue regeneration (13).

A similar argument can be made for *C. rectus*, which, like *A. actinomycetemcomitans*, produces a toxin that can destroy host immune cells (10, 11). This species is the most prevalent subgingival periodontal pathogen in the predominant cultivable microflora from subjects having diabetes mellitus or AIDS, diseases associated with significant immune suppression (22, 23). Again, these diseases were not found in a large number of subjects in our cohort, so bacteria associated with subgingival infections in these subjects would not likely enter the logistic regression model. Obviously, for subjects with diabetes mellitus or AIDS, *C. rectus* could be an important subgingival pathogen, contributing significantly to the pathogenesis of their periodontal disease.

TRANSMISSION OF PERIODONTAL PATHOGENS

The initial step in the pathogenesis of periodontal disease is the transmission of periodontal pathogens into the human oral cavity. To examine questions of transmission, it is necessary to detect periodontal pathogens in clinical samples of subgingival dental plaque and also to differentiate one strain from another. In this way, strains can be tracked from subject to subject, particularly in longitudinal studies.

Determination of strain variability can also be used to examine the clonality of oral pathogens. There are in most bacterial species a number of discrete clonal lineages, and in most of the pathogenic bacterial species examined to date, a small number of clonal lineages within a species are responsible for most cases of the disease. As Socransky and Haffajee proposed (18), only a small number of clones within species of periodontal pathogens may actually be responsible for attachment loss in periodontitis. This could explain why many patients harbor periodontal pathogenic species in the absence of demonstrable attachment loss, i.e., infection with a nonpathogenic clone.

Previously, strain differences were examined using phenotypic methods such as biotyping or antigenic determinations such as serotyping (16, 21, 25). However, as in the clinical detection of target microorganisms, techniques in molecular biology are becoming increasingly useful in examining variations between strains within periodontal pathogenic species. These methods include restriction endonuclease analysis (4, 26), restriction fragment length polymorphisms (7), RNA analysis (6), and a new method, the arbitrarily primed PCR (19).

The arbitrarily primed PCR procedure involves DNA extraction from a bacterial isolate and amplification of DNA segments by means of synthetic oligonucleotide primers. The amplified DNA segments can then be visualized by agarose gel electrophoresis, resulting in an "amplitype" for each strain. A critical step in the process is the selection of a DNA primer that will discriminate among strains within the target species. For example, from a battery of over 20 primers, we have found only one primer (AGTCAGCCAC) that is useful in discriminating among strains of *P. gingivalis*.

Arbitrarily primed PCR has been used by Preus et al. (15) to study the distribution and transmission of *A. actinomycetemcomitans* in families of patients with established periodontitis. Using a synthetic oligonucleotide (GGGTAACGCC), they distinguished 14 different amplification patterns in 20 *A. actinomycetemcomitans* strains (14). In six of seven families, husbands and wives had different types of *A. actinomycetemcomitans* in their oral cavities. Nine of 10 children harbored an *A. actinomycetemcomitans* strain identical to that found in one parent, while one child had two different types of *A. actinomycetemcomitans*, both of which were identical to those found in the parents (15).

Clearly, a great deal of progress has been made in identifying periodontal pathogens, in determining their importance as risk factors for attachment loss, and in developing methods to study the transmission of these pathogens into the oral cavity. As techniques in molecular biology are increasingly applied to the problem of periodontal infection, it is likely that effective clinical strategies will soon be available to prevent initial infection, reinfection, and recurrence of periodontal disease.

REFERENCES

1. **Beck, J. D., G. G. Koch, J. J. Zambon, R. J. Genco, and G. E. Tudor.** 1992. Evaluation of oral bacteria as risk indicators for periodontitis in older adults. *J. Periodontol.* **63:**93–99.
2. **Bonta, Y., J. J. Zambon, M. Neiders, and R. J. Genco.** 1985. Rapid identification of periodontal pathogens in subgingival dental plaque: comparison of indirect immunofluorescence microscopy with bacterial culture for detection of *Actinobacillus actinomycetemcomitans. J. Dent. Res.* **64:**793–798.
3. **Carlos, J. P., M. D. Wolfe, J. J. Zambon, and A. Kingman.** 1988. Periodontal disease in adolescents: some clinical and microbiologic correlates of attachment loss. *J. Dent. Res.* **67:**1510–1514.
4. **Chen, C.-K., G. J. Sunday, J. J. Zambon, and M. E. Wilson.** 1990. Restriction endonuclease analysis of *Eikenella corrodens. J. Clin. Microbiol.* **28:**1265–1270.
5. **Christersson, L. A., B. Albini, J. J. Zambon, and R. J. Genco.** 1987. Tissue localization of *Actinobacillus actinomycetemcomitans* in human periodontitis. I. Light, immunofluorescence, and electron microscopic results. *J. Periodontol.* **58:**529–539.
6. **Dewhirst, F. E., C.-K. Chen, B. J. Paster, and J. J. Zambon.** 1993. Phylogenetic position of *Eikenella corrodens*-like human oral isolates in the family *Neisseriaceae* and description of *Kingella orale* sp. nov. *Int. J. Syst. Bacteriol.* **43:**490–499.
7. **DiRienzo, J., and J. Slots.** 1990. Genetic approach to the study of epidemiology and pathogenesis of *Actinobacillus actinomycetemcomitans* in localized juvenile periodontitis. *Arch. Oral Biol.* **35**(Suppl.):79S–84S.
8. **Drake, C. W., J. Hunt, J. D. Beck, and J. J. Zambon.** 1993. The distribution and interrelationship of *Actinobacillus actinomycetemcomitans, Porphyromonas gingivalis, Prevotella intermedia,* and BANA scores among older adults. *J. Periodontol.* **64:**89–94.

9. **Genco, R. J., L. A. Christersson, and J. J. Zambon.** 1986. Juvenile periodontitis. *Int. Dent. J.* **36:**169–176.

10. **Gillespie, J., E. DeNardin, S. Radel, J. Kuracina, J. Smutko, and J. J. Zambon.** 1992. Production of an extracellular toxin by the oral pathogen *Campylobacter rectus*. *Microb. Pathog.* **12:**69–77.

11. **Gillespie, M. J., J. Smutko, G. G. Haraszthy, and J. J. Zambon.** 1993. Isolation and characterization of the *Campylobacter rectus* cytotoxin. *Microb. Pathog.* **14:**203–215.

12. **Grossi, S. G., J. J. Zambon, A. W. Ho, G. Koch, R. G. Dunford, E. E. Machtei, O. M. Norderyd, and R. J. Genco.** 1994. Assessment of risk for periodontal disease. I. Risk indicators for attachment loss. *J. Periodontol.* **65:**260–267.

13. **Machtei, E. E., M. I. Cho, R. Dunford, J. Norderyd, J. J. Zambon, and R. J. Genco.** Clinical, microbiological, and histological factors which influence the success of regenerative periodontal therapy. *J. Periodontol.,* in press.

14. **Preus, H. R., V. I. Haraszthy, J. J. Zambon, and R. J. Genco.** 1993. Differentiation of *Actinobacillus actinomycetemcomitans* by arbitrarily primed polymerase chain reaction. *J. Clin. Microbiol.* **31:**2773–2776.

15. **Preus, H. R., J. J. Zambon, E. E. Machtei, R. G. Dunford, and R. J. Genco.** 1994. The distribution and transmission of *Actinobacillus actinomycetemcomitans* in families with established adult periodontitis. *J. Periodontol.* **65:**2–7.

16. **Saarela, M., S. Asikainen, S. Alaluusua, L. Pyhala, C.-H. Lai, and H. Jousimies-Somer.** 1992. Frequency and stability of mono- or poly-infection by *Actinobacillus actinomycetemcomitans* serotypes a, b, c, d or e. *Oral Microbiol. Immunol.* **7:**277–279.

17. **Schenkein, H. A., J. A. Burmeister, T. E. Koertge, C. N. Brooks, A. M. Best, L. V. H. Moore, and W. E. C. Moore.** 1993. The influence of race and gender on periodontal microflora. *J. Periodontol.* **64:**292–296.

18. **Socransky, S. S., and A. D. Haffajee.** 1992. The bacterial etiology of destructive periodontal disease: current concepts. *J. Periodontol.* **63**(Suppl.)**:**322–331.

19. **Welsh, J., and M. McClelland.** 1990. Fingerprinting genomes using PCR with arbitrary primers. *Nucleic Acids Res.* **18:**7213–7218.

20. **Zambon, J. J.** 1985. *Actinobacillus actinomycetemcomitans* in human periodontal disease. *J. Clin. Periodontol.* **12:**1–20.

21. **Zambon, J. J., L. A. Christersson, and J. Slots.** 1983. *Actinobacillus actinomycetemcomitans* in human periodontal disease: prevalence in patient groups and distribution of biotypes and serotypes. *J. Periodontol.* **54:**707–711.

22. **Zambon, J. J., H. Reynolds, J. Fisher, M. Shlossman, R. Dunford, and R. J. Genco.** 1988. Microbiological and immunological studies of adult periodontitis in patients with non-insulin dependent diabetes mellitus. *J. Periodontol.* **59:**23–31.

23. **Zambon, J. J., H. Reynolds, and R. J. Genco.** 1990. Studies of the subgingival microflora in patients with acquired immunodeficiency syndrome. *J. Periodontol.* **61:**699–705.

24. **Zambon, J. J., H. S. Reynolds, P. Chen, and R. J. Genco.** 1985. Rapid identification of periodontal pathogens in subgingival dental plaque: comparison of indirect immunofluorescence microscopy with bacterial culture for detection of *Bacteroides gingivalis*. *J. Periodontol.* **56:**(Suppl.)**:**32–40.

25. **Zambon, J. J., J. Slots, and R. J. Genco.** 1983. Serology of oral *Actinobacillus actinomycetemcomitans* and serotype distribution in human periodontal disease. *Infect. Immun.* **41:**19–27.

26. **Zambon, J. J., G. J. Sunday, and J. Smutko.** 1990. Molecular genetic analysis of *Actinobacillus actinomycetemcomitans* epidemiology. *J. Periodontol.* **61:**75–80.

Molecular Pathogenesis of Periodontal Disease
Edited by Robert Genco et al.
© 1994 American Society for Microbiology, Washington, DC 20005

Chapter 2

Porphyromonas gingivalis Fimbriae: Structure, Function, and Insertional Inactivation Mutants

Robert J. Genco, Hakimuddin Sojar, Jin-Yong Lee, Ashu Sharma, Gurrinder Bedi, Moon-Il Cho, and David W. Dyer

Various forms of periodontal diseases are closely associated with specific subgingival bacteria (36). *Porphyromonas gingivalis*, a gram-negative anaerobe, has been implicated as an important etiologic agent in adult periodontitis (25, 32, 35, 43). *P. gingivalis* is often detected in large numbers in subgingival lesions of periodontal patients and is rarely detectable in patients or sites exhibiting periodontal health (2, 3, 34, 40). *P. gingivalis* produces a large number of putative virulence factors, including several proteases (9, 11, 38, 39), hemolysins (4, 30; also see chapter 3 of this volume), and collagenases (1, 18; also see chapter 3). *P. gingivalis* fimbriae are thought to be important in mediating adherence to host tissues and hence are likely to be important virulence determinants (15, 41). Bacterial coherence may also be important for colonization of the oral cavity (5, 10, 28, 33), and *P. gingivalis* has been found to adhere to several bacteria, including *Actinomyces viscosus* (21), *Actinomyces naeslundii* and *Actinomyces israelii* (23, 29), *Streptococcus sanguis* (38), and *Streptococcus mitis* (26). In this communication we report on studies elucidating the role of *P. gingivalis* fimbriae in virulence.

P. GINGIVALIS FIMBRIAE: STRUCTURE OF THE FIMBRILLIN SUBUNIT

The subunit of *P. gingivalis* fimbriae, fimbrillin, was purified and found by gel electrophoresis to be approximately 43 kDa (41). The subunit was characterized as a polypeptide with little or no detectable carbohydrate. The N-terminal sequence

Robert J. Genco, Hakimuddin Sojar, Jin-Yong Lee, Ashu Sharma, and Moon-Il Cho • Department of Oral Biology, School of Dental Medicine, State University of New York at Buffalo, 3435 Main Street, Foster Hall, Buffalo, New York 14214-3092. *Gurrinder Bedi* • Maginin Pharmaceuticals Inc., 5110 Campus Drive, Plymouth Meeting, Pennsylvania 19462. *David W. Dyer* • Department of Microbiology and Immunology, Health Sciences Center, University of Oklahoma, Oklahoma City, Oklahoma 73190.

of the subunit protein appears to be unrelated to the N termini of fimbriae from other species including *Bacteroides nodosus* (42).

Molecular cloning and sequencing of the gene encoding the fimbrial subunit of *P. gingivalis* were carried out by Dickinson and coworkers (7), and the gene was found to be present as a single copy on the chromosome. The predicted size of the mature protein was 35,924 Da, and the deduced protein sequence had no marked similarity to known fimbrial sequences. Furthermore, no homologous sequences could be found on other black-pigmented *Bacteroides* species, which suggests that the *P. gingivalis* fimbriae represent a unique class of fimbrial subunit proteins (7).

FIMBRIAE OF *P. GINGIVALIS*

A rapid and reproducible method for purification of the 43-kDa fimbriae from *P. gingivalis* was described by Sojar et al. (37). This method results in purification of the 43-kDa fimbrial protein to homogeneity in native-like form, free from other components, including a commonly found contaminant, the 75-kDa protein. Briefly, after *P. gingivalis* have been grown and harvested, fimbriae are sheared from the cells by mild sonication and the fimbrial proteins are separated by differential centrifugation. Fimbriae are selectively precipitated in the presence of 1% sodium dodecyl sulfate (SDS) and 0.2 M $MgCl_2$ at pH 6.5. The precipitation step is repeated three or four times to obtain native-like 43-kDa protein free of 75-kDa protein or other contaminants as assessed by silver-stained gel electrophoresis, immunoblotting, amino acid analysis, and N-terminal sequencing.

The purified 43-kDa protein showed fimbria-like morphology under the electron microscope, and specific antibodies to this material localized the 43-kDa protein by immunogold labeling to unbranched fimbriae with diameters of approximately 3 to 5 nm on whole cells of *P. gingivalis* (Fig. 1). Circular dichroism studies suggested high levels of β-sheet structure (37).

Fimbriae purified from various *P. gingivalis* strains differ in size (ranging from 41 to 49 kDa), amino-terminal sequences, and antigenicity (19). In addition, heterogeneity was shown by restriction fragment length polymorphism (RFLP) using the N-terminal coding sequence of *fimA* from *P. gingivalis* 381 as a probe for the gene encoding the fimbrial subunit protein (22). There were 25 RFLP patterns among 39 strains, suggesting considerable heterogeneity at the *fimA* locus. The fimbriae from *P. gingivalis* 14-7-K appear to be significantly different from those of other strains studied by RFLP. Studies are in progress to determine whether this strain represents another fimbrial type on *P. gingivalis* or whether it is simply a variant showing greater differences than the other strains studied have shown.

ROLE OF FIMBRIAE IN ADHERENCE

Fimbriae are important in mediating adherence of many bacterial species to the surfaces they colonize and infect. *P. gingivalis* fimbriae appear to mediate adherence to oral epithelial cells and salivary pellicle-coated tooth surfaces. For example, Isogai et al. (15) showed that monoclonal antibodies to *P. gingivalis*

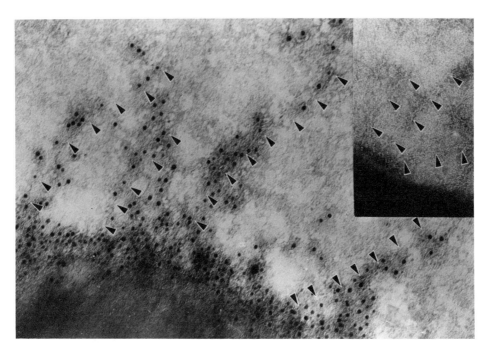

Figure 1. Immunogold localization of the 43-kDa fimbrial protein on whole cells of *P. gingivalis* 2561. Note the heavy labeling of fimbriae (arrowheads) with gold particles and the lack of labeling in the control (inset). Reprinted with permission from reference 37.

fimbriae block adhesion of *P. gingivalis* to human buccal epithelial cells. Studies from our laboratories show that *P. gingivalis* can bind to saliva-coated hydroxyapatite beads in a concentration-dependent manner and that this binding is inhibited by purified fimbriae (20). Furthermore, synthetic peptides corresponding primarily to the C-terminal region of the protein inhibit the binding of *P. gingivalis* to saliva-coated hydroxyapatite, suggesting that this is the region of the active site(s) responsible for binding of *P. gingivalis* to surface-bound salivary components (Fig. 2).

Recently, *P. gingivalis* fimbrillin polypeptide has been expressed and purified from *Escherichia coli* (31). To produce recombinant fimbrillin, a coding region of the fimbrillin gene from *P. gingivalis* 2561 was amplified by the PCR and cloned into the pET-11d vector. The recombinant plasmid was transformed into *E. coli* BL21, and protein expression was induced with isopropyl β-D-thiogalactopyranoside. The expressed protein was purified by gel filtration from inclusion bodies after solubilization with urea. The purified recombinant fimbrillin polypeptide (r-fim 10-337) corresponded to amino acids 10 to 337 of the deduced amino acid sequence of the fimbrillin. The r-fim 10-337 reacted with antibodies to fimbrillin as well as with antibodies to synthetic peptides corresponding to amino acid sequence of the fimbrillin. Recombinant fim 10-337 was capable of inhibiting the binding of *P. gingivalis* 2561 cells to saliva-coated hydroxyapatite beads (Fig. 3).

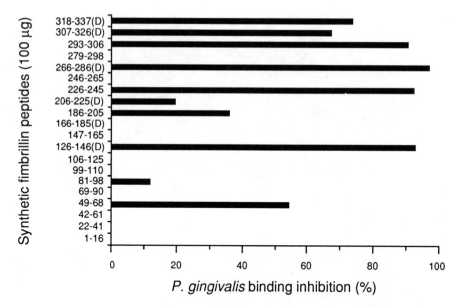

Figure 2. Inhibition of *P. gingivalis* binding to saliva-coated hydroxyapatite beads exerted by synthetic peptides corresponding to segments of the fimbrillin amino acid sequence. Saliva-coated hydroxyapatite beads (2 mg) were incubated with 1.2×10^8 cells of *P. gingivalis* 2561 in the presence of 100 µg of each peptide. Inhibition is expressed as percent inhibition of the control binding that occurred when no peptide was added to the system (taken to be 100%). Reprinted with permission from reference 20.

This confirms the role of the fimbrillin polypeptide in the binding of *P. gingivalis* to saliva-coated hydroxyapatite. The availability of large amounts of recombinant fimbrillin will allow detailed studies of the epitopes, binding domains, and receptor molecules important in fimbria-mediated interactions.

IN VIVO SIGNIFICANCE OF FIMBRIAE

The in vitro studies discussed above strongly suggest that the fimbrillin polypeptide subunits of *P. gingivalis* fimbriae play a major role in the binding and adherence of *P. gingivalis* to salivary pellicle-coated surfaces. The in vitro evidence for the role of *P. gingivalis* fimbrillin subunit in adherence can be summarized as follows: (i) the binding of *P. gingivalis* whole cells to the saliva-coated hydroxyapatite model for pellicle-coated surfaces is inhibited by highly purified, native-like *P. gingivalis* fimbriae; (ii) the binding of *P. gingivalis* cells to saliva-coated hydroxyapatite is inhibited by synthetic peptides corresponding to linear sequences mainly in the C-terminal one-third of the fimbrillin subunit; and (iii) the binding of *P. gingivalis* cells to saliva-coated hydroxyapatite is inhibited by the purified recombinant fimbrillin subunit corresponding to amino acids 10 to 337, showing that the fimbrillin polypeptide is important in *P. gingivalis* binding to saliva-coated surfaces.

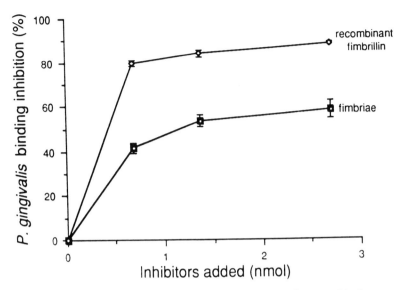

Figure 3. Inhibition of the binding of *P. gingivalis* whole cells to saliva-coated hydroxyapatite. The conditions of the assay were the same as for Fig. 2. The inhibitors include fimbriae, which is highly purified native fimbriae purified from cells, and recombinant fimbrillin purified from *E. coli*, which was induced to express the fimbrillin. The percent inhibition is compared to binding of *P. gingivalis* cells to saliva-coated hydroxyapatite in the absence of inhibitor, which is taken to be 100%. Reprinted with permission from reference 31.

It was then necessary to evaluate the extent to which the fimbrillin subunit contributes to *P. gingivalis* colonization and subsequent periodontal destruction in vivo. For these experiments we developed an insertional inactivation mutant of the *P. gingivalis* fimbrillin subunit. This mutant was then inoculated as a monoinfectant in a gnotobiotic rat model of *P. gingivalis*-induced periodontal destruction (16, 17). Previous experiments had shown that infection with wild-type strains of *P. gingivalis* would cause a periodontal disease-like syndrome involving loss of periodontal bony support (17). Furthermore, Klausen and coworkers (16) showed that gnotobiotic rats preimmunized with *P. gingivalis* whole cells were protected against periodontal destruction. In a further study, Evans and coworkers (8) found that preimmunization with purified native fimbriae protects against periodontal destruction in *P. gingivalis*-infected gnotobiotic rats, while comparable immunization with the 75-kDa component does not reduce the level of periodontal disease in these animals. These studies suggest that the *P. gingivalis*-infected gnotobiotic rat is a useful model in which to directly assess the role of the fimbrillin subunit in *P. gingivalis*-mediated periodontal destruction.

INACTIVATION OF THE *P. GINGIVALIS fimA* GENE

Targeted insertional mutagenesis was used to inactivate the *P. gingivalis fimA* gene. A fragment containing an internal coding sequence of the *fimA* gene of *P.*

gingivalis 381 was cloned into a plasmid and conjugally mobilized into *P. gingivalis* 381. This resulted in disruption of the *fimA* gene (24). Briefly, the method was as follows. Genomic DNA was isolated from *P. gingivalis*, digested with RNase, and precipitated with isopropanol. Plasmids pVAL-7 and pUC13Bg12.1 were purified from *E. coli*. The *Hinc*II fragment of the *fimA* 381 clone contained most of the internal *fimA* coding sequence. This fragment was inserted into the *EcoRV* site of pVAL-7. This construct was conjugally mobilized into *P. gingivalis* 381, and the resident *fimA* gene was found to be inactivated by homologous recombination.

Two mutants, DPG3 and DPG5, were obtained, and their stabilities were initially tested by growth in the absence of antibiotic selection. The growth kinetics of the two strains were identical to that of wild-type strain 381. In the absence of antibiotic selection, 100% of the clones examined remained erythromycin resistant. These data suggest that DPG3 and DPG5 are relatively stable; however, each insertion mutant was maintained on agar plates containing erythromycin for all subsequent experiments. SDS-polyacrylamide gel electrophoresis (PAGE) and Western immunoblots of total membrane preparations of the wild types of *P. gingivalis* 381, DPG3, and DPG5 were analyzed for alterations in membrane protein profile. Immunoblot analysis indicated that membrane preparations of both DPG3 and DPG5 lacked the 43-kDa fimbrial protein present in strain 381 (Fig. 4). Minor differences in other components between wild-type and mutant membrane protein profiles also were seen. For example, both mutants showed a slight decrease in the amounts of 29-kDa protein and an increase in the amounts of 32-kDa protein compared to wild type. Neither the 29- nor the 32-kDa protein reacted with anti-fimbrillin or antifimbria antisera.

It appears from these studies that insertion by homologous recombination of a large segment of the internal coding region of the *fimA* gene blocked production of fimbrillin by the mutant strain. Structurally, strain DPG3 also appeared to be devoid of fimbriae compared to the wild-type strain (Fig. 5). Electron micrographs of cross-sections of strain DPG3 indicated that the cell walls were typical

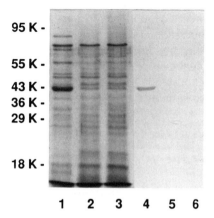

95 K -
55 K -
43 K -
36 K -
29 K -

18 K -

1   2   3   4   5   6

Figure 4. SDS-PAGE and Western immunoblot using SUNY extract of *P. gingivalis* 381 wild-type DPG3 and DPG5 mutants. For lanes 1 to 3, sonicates were separated on a 10% polyacrylamide gel and visualized by Coomasie blue. Lane 1, strain 381; lane 2, DPG3; lane 3, DPG5; lanes 4 through 6, immunoblot using antifimbrillin.

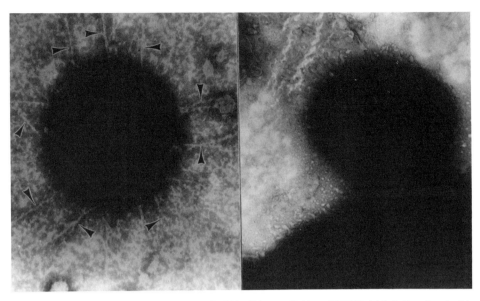

Figure 5. Electron microscopy of *P. gingivalis* 381 wild type (left) and DPG3 (right). Reprinted with permission from reference 24.

for a gram-negative bacterium and were not significantly different from the cell walls of the wild-type strain. Strain DPG5 showed electron-translucent ghosts in several preparations, suggesting that major alterations in cell surface had occurred, and we did not study this mutant further. In contrast, DPG3 lacked detectable fimbriae, but no major structural or biochemical alterations were seen in the cell wall. Therefore, *P. gingivalis* DPG3 was used as the *fimA* mutant for further study.

In vitro characterization of the *fimA* mutant of *P. gingivalis* DPG3 showed that DPG3 was not impaired in its ability to hemagglutinate sheep erythrocytes; i.e., it showed hemagglutinating activity comparable to that of the wild type. Furthermore, *P. gingivalis* DPG3 exhibited coadherence to *Streptococcus gordonii* G9B comparable to that of *P. gingivalis* 381 wild type. Thus, it appears that an intact fimbrial structure is not required for *P. gingivalis* hemagglutination or coadhesion with *S. gordonii*. In contrast, binding of DPG3 to saliva-coated hydroxyapatite beads was markedly reduced compared to wild-type binding. These data indicate that the fimbrial structure is important for adherence of *P. gingivalis* to this saliva-coated substrate. Having characterized the DPG3 mutant, we next proceeded to test the mutant in vivo in a gnotobiotic rat experiment.

INFECTION OF GNOTOBIOTIC RATS WITH *P. GINGIVALIS fimA* MUTANT

Male, germfree, Sprague-Dawley rats were kept under gnotobiotic conditions. Three groups were kept in separate isolators and at 4 weeks were either left uninfected, infected with wild-type *P. gingivalis*, or infected with the *P. gingivalis* *fimA* mutant DPG3. Each rat infected with *P. gingivalis* received 0.5 ml of 1.5 × 10^{12} cells per ml three times by gavage at 48-h intervals. Forty-two days after this last inoculation, the rats were sacrificed and exsanguinated under anesthesia. The periodontal bone level was assessed as defleshed rat jaws by both morphometric and radiographic methods as decribed by Klausen et al. (17). The morphometric method assesses bone loss from the cementoenamel junction to the alveolar bone crest on buccal root surfaces of maxillary teeth. The radiographic technique estimates the interproximal bone loss on the first mandibular molars.

Uninfected rats showed little or no bone loss over the 42-day period, and rats infected with wild-type *P. gingivalis* showed significant bone loss compared to that in the uninfected controls, findings consistent with those of previous experiments. However, the DPG3-inoculated rats failed to demonstrate bone loss by either method that was statistically greater than the level seen in the uninoculated controls (see Fig. 4 of chapter 23 in this volume). This study strongly suggests that the *fimA* DPG3 mutant of *P. gingivalis* is unable to induce periodontal destruction when inoculated into the germfree rat model system under conditions identical to that in which the wild type is able to cause periodontal destruction. Hence, the fimbriae of *P. gingivalis* likely play an important role in the virulence of this organism, possibly by mediating initial adherence and colonization of the oral cavity, a necessary first step in the initiation of periodontal destruction.

Although these studies suggest the fimbrillin polypeptide as a major determinant of in vitro and in vivo adherence of *P. gingivalis*, further studies will be needed to determine whether there are other proteins, such as adherence ligands, associated with the fimbrillin or other fimbrial-associated adhesins that are important in modulating this effect. For example, these other putative adhesins may be part of the fimbrial structure, but in the absence of a functional *fimA* protein, e.g., in the insertional inactivation mutant, they would then not be part of the fimbrial structure and would therefore not be able to function. Biochemical studies searching for fimbria-associated adhesins and genetic studies of the *fimA* gene cluster may help identify fimbria-associated proteins important in the synthesis and function of this important organelle.

SUMMARY

Adhesion is likely an important role for *P. gingivalis* fimbriae; however, these fimbriae have been reported to have other biologic activities. For example, *P. gingivalis* fimbriae function as antigens inducing a B-cell response after oral or parenteral immunization (27). *P. gingivalis* fimbriae also have been reported to stimulate fibroblast-derived cytokine FTAF (14), to induce expression of the neutrophil chemotactic factor KC gene of mice (13), and to induce interleukin-1 production from macrophages (12).

The exact role of *P. gingivalis* fimbriae in virulence awaits further study; however, the in vivo significance of *P. gingivalis* fimbriae as an important virulence factor is strongly suggested by the rat model. This model does not fully assess other interactions that may be occurring in human periodontal disease. These include preexisting immunity and *P. gingivalis* interactions with other oral organisms. Humans with periodontal disease make antibodies to the 43-kDa fimbrial subunit of *P. gingivalis* during the course of disease (6), which suggests that the fimbriae are expressed during periodontal infection and that the host reacts to the fimbrial molecule. Taken together, the evidence gives us confidence that designing experiments to further assess the mechanism of action of *P. gingivalis* fimbrillin in human periodontal disease is a useful endeavor.

REFERENCES

1. **Birkedal-Hansen, H., R. E. Taylor, J. J. Zambon, P. K. Barua, and M. E. Neiders.** 1988. Characterization of collagenolytic activity from strains of *Bacteroides gingivalis*. *J. Periodontal Res.* **23:**258–264.
2. **Christersson, L. A., B. G. Rosling, R. G. Dunford, U. M. E. Wikesjö, J. J. Zambon, and R. J. Genco.** 1988. Monitoring of subgingival *Bacteroides gingivalis* and *Actinobacillus actinomycetemcomitans* in the management of advanced periodontitis. *Adv. Dent. Res.* **2**(2)**:**382–388.
3. **Christersson, L. A., J. J. Zambon, R. G. Dunford, S. G. Grossi, and R. J. Genco.** 1989. Specific subgingival bacteria and diagnosis of gingivitis and periodontitis. *J. Dent. Res.* **68:**1633–1639.
4. **Chu, L., T. E. Bramanti, S. C. Holt, and J. L. Ebersole.** 1991. Hemolytic activity in the periodontopathogen, *Porphyromonas gingivalis:* kinetics of enzyme formation and localization. *Infect. Immun.* **58:**1932–1940.
5. **Clark, W. B., L. L. Baumann, and R. J. Genco.** 1978. Comparative estimates of bacterial affinities and adsorption sites on hydroxyapatite surfaces. *Infect. Immun.* **19:**846–853.
6. **De Nardin, A. M., H. T. Sojar, S. G. Grossi, L. A. Christersson, and R. J. Genco.** 1991. Humoral immunity of older adults with periodontal disease to *Porphyromonas gingivalis*. *Infect. Immun.* **59:**4363–4370.
7. **Dickinson, D. P., M. A. Kubiniec, F. Yoshimura, and R. J. Genco.** 1988. Molecular cloning and sequencing of the gene encoding the fimbrial subunit protein of *Bacteroides gingivalis*. *J. Bacteriol.* **170:**1658–1665.
8. **Evans, R. T., B. Klausen, H. T. Sojar, G. S. Bedi, C. Sfintescu, N. S. Ramamurthy, L. M. Golub, and R. J. Genco.** 1992. Immunization with *Porphyromonas* (*Bacteroides*) *gingivalis* fimbriae protects against periodontal destruction. *Infect. Immun.* **60:**2926–2935.
9. **Fujimura, S., and T. Nakamura.** 1987. Isolation and characterization of a protease from *Bacteroides gingivalis*. *Infect. Immun.* **55:**716–720.
10. **Gibbons, R. J., and I. Etherden.** 1983. Comparative hydrophobicities of oral bacteria and their adherence to salivary pellicles. *Infect. Immun.* **41:**1190–1196.
11. **Grenier, D., and B. C. MacBride.** 1987. Isolation of a membrane-associated *Bacteroides gingivalis* glycylprolyl protease. *Infect. Immun.* **55:**3131–3136.
12. **Hanazawa, S., Y. Murakami, K. Hirose, S. Amano, Y. Ohmori, H. Higuchi, and S. Kitano.** 1991. *Bacteroides* (*Porphyromonas*) *gingivalis* fimbriae activate mouse peritoneal macrophages and induce gene expression and production of interleukin-1. *Infect. Immun.* **59:**1972–1977.
13. **Hanazawa, S., Y. Murakami, A. Takeshita, H. Kitami, K. Ohta, S. Amano, and S. Kitano.** 1992. *Porphyromonas gingivalis* fimbriae induce expression of the neutrophil chemotactic factor KC gene of mouse peritoneal macrophages: role of protein kinase C. *Infect. Immun.* **60:**1544–1549.
14. **Hirose, K.** 1990. Stimulatory effects of *Bacteroides gingivalis* fimbriae on production of fibroblasts-derived thymocyte activating factor (FTAF) by human gingival fibroblasts. *J. Meikai Univ. School Dent.* **19**(1)**:**127–136.
15. **Isogai, H., E. Isogai, F. Yoshimura, T. Suzuki, W. Kogota, and K. Takano.** 1988. Specific inhibition of adherence of an oral strain of *Bacteroides gingivalis* 381 to epithelial cells by monoclonal antibodies against the bacterial fimbriae. *Arch. Oral Biol.* **33:**479–485.
16. **Klausen, B., R. T. Evans, N. S. Ramamurthy, L. M. Golub, C. Sfintescu, J.-Y. Lee, G. S. Bedi,**

J. J. Zambon, and R. J. Genco. 1991. Periodontal bone level and gingival proteinase activity in gnotobiotic rats immunized with *Bacteroides gingivalis*. *Oral Microbiol. Immunol.* **6:**193–201.

17. Klausen, R., R. T. Evans, and C. Sfintescu. 1989. Two complementary methods of assessing periodontal bone level in rats. *Scand. J. Dent. Res.* **97:**494–499.

18. Lawson, D. A., and T. F. Meyer. 1992. Biochemical characterization of *Porphyromonas (Bacteroides) gingivalis* collagenase. *Infect. Immun.* **60:**1524–1529.

19. Lee, J.-Y., H. T. Sojar, G. S. Bedi, and R. J. Genco. 1991. *Porphyromonas gingivalis* fimbrillin: size, amino-terminal sequence, and antigenic heterogeneity. *Infect. Immun.* **59:**383–389.

20. Lee, J.-Y., H. T. Sojar, G. S. Bedi, and R. J. Genco. 1992. Synthetic peptides analogous to the fimbrillin sequence inhibit adherence of *Porphyromonas gingivalis*. *Infect. Immun.* **60:**1662–1670.

21. Li, J., and R. P. Ellen. 1989. Relative adherence of *Bacteroides* species and strains to *Actinomyces viscosus* on saliva-coated hydroxyapatite. *J. Dent. Res.* **68:**1308–1312.

22. Loos, B. G., and D. W. Dyer. 1992. Restriction fragment length polymorphism analysis of the fimbrillin locus, *fimA*, of *Porphyromonas gingivalis*. *J. Dent. Res.* **71:**1173–1181.

23. Lund, B., F. Lindberg, M. Baga, and S. Normark. 1985. Globoside-specific adhesins of uropathogenic *Escherichia coli* are encoded by similar *trans*-complementable gene clusters. *J. Bacteriol.* **162:**1293–1301.

24. Malek, R., J. G. Fisher, A. Caleca, M. Stinson, C. J. van Oss, J.-Y. Lee, M.-I. Cho, R. J. Genco, R. T. Evans, and D. W. Dyer. 1994. Inactivation of the *Porphyromonas gingivalis fimA* gene blocks periodontal damage in gnotobiotic rats. *J. Bacteriol.* **176:**1052–1059.

25. Mayrand, D., and S. C. Holt. 1988. Biology of asaccharolytic black-pigmented *Bacteroides* species. *Microbiol. Rev.* **52:**134–152.

26. Murakami, Y., H. Nagata, A. Amano, M. Takagaki, S. Shizukuishi, A. Tsunemitsu, and S. Aimoto. 1991. Inhibitory effects of human salivary histatins and lysozyme on coaggregation between *Porphyromonas gingivalis* and *Streptococcus mitis*. *Infect. Immun.* **59:**3284–3286.

27. Ogawa, T., Y. Kusumoto, H. Kiyono, J. R. McGhee, and S. Hamada. 1992. Occurrence of antigen specific B cells following oral or parenteral immunization with *Porphyromonas gingivalis*. *Int. Immunol.* **4:**1003–1010.

28. Okuda, K., J. Slots, and R. J. Genco. 1981. *Bacteroides gingivalis, Bacteroides asaccharolyticus* and *Bacteroides melaninogenicus:* cell surface morphology and adherence to erythrocytes and human buccal epithelial cells. *Curr. Microbiol.* **6:**7–12.

29. Schwarz, S., R. P. Ellen, and D. A. Grove. 1987. *Bacteroides gingivalis-Actinomyces viscosus* cohesive interactions as measured by a quantitative binding assay. *Infect. Immun.* **55:**2391–2397.

30. Shah, H. N., and S. E. Charbia. 1989. Lysis of erythrocytes by the secreted cysteine proteinase from *Porphyromonas gingivalis* W83. *FEMS Microbiol. Lett.* **61:**213–218.

31. Sharma, A., H. T. Sojar, J.-Y. Lee, and R. J. Genco. 1993. Expression of a functional *Porphyromonas gingivalis* fimbrillin polypeptide in *Escherichia coli:* purification, physiochemical and immunochemical characterization, and binding characteristics. *Infect. Immun.* **61:**3570–3573.

32. Slots, J., and R. J. Genco. 1984. Black-pigmented *Bacteroides* species, *Capnocytophaga* species and *Actinobacillus actinomycetemcomitans* in human periodontal disease: virulence factors in colonization, survival and tissue destruction. *J. Dent. Res.* **63:**412–421.

33. Slots, J., and R. J. Gibbons. 1987. Attachment of *Bacteroides melaninogenicus* subsp. *asaccharolyticus* to oral surfaces and its possible role in colonization of the mouth and of periodontal pockets. *Infect. Immun.* **19:**254–264.

34. Slots, J., and M. A. Listgarten. 1988. *Bacteroides gingivalis, Bacteroides intermedius* and *Actinobacillus actinomycetemcomitans* in human periodontal diseases. *J. Clin. Periodontol.* **15:**85–93.

35. Socransky, S. S. 1977. Microbiology of periodontal disease—present status and future considerations. *J. Periodontol.* **48:**497–504.

36. Socransky, S. S., and A. D. Haffajee. 1991. Microbial mechanisms in the pathogenesis of destructive periodontal diseases: a critical assessment. *J. Periodontal Res.* **26:**195–212.

37. Sojar, H. T., J.-Y. Lee, G. S. Bedi, M.-I. Cho, and R. J. Genco. 1991. Purification, characterization and immunization of fimbrial protein from *Porphyromonas (Bacteroides) gingivalis*. *Biochem. Biophys. Res. Commun.* **175:**713–719.

38. **Suido, H., M. E. Neiders, P. K. Barus, M. Nakamura, P. A. Mashimo, and R. J. Genco.** 1987. Characterization of N-CBz-glycyl-glycyl-arginyl peptidase and glycyl-glycyl peptidase of *Bacteroides gingivalis. J. Periodontal Res.* **22:**412–418.

39. **Tsutsui, H., T. Kinorchi, Y. Wakano, and Y. Ohnishi.** 1987. Purification and characterization of a protease from *Bacteroides gingivalis* 381. *Infect. Immun.* **55:**420–427.

40. **Wennström, J. L., G. Dahlén, J. Svensson, and S. Nyman.** 1987. *Actinobacillus actinomycetemcomitans, Bacteroides gingivalis* and *Bacteroides intermedius:* predictors of attachment loss? *Oral Microbiol. Immunol.* **2:**158–163.

41. **Yoshimura, F., K. Takahashi, Y. Nodasaka, and T. Suzuki.** 1984. Purification and characterization of a novel type of fimbriae from the oral anaerobe *Bacteroides gingivalis. J. Bacteriol.* **160:**949–957.

42. **Yoshimura, F., T. Takasawa, M. Yoneyama, T. Yamaguchi, H. Shiokawa, and T. Suzuki.** 1985. Fimbriae from the oral anaerobe *Bacteroides gingivalis:* physical, chemical, and immunological properties. *J. Bacteriol.* **163:**730–734.

43. **Zambon, J. J., H. S. Reynolds, and J. Slots.** 1981. Black-pigmented *Bacteroides* spp. in the human oral cavity. *Infect. Immun.* **32:**198–203.

Molecular Pathogenesis of Periodontal Disease
Edited by Robert Genco et al.
© 1994 American Society for Microbiology, Washington, DC 20005

Chapter 3

Molecular Genetic Approaches for Identifying Virulence Factors of *Porphyromonas gingivalis*

Howard K. Kuramitsu

Periodontal diseases are now recognized as mixed bacterial infections of the supporting tissues of teeth (37). A variety of approaches have implicated *Porphyromonas gingivalis* and several other gram-negative anaerobic bacteria as important etiologic agents in these diseases (35). Therefore, it is important to identify the virulence factors used by these organisms to initiate tissue destruction in the gingival margin. One effective strategy for identifying these factors is to utilize isogenic mutants of *P. gingivalis* altered in a potential virulence factor in conjunction with appropriate animal model systems (Fig. 1). The introduction of molecular genetic approaches for isolating and characterizing the genes encoding some of these factors represents an initial step in such a general strategy. In addition, the recent development of gene transfer systems for *P. gingivalis* (8, 31) has allowed the construction of defined mutants in these organisms (7, 21).

A comparison of *P. gingivalis* strains with other microorganisms present in the human oral cavity revealed that the former organisms expressed several potential virulence traits (14, 35, 37). Among these were the elaboration of collagenases capable of degrading the type I and IV collagens abundant in periodontal tissue (10, 20, 33) and a relatively high proteolytic activity (14). In addition, since hemin appears to be growth limiting for these organisms (3) and also influences their pathogenicity in animal model systems (14, 24), the expression of both hemagglutinating and hemolytic activities by *P. gingivalis* can also be rationalized as potential virulence factors. It has been proposed that these factors, as well as others that will not be discussed in this review, may be important in the virulence of these organisms (14, 37). However, this possibility has not yet been rigorously tested for each of these properties. In this regard, the isolation of several genes from *P. gingivalis* expressing some of these activities (Table 1) and their utilization in constructing isogenic mutants will be crucial in identifying periodontopathic virulence factors.

Howard K. Kuramitsu • Department of Oral Biology, State University of New York, Buffalo, New York, 14214-3092.

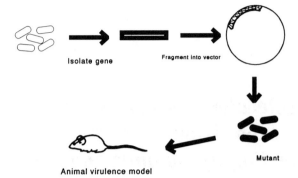

Isolate gene

Fragment into vector

Mutant

Animal virulence model

Figure 1. General strategy to test for virulence factors in *P. gingivalis*. Internal DNA fragments from isolated genes are transferred into conjugative plasmids in *Escherichia coli*. Following conjugation with *P. gingivalis* recipients, homologous recombination of the plasmid into the chromosome results in mutants containing two incomplete copies of the cloned gene. The mutants can then be compared with the parental organisms for virulence in appropriate animal models.

COLLAGENASES

The expression of collagenases capable of degrading human collagens is limited to only a small group of bacteria (34). In the human oral cavity, only *P. gingivalis* strains appear to be capable of degrading type I collagen (22). Since extensive breakdown of type I collagen occurs during periodontal destruction (27), it is reasonable to think that the collagenases of these organisms may act as virulence factors in periodontitis. More recent investigations have demonstrated that the majority of the collagenase activity present in the gingival fluids from periodontal inflammatory sites is host derived (39). Nevertheless, evidence that some of this destruction may be due to bacterial collagenases that degrade type I collagen into small peptide fragments was also obtained (38). It therefore appears that the *P. gingivalis* collagenases may play only a minor role in the direct destruction of type I collagen in periodontal tissue. However, these enzymes together with other pro-teases elaborated by *P. gingivalis* may aid in providing these organisms and other

Table 1. Genes for putative virulence factors from *P. gingivalis*

Gene	Proposed function	Reference
Fimbrillin	Colonization	6
prtT	Protease	26
tpr	Protease	2
44-kDa protease	Protease and hemagglutinin	16
150-kDa protease	Protease; fibronectin and fibrinogen binding	29
hagA	Hemagglutinin	32
hagB	Hemagglutinin	21
hagC	Hemagglutinin	21
hagD	Hemagglutinin	30
Hemolysin	Erythrocyte lysis	17

subgingival plaque bacteria with a source of amino acids following hydrolysis of collagens.

Recently, a gene coding for a collagenase activity has been isolated from *P. gingivalis* ATCC 53977 and characterized (18). This gene, *prtC*, appears to be present in each of the three major serotypes of this organism. Unlike mammalian collagenases, the *prtC* gene product appears to degrade type I collagen into small peptide fragments. The enzyme is also somewhat unusual in that it does not express detectable gelatinase activity. More recently, this same gene has been isolated from strain 381 by PCR in conjunction with oligonucleotide primers derived from the sequence of the *prtC* gene, and the enzyme product was demonstrated to weakly degrade type I collagen but not type III collagen (1). In this regard, another collagenase activity distinct from the *prtC* gene product has been isolated from strain 53977 (40). This enzyme appears to be similar to that described earlier (35) in that the enzyme also displays trypsinlike activity, and it will be of interest to isolate and characterize the gene encoding this activity.

The availability of genes from *P. gingivalis* that express collagenase activity now makes it possible to construct isogenic mutants defective in this activity for evaluation in animal model systems (4, 15). In addition, these mutants will help determine how many collagenase genes are expressed by each strain of *P. gingivalis* and what the potential role in virulence of each of these genes is.

PROTEASES

Since *P. gingivalis* strains are highly proteolytic (14, 37), it has been proposed that this property may contribute to the virulence of these organisms. In this regard, proteases may be important for (i) hydrolysis of proteins to provide essential amino acids for the organisms, (ii) neutralization of host immune systems, (iii) degradation of supporting tissue in the periodontal pocket, (iv) activation of host procollagenases (38), (v) exposure of cryptotopes for bacterial attachment to host tissue, and (vi) mediation of attachment to erythrocytes and other cells. Therefore, it would not be surprising if one or more proteases produced by *P. gingivalis* turn out to play a significant role in the overall virulence of this organism.

A number of laboratories have reported the purification and characterization of proteases from *P. gingivalis* (reviewed in reference 14). With this approach alone, it has been difficult to determine the number of specific proteases produced by each strain, since multiple proteases are present (11) and the highly proteolytic nature of the organism makes differentiation of the enzymes problematic (i.e., partial degradation of a protease can alter the molecular size as well as some of the properties of the enzyme). Nevertheless, this organism likely expresses two major classes of proteases: serine and cysteine proteases, with several of the highly purified enzymes from *P. gingivalis* belonging to the latter group (5, 13, 25). Despite the fact that proteases capable of hydrolyzing the synthetic trypsin substrates BANA (*N*-α-benzoyl-D,L-arginine-β-naphthylamide) or BAPNA (*N*-α-benzoyl-D,L-arginine-*p*-nitroanilide) can be readily detected in *P. gingivalis* and have been designated "trypsinlike proteases" (11), more recent results with highly purified enzyme preparations indicate that some of these enzymes are specific for arginine- and not lysine-containing peptide bonds (5, 13, 25, 26).

In order to further characterize the *P. gingivalis* proteases, several laboratories have reported the isolation of the genes encoding protease activity (Table 1). Bourgeau and colleagues (2) characterized the *tpr* gene from *P. gingivalis* W83 expressing a 62.5-kDa protease. The deduced amino acid sequence of the gene revealed a domain with homology to cysteine proteases from both eucaryotes and procaryotes. Independently, Park and McBride (28) isolated the same gene from strain W83. Preliminary evidence from this laboratory (23) has further suggested that this gene may be important for the growth of the organism, since isogenic mutants of this strain that are defective in the protease gene grew very poorly in vitro. This suggests that the *tpr* gene might play an important role in the growth of *P. gingivalis* in the environment of the periodontal pocket.

Another protease gene, *prtT*, has recently been isolated from strain 53977 and characterized (26). This gene encodes a protein of 53.9 kDa whose activity was strongly dependent on the presence of thiol agents, a feature characteristic of most proteases from *P. gingivalis*. In contrast to the *tpr* gene product, the *prt* protease hydrolyzed the synthetic trypsin substrate BAPNA. Although similar in size to the gingipain-1 protease recently described (5), the *prtT* protease does not contain any peptide sequences analogous to the partial amino acid sequence determined for the former enzyme (5). Since the *prtT* gene is located near the *prtC* collagenase gene on the strain 53977 chromosome and since the protease degrades gelatin, it has been proposed that the protease may play a role in the further cleavage of degradation products of type I collagen produced by the action of collagenases.

Previously, Yoshimura's laboratory (25) described a trypsinlike cysteine protease that also appears to act as a hemagglutinin with human erythrocytes. The gene for this enzyme has recently been isolated (16), and its characterization is currently in progress. The N-terminal amino acid sequence of this 44-kDa protease has been determined (Table 2); it is not homologous to sequences present in either the *tpr* or the *prtT* gene products. A protease that appears to mediate binding of *P. gingivalis* W12 to fibrinogen as well as to fibronectin has also been previously characterized (19). The gene for this 150-kDa protease has recently been isolated and partially sequenced (29). Interestingly, several domains of this enzyme share extensive amino acid sequence homology with the *hagA* hemagglutinin isolated from strain 381 (32). The significance of such homology remains to be determined, but the homology suggests a possible evolutionary relationship between adhesins and proteases from *P. gingivalis*.

Table 2. N-terminal amino acid sequences of *P. gingivalis* proteases

Name	N-terminal sequence	Source[a]	Reference
44 kDa	SGQAEIVLEAHDV	Protein	41
Gingipain	YTPVEEKQNGRMIVIVAKKYEGDIKDFVDW	Protein	5
PRTT	MVYHVYPNWFRLQNFPVCYSSK	Nucleic acid	26
TPR	MEKKLVPQSISKERLQKLEAQATLT	Nucleic acid	2

[a] Protein, N-terminal amino acid sequencing of protein; Nucleic acid, amino acid sequence deduced from nucleotide sequence.

It is of interest that the 44-kDa protease appears to act as a bifunctional molecule at the same substrate site (i.e., inhibitors such as leupeptin interfere with both protease and hemagglutinating activities [25]). In contrast, the 150-kDa protease contains a fibrinogen-fibronectin binding site that appears to be distinct from the protease catalytic site (19). It will be of interest to determine whether other cell surface adhesins of *P. gingivalis* are also bifunctional molecules.

Assuming that no strain-specific proteases are elaborated by *P. gingivalis*, it is likely that at least four distinct proteases are expressed by these organisms (Table 2). This speculation is based on the known amino acid sequences derived from N-terminal amino acid sequencing of the proteins or on the sequences deduced by nucleotide sequencing of the cloned genes. The specific role of each of these proteases in the physiology or virulence of *P. gingivalis* remains to be determined, but the problem can now be addressed by utilizing isogenic protease mutants.

PROSPECTS FOR DEFINING *P. GINGIVALIS* VIRULENCE FACTORS

Both biochemical and molecular genetic approaches indicate that each *P. gingivalis* strain can very likely express multiple collagenases, proteases, and hemagglutinins (30). One practical consequence of this situation is that it may be difficult to isolate some of the genes involved in these activities by utilizing the transposon mutagenesis strategy (9). Since this strategy relies on screening for the loss of a particular function, the presence of multiple genes expressing identical phenotypes may make it difficult to isolate such mutants by this approach.

Because there appear to be two collagenase genes in each strain, it should be feasible to construct both single and double collagenase mutants following conjugation with plasmids containing internal DNA fragments of each gene. Such mutants could be tested in both animal inflammation (4) and periodontitis (15) model systems to assess the relative contributions of each gene product to virulence (of course recognizing the limitations of such models relative to human periodontitis). It may be more difficult to make such an assessment for the proteases and hemagglutinins of these organisms, since more than two structural genes are capable of expressing each property. However, judging from testing in animal model systems, it may be that not all of these genes are equally important for virulence, and it may be possible to identify those that play a critical role in the pathogenicity of *P. gingivalis*. It will also be necessary to examine the regulation of expression of these potential virulence factors to determine whether local environmental conditions (presence of other microorganisms, pH, extent of bleeding, etc.) in the animal models are relevant to those in the human periodontal pocket. This condition is one potential confounding factor in relating the results of animal models to the human situation.

The virulence potentials of individual *P. gingivalis* gene products could also be assessed by transferring the relevant genes to other less pathogenic organisms by utilizing novel gene transfer techniques (12). These chimeric organisms could then be tested in appropriate animal models for confirmation of their important roles in pathogenicity. It is likely that the availability of isolated genes coding for potential periodontopathic virulence factors and their utilization in a variety of approaches will lead to a more rigorous testing of their possible roles in periodontitis.

Acknowledgments. I thank D. Dyer, S. C. Holt, M. Lantz, D. Mayrand, B. C. McBride, A. Progulske-Fox, and F. Yoshimura for communicating unpublished results to me.

Work from my laboratory that is cited in this review was supported in part by U.S. Public Health Service grant DE08293.

REFERENCES

1. **Aoki, T., T. Kuroki, S. Otake, and T. F. Meyer.** 1993. Program Abstr. 71st Int. Assoc. Dent. Res., abstr. 415.
2. **Bourgeau, G., H. Lapointe, P. Peloquin, and D. Mayrand.** 1992. Cloning, expression, and sequencing of a protease gene *(tpr)* from *Porphyromonas gingivalis* W83 in *Escherichia coli. Infect. Immun.* **60:**3186–3192.
3. **Bramanti, T. E., and S. C. Holt, 1991.** Roles of porphyrins and host iron transport proteins in regulation of growth of *Porphyromonas gingivalis. J. Bacteriol.* **173:**7330–7339.
4. **Chen, P. B., M. E. Neiders, S. J. Millar, H. S. Reynolds, and J. J. Zambon.** 1987. Effect of immunization on experimental *Bacteroides gingivalis* infection in a murine model. *Infect. Immun.* **55:**2534–2537.
5. **Chen, Z., J. Potempa, A. Polanowski, M. Wikstrom, and J. Travis.** 1992. Purification and characterization of a 50-kDa cysteine protease (gingipain) from *Porphyromonas gingivalis. J. Biol. Chem.* **267:**18896–18901.
6. **Dickenson, D. P., M. A. Kubiniec, F. Yoshimura, and R. J. Genco.** 1988. Molecular cloning and sequencing of the gene encoding the fimbrial subunit protein of *Bacteroides gingivalis. J. Bacteriol.* **170:**1658–1665.
7. **Dyer, D.** Personal communication.
8. **Dyer, D. W., G. Bilalis, J. H. Michel, and R. Malek.** 1992. Conjugal transfer of plasmid and transposon DNA from *Escherichia coli* into *Porphyromonas gingivalis. Biochem. Biophys. Res. Commun.* **186:**1012–1019.
9. **Genco, C. A., M. Lassiter, R. Y. Forng, and M. O. Basil.** 1993. Program Abstr. 71st Int. Assoc. Dent. Res., abstr. 123.
10. **Gibbons, R. J., and B. MacDonald.** 1961. Degradation of collagenous substrates by *Bacteroides melaninogenicus. J. Bacteriol.* **81:**614–621.
11. **Grenier, D., G. Chao, and B. C. McBride.** 1989. Characterization of sodium dodecyl sulfate-stable *Bacteroides gingivalis* proteases by polyacrylamide gel electrophoresis. *Infect. Immun.* **57:**95–99.
12. **Guiney, D. G., and K. Bouic.** 1990. Detection of conjugal transfer systems in oral, black-pigmented *Bacteroides* spp. *J. Bacteriol.* **172:**495–497.
13. **Hinode, D., H. Hayashi, and R. Nakamura.** 1991. Purification and characterization of three types of proteases from culture supernatants of *Porphyromonas gingivalis. Infect. Immun.* **59:**3060–3068.
14. **Holt, S. C., and T. E. Bramanti.** 1991. Factors in virulence expression and their role in periodontal disease pathogenesis. *Crit. Rev. Oral Biol. Med.* **2:**177–281.
15. **Holt, S. C., J. Ebersole, J. Felton, M. Brunsvold, and K. S. Kornman.** 1988. Implantation of *Bacteroides gingivalis* in nonhuman primates initiates progression of periodontitis. *Science* **239:**55–57.
16. **Ikeda, T., F. Yoshimura, and M. Nishikata.** 1993. Program Abstr. 71st Int. Assoc. Dent. Res., abstr. 119.
17. **Karunakaran, T., and S. C. Holt.** Personal communication.
18. **Kato, T., N. Takahashi, and H. K. Kuramitsu.** 1992. Sequence analysis and characterization of the *Porphyromonas gingivalis prtC* gene, which expresses a novel collagenase activity. *J. Bacteriol.* **174:**3889–3895.
19. **Lantz, M. S., R. D. Allen, L. W. Duck, J. L. Blume, L. M. Switalski, and M. Hook.** 1991.

Identification of *Porphyromonas gingivalis* components that mediate its interaction with fibronectin. *J. Bacteriol.* **173:**4263–4270.

20. **Lawson, D. A., and T. F. Meyer.** 1992. Biochemical characterization of *Porphyromonas (Bacteroides) gingivalis* collagenase. *Infect. Immun.* **60:**1524–1529.

21. **Lepine, G., and A. Progulske-Fox.** 1993. Program Abstr. 71st Int. Assoc. Dent. Res., abstr. 120.

22. **Mayrand, D., and D. Grenier.** 1985. Detection of collagenase activity in oral bacteria. *Can. J. Microbiol.* **31:**134–138.

23. **McBride, B. C.** Personal communication.

24. **McKee, A. S., A. S. McDermid, A. Baskerville, A. B. Dowsett, D. C. Ellwood, and P. D. Marsh.** 1986. Effect of hemin on the physiology and virulence of *Bacteroides gingivalis* W50. *Infect. Immun.* **52:**349–355.

25. **Nishikata, M., and F. Yoshimura.** 1991. Characterization of *Porphyromonas (Bacteroides) gingivalis* hemagglutinin as a protease. *Biochem. Biophys. Res. Commun.* **178:**336–342.

26. **Otogoto, J., and H. K. Kuramitsu.** 1993. Isolation and characterization of the *Porphyromonas gingivalis prtT* gene, coding for protease activity. *Infect. Immun.* **61:**117–123.

27. **Page, R. E., and H. E. Schroeder.** 1981. Current status of the host response in chronic marginal periodontitis. *J. Periodontol.* **52:**477–491.

28. **Park, Y., and B. C. McBride.** 1992. Cloning of a *Porphyromonas (Bacteroides) gingivalis* protease gene and characterization of its product. *FEMS Microbiol. Lett.* **92:**273–278.

29. **Patti, J. M., M. S. Lantz, and A. Progulske-Fox.** 1993. Program Abstr. 71st Int. Assoc. Dent. Res., abstr. 418.

30. **Progulske-Fox, A.** Personal communication.

31. **Progulske-Fox, A., A. Oberste, and W. P. McArthur.** 1989. Transfer of plasmid pE5-2 from *Escherichia coli* to *Bacteroides gingivalis* and transposition of Tn4351 to the *B. gingivalis* chromosome. *Oral Microbiol. Immunol.* **4:**132–134.

32. **Progulske-Fox, A., S. Tumwasorn, and S. C. Holt.** 1989. The expression and function of a *Bacteroides gingivalis* hemagglutinin gene in *Escherichia coli. Oral Microbiol. Immunol.* **4:**121–131.

33. **Robertson, P., M. Lantz, P. T. Marucha, K. S. Kornman, C. L. Trummel, and S. C. Holt.** 1982. Collagenolytic activity associated with *Bacteroides* species and *Actinobacillus actinomycetemcomitans. J. Periodontal. Res.* **17:**275–283.

34. **Seifter, S., and E. Harper.** 1970. Collagenases. *Methods Enzymol.* **16:**613–635.

35. **Slots, J., and R. J. Genco.** 1984. Black-pigmented *Bacteroides* species, *Capnocytophaga* species, and *Actinobacillus actinomycetemcomitans* in human periodontal diseases: virulence factors in colonization, survival, and tissue destruction. *J. Dent. Res.* **63:**412–421.

36. **Smalley, J. W., A. J. Birss, and C. A. Shuttleworth.** 1988. The degradation of type I collagen and human plasma fibronectin by the trypsin-like enzyme and extracellular membrane vesicles of *Bacteroides gingivalis* W50. *Arch. Oral Biol.* **33:**323–329.

37. **Socransky, S. S., and A. D. Haffajee.** 1991. Microbial mechanisms in the pathogenesis of destructive periodontal diseases: a critical assessment. *J. Periodontal Res.* **26:**195–212.

38. **Sorsa, T., T. Ingman, K. Suomalainen, M. Haapasalo, Y. T. Konttinen, O. Lindy, H. Saari, and V.-J. Uitto.** 1992. Identification of proteases from periodontopathic bacteria as activators of latent neutrophil and fibroblast-type interstitial collagenase. *Infect. Immun.* **60:**4491–4495.

39. **Uitto, V.-J., K. Suomalainen, and T. Sorsa.** 1990. Salivary collagenase. Origin, characteristics, and relationship to periodontal health. *J. Periodontal. Res.* **25:**135–142.

40. **Yoneda, M., and H. K. Kuramitsu.** Unpublished data.

41. **Yoshimura, F.** Personal communication.

Molecular Pathogenesis of Periodontal Disease
Edited by Robert Genco et al.
© 1994 American Society for Microbiology, Washington, DC 20005

Chapter 4

DNA Fingerprinting of *Porphyromonas gingivalis* by Arbitrarily Primed PCR

Christian Mouton and Christian Ménard

RATIONALE FOR TYPING

There is a need for objective and easy-to-use tests to delineate taxa within a single pathogenic species and to provide a basis for the identification of individual strains. Quick and easy means of distinguishing between isolates that are morphologically and biochemically identical are particularly desirable for sorting epidemiological information concerning bacterial reservoirs and transmission of any given pathogen. Such procedures are usually called typing. In particular, in those infectious diseases for which the probable etiologic agent is a member of the normal flora, simple species identification is not useful in distinguishing between infection and colonization or in tracing the source of the infecting organism. Such is the case for *Porphyromonas (Bacteroides) gingivalis,* a suspected etiologic agent of chronic adult periodontitis. There is experimental evidence for virulent and nonvirulent strains of this species (reviewed in reference 19) that awaits being verified in vivo, pending the availability of a suitable typing method.

Conventional typing methods in general bacteriology include biochemical profiling, serotyping, antibiotic susceptibility patterns, bacteriocin production and susceptibility, electrophoretic profiling of whole-cell proteins, and phage typing, to name a few. Of these, only a few have been applied to *P. gingivalis*. Serotyping remains the most widely used typing method in bacteriology, and several serological studies of *P. gingivalis* have, to a certain extent, paved the way for a serotyping scheme for the species. Indeed, serological studies of the former taxon *Bacteroides melaninogenicus* were useful in revealing the heterogeneity of the group (15), and the serological analysis of asaccharolytic black-pigmented *Bacteroides* spp. was instrumental in the emergence of *Bacteroides gingivalis* (now *Porphyromonas gingivalis*) as a species distinct from *Bacteroides asaccharolyticus* (now *Porphyromonas asaccharolytica*) (23, 24). Other studies (7, 26, 29) have further shown that *P. gingivalis* is antigenically distinct from other black-pigmented *Bacteroides* spp. Our

Christian Mouton and Christian Ménard • Groupe de Recherche en Ecologie Buccale, Faculté de Médecine Dentaire, Université Laval, Québec, Québec, Canada G1K 7P4.

own work (27) has suggested serological heterogeneity within *P. gingivalis* by differentiating between human and animal serotypes. Three serogroups that seem to correlate with pathogenicity have been described (8). Gmür et al. (11) reported two serotypes, Umemoto et al. (33) reported three, Nagata et al. (25) reported four, and Fujiwara et al. (9) reported two lipopolysaccharide serogroups. Overall, these studies revealed that *P. gingivalis* is a relatively diverse species, but success in providing a reliable and discriminatory typing method has been only limited, beyond the fact that in four of the above-mentioned studies in which strains 381 and W50 were analyzed, they were reported as belonging to separate serovars. Obviously, there is a need to establish an alternative and more accurate test method to epidemiologically group related strains within the species and possibly to establish correlations with pathogenicity. A review of the conventional typing methods for black-pigmented gram-negative anaerobes, including *P. gingivalis,* has recently been presented elsewhere (37).

Each of the conventional typing methods has specific applications and advantages; however, these methods rarely allow strain identification, thus hampering detailed epidemiological studies. It is now generally accepted that the low level of discrimination and the poor reproducibility associated with conventional methods (which, because they use phenotypic properties that rely on gene expression, may be unstable) can be circumvented by using DNA-based characterization. Indeed, with advances in molecular biology, new techniques for systematics and for strain-specific epidemiological studies have become available, and they offer the potential of determining whether clinical significance can be related to typing results.

FINGERPRINTING TOOLS

Since the discrimination of microorganisms by a detectable phenotype implies that the organisms should also be distinguishable at the DNA level, working with DNA has become part of the mainstream of molecular typing methods. Few studies have applied the molecular typing methods presently available to investigations of the genetic heterogeneity and/or transmission of periodontal pathogens: restriction endonuclease analysis (REA) (4, 18, 28, 35, 36, 42), restriction fragment length polymorphism (RFLP) (6, 16), ribotyping (30, 32, 34), and multilocus enzyme electrophoresis (MLEE) (17). A brief review of the advantages and limits of these techniques and the data on *P. gingivalis* they have generated is presented below.

Whole-Cell DNA REA

Restriction endonucleases recognize and cleave double-stranded DNA at specific base pair sequences. The DNA fragments generated are separated by electrophoresis, stained with ethidium bromide, and visualized with UV light. The genetic heterogeneity and homogeneity of strains can then be evaluated by comparing the numbers and sizes of the DNA fragments obtained. These DNA fragment patterns constitute a specific fingerprint for characterizing each strain. However, patterns are complex and often difficult to compare, as they suffer from the large number of generated fragments that often hinder the resolution of individual fragments. Ideally, no more than 30 bands should be obtained, which involves screening a large number of endonucleases. The REA fingerprinting study of Loos et al. (18) distinguished 25 patterns among 39 strains, indicative of a considerable genetic heterogeneity within the species *P. gingivalis*. Using REA, van Steenbergen et al. (36) showed that two *P. gingivalis* isolates from two schoolgirls were distinct, as they were also distinct from laboratory strains: only the two well-known laboratory strains W50 and W83 had similar patterns. This observation suggested that bacterial cross-infection did not occur between subjects in this school. The same authors (35) recently provided evidence for the transmission of *P. gingivalis* between spouses. It was shown that with one exception, each individual was colonized with only one clonal type of *P. gingivalis* and that the REA patterns of all isolates from unrelated individuals were distinct; in contrast, both husband and wife in six of the eight couples under study shared identical REA types.

RFLP

DNA fragments first obtained by the restriction enzyme technique can be identified by a DNA probe, so that polymorphism in the DNA probe region can be detected. Geneticists have coined the name RFLP for this procedure. DNA is digested by restriction enzymes, and the fragments generated are separated by size on a gel, transferred to a solid support, and hybridized to a labeled cloned sequence acting as a probe. Small structural differences in the vicinity of the genomic DNA regions homologous to the probe affect the length of the fragment containing the probe-homologous sequence. This can be detected as a change, termed an RFLP, in the location of the hybridization band. Although this procedure yields patterns easier to analyze than those of REA and has been applied with success to differentiate organisms that appear homogeneous by biochemical or serological tests, it remains cumbersome to use. RFLPs suffer from the additional limitations of high cost when large populations are examined and of monitoring only that part of the genome that possesses the restriction sites under scrutiny. RFLP analysis with the *fimA* gene distinguished 25 patterns among 39 strains of *P. gingivalis* that could be divided into nine groups (16).

rRNA Restriction Patterns, Ribotyping

Ribotyping is another approach to reducing the number of interpretable DNA fragments. It consists of the hybridization of standard REA fragments with *Escherichia coli* rDNA (DNA coding for rRNA), which is well conserved among bacteria and thus provides a broad spectrum of hybridization with virtually all bacterial genomes. Only DNA fragments that contain a portion of the rRNA genes are visualized, thus decreasing the number of stained fragments to 20 or less. A preliminary study (32) of 25 strains of *P. gingivalis* from six subjects revealed that all isolates from the same subject were of the same ribotype, but ribotypes differed from subject to subject. Saarela et al. (30) also observed that the ribotypes of intraindividual strains were identical in eight of nine subjects, whereas those of interindividual strains were different. In a recent study (34), ribotyping was used as one of three different methods for typing *P. gingivalis*. Eleven ribotypes were identified among 32 isolates obtained from eight couples; the ribotypes from unrelated subjects were all distinct, whereas both husband and wife in six couples shared identical ribotypes.

MLEE

MLEE is based on the principle that mutations in genes coding for bacterial metabolic enzymes can be detected by differences in the electrophoretic migration patterns of the enzymes. Since many bacterial enzymes are polymorphic, MLEE provides a highly discriminatory method for the detection of bacterial clones, which are identified by distinctive electrophoretic enzyme profiles, called electrophoretic types (ET). A comprehensive study of 100 strains of *P. gingivalis* (17) revealed a total of 78 ETs, and cluster analysis placed these strains in three phylogenetic divisions. The average genetic distances between ETs of division I (88 human and 4 monkey strains) and those of division II (5 animal strains) and division III (3 isolates from sheep with broken-mouth periodontitis) were 0.70 and 0.87, respectively. The major outcome of this work was to indicate that periodontal patients are infected by strains with a wide variety of chromosomal genotypes and, accordingly, that interclonal variation in pathogenicity is small.

Another chromosome-based epidemiological marker system used to generate polymorphisms is pulsed-field gel electrophoresis (PFGE). It has been proposed that PFGE offers greater discriminatory power than ribotyping, since this method detects the distribution of restriction sites throughout the chromosome. This method has not yet been applied to studies of *P. gingivalis*. Plasmid typing, another molecular method, has the shortcoming of not being useful with plasmid-free species, of which *P. gingivalis* is one.

AP-PCR Generates Fingerprints: the RAPD Pattern

A novel procedure, called arbitrarily primed PCR (AP-PCR), can be used for generating fingerprints of the bacterial genome. Only a few studies have applied this new method to studies of *P. gingivalis* (3, 5). The most complete set of data comes from our own study (21), which will be summarized below.

In 1990, Williams et al. (40) and Welsh and McClelland (38) simultaneously proposed that amplification of random segments of genomic DNA can be directed by a single oligonucleotide primer of arbitrary sequence by using AP-PCR. A characteristic spectrum of short DNA products of various complexities is thus generated, and polymorphisms in the lengths of the amplified sequences obtained, or randomly amplified polymorphic DNA (RAPD), can be used to compare bacterial strains. This method is distinct from the classic PCR in the use of a single primer instead of two and a low-stringency annealing temperature. An AP-PCR protocol that produces DNA fingerprints both from pure genomic DNA and directly from *P. gingivalis* colonies has been developed.

Pure genomic DNA is obtained by using the minipreparation procedure of

Wilson (41), to which an RNase treatment done according to Smith et al. (31) is added.

From a bank of synthetic oligonucleotide primers, each 9 nucleotides long, four primers were selected (910-05 [5′ CCGGCGGCG], 910-09 [5′ CCGGGCCGC], 940-11 [5′ GTCTCGGGG], and 970-11 [5′ GTAAGGCCG]). The nucleotide sequence of each primer in this bank had been determined by a computer using a random sequence generator program in the absence of any nucleotide sequence information for the species tested. Amplification reactions were performed by using a modification of the method described by Williams et al. (40). We used a reaction mixture of 25 μl containing 20 mM Tris-HCl (pH 8.3); 50 mM KCl; 3 mM MgCl$_2$; 0.001% gelatin; 200 μM each dATP, dCTP, dGTP, and dTTP; 0.4 μM primer; 25 ng of genomic DNA; and 1.25 U of *Taq* DNA polymerase (Pharmacia, Baie d'Urfé, Canada). Amplification was performed in a DNA Thermal Cycler (Perkin Elmer Cetus, Montreal, Canada) programmed for 25 cycles of 1 min at 94°C, 1 min at 32°C, and 2 min at 72°C, with the fastest available transitions between temperatures. A negative control without template DNA was included in each AP-PCR run. All such controls were negative. Amplification products were compared by electrophoresis in 1.5% SeaKem GTG agarose gels (FMC, Rockland, Maine) in 0.04 M Tris-acetate–0.002 M EDTA (pH 8.5), stained with ethidium bromide, and photographed on a UV transilluminator. A 1-kb DNA ladder (GIBCO BRL, Burlington, Canada) was included as a size marker.

That fingerprints from *P. gingivalis* can be obtained by AP-PCR is shown in Fig. 1. A characteristic spectrum of short DNA products of various complexities is generated, and polymorphisms in the lengths of amplified sequences obtained (RAPD) can be used to compare bacterial strains. In the experiment resulting in Fig. 1, the RAPD patterns of five well-known laboratory strains were compared, allowing us to explore the discriminative powers of two primers.

Banding patterns containing from one to eight amplicons located between the 298- and 3,054-bp markers are indicative of genetic polymorphism. The small number of amplicons obtained in each pattern attests that AP-PCR fingerprinting does not suffer from a large number of generated fragments, a significant improvement over REA. When primer 970-11 is used, there is an obvious genetic relatedness, as indicated by a set of three amplicons common to all strains. Pairwise comparison of strains W50 and W83 suggests that these two strains are identical. Pairwise

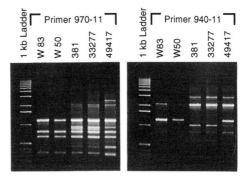

Figure 1. AP-PCR fingerprints of five key laboratory strains of *P. gingivalis* obtained with two primers. Genetic polymorphism within the species is shown by the distinct banding patterns (number and position of the amplicons); each pattern is called an RAPD fingerprint. Genetic relatedness is indicated by amplicons common to strains, and distinct RAPD fingerprints attest to the discriminative power of each primer.

comparison of 381 and 33277 suggests that these two strains are also identical, although they clearly belong to a type different from that of W50 and W83. Strain 49417, which has amplicons in common with both previous types, is nevertheless distinct and thus represents yet another type. When primer 940-11 is used, strains 381 and 33277 once again appear identical, a confirmation of the observation by Loos et al. (18) that the reference American Type Culture Collection strain is most likely derived from strain 381. In contrast, W50 and W83 are no longer identical, suggesting that these two closely related strains may in fact belong to two distinct clonal types. The fingerprint of strain 49417 is once again distinct from that of the other types. It is noticeable that the discriminatory power of AP-PCR does not rely on the complexity of fingerprints, since, when a primer such as 940-11 is used, patterns of one to three amplicons do the job.

Interassay reproducibility of the AP-PCR procedure is excellent, since it has been shown that the patterns and sizes of amplified fragments remain the same for 10 DNA preparations obtained from our standard strain (ATCC 49417) over a period of 8 months (data not shown). It has also been shown that the use of four preparations ([i] pure genomic DNA, [ii] DNA and protein, [iii] DNA and RNA, and [iv] DNA, protein, and RNA) as templates for amplification by AP-PCR resulted in identical banding patterns. This result clearly demonstrates that contamination of DNA by foreign material does not alter the profile of the amplification products (amplicons). This observation leads us to the conclusion that for AP-PCR, as for PCR, there is no requirement for highly purified DNA.

Accordingly, the AP-PCR protocol can be simplified by using cells harvested directly from colonies to provide template DNA (Fig. 2). A single colony at least 1 mm in diameter harvested with a disposable sterile needle is added to a sterile

TEMPLATE DNA FROM WHOLE CELLS

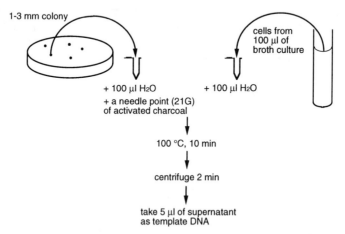

Figure 2. Template DNA to be used for AP-PCR amplification can be obtained directly from bacterial cells; using colonies, a soluble product, probably derived from porphyrin, acted as a PCR inhibitor. Inhibition was overcome by adsorbing the suspension of boiled cells with activated charcoal. AP-PCR of a single colony is a quick, simple, reproducible procedure.

microcentrifuge tube containing 100 µl of sterile water and a needlepoint (21 gauge) of activated charcoal (Sigma, St. Louis, Mo.) to absorb porphyrin-derived pigment. Samples are then placed in a 100°C water bath for 10 min and centrifuged for 2 min in a microcentrifuge, and 5-µl aliquots of supernatant are used as templates for AP-PCR. Such preparations, as well as cells from a broth culture, yield fingerprints similar to those made with pure genomic DNA (21), thus allowing the bypassing of a time-consuming and expensive procedure. All controls and samples should be prepared in the same manner (genomic DNA or whole cell) to avoid any possible inconsistency.

To summarize, AP-PCR can generate DNA fingerprints for *P. gingivalis,* and AP-PCR using colonies is a quick, simple, reproducible procedure.

RAPD MARKERS AS POTENTIAL TOOLS FOR TAXONOMIC STUDIES

Traditional methods for identifying bacteria have mostly relied on phenotypic characteristics. In recent years, considerable taxonomic improvements have emerged with the introduction of molecular characterization, e.g., DNA base composition, DNA-DNA hybridization, gene sequencing, etc. (reviewed in reference 2). Nucleic acid hybridization has contributed significantly to the progress of bacterial classification, but its use in identification requires that the nucleic acid of the unknown strain be tested against numerous reference nucleic acids. This can be done only in a laboratory with a collection of reference nucleic acids. Over the last 10 years, considerable effort has been expended to develop genotypic taxonomic methods for rapid identification of microorganisms. A variety of novel procedures have been tried. One of the most promising of these has been the use of DNA probes that are specific for an organism's 16S rRNA or 16S + 23S rRNA (13). Since genomic DNA amplification by AP-PCR can produce DNA fingerprints, it is conceivable that differences between fingerprints would reflect differences in the sequences of the templates that might be of taxonomic value. AP-PCR thus appears to be a promising simple tool for estimating relatedness among species of bacteria that are closely related and among divergent strains within species. Up to now, studies involving genetic characterization by AP-PCR for bacterial or fungal systematics have included only a few taxa, e.g., *Leptosphaeria maculans* (12), *Listeria* spp. (20), *Streptococcus uberis* (14), and *Aspergillus fumigatus* (1).

At present, only a single published study reports the use of AP-PCR for taxonomic purposes (39). In this study, Welsh and coworkers resolved into three distinct phyletic groups a collection of 30 Eurasian and North American isolates of spirochetes that are generally categorized as *Borrelia burgdorferi.* The groups identified were the same as those determined by MLEE, DNA-DNA hybridization, and RFLP, thus suggesting three related groups of potential species. Three primers (33, 20, and 18 nucleotides long) were used to generate fingerprints. Amplicons were scored as 1 (present) or 0 (absent) in order to construct a data matrix that was analyzed by phylogenetic analysis using parsimony (PAUP) and the PHYLIP phylogeny inference package.

Data from a first study of the taxon *P. gingivalis* by AP-PCR are available

(22). In it, cluster analysis of the RAPD fingerprints revealed two major genetic groups that matched the human and animal biotypes.

Nine human strains and 7 animal strains of *P. gingivalis* as well as 17 strains other than *P. gingivalis* were analyzed with four nanomer primers. Three primers generated RAPD fingerprints that allowed the 16 *P. gingivalis* strains to be differentiated, two of the primers yielded species-specific markers, and two of the primers permitted biotype distinction. Figure 3 shows fingerprints obtained from human strains with primer 970-11 (5′ GTAAGGCCG). These are highly polymorphic, with nine distinct banding patterns, each unique signature corresponding to a strain containing 5 to 12 amplicons. Two amplicons at 1,146 and 756 bp were common to all human strains tested, whereas others were shared by only some strains, and still others were unique. This primer offers potential as a tool for identification at the species level because of the species-specific markers and at the strain level because of the unique amplicons. Use of primer 970-11 on DNA from animal strains yielded highly polymorphic fingerprints with as many banding patterns as there were isolates. This primer thus allows a clear-cut distinction between the two biotypes, mostly because of the absence in the animal fingerprints of the 1,146-bp amplicon common to all human strains. No such amplicons could be detected when AP-PCR with the same primer was run on a set of 17 strains from related and unrelated species. This indicates that common amplicons at 1,146 and 756 bp are restricted to *P. gingivalis* strains, suggesting their use as valuable genetic markers for this species.

To further investigate the affinity between individual strains and groups of strains, data obtained with the four primers were combined for a cluster analysis assuming equal weight for all characters. A genetic distance of 0.67 revealed two major phenetic groups (Fig. 4) that matched the already established human and animal biotypes. The Mantel statistic was computed and found to be significant at $P < 0.001$, indicating the validity of recognition of the two groups. This observation raises the question of whether the biotypes represent different species, as recently

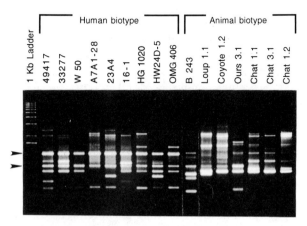

Figure 3. RAPD fingerprints of 16 strains of *P. gingivalis* obtained with primer 970-11. Arrowheads on the left indicate the 1,146- and 756-bp amplicons shared by all strains of the human biotype.

Genetic distance

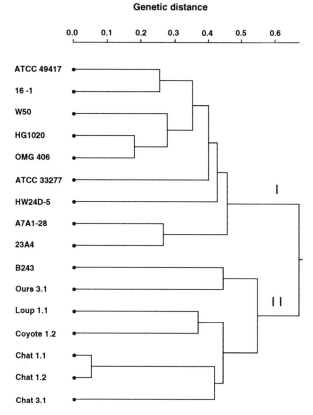

Figure 4. Dendrogram of cluster analysis based on RAPD fingerprints. The number on the horizontal axis indicates genetic distance as determined by the genetic similarity of Nei and UPGMA clustering. Group I and group II, separated by a genetic distance of 0.67, match the human and animal biotypes, respectively, of *P. gingivalis*.

suggested by Loos et al. (17), who reported a genetic distance of 0.7 by MLEE. The genetic distance of 0.67 observed between human and animal biotypes reflects a sufficient genetic difference, albeit based on a limited number of strains, to support the hypothesis that the *P. gingivalis* nomenspecies actually consists of two genospecies. Clearly, only an analysis of a larger sample of *P. gingivalis* strains by AP-PCR reinforced by more traditional methods such as DNA-DNA hybridization used as a gold standard can provide the answer.

The two preliminary studies discussed above clearly indicate that AP-PCR has the potential to create genetic frameworks useful in taxonomy. Whether a cladistic approach, using a maximum parsimony method (the *B. burgdorferi* study), or a phenetic approach, using UPGMA (the *P. gingivalis* study), should be chosen for evaluating the diversity and relationships among a collection of organisms is beyond the scope of this chapter. The objective of classification in clinical bacteriology is to distinguish natural groups. This can now be done by methods that yield quantitative estimates of genetic relationships to ultimately correlate the grouping data obtained with host occurrence, disease associations, and putative virulence factors, a most desirable goal (17).

It should be noted that the above-mentioned studies assigned molecular weights to the DNA fragments by hand, a time-consuming and somewhat difficult, imprecise procedure. Because RAPD patterns are simple, they lend themselves to image-scanning devices. It would therefore be possible to use the digitized information for automatic band quantification (molecular weight, optical density), which in turn would allow computer-assisted fingerprint storage and comparison by means of numerical analysis. It thus can be anticipated that AP-PCR-based molecular systematics studies have a bright future, since the possibility of using the power of computer-assisted analysis will allow us to combine phenotypic and phylogenetic approaches. In Fig. 5, a flow chart describing a tentative protocol for the numerical analysis of RAPD patterns is shown.

RAPD MARKERS AS POTENTIAL TOOLS FOR TRANSMISSION STUDIES

A major objective of molecular epidemiology is to study the transmission of bacteria, which can be done by various methods; the method of choice is the one that can ascertain that two recovered organisms originate from the same clone. For this purpose, the traditional technique used by molecular epidemiologists has been REA; more recently, RFLP or ribotyping has been used, and occasionally MLEE has been used. The technical complexity and the costs involved in examining large populations have probably limited the use of these techniques, as reflected by the few studies of periodontopathogens. Data from a recent study (34) have demonstrated that AP-PCR typing can be of great value in epidemiological studies of *P. gingivalis* and may also be useful with other bacteria. In this study, three different methods for typing *P. gingivalis,* i.e., REA typing, ribotyping, and AP-PCR, were compared for 32 isolates obtained from eight patients with severe periodontitis and from their spouses. The data obtained with the three methods were in agreement: unrelated individuals all had distinct DNA patterns, whereas both husband and wife in six couples shared types indistinguishable by all three methods. Isolates from different sites in the same individual gave indistinguishable patterns. A typical example of the different RAPD patterns that can be obtained by AP-PCR from a single colony is shown in Fig. 6. Although all three methods demonstrate that *P. gingivalis* can be transmitted between spouses, AP-PCR was the easiest and least time-consuming method.

CONCLUDING REMARKS

AP-PCR is a powerful tool with many potential applications for more efficient management of periodontal diseases through a better knowledge of the molecular basis of pathogenesis. Among the modern tools of molecular systematics, the novel AP-PCR technology allows for a definition of groups within a bacterial species and

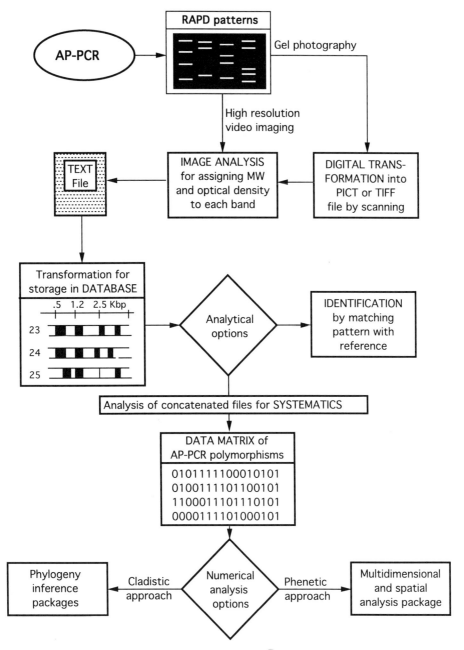

Figure 5. Flow chart describing a tentative protocol for the numerical analysis of RAPD patterns. MW, molecular weight.

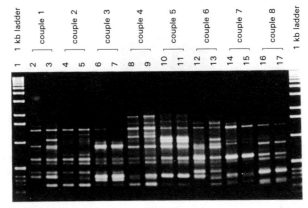

Figure 6. RAPD fingerprints of colonies of 16 *P. gingivalis* isolates from eight married couples. Primer 910-05 was used. Unrelated individuals all show distinct RAPD fingerprints; patient (lane 2) and spouse (lane 3) in couple 1 and patient (lane 12) and spouse (lane 13) in couple 6 show distinct RAPD fingerprints. Husband and wife in couples 2, 3, 4, 5, 7, and 8 share identical RAPD fingerprints. (From reference 34 with permission.)

provides a method for their identification. Such groupings allow the diversity of the strains within a habitat to be defined and offer the potential to facilitate the definition of pathogenic and virulent strains. Whether only virulent pathotypes of *P. gingivalis* can cause disease or whether the microorganism is an opportunistic pathogen (10, 17) remains an open question. The recent data of Loos et al. (17) indicate that no specific genetic lineages or clusters of clones of *P. gingivalis* are associated with distinct types of infections. AP-PCR fingerprinting of DNA of large numbers of *P. gingivalis* strains recovered from both healthy subjects and patients with periodontitis, including strains from sites with various degrees of inflammation and pocket formation, will help disprove or confirm that *P. gingivalis* is an opportunistic pathogen.

Acknowledgments. We thank R. Brousseau for providing the synthetic primers and for helpful discussions, P. Legendre for help with data analysis, D. Fournier for managing our collection of strains, and M. van Steenbergen for a fruitful collaboration.

This work was financed by Medical Research Council of Canada grant MA-8761.

REFERENCES

1. **Aufauvre-Brown, A., J. Cohen, and D. W. Holden.** 1992. Use of randomly amplified polymorphic DNA markers to distinguish isolates of *Aspergillus fumigatus. J. Clin. Microbiol.* **30:**2991–2993.
2. **Brenner, D. J.** 1991. Taxonomy, classification, and nomenclature of bacteria, p. 209–215. *In* A. Balows, W. J. Hausler, Jr., K. L. Herrmann, H. D. Isenberg, and H. J. Shadomy, (ed.), *Manual of Clinical Microbiology,* 5th ed. American Society for Microbiology, Washington, D.C.
3. **Chen, C.-K. C., Y. Liu, and J. Slots.** 1993. DNA fingerprinting of *Porphyromonas gingivalis* by arbitrary primers/polymerase chain reaction (AP-PCR), abstr. 397. *J. Dent. Res.* **72**(Special Issue)**:**153.
4. **Chen, C.-K. C, T. V. Potts, and M. E. Wilson.** 1990. DNA homologies shared among *E. corrodens* isolates and other corroding bacilli from the oral cavity. *J. Periodontal Res.* **25:**106–112.

5. **Dewhirst, F. E., B. J. Paster, G. J. Fraser, L. M. Martin, and S. S. Socransky.** 1993. Clonal typing of *Campylobacter rectus* and *Porphyromonas gingivalis* by APPCR, abstr. 394. *J. Dent. Res.* **72**(Special Issue):153.

6. **DiRienzo, J. M., S. Cornell, L. Kazoroski, and J. Slots.** 1990. Probe-specific DNA fingerprinting applied to the epidemiology of localized juvenile periodontitis. *Oral Microbiol. Immunol.* **5**:49–56.

7. **Ebersole, J. L., M. A. Taubman, D. J. Smith, and D. E. Frey.** 1988. Serological classification of *Bacteroides* from the human oral cavity. *J. Periodontol.* **23**:22–27.

8. **Fisher, J. G., J. J. Zambon, P. Chen, and R. J. Genco.** 1987. *Bacteroides gingivalis* serogroups and correlation with virulence, abstr. 817. *J. Dent. Res.* **65**(Special Issue):816.

9. **Fujiwara, T., T. Ogawa, S. Sobue, and S. Hamada.** 1990. Chemical immunobiological and antigenic characterizations of lipopolysaccharides from *Bacteroides gingivalis* strains. *J. Gen. Microbiol.* **136**:319–326.

10. **Genco, R. J., J. J. Zambon, and L. A. Christersson.** 1988. The origin of periodontal infections. *Adv. Dent. Res.* **2**:245-259.

11. **Gmür, R., G. Werner-Felmayer, and B. Guggenheim.** 1988. Production and characterization of monoclonal antibodies specific for *Bacteroides gingivalis*. *Oral Microbiol. Immunol.* **3**:181–186.

12. **Goodwinn, P. H., and S. L. Annis.** 1991. Rapid identification of genetic variation and pathotype of *Leptosphaeria maculans* by random amplified polymorphic DNA assay. *Appl. Environ. Microbiol.* **57**:2482–2486.

13. **Grimont, F., and P. A. D. Grimont.** 1986. Ribosomal ribonucleic acid gene restriction patterns as potential taxonomic tools. *Ann. Inst. Pasteur/Microbiol.* **137B**:165–175.

14. **Jayarao, B. M., B. J. Bassam, G. Caetanno-Anollés, P. M. Gresshoff, and S. P. Oliver.** 1992. Subtyping of *Streptococcus uberis* by DNA amplification fingerprinting. *J. Clin. Microbiol.* **30**:1347–1350.

15. **Lambe, D. W. J.** 1974. Determination of *Bacteroides melaninogenicus* serogroups by fluorescent antibody staining. *Appl. Microbiol.* **28**:561.

16. **Loos, B. G., and D. W. Dyer.** 1992. Restriction fragment length polymorphism analysis of the fimbrillin locus, *fimA*, of *Porphyromonas gingivalis*. *J. Dent. Res.* **71**:1173–1181.

17. **Loos, B. G., D. W. Dyer, T. S. Whittam, and R. K. Selander.** 1993. Genetic structure of populations of *Porphyromonas gingivalis* associated with periodontitis and other oral infections. *J. Clin. Microbiol.* **61**:204–212.

18. **Loos, B. G., D. Mayrand, R. J. Genco, and D. P. Dickinson.** 1990. Genetic heterogeneity of *Porphyromonas (Bacteroides) gingivalis* by genomic DNA fingerprinting. *J. Dent. Res.* **69**:1488–1493.

19. **Mayrand, D., and S. C. Holt.** 1988. Biology of asaccharolytic black-pigmented *Bacteroides* species. *Microbiol. Rev.* **52**:134–152.

20. **Mazurier, S. I., and K. Wernars.** 1992. Typing of *Listeria* strains by random amplification of polymorphic DNA. *Res. Microbiol.* **143**:499–505.

21. **Ménard, C., R. Brousseau, and C. Mouton.** 1992. Application of polymerase chain reaction with arbitrary primer (AP-PCR) to strain identification of *Porphyromonas (Bacteroides) gingivalis*. *FEMS Microbiol. Lett.* **95**:163–168.

22. **Ménard, C., and C. Mouton.** 1993. Randomly amplified polymorphic DNA analysis confirms the biotyping scheme of *Porphyromonas gingivalis*. *Res. Microbiol.* **144**:445–455.

23. **Mouton, C., P. G. Hammond, J. Slots, and R. J. Genco.** 1981. Identification of *Bacteroides gingivalis* by fluorescent antibody staining. *Ann. Inst. Pasteur/Microbiol.* **132B**:69–83.

24. **Mouton, C., J. Slots, and R. J. Genco.** 1979. Serology of *Bacteroides melaninogenicus* by the indirect fluorescent antibody technique, abstr. 56. *J. Dent. Res.* **58A**:106.

25. **Nagata, A., T. Man-Yoshi, M. Sato, and R. Nakamura.** 1991. Serological studies of *Porphyromonas (Bacteroides) gingivalis* and correlation with enzyme activity. *J. Periodontal Res.* **26**:184–190.

26. **Okuda, K., K. Ohta, T. Kato, I. Takazoe, and J. Slots.** 1986. Antigenic characteristics and serological identification of 10 black-pigmented *Bacteroides* species. *J. Clin. Microbiol.* **24**:89–95.

27. **Parent, R., C. Mouton, L. Lamonde, and D. Bouchard.** 1986. Human and animal serotypes of *Bacteroides gingivalis* defined by crossed immunoelectrophoresis. *Infect. Immun.* **51**:909–918.

28. **Preus, H. R., and I. Olsen.** 1988. Possible transmittance of *A. actinomycetemcomitans* from a dog to a child with rapidly destructive periodontitis. *J. Periodontal Res.* **23:**68–71.
29. **Reed, M. J., J. Slots, C. Mouton, and R. J. Genco.** 1980. Antigenic studies of oral and nonoral black-pigmented *Bacteroides* strains. *Infect. Immun.* **51:**286–293.
30. **Saarela, M., A.-M. Stucki, B. von Troil-Linden, S. Alaluusua, H. Jousimies-Somer, and S. Asikainen.** 1993. Intra- and inter-individual comparison of *Porphyromonas gingivalis* genotypes. *FEMS Immunol. Med. Microbiol.* **6:**99–102.
31. **Smith, G. L. F., C. Sansone, and S. S. Socransky.** 1989. Comparison of two methods for the small-scale extraction of DNA from subgingival microorganisms. *Oral Microbiol. Immunol.* **4:**135–140.
32. **Socransky, S. S., and L. Martin.** 1992. Ribotyping of *P. gingivalis* and *C. rectus,* abstr. 1127. *J. Dent. Res.* **71**(Special Issue):246.
33. **Umemoto, T., E. Ishikawa, K. Watanabe, N. Hamada, and M. Iida.** 1989. Serological classification of *Bacteroides gingivalis* and serogroup distribution in periodontitis patients. *Dent. Jpn.* **26:**27–30.
34. **van Steenbergen, T. J. M., C. Ménard, C. J. Tijhof, C. Mouton, and J. de Graaff.** 1993. Comparison of 3 molecular typing methods in studies of transmission of *Porphyromonas gingivalis. J. Med. Microbiol.* **39:**416–421.
35. **van Steenbergen, T. J. M., M. D. A. Petit, L. H. M. Scholte, U. van der Velden, and J. de Graaff.** 1993. Transmission of *Porphyromonas gingivalis* between spouses. *J. Clin. Periodontol.* **20:**340–345.
36. **van Steenbergen, T. J. M., U. van der Velden, F. Abbas, and J. de Graaff.** 1991. Microflora and bacterial DNA restriction enzyme analysis in young adults with periodontitis. *J. Periodontol.* **62:**235–241.
37. **van Steenbergen, T. J. M., A. J. van Winkelhoff, and J. de Graaff.** 1993. Classification and typing methods of black-pigmented gram-negative anaerobes. *FEMS Immunol. Med. Microbiol.* **6:**83–88.
38. **Welsh, J., and M. McClelland.** 1990. Fingerprinting genomes using PCR with arbitrary primers. *Nucleic Acids Res.* **18:**7213–7218.
39. **Welsh, J., C. Pretzman, D. Postic, I. Saint Girons, G. Baranton, and M. McClelland.** 1992. Genomic fingerprinting by arbitrarily primed polymerase chain reaction resolves *Borrelia burgdorferi* into three distinct phyletic groups. *Int. J. Syst. Bacteriol.* **42:**370–377.
40. **Williams, J. G. K., A. R. Kubelik, K. J. Livak, J. A. Rafalski, and S. V. Tingey.** 1990. DNA polymorphisms amplified by arbitrary primers are useful as genetic markers. *Nucleic Acids Res.* **18:**6531–6535.
41. **Wilson, K.** 1991. Preparation of genomic DNA from bacteria, p. 2.4.1–2.4.2. *In* F. M. Ausubel, R. Brent, R. E. Kingston, D. D. Moore, J. G. Seidman, J. A. Smith, and K. Struhl (ed.), *Current Protocols in Molecular Biology.* Wiley-Interscience, Philadelphia.
42. **Zambon, J. J., G. J. Sunday, and J. S. Smutko.** 1990. Molecular genetic analysis of *Actinobacillus actinomycetemcomitans* epidemiology. *J. Periodontol.* **61:**75–80.

Molecular Pathogenesis of Periodontal Disease
Edited by Robert Genco et al.
© 1994 American Society for Microbiology, Washington, DC 20005

Chapter 5

Relationship between Iron Availability and Periodontal Disease Associated with *Porphyromonas gingivalis* and *Actinobacillus actinomycetemcomitans*

J. Leslie Winston and David W. Dyer

Human physiology has evolved a variety of mechanisms to combat microbial pathogens, including acute-phase and specific, long-term immune mechanisms for actively seeking out and eliminating invading microbes. The human body also uses several nonspecific defenses to control bacterial infections. Among these, nutritional deprivation by controlling iron (Fe) availability is extremely important in limiting microbial growth. Fe sequestration in the host also prevents cell and tissue damage caused by free Fe. In an aqueous, neutral-pH, oxidized environment (that is, biological conditions), Fe exists in the $+3$ (ferric) oxidation state. Free Fe^{3+} ions are extremely insoluble ($K_{sp} = 10^{-38}$ M) and toxic; Fe^{3+} ions generate free radicals that cause tissue damage (32). To control Fe toxicity, Fe in body fluids is very carefully managed. Extracellularly, Fe^{3+} is bound to transferrin (Tf; in plasma) and lactoferrin (Lf; in saliva and other mucosal secretions). These proteins hold free extracellular Fe at extremely low levels; the free-Fe concentration in plasma has been estimated at $<10^{-18}$ M, far below the 100 to 400 nM Fe needed for bacterial growth (8). Intracellular Fe is contained within heme, hemosiderin, and ferritin (34). This Fe sequestration holds freely available Fe to such low levels that most microbes are growth limited, a situation termed "nutritional immunity" (47).

Since Fe is an essential nutrient for virtually all bacteria, pathogens have evolved specific Fe acquisition systems to circumvent nutritional immunity. One method of Fe transport used by pathogenic bacteria depends on siderophores. Siderophores are low-molecular-weight (600- to 900-Da) Fe-chelating compounds produced in copious quantities by Fe-starved bacteria. The excreted siderophore removes Fe^{3+} from Tf or Lf, and the Fe^{3+}-siderophore complex is then used as an Fe source by the bacterium (33). Alternatively, some bacteria have specific cell

J. Leslie Winston • Department of Oral Biology, School of Dental Medicine, State University of New York at Buffalo, Buffalo, New York 14214. *David W. Dyer* • Department of Microbiology and Immunology, Health Sciences Center, University of Oklahoma, Oklahoma City, Oklahoma 73190.

surface receptors for Tf and Lf (4, 44). The Fe-carrying protein binds to the bacterial receptor, and Fe is removed and then internalized without siderophore production.

Our laboratory is interested in the possibility of interfering with periodontitis associated with *Porphyromonas gingivalis* and *Actinobacillus actinomycetemcomitans* by targeting the cell surface components of the Fe transport systems of these pathogens. Experimentally, the identities of these proteins are easily revealed: when bacteria are starved for Fe, cell surface components of the Fe transport system are synthesized in highly increased amounts. We therefore use the term FeRP (Fe-repressible protein) to describe these proteins.

The development of vaccines that target Fe transport shows great promise. Cell surface FeRPs of bacterial Fe transport systems are commonly immunogenic in humans (3, 16, 38, 40, 49). Fe transport is essential for disease (23, 48) and the immune response against FeRPs can block iron transport (25, 40). To obtain Fe from a siderophore, Tf, Lf, or other host Fe source, the cell surface Fe transport FeRPs must interact intimately (bind) with the Fe carrier. This requires significant structural (and therefore antigenic) conservation among Fe transport proteins from different strains of a pathogenic species. Thus, antibodies directed against conserved bacterial Fe transport surface components may block Fe uptake and suppress microbial growth by reexerting nutritional immunity. Such an immune response would be equally effective if the antibodies elicited were capable of activating complement-mediated killing or were opsonophagocytic antibodies.

Development of chemotherapeutic agents that compete with or exploit bacterial Fe transport may also control infection. For example, phenolate-substituted cephalosporins are being investigated for use as antimicrobial agents (12, 46). The phenolate moiety is recognized as a siderophore by several gram-negative bacteria and can allow the cephalosporin to gain access to the bacterial periplasmic space. Several phenolate-substituted cephalosporins inhibit *Escherichia coli* strains that were previously resistant to the cephalosporin alone (12, 46).

HOW DO BACTERIAL Fe TRANSPORT SYSTEMS WORK?

Most pathogens synthesize and excrete siderophores for obtaining Fe in vivo (32, 33). For instance, aerobactin is absolutely essential for virulence of invasive *E. coli* (23). Aerobactin removes Fe^{3+} from Tf, and the Fe^{3+}-aerobactin complex then binds to a specific cell surface receptor (33). The Fe^{3+}-aerobactin complex is processed to remove the Fe, while the siderophore is released for another round of Fe acquisition (33). Many microbes can use exogenous siderophores produced by other bacteria. Although a bacterium may be incapable of directly obtaining Fe from Tf or Lf, a pathogen may indirectly obtain Tf- or Lf-bound Fe by utilizing a siderophore produced by a coinfecting microorganism. This may have special relevance for periodontal lesions, which often contain a mixed microbial flora. Other pathogens have FeRP receptors that specifically bind Tf, Lf, and possibly hemoglobin (4, 26, 44). However, the mechanisms by which Fe is removed from these Fe carriers after binding to bacterial receptors are unknown.

Heme is an important Fe source for many pathogens, so heme uptake systems are currently a subject of considerable interest. Heme Fe is typically found intra-cellularly, bound to hemoglobin, myoglobin, and cytochromes (34). Intravascular hemolysis may liberate heme or hemoglobin, but circulating hemoglobin is rapidly bound by plasma haptoglobin, while heme is bound tightly by hemopexin and serum albumin and then rapidly cleared by the liver (31). Bacterial pathogens utilize both free heme and heme-containing proteins as sole sources of Fe (13, 14, 35).

Extracellular Fe reduction may represent an additional mechanism that allows pathogens to overcome nutritional immunity (1, 15). Fe^{2+} may be easier to transport and assimilate because of its solubility properties (8). In this context, the reduced anaerobic milieu in which *P. gingivalis* and *A. actinomycetemcomitans* reside may provide sufficient Fe^{2+} for growth.

Fe IN THE PERIODONTAL ENVIRONMENT

Little is known about the Fe sources that exist in the periodontal environment (Fig. 1). Mukherjee (30) found 26 to 170 μM Fe in the gingival crevicular fluid (GCF) of patients with adult periodontitis. This is far greater than the concentration of Tf-bound Fe normally found in plasma, suggesting that the Fe content is probably derived from many sources. Lf, Tf, and hemoglobin (11) are known constituents of GCF and probably support the growth of periodontopathogens in vivo. Further, the cellular composition of the periodontium in health and disease likely provides a rich supply of intracellular Fe, including ferritin and cytochromes.

EXPRESSION OF BACTERIAL VIRULENCE DETERMINANTS IS OFTEN REGULATED BY Fe AVAILABILITY

In addition to Fe transport proteins, several bacterial virulence determinants are regulated by Fe availability. The *E. coli* α-hemolysin (45) and the *Pasteurella*

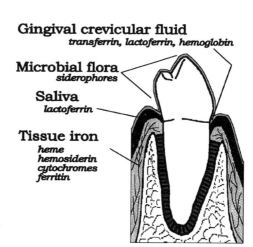

Figure 1. Potential iron sources in the oral cavity that may support periodontal infections.

haemolytica leukotoxin (LTX; 42) are two toxins whose expression is enhanced when bacteria are starved for Fe. These toxins share a similar structural motif, placing these proteins in the RTX (repeat in structural toxin) family. Since the *A. actinomycetemcomitans* LTX shows considerable nucleotide sequence similarity to other RTX proteins (24), the Fe regulation of this LTX is currently being investigated (see below).

Fe TRANSPORT BY *P. GINGIVALIS* AND *A. ACTINOMYCETEMCOMITANS*

It is important to study the putative periodontopathogens *P. gingivalis* and *A. actinomycetemcomitans* because their association with the periodontal disease process is well documented (38, 50). However, little is currently known about Fe transport by these organisms. The ecological niches of *P. gingivalis* and *A. actinomycetemcomitans* in the oral cavity share some similarities, but their differences may have led to the evolution of different Fe transport systems. *P. gingivalis* is an obligate anaerobe (17) whose main oral niche appears to be dental plaque (38). *A. actinomycetemcomitans*, on the other hand, is a facultative anaerobe demonstrated to persist both in the anaerobic periodontal pocket (39) and in highly oxygenated gingival tissues (10, 29). Under anaerobic (reduced) conditions, Fe^{2+} is probably available, suggesting that these organisms can transport Fe^{2+}. Additional Fe sources that each bacterium may utilize likely mirror the Fe sources that are available in the ecological niche occupied by that organism.

P. gingivalis

The majority of studies of Fe and *P. gingivalis* pathogenesis have focused on the utilization of heme as an Fe source. The requirement of heme as a growth supplement was first noted by Gibbons and MacDonald (17). Shah et al. (37) demonstrated that *P. gingivalis* transports the intact heme molecule into the cell and presumably extracts Fe from the porphyrin ring intracellularly. We established Fe-limited growth conditions (2) for *P. gingivalis* in a complex medium (half-strength brain heart infusion broth) supplemented with protoporphyrin IX (PIX). The growth of *P. gingivalis* A7A1-28, 381, Bowden 18/10, and W50 was significantly reduced by 125 μM 2,2′-dipyridyl (DP), a chelator with Fe^{2+} specificity (2). DP-induced growth inhibition was reversible if ferrous ammonium sulfate, but not ferric nitrate, was added to the medium. This observation demonstrates that growth inhibition was due to chelation of Fe^{2+} and suggests that Fe^{2+} can be utilized for growth. Under these growth conditions, heme, hemoglobin, and cytochrome *c* also served as Fe sources for *P. gingivalis* (36). Utilization of cytochrome *c* as an Fe source was significant, as the heme bound to this protein is covalently attached. In order to remove Fe from cytochrome *c, P. gingivalis* would need to first proteolytically degrade the protein. Proteolysis may be an important nonspecific step in the utilization of other protein-bound Fe sources by *P. gingivalis*. Bramanti and Holt (6) obtained similar results and also showed that *P. gingivalis* uses myoglobin and catalase as porphyrin sources.

Bramanti and Holt (5) examined membranes from *P. gingivalis* W50. The organism produced at least 10 new proteins, ranging from 26 to 80 kDa, under conditions of heme starvation. With heme limitation, the 26-kDa protein was found on the *P. gingivalis* surface. When these cultures were shifted to a hemin-rich environment, the 26-kDa protein seemed to disappear from the cell surface, becoming embedded in the deeper recesses of the outer membrane. Bramanti and Holt (7) suggested that this translocation event may be responsible for heme transport. This model is quite unique, as bacterial outer membrane proteins have not been shown to physically move across the outer membrane.

We demonstrated that PIX could substitute for heme in supporting the growth of *P. gingivalis* (2). In subsequent experiments, we showed that PIX was not a classical Fe chelator (28) and therefore was not acting to carry Fe into *P. gingivalis*. These findings suggest that *P. gingivalis* utilizes heme and PIX as porphyrin sources in addition to using heme as an Fe source. We further demonstrated that a variety of Fe-free porphyrins will substitute for heme in stimulating the growth of *P. gingivalis* (28).

Since Tf is present in GCF, several investigators have considered the possibility that Tf serves as an Fe source for *P. gingivalis*. Tf supported *P. gingivalis* growth in mycoplasma broth (6) and Trypticase soy broth–0.5% yeast extract (21). In these studies, Tf substituted for heme in supporting the growth of *P. gingivalis*. It is presently unclear how Tf acts as a heme substitute: Tf is not a heme-containing protein but carries Fe^{3+} bound directly to the protein and coordinated with a carbonate or bicarbonate anion (19). Apo-Tf (Fe free) did not support the growth of *P. gingivalis*. Thus, it appears that this growth stimulation effect is somehow associated with Fe, but the mechanism is unclear, since Fe is plentiful in the complex media used for these experiments. Complex media contain between 600 nM and 30 μM endogenous Fe (6; our unpublished data). These values exceed the 100 to 400 nM Fe required for bacterial growth (8). Thus, although Tf and heme may provide Fe to *P. gingivalis* in these experiments, the growth assays performed in the absence of Fe chelators probably did not assay specifically for Fe transport, as the endogenous Fe in these complex media would be sufficient to support growth.

When starved for Fe by DP, *P. gingivalis* A7A1-28 expressed two outer membrane FeRPs of 24 and 43 kDa (2). Six strains of *P. gingivalis* (381, Bowden 18/10, RB22D-1, EM-3, AJW-1, and AJW-3) produced a 24-kDa FeRP that comigrated with that from A7A1-28 on sodium dodecyl sulfate-polyacrylamide gel electrophoresis (43). The 24-kDa FeRPs from all these strains reacted with antisera obtained from mice immunized with Fe-starved A7A1-28, suggesting that this FeRP is a conserved outer membrane antigen and presumably has a conserved function. However, strain W50 did not produce this protein (43).

Four *P. gingivalis* FeRPs reacted with antibodies found in sera from patients with adult periodontitis (9). These proteins elicited a significant proportion of the human immunoglobulin G response against *P. gingivalis* membrane proteins, and ~40% of the patient sera reacted with the 24-kDa FeRP. Studies to determine whether the 24-kDa FeRP represents a possible vaccine antigen and whether the 24-kDa protein is involved in *P. gingivalis* Fe transport are under way.

As mentioned above, Fe availability often controls the expression of many bacterial virulence determinants. Since *P. gingivalis* proteolytic enzymes are thought to enhance in vivo survival (38), we examined the role of Fe availability in the expression of protease activity by *P. gingivalis* (2). However, when starved for Fe by DP, *P. gingivalis* protease activity did not appear to be affected (2).

A. actinomycetemcomitans

We established Fe-limited growth conditions for *A. actinomycetemcomitans* by using chemical chelators in Trypticase soy broth containing 0.5% yeast extract and 0.1% sodium bicarbonate (49). DP (Fe^{2+} chelator) inhibited the growth of *A. actinomycetemcomitans* Y4, JP2, and 75 when the cultures were grown under anaerobiosis. DP did not inhibit the aerobic growth of the organism, but ethylene-di-(*o*)-hydroxyphenylacetic acid (EDDA; a Fe^{3+} chelator) inhibited *A. actinomycetemcomitans* under aerobiosis. This chelator-induced growth inhibition was reversible by both Fe^{3+} and Fe^{2+} sources, indicating that the growth inhibition was due to Fe chelation rather than to the binding of other metal ions by these chelators. Consequently, *A. actinomycetemcomitans* appears to utilize both Fe^{2+} and Fe^{3+} sources. This is not unexpected, as the organism is capable of metabolizing in anaerobic (reduced) and aerobic (oxidized) environments. However, experiments designed to identify potential in vivo Fe sources for *A. actinomycetemcomitans* in the presence of DP and EDDA have proven difficult. Chelators can shuttle Fe between themselves and Tf and Lf, making it difficult to assess whether a bacterium can use these Fe carrier proteins as Fe sources (20). We are currently focusing on the development of an Fe-limited growth medium that lacks chelating agents for use in performing these experiments in the absence of chelators.

A. actinomycetemcomitans may not produce siderophores. We could not detect phenolate or hydroxamate siderophores in culture supernatants of Fe-starved *A. actinomycetemcomitans* (49). However, *A. actinomycetemcomitans* may produce a siderophore of novel chemical structure, which would not react with the chemical assays used to identify phenolate or hydroxamate compounds (22). Preliminary experiments suggest that *A. actinomycetemcomitans* may use the Fe^{3+} chelator Desferal, as an iron source; Desferal is a semisynthetic derivative of the microbial siderophore ferrioxamine B (22). Thus, *A. actinomycetemcomitans* may use exogenous siderophores rather than producing its own siderophore. We also attempted to assess whether *A. actinomycetemcomitans* possessed a Tf and/or Lf receptor, as has been reported for *Actinobacillus pleuropneumoniae* (18). However, Tf- or Lf-horseradish peroxidase conjugates did not bind to iron-starved *A. actinomycetemcomitans* cells immobilized on nitrocellulose, suggesting that *A. actinomycetemcomitans* does not have receptors for Tf or Lf (49). It is currently unclear whether the organism can use these proteins as Fe sources and what the possible mechanism of Fe removal from these proteins may be. No studies have yet examined whether *A. actinomycetemcomitans* uses heme or other Fe sources for growth in vivo.

Under Fe-limited anaerobic growth conditions (49), strains Y4, JP2, and 75 expressed a common FeRP of ~70 kDa. Strain Y4 produced an additional FeRP of 53 kDa, while strain JP2 produced strain-specific FeRPs of 40 and 60 kDa. The expression of the 70-kDa FeRP by all three strains suggests that this protein has a conserved function and possibly is involved in Fe transport by *A. actinomycetem-comitans*. Anti-70-kDa FeRP antibodies reacted in sera from five of six patients with localized juvenile periodontitis and in one of three periodontally healthy patients (49), suggesting that this protein may be antigenically conserved among serologic variants of *A. actinomycetemcomitans*.

Since the synthesis of at least two other members of the RTX family is regulated by Fe availability (42, 45), we (see below) and others (41) have examined whether LTX expression is influenced by Fe starvation. The leukotoxic activities of strains Y4, JP2, and 75 were examined using trypan blue exclusion bioassays on HL-60 cells, an acute promyelocytic leukemia cell line that responds to the *A. actino-mycetemcomitans* LTX much as human polymorphonuclear leukocytes do (51). Strain Y4 did not produce LTX (data not shown). Strain JP2 demonstrated killing values (expressed as mean percent viability of control) at 60 min of 55.25 ± 20.91 for iron-limited cultures and 60.00 ± 12.92 for iron-sufficient cultures, indicating no difference in leukotoxic activity with Fe availability. Strain 75 produced similar results (data not shown). We also measured LTX concentrations by an enzyme-linked immunosorbent assay (ELISA) with antiserum kindly provided by T. Inzana (27) against the *A. pleuropneumoniae* hemolysin, which cross-reacts with the *A. actinomycetemcomitans* LTX (data not shown). However, we could not detect any appreciable differences in the LTX activities of strains Y4, JP2, and 75 when they were Fe starved (Table 1). Recently, Spitznagel and Kolodrubetz (41) examined whether *A. actinomycetemcomitans* LTX was Fe regulated. They used a *lacZ-lktA* gene fusion, which allowed *lktA* expression to be quantitated by measuring β-galactosidase activity. In these assays, *lktA* expression was not Fe regulated. Lally and Kieba (see chapter 7) described a segment within the JP2 *lktA* promoter with homology to the *fur* (ferric uptake regulator) consensus sequence. At present, it is not known whether this region is functional in *A. actinomycetemcomitans*.

Table 1. Comparison by ELISA of the effect of iron availability on *A. actinomycetemcomitans* LTX

Strain	A_{490}[a]		Ratio, $+Fe/-Fe$	
	$+O_2$	$-O_2$	$+O_2$	$-O_2$
Y4 $+Fe$	0.409	0.432	0.95	0.96
Y4 $-Fe$	0.423	0.442		
JP2 $+Fe$	0.422	0.416	1.06	0.99
JP2 $-Fe$	0.441	0.447		
75 $+Fe$	0.451	0.441	1.02	1.01
75 $-Fe$	0.435	0.430		

[a] A_{490} of *A. pleuropneumoniae* J45 positive control = 0.669; A_{490} of conjugate control = 0.068. $+O_2$, aerobic growth conditions; $-O_2$, anaerobic growth conditions.

CONCLUDING REMARKS

While we have begun to develop a better understanding of the importance of Fe transport in the pathogenesis of *P. gingivalis* and *A. actinomycetemcomitans*, we are still naive regarding the true availability of Fe in the periodontal environment. We have begun accumulating evidence that both putative periodontopathogens may be significantly affected by nutritional immunity. In addition, we are aware of the total Fe concentration in the GCF (30), but it is not clear whether this Fe is freely available to periodontal pathogens or is sequestered, as it is elsewhere in the human host. There is a great necessity for accurate determination of the Fe sources available in the GCF and surrounding periodontal tissues so that we can more intelligently describe Fe transport for all periodontal pathogens.

Acknowledgments. We thank Mirdza Neiders, Mary Bayers-Thering, M. Jane Gillespie, and Renae Malek for their contributions to these studies.

This work was supported by U.S. Public Health Service grants DE08240, DE00158, and DE05640-01.

REFERENCES

1. **Adams, T. J., S. Vartivarian, and R. E. Cowart.** 1990. Iron acquisition systems of *Listeria monocytogenes. Infect. Immun.* **58:**2715–2718.
2. **Barua, P. K., D. W. Dyer, and M. E. Neiders.** 1990. Effect of iron limitation on *Bacteroides gingivalis. Oral Microbiol. Immunol.* **5:**263–268.
3. **Black, J. R., D. W. Dyer, M. K. Thompson, and P. F. Sparling.** 1986. Human immune response to iron-repressible outer membrane proteins of *Neisseria meningitidis. Infect. Immun.* **54:**710–713.
4. **Blanton, K. J., G. D. Biswas, J. Tsai, J. Adams, D. W. Dyer, S. M. Davis, G. G. Koch, P. K. Sen, and P. F. Sparling.** 1990. Genetic evidence that *Neisseria gonorrhoeae* produces specific receptors for transferrin and lactoferrin. *J. Bacteriol.* **172:**5225–5235.
5. **Bramanti, T. E., and S. C. Holt.** 1990. Iron-regulated outer membrane proteins in the periodontopathic bacterium, *Bacteroides gingivalis. Biochem. Biophys. Res. Commun.* **166:**1146–1154.
6. **Bramanti, T. E., and S. C. Holt.** 1991. Roles of porphyrins and host iron transport proteins in regulation of growth of *Porphyromonas gingivalis* W50. *J. Bacteriol.* **173:**7330–7339.
7. **Bramanti, T. E., and S. C. Holt.** 1992. Localization of a *Porphyromonas gingivalis* 26-kilodalton heat-modifiable, hemin-regulated surface protein which translocates across the outer membrane. *J. Bacteriol.* **174:**5827–5893.
8. **Bullen, J. J., H. J. Rogers, and E. Griffiths.** 1978. Role of iron in bacterial infection. *Curr. Top. Microbiol. Immunol.* **80:**1–35.
9. **Chen, C.-K. C., A. DeNardin, D. W. Dyer, R. J. Genco, and M. E. Neiders.** 1991. Human immunoglobulin G antibody response to iron-repressible and other outer membrane proteins of *Porphyromonas (Bacteroides) gingivalis. Infect. Immun.* **59:**2427–2433.
10. **Chrissterson, L. A., B. Albini, J. J. Zambon, U. M. E. Wikesjo, and R. J. Genco.** 1987. Tissue localization of *Actinobacillus actinomycetemcomitans* in human periodontitis. I. Light, immunofluorescence, and electron microscopic studies. *J. Periodontol.* **58:**529–539.
11. **Cimasoni, G.** 1983. Crevicular fluid updated. *Monogr. Oral Sci.* **12:**112–117.
12. **Curtis, N. A. C., R. L. Eisenstadt, S. J. East, R. J. Cornford, L. A. Walker, and A. J. White.** 1988. Iron-regulated outer membrane proteins of *Escherichia coli* K-12 and mechanism of action of catechol-substituted cephalosporins. *Antimicrob. Agents Chemother.* **32:**1879–1886.
13. **Dyer, D. W., E. P. West, and P. F. Sparling.** 1987. Effects of serum carrier proteins on the growth of pathogenic neisseriae with heme-bound iron. *Infect. Immun.* **55:**2171–2175.

14. **Eaton, J. W., P. Brandt, J. R. Mahoney, and J. T. J. Lee.** 1982. Haptoglobin: a natural bacteriostat. *Science* **215:**691–693.

15. **Evans, S. L., J. E. L. Arceneaux, B. R. Byers, M. E. Martin, and H. Aranha.** 1986. Ferrous iron transport in *Streptococcus mutans*. *J. Bacteriol.* **168:**1096–1099.

16. **Fernandez-Beros, M. E., C. Gonzalez, M. A. McIntosh, and F. C. Cabello.** 1989. Immune response to the iron-deprivation-induced proteins of *Salmonella typhi* in typhoid fever. *Infect. Immun.* **57:**1271–1275.

17. **Gibbons, R. J., and J. B. MacDonald.** 1960. Hemin and vitamin K compounds as required factors for the cultivation of certain strains of *Bacteroides melaninogenicus*. *J. Bacteriol.* **80:**164–170.

18. **Gonzalez, G. C., D. L. Caamano, and A. B. Schryvers.** 1990. Identification and characterization of a porcine-specific transferrin receptor in *Actinobacillus pleuropneumoniae*. *Mol. Microbiol.* **4:**1173–1179.

19. **Graham, G. A., and G. W. Bates.** 1977. Factors influencing the rate of release of iron from Fe^{3+} transferrin-CO_3^{2-}, p. 273–290. *In* E. B. Brown, P. A. J. Fielding, and R. R. Crichton (ed.), *Proteins of Iron Metabolism*. Grune and Stratton, New York.

20. **Harmuth-Hoene, A.-E., M. Vladar, and R. Ohrtmann.** 1969. Fe(III)-exchange between transferrin and chelates in vitro. *Chem. Biol. Interactions.* **1:**271–283.

21. **Inoshita, E., K. Iwakura, A. Amano, T. Tamagawa, and S. Shizukuishi.** 1991. Effect of transferrin on the growth of *Porphyromonas gingivalis*. *J. Dent. Res.* **70:**1258–1261.

22. **Keller-Schierlein, W.** 1976. *In* W. F. Anderson and M. C. Hiller (ed.), *Development of Iron Chelators for Clinical Use,* p. 53–82. Department of Health, Education, and Welfare publication no. 76-994. National Institutes of Health, Bethesda, Md.

23. **Konopka, K., A. Bindereif, and J. B. Neilands.** 1982. Aerobactin-mediated utilization of transferrin iron. *Biochemistry* **21:**6503–6508.

24. **Kraig, E., T. Dailey, and D. Kolodrubetz.** 1990. Nucleotide sequence of the leukotoxin gene from *Actinobacillus actinomycetemcomitans*: homology to the alpha-hemolysin/leukotoxin gene family. *Infect. Immun.* **58:**920–929.

25. **Lee, B. C., and P. Hill.** 1992. Identification of an outer-membrane haemoglobin-binding protein in *Neisseria meningitidis*. *J. Gen. Microbiol.* **138:**2647–2656.

26. **LeRoy, D., D. Expert, A. Razafindratsita, A. Deroussent, J. Cosme, C. Bohuon, and A. Andremont.** 1992. Activity and specificity of a mouse monoclonal antibody to ferric aerobactin. *Infect. Immun.* **60:**768–772.

27. **Ma, J., and T. J. Inzana.** 1990. Indirect enzyme-linked immunosorbent assay for detection of antibody to a 110,000-molecular-weight hemolysin of *Actinobacillus pleuropneumoniae*. *J. Clin. Microbiol.* **28:**1356–1361.

28. **Malek, R. L., and D. W. Dyer.** Growth stimulation of *Porphyromonas gingivalis* by porphyrins and transferrin, abstr. B-299, p. 79. *Abstr. 93rd Gen. Meet Am. Soc. Microbiol.*, *1993*.

29. **Meyer, D., P. Sreenivasan, and P. Fives-Taylor.** 1991. Evidence for invasion of a human oral cell line by *Actinobacillus actinomycetemcomitans*. *Infect. Immun.* **59:**2719–2726.

30. **Mukherjee, S.** 1985. The role of crevicular fluid iron in periodontal disease. *J. Periodontol.* **56:**22–27.

31. **Muller-Eberhard, U.** 1970. Hemopexin. *N. Engl. J. Med.* **283:**1090–1094.

32. **Neilands, J. B.** 1981. Microbial iron compounds. *Annu. Rev. Biochem.* **50:**715–731.

33. **Neilands, J. B.** 1982. Microbial envelope proteins related to iron. *Annu. Rev. Microbiol.* **36:**285–309.

34. **Payne, S. M.** 1988. Iron and virulence in the family *Enterobacteriaceae*. *Crit. Rev. Microbiol.* **16:**81–111.

35. **Pidcock, K. A., J. A. Wooten, B. A. Daley, and T. L. Stull.** 1988. Iron acquisition by *Haemophilus influenzae*. *Infect. Immun.* **56:**721–725.

36. **Sandele, P. J., D. W. Dyer, and M. E. Neiders.** 1990. Utilization by *Bacteroides gingivalis* of heme-compounds as an iron source, abstr. 1513. *J. Dent. Res.* **69:**298.

37. **Shah, H. N., R. Bonnett, B. Mateen, and R. A. D. Williams.** 1979. The porphyrin pigmentation of subspecies of *Bacteroides melaninogenicus*. *Biochem. J.* **180:**45–50.

38. **Slots, J., and M. A. Listgarten.** 1988. *Bacteroides gingivalis, Bacteroides intermedius,* and *Actinobacillus actinomycetemcomitans* in human periodontal diseases. *J. Clin. Periodontol.* **15:**85–93.

39. **Smith, L. D., and B. L. Williams.** 1984. Anaerobes in the microflora of the human body, p. 262–279. *In* A. Balows (ed.), *The Pathogenic Bacteria.* Charles C. Thomas, Publisher, Springfield, Ill.

40. **Sokol, P. A.** 1987. Surface expression of ferripyochelin-binding protein is required for virulence of *Pseudomonas aeruginosa. Infect. Immun.* **55:**2021–2025.

41. **Spitznagel, J., and D. Kolodrubetz.** 1993. Environmental regulation of leukotoxin production in *Actinobacillus actinomycetemcomitans,* abstr. 388. *J. Dent. Res.* **72:**152.

42. **Strathdee, C. A., and R. Y. C. Lo.** 1989. Regulation of expression of the *Pasteurella haemolytica* leukotoxin determinant. *J. Bacteriol.* **171:**5955–5962.

43. **Thering, M. B., D. W. Dyer, and M. E. Neiders.** 1991. Iron-repressible membrane proteins of *Porphyromonas gingivalis,* abstr. 1917. *J. Dent. Res.* **70:**506.

44. **Tsai, J., D. W. Dyer, and P. F. Sparling.** 1988. Loss of transferrin receptor activity in *Neisseria meningitidis* correlates with inability to use transferrin as an iron source. *Infect. Immun.* **56:**3132–3138.

45. **Waalwijk, C., D. M. MacLaren, and J. DeGraaff.** 1983. In vivo function of hemolysin in the nephropathogenicity of *Escherichia coli. Infect. Immun.* **42:**245–249.

46. **Watanabe, N.-A., T. Nagasu, K. Katsu, and K. Kitoh.** 1987. E-0702, a new cephalosporin, is incorporated into *Escherichia coli* cells via the *tonB*-dependent iron transport system. *Antimicrob. Agents Chemother.* **31:**497–504.

47. **Weinberg, E. D.** 1978. Iron and infection. *Microbiol. Rev.* **42:**45–66.

48. **Williams, P. K., and H. K. George.** 1979. ColV plasmid-mediated iron uptake and the enhanced virulence of invasive strains of *Escherichia coli,* p. 161–172. *In* K. Timmis and A. Puhler (ed.), *Plasmids of Medical and Environmental and Commercial Importance.* Elsevier/North-Holland Publishing Co., Amsterdam.

49. **Winston, J. L., C.-K. C. Chen, M. E. Neiders, and D. W. Dyer.** 1993. Membrane protein expression by *Actinobacillus actinomycetemcomitans* in response to iron availability. *J. Dent. Res.,* **72:**1366–1373.

50. **Zambon, J. J.** 1988. *Actinobacillus actinomycetemcomitans* in the pathogenesis of human periodontal disease. *Adv. Dent. Res.* **2:**269–274.

51. **Zambon, J. J., C. DeLuca, J. Slots, and R. J. Genco.** 1983. Studies of leukotoxin from *Actinobacillus actinomycetemcomitans* using the promyelocytic HL-60 cell line. *Infect. Immun.* **40:**205–212.

Molecular Pathogenesis of Periodontal Disease
Edited by Robert Genco et al.
© 1994 American Society for Microbiology, Washington, DC 20005

Chapter 6

Invasion of Cultured Epithelial Cells by Periodontopathogens

Paula M. Fives-Taylor, Diane H. Meyer, Prem K. Sreenivasan, and Keith P. Mintz

A variety of pathogenic mechanisms (28, 70) and virulence factors (65) are involved in the periopathic potential of organisms implicated in periodontal disease. Currently, there is neither agreement nor consensus as to the actual mechanism(s) of pathogenesis or the specific virulence factors among these organisms (77). It is our hypothesis that invasion of the epithelial-cell barrier by periodontopathogens is a virulence mechanism and plays a basic role in the pathogenesis and progression of periodontal disease. Epithelial cells act as the first barrier against bacterial infection. Attachment to and penetration of these cells is an early step in establishing infection by a variety of microorganisms. Pathogens such as *Shigella* spp. (20), *Salmonella* spp. (19, 29), *Listeria* spp. (24, 42), *Yersinia* spp. (74), enteroinvasive *Escherichia coli* (40), *Neisseria gonorrhoeae* (64), and *Bordetella pertussis* and *Bordetella parapertussis* (14) can be internalized by cells that are not usually phagocytic (nonprofessional phagocytes) (16). This pathogenic strategy, called cell invasion, confers upon the microorganism unique survival strategies (17, 21, 49). Organisms are safely sequestered from host defenses and the antimicrobial substances present on mucosal surfaces. Nonspecific host defense clearing mechanisms are avoided, and these organisms grow and multiply in a nutritionally rich environment free of other competing organisms. Some intracellular bacteria, e.g., *Shigella* and *Listeria* spp., utilize their intracellular niche to transmit themselves to uninfected neighboring cells (15).

The ability to invade epithelial cells is a virulence factor for *Salmonella* spp. (25). In that study (25), noninvasive mutants were constructed by transposon mutagenesis of a cloned invasion gene. The noninvasive mutants were unable to es-

Paula M. Fives-Taylor, Diane H. Meyer, and Keith P. Mintz • Department of Microbiology and Molecular Genetics, University of Vermont, Stafford Hall, Burlington, Vermont 05405.   ***Prem K. Sreenivasan*** • Laboratory of Molecular Infectious Diseases, The Rockefeller University, 1230 York Avenue, New York, New York 10021.

tablish infection when the organisms entered by the normal fecal-oral route of entry but remained virulent when injected directly into the blood stream.

MECHANISMS OF INVASION

Two distinct modes of invasion have been reported. *E. coli* and *Shigella* spp. enter the epithelial cell, grow within the cell, and cause localized infection and destruction of the underlying mucosa. In contrast, *Salmonella* and *Yersinia* spp. penetrate the epithelial cell and transit through it to deeper tissues without causing major damage to the mucosa. The complexity of the process and the number of genes required for invasion vary considerably among bacterial genera and even among bacterial species within the same genus (18). *Shigella flexneri* requires at least three genes located on a 70-kb plasmid (39, 61, 63). *Yersinia enterocolitica* has a single chromosomal gene that is responsible for invasion (34). The closely related species *Yersinia pseudotuberculosis* requires at least two chromosomal genes to confer the same phenotype (47). While no generalizations about bacterial invasion can be made, successful pathogens have some mechanisms in common. Most invasive bacteria adhere to eucaryotic cell surfaces and cellular structures such as microvilli prior to localization within the eucaryotic cells. Internalization requires the participation of eucaryotic cytoskeletal components. For example, invasive *Salmonella* species and invasive *E. coli* result in a loss of the eucaryotic microvilli that regenerate following invasion (11, 18). After entering the eucaryotic cell, *Salmonella* spp., *Shigella* spp., and *E. coli* are surrounded by polymerized actin. Many invasive bacteria such as *Shigella* species and invasive *E. coli* multiply in target cells. *Yersinia* and *Salmonella* species use their intracellular locations as waystations for entering underlying deeper tissue. A host cell-derived membrane vacuole usually surrounds invasive bacteria upon entry. Some invasive bacteria (*Shigella* and *Listeria* spp.) lyse the vacuole and multiply in the cytoplasm. In contrast, *Yersinia* and *Salmonella* spp. remain localized within the vacuole. In general, bacteria capable of extensive multiplication within the host cell usually lyse the vacuole.

A common theme is emerging concerning the differences between invasive and noninvasive bacteria that enable the former to localize within eucaryotic cells. Bacterial attachment to eucaryotic cells alone does not end in invasion (15). Environmental stimuli such as O_2 tension, temperature, presence of specific bacterial gene products, and density and distribution of eucaryotic receptors may influence invasion. *Salmonella typhimurium* invasion is stimulated by growth under O_2 limitation. Furthermore, there is a 50-fold increase in a particular *S. typhimurium* gene following bacterial contact with eucaryotic cells but not after contact with inert substances such as agarose or Matrigel (75). The invasion genes of *Yersinia* spp. and *S. flexneri* (46) are temperature regulated. *Yersinia* spp. grown at 28°C synthesize the invasin protein that efficiently promotes entry into eucaryotic cells but express a different and less efficient invasion protein, Ail, at 37°C (33). On the other hand, *S. flexneri* synthesizes several proteins responsible for invasion when grown at 37°C but not at 30°C (63). Expression of a *S. typhimurium* gene required for invasion is regulated by changes in DNA supercoiling (26).

MECHANISM OF BACTERIAL UPTAKE BY HOST CELLS

Invasion of epithelial cells by bacteria involves subversion of normal host cell function by the bacteria. Invasive bacteria trigger host cell signal transduction mechanisms that induce cytoskeletal rearrangements leading to bacterial uptake (53). The invasin protein of *Yersinia* spp. binds to and clusters integrin receptors at the bacterium–epithelial-cell boundaries. The integrin clustering activates host tyrosine protein kinase, which in turn triggers cytoskeletal rearrangements that facilitate uptake (54). *S. typhimurium* phosphorylates the tyrosine receptor for epidermal growth factor (EGF) before invasion (27). In contrast, noninvasive mutants that adhere to eucaryotic cells do not phosphorylate the EGF receptor and are not internalized. Interestingly, the addition of EGF to eucaryotic cells causes membrane ruffling, stimulation of pinocytosis, redistribution of cell surface receptors, generation of a calcium flux, and changes in intracellular pH (55). Similar effects on the eucaryotic cytoskeleton are a feature typical of *Salmonella* invasion (27). The precise mechanisms for protein phosphorylation and the role of EGF in invasion remain unclear. However, tyrosine phosphorylation plays a central role in transducing signals to the host cell and communicating with the host cell. A role for calcium in intracellular signaling has been noted in the entry of enteropathogenic *E. coli* into eucaryotic cells (2). Calcium may play a role in activating actin and dissociating proteins such as villin that result in microvilli disruption. These examples, while interesting, probably are but a few of the sophisticated ways that pathogens exploit host cell functions. The close interaction between organism and host apparently resulted in an evolutionary pressure for the organism to subvert these host cell functions for survival.

CLINICAL OBSERVATIONS OF INVASION
BY PERIODONTOPATHOGENS

The suggestion that periodontopathogens invade the gingival tissue has received intermittent support since the beginning of the century. The hypothesis, like the disease, has bursts of activity. As early as 1907, Goadby (30) postulated that bacteria invade oral tissues. Turner and Drew (73) presented histological evidence supporting the presence of diphtheroids in the gingiva of patients with periodontal disease. Beckwith et al. (3) cultivated bacteria from tissue aspirates of patients with gingivitis. They located bacteria in the periodontal membrane that were moving toward the apex. Twenty-two years later, Ray and Orban (51) reported seeing degradation and necrosis of tissue in gingival biopsy samples. They hypothesized that the necrosis was caused by bacterial invasion. During the 1960s and 1970s, the idea that gingival tissues were invaded by bacteria was reintroduced. Again, investigators thought that invasion might play a role in periodontal disease (22, 23, 38, 41, 45, 56, 72). Bacteria were noted in the gingival epithelium, basal stratum, and connective tissue beneath the basal lamina and along the alveolar bone. The number of microorganisms in the tissues was always higher in samples from diseased sites. Saglie and coworkers were the leading proponents of this bacterial-invasion hypothesis in the 1980s. They found that bacterial invasion of soft periodontal tissue and bone was common in advanced periodontitis and localized juvenile periodontitis (58). Bacteria were found invading the epithelial wall (57), in the enlarged intra-cellular spaces of the pocket epithelium surface, on the epithelial side of the basal lamina, and in the connective tissue. Bacteria were in specific intracellular locations and presented a definite pattern of penetration.

Gram-negative rods were observed in the underlying tissues, but their species were not determined. By means of immunofluorescence (6, 57), some internal bacteria were identified as *Actinobacillus actinomycetemcomitans*, pinpointing *A. actinomycetemcomitans* as an invading pathogen. In subsequent studies, this organism was found in distinct arrangements within diseased gingiva (59). It was arranged in a pyramid-like formation with the base toward the outer epithelium, and its presence was correlated with decreased keratin in the epithelium. The consistent presence of *A. actinomycetemcomitans* in the oral epithelium in the absence of bacteria in the deeper tissue led the investigators to propose that the oral epithelium may serve as the port for bacterial entry into the underlying gingiva (59). Christersson et al. (6) were able to cultivate *A. actinomycetemcomitans* from 8 of 11 gingival biopsy samples. The presence of large numbers of the organism in the infected periodontal pocket was correlated with the number of viable organisms localized within the gingiva. These results correlate well with those of earlier microscopic and immunochemical techniques that were used to implicate bacterial invasion during periodontitis. In an attempt to explain the bursts of disease activity associated with periodontitis, immunochemical studies were performed with gingival biopsy samples obtained from patients with active and inactive periodontal lesions. A positive correlation was made between the intragingival presence of *A. actinomycetemcomitans* and active periodontal disease (57). Although *A. actinomycetemcomitans* was localized in the gingival tissue of inactive sites, it was consistently present in significantly large numbers in the sites undergoing active periodontal destruction.

In the only animal study (germfree rats) of invasion, *Bacteroides melaninogenicus* invaded the cells and connective tissues, while *Actinomyces viscosus* did not (1).

IN VITRO MODELS FOR CELL INVASION
BY PERIODONTOPATHOGENS

To date, there have been four reports on the development of in vitro invasion models for periodontopathogens. Three models assay for different aspects of invasion. An in vitro invasion model utilizing PF-HR-9 cells was developed (76). This model utilized a basement membrane-like matrix that was composed of the same components as in vivo basement membranes. Winkler et al. (76) studied attachment to and solubilization of the isolated matrix by *P. gingivalis*. This model is an example of tissue invasion and proposes that bacteria may reach even deeper tissues by solubilization of the basement membrane matrix. The investigators did not look at cell invasion or propose a model to explain how the organisms got to the basement membrane. Saglie et al. (60) reported a gingival equivalent model for invasion by *A. actinomycetemcomitans* whereby gingival fibroblasts were mixed with collagen to form a lattice. Gingival epithelial cells seeded on the lattices developed into a multilayered epithelium. A strain of *A. actinomycetemcomitans* was used to infect the model. Analysis by electron microscopy revealed that bacteria were deep within the collagen lattice, and they were seen attached to and within some of the epithelial cells. This model allows one to look at tissue invasion but is not easily amenable to the study of cell invasion. An in vitro assay for invasion of the KB cell line (derived from a human oral epidermoid carcinoma) by *A. actinomycetemcomitans* has been developed (44, 67). This model allows us to study cell invasion and is quantitative, permitting the kinetics of invasion to be studied. Briefly, *A. actinomycetemcomitans* is incubated with confluent monolayers of KB cells for 2 h before the addition of gentamicin, which kills all extracellular *A. actinomycetemcomitans*. The monolayers are then extensively washed and lysed with added Triton X-100. The released internalized bacteria are enumerated by plate counting.

INVASION OF EPITHELIAL CELLS BY PERIODONTOPATHOGENS

Little information is available about the mechanisms that periodontopathogens use to maintain themselves in the oral cavity. Several options are attachment to epithelial cells, attachment to the tooth enamel, penetration of the epithelial cell either directly or through the intracellular spaces at the junction of the epithelial cells, and coaggregation with other microorganisms that are firmly attached to oral tissues. In the last few years, investigations have centered around the attachment to and penetration of epithelial cells by three of the important periodontopathogens: *A. actinomycetemcomitans*, *Porphyromonas gingivalis*, and *Treponema denticola*. The criteria for invasion generally include the following: (i) the organisms can be seen internalized by electron microscopy; (ii) eucaryotic microfilaments are required, so the process is sensitive to cytochalasin-D; and (iii) metabolically active bacteria and eucaryotic cells are required.

A. actinomycetemcomitans

A. actinomycetemcomitans is capable of attaching to hydroxyapatite, a natural tooth material (36, 52), and to cultured epithelial cells (43, 48, 71). The ability of this organism to localize within cultured epithelial cells has been demonstrated by the quantitative cell culture assay, with gentamicin used to kill external cells (44). *Haemophilus aphrophilus,* a closely related oral species, did not invade cultured cells and served as a negative control in all experiments. Results are presented as an invasion index of *A. actinomycetemcomitans* compared to that of *H. aphrophilus*. Twenty-four percent of *A. actinomycetemcomitans* isolates (44) screened from different areas in the United States were invasive (Fig. 1). Invasion by *A. actinomycetemcomitans* was comparable to that by other invasive members of the family *Enterobacteriaceae*. Invasiveness was associated with smooth colony morphology

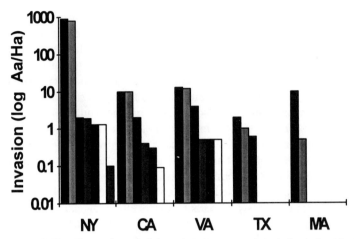

Figure 1. Invasion of KB cell monolayers of *A. actinomycetemcomitans* isolates. Invasion assays were carried out for 2 h at a multiplicity of infection of 2,500:1. Results were normalized to invasion by *H. aphrophilus,* which was run concurrently, and are presented as an invasion index on a log scale. NY, New York; CA, California; VA, Virginia; TX, Texas; MA, Massachusetts.

(44), although smoothness alone was not sufficient for invasion, as some smooth isolates did not invade (Fig. 2).

A. *actinomycetemcomitans* enters the cells by receptor-mediated endocytosis and in early stages is surrounded by endosomal membranes that disintegrate with time (68). Invasion by *A. actinomycetemcomitans* is an active process requiring energy and both bacterial and eucaryotic de novo protein synthesis. Eucaryotic microfilaments but not microtubules are required for internalization. KB cells form numerous microvilli in the presence of *A. actinomycetemcomitans* and many bac-

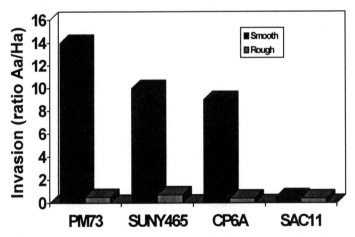

Figure 2. Invasion of KB cell monolayers of *A. actinomycetemcomitans* rough and smooth phenotype pairs. Invasion assays were carried out for 2 h at a multiplicity of infection of 2,500:1. Results are normalized to invasion by *H. aphrophilus,* which was run concurrently.

teria are observed attached to the microvilli. *A. actinomycetemcomitans* require-ments for entry into eucaryotic cells are similar to the requirements of many but not all invasive bacteria (Table 1).

Sodium dodecyl sulfate-polyacrylamide gel electrophoresis of SUNY 465 (19a) revealed three peptide differences between the rough and smooth isolates. Ab-sorption of anti-SUNY 465 smooth whole-cell sera with the rough isolate removed all the antigenic reaction of the smooth isolate. These data suggested that the peptide differences were modifications of the peptides present in the smooth isolate. The switch from smooth to rough morphology was associated with loss of inva-siveness but increased ability to adhere to KB cells. Preliminary data suggest that a low-molecular-weight model molecule induces the switching of SUNY 465 be-tween smooth and rough forms.

To identify the macromolecules involved in the adhesion to and invasion of epithelial cells by *A. actinomycetemcomitans*, monoclonal antibodies to the whole bacterium were developed. Two antibodies have been isolated (this study). These antibodies inhibit *A. actinomycetemcomitans* invasion of (Fig. 3) but not attachment to epithelial cells. These data imply that different macromolecules are involved in

Table 1. Comparision of requirements and strategies used by facultatively intracellular bacteria to invade epi-thelial cells[a]

Organism	Attachment	Microfilament	Microtubule	Energy metabolism	Endosomal acidification	Protein synthesis	Nucleic acid synthesis	Receptor-mediated endocytosis	Reference(s)
Actinobacillus actinomycetemcomitans	+	+	−	+	−	+	−	+	44, 68
Haemophilus spp.	+	+	+	ND	ND	ND	ND	ND	69
Enterobacteriaceae									
Shigella spp.	?	+	−	+	−	ND	ND	−	17, 31, 32
EPEC	+	+	+	ND	ND	ND	ND	ND	10
EIEC	+	+	−	ND	ND	ND	ND	ND	10, 66
Yersinia pseudotuberculosis	+	+	−	−	−	−	−	+	17, 18, 47, 66
Yersinia enterocolitica	+	+	−	−	−	−	−	+	17, 18, 47, 66
Salmonella spp.	+	+	−	+	−	+	±	+	17, 18
Edwardsiella spp.	ND	+	+	ND	−	ND	ND	−	35
Brucella abortus	+	ND	ND	ND	ND	ND	ND	ND	9
Neisseria gonorrhoeae	+	+	+	ND	ND	+	ND	ND	5
Campylobacter spp.	+	+	ND	+	ND	ND	ND	ND	8
Bordetella spp.	+	+	ND	ND	ND	ND	ND	−	14
Listeria monocytogenes	+	+	−	ND	ND	−	ND	ND	24, 42
Listeria pneumophila	+	ND	ND	ND	ND	ND	ND	ND	50
Aeromonas spp.	+	ND	ND	ND	ND	ND	ND	ND	4
Borrelia burgdorferi	+	−	−	+	ND	−	ND	ND	7

[a]+, required; −, not required; ND, not determined; ?, unknown; ±, RNA synthesis required; EPEC, enteropathogenic *E. coli*; EIEC, enteroinvasive *E. coli*.

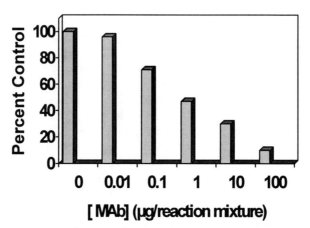

Figure 3. Inhibition of invasion of *A. actinomycetemcomitans* SUNY 465 by monoclonal antibody K26. Various concentrations of antibody were incubated with 10^8 bacteria for 1 h before being added to the monolayer of KB cells.

invasion and attachment of this organism to epithelial cells. The identity of the macromolecules to which these antibodies bind is currently under investigation.

P. gingivalis

Approximately 10% of attached *P. gingivalis* invade KB epithelial cells in a cell culture assay including trospectomycin (12) or a combination of metronidazole and gentamicin (62) to kill external cells. *P. gingivalis* penetration of primary cultures of human gingival epithelial cells also has been demonstrated (37). Initial contact between *P. gingivalis* and the epithelial cell stimulates the cellular membrane to form coated pits, implying that internalization occurs through the receptor-mediated endocytosis pathway (62). *P. gingivalis* enters the cells in a membrane-derived vacuole that disintegrates in time. Interestingly, intracellular *P. gingivalis* still releases the outer membrane vesicles called blebs, which suggests that *P. gingivalis* remains metabolically active within the eucaryotic cell.

T. denticola

T. denticola binds to both human fibroblast and epithelial cells in culture (13). Effective binding occurs at the tip of the treponeme. In in vitro models, the tip binding can be produced with fibronectin membranes and/or laminin. Fibronectin receptors located throughout the sheath of the organism orient toward the tip of the organism in response to fibronectin. The treponemes can also bind to RGD peptides, suggesting that the receptor may be integrin-like. Preincubation of the treponemes with the RGD peptides, however, is not sufficient to inhibit subsequent binding. Hence, other adhesins may be involved. Actin filaments are reorganized in fibroblasts in response to the treponemal attachment, suggesting that the organism can invade the fibroblast cell. However, very few cells with internalized treponemes can be seen. Therefore, invasion of fibroblasts may be a rare event for this organism. Invasion of KB cells by *T. denticola* has not been noted. Rapid F actin depolymerization, shrinkage of the KB cell, and loss of volume regulation occur in response to *T. denticola* attachment. However, *T. denticola* has been internalized in HEp-2 cells (32a). Whether this discrepancy reflects differences between KB and HEp-2 cell surface receptors or the experimental protocols awaits further experimentation.

In summary, at least some periodontal pathogens invade cultured oral epithelial cells and primary gingival tissue. The next 10 years should see an explosion in our understanding of the mechanisms and macromolecules used by these invasive periodontopathogens.

Acknowledgments. This work was supported in part by Public Health Service grants RO1DE09760 and F23DE05626 from the National Institute of Dental Research.

REFERENCES

1. **Allenspach-Petrzilka, G. E., and B. Gugenheim.** 1982. *Bacteroides melaninogenicus* subsp. *intermedius* invasion of rat gingival tissue. *J. Periodontal Res.* **17:**456–459.
2. **Baldwin, T. J., W. Ward, A. Aitken, S. Knutton, and P. H. Williams.** 1991. Elevation of intracellular free calcium levels in HEp-2 cells infected with enteropathogenic *Escherichia coli*. *Infect. Immun.* **59:**1599–1604.
3. **Beckwith, T. D., F. V. Simonton, and E. J. Rose.** 1927. The presence of bacterial microorganisms in human gingival tissue in gingivitis. *Dent. Cosmos* **6:**164–167.
4. **Carrello, A. K., K. A. Silburn, J. R. Budden, and B. J. Chang.** 1988. Adhesion of clinical and environmental *Aeromonas* isolates to HEp-2 cells. *J. Med. Microbiol.* **26:**19–27.
5. **Chen, J., P. Bavoil, and V. L. Clark.** 1990. Enhancement of invasive ability of *Neisseria gonorrhoeae* by contact with HecIB, an adenocarcinoma endometrial cell line. *Mol. Microbiol.* **5:**1531–1538.
6. **Christersson, L. A., B. Albini, J. J. Zambon, U. M. E. Wikesjo, and R. J. Genco.** 1987. Tissue localization of *Actinobacillus actinomycetemcomitans* in human periodontitis. I. Light immunofluorescence and culture techniques. *J. Periodontol.* **58:**529–539.
7. **Comstock, L. E., and D. D. Thomas.** 1991. Characterization of *Borrelia burgdorferi* invasion of cultured endothelial cells. *Microb. Pathog.* **10:**137–148.
8. **De Melo, M. A., G. Gabbiani, and J. C. Pechere.** 1989. Cellular events and intracellular survival of *Campylobacter jejuni* during infection of HEp-2 cells. *Infect. Immun.* **57:**2214–2222.
9. **Detilleux, P. G., B. L. Deyoe, and N. F. Cheville.** 1990. Penetration and intracellular growth of *Brucella abortus* in nonphagocytic cells in vitro. *Infect. Immun.* **58:**2320–2328.
10. **Donnenberg, M. S., A. Donohue Rolfe, and G. T. Keusch.** 1990. A comparison of HEp-2 cell invasion of enteropathogenic and enteroinvasive *Escherichia coli*. *FEMS Microbiol. Lett.* **57:**83–86.
11. **Donnenberg, M. S., and J. B. Kaper.** 1992. Enteropathogenic *Escherichia coli*. *Infect. Immun.* **60:**3953–3961.
12. **Duncan, M. J., S. Nakao, Z. Skobe, and H. Xie.** 1993. Interactions of *Porphyromonas gingivalis* with epithelial cells. *Infect. Immun.* **61:**2260–2265.
13. **Ellen, R. P., I. A. Song, S. Buivids, and C. A. G. McCullock.** 1993. Morphology of *Treponema denticola* invasion in gingival fibroblasts. *J. Dent. Res.* **72** (special issue)**:** 411.
14. **Ewanowich, C. A., A. R. Melton, A. A. Weiss, R. K. Sherburne, and M. S. Peppler.** 1989. Invasion of HeLa 229 cells by virulent *Bordetella pertussis*. *Infect. Immun.* **57:**2698–2704.
15. **Falkow, S.** 1991. Bacterial entry into eukaryotic cells. *Cell* **65:**1099–1102.
16. **Falkow, S., P. Small, R. Isberg, S. F. Hayes, and D. Corwin.** 1987. A molecular strategy for the study of bacterial invasion. *Rev. Infect. Dis.* **9:**S450–S455.
17. **Finlay, B. B., and S. Falkow.** 1988. A comparison of microbial invasion strategies used by *Salmonella*,

Shigella, and *Yersinia* species, p. 277–243. *In* M. A. Horowitz (ed.), *Bacterial-Host Cell Interaction.* Alan R. Liss, Inc., New York.

18. **Finlay, B. B., and S. Falkow.** 1989. Common themes in microbial pathogenicity. *Microbiol. Rev.* **53:**210–230.

19. **Finlay, B. B., F. Heffron, and S. Falkow.** 1989. Epithelial cell surfaces induce Salmonella proteins required for bacterial adherence and invasion. *Science* **243:**940–943.

19a. **Fives-Taylor, P. M., and P. K. Sreenivasan.** Unpublished data.

20. **Formal, S. B., H. L. Dupont, R. B. Hornick, M. J. Snyder, J. Libonati, and E. H. LaBrec.** 1971. Experimental models in the investigation of the virulence of dysentery bacilli and *Escherichia coli.* *Ann. N.Y. Acad. Sci.* **176:**190–196.

21. **Formal, S., T. L. Hale, and P. J. Sansonetti.** 1983. Invasive enteric pathogens. *Rev. Infect. Dis.* **5:**S702–S707.

22. **Frank, R. M.** 1980. Bacterial penetration in the apical pocket wall of advanced human periodontitis. *J. Periodontal Res.* **15:**563–573.

23. **Frank, R. M., and J. C. Voegel.** 1978. Bacterial bone resorption in advanced cases of human periodontitis. *J. Periodontal Res.* **13:**251–261.

24. **Gaillard, J. L., P. Berche, J. Mounier, S. Richard, and P. Sansonetti.** 1987. In vitro model of penetration and intracellular growth of *Listeria monocytogenes* in the human enterocyte-like cell line Caco-2. *Infect. Immun.* **55:**2822–2829.

25. **Galan, J. E., and R. C. Curtiss III.** 1989. Cloning and molecular characterization of genes whose products allow Salmonella typhimurium to penetrate tissue culture cells. *Proc. Natl. Acad. Sci. USA* **86:**6383–6387.

26. **Galan, J. E., and R. C. Curtiss III.** 1990. Expression of *Salmonella typhimurium* genes required for invasion is regulated by changes in DNA supercoiling. *Infect. Immun.* **58:**1879–1885.

27. **Galan, J. E., J. Pace, and M. J. Hayman.** 1992. Involvement of the epidermal growth factor receptor in the invasion of cultured mammalian cells by *Salmonella typhimurium. Nature* (London) **357:**588–589.

28. **Genco, R. J., and J. Slots.** 1984. Host responses in periodontal diseases. *J. Dent. Res.* **63:**441–451.

29. **Gianella, R. A., O. Washington, P. Gemski, and S. B. Formal.** 1973. Invasion of HeLa cells by *Salmonella typhimurium*: a model for study of invasiveness of *Salmonella. J. Infect. Dis.* **128:**69–75.

30. **Goadby, K. W.** 1907. Erasmus Wilson's lecture on pyorrhea alveolaris. *Lancet* **i:**633.

31. **Hale, T. L., and P. F. Bonventre.** 1979. *Shigella* infection of Henle intestinal epithelial cells: role of the bacterium. *Infect. Immun.* **24:**879–886.

32. **Hale, T. L., R. E. Morris, and P. F. Bonventre.** 1979. *Shigella* infection of Henle intestinal epithelial cells: role of the host cell. *Infect. Immun.* **24:**887–894.

32a. **Holt, S. C.** Personal communication.

33. **Isberg, R. R.** 1991. Discrimination between intracellular uptake and surface adhesion of bacterial pathogens. *Science* **252:**934–938.

34. **Isberg, R. R., and S. Falkow.** 1985. A single genetic locus encoded by *Y. pseudotuberculosis* permits invasion of cultured animal cells by *Escherichia coli* K-12. *Nature* (London) **317:**262–264.

35. **Janda, J. M., S. L. Abbott, and L. S. Oshino.** 1991. Penetration and replication of *Edwardsiella* subsp. in HEp-2 cells. *Infect. Immun.* **59:**154–161.

36. **Kagermeier, A. S., and J. London.** 1985. *Actinobacillus actinomycetemcomitans* strains Y4 and N27 adhere to hydroxyapatite by distinctive mechanisms. *Infect. Immun.* **47:**654–658.

37. **Lamont, R. J., D. Oda, R. E. Persson, and G. R. Persson.** 1992. Interaction of *Porphyromonas gingivalis* with epithelial cells maintained in culture. *Oral Microbiol. Immunol.* **7:**364–367.

38. **Listgarten, M. A.** 1965. Electron microscopic observations on the bacterial flora of acute necrotizing ulcerative gingivitis. *J. Periodontol.* **36:**328–339.

39. **Maurelli, A. T., B. Baudry, H. d'Hauteville, T. L. Hale, and P. J. Sansonetti.** 1985. Cloning of plasmid DNA sequences involved in invasion of HeLa cells by *Shigella flexneri. Infect. Immun.* **49:**164–171.

40. **Mehlman, I. J., E. L. Eide, A. C. Sanders, M. Fishbein, and C. G. Aulisio.** 1977. Methodology for recognition of invasive potential of *Escherichia coli. J. Assoc. Off. Anal. Chem.* **60:**546–551.

41. **Merrell, B. R., S. W. Joseph, L. J. Casazza, and J. F. Duncan.** 1981. Bacterial bone resorption in noma (gangrenous stomatitis). *J. Oral Pathol.* **10:**173–177.

42. **Meyer, D. H., M. Bunduki, D. M. Beliveau, and C. W. Donnelly.** 1992. Differences in invasion and adherence of *Listeria monocytogenes* with mammalian gut cells. *Food Microbiol.* **9:**115–126.

43. **Meyer, D. H., and P. M. Fives-Taylor.** 1993. Evidence that extracellular components function in adherence of *Actinobacillus actinomycetemcomitans* to epithelial cells. *Infect. Immun.* **61:**4933–4936.

44. **Meyer, D. H., P. K. Sreenivasan, and P. M. Fives-Taylor.** 1991. Evidence for the invasion of a human oral cell line by *Actinobacillus actinomycetemcomitans. Infect. Immun.* **59:**2719–2726.

45. **Michel, C.** 1969. Etude ultrastructurale de l'os alveolaire au cours des parodontolyses. *Paradontologie* **23:**191–210.

46. **Miller, J. F., J. J. Mekalanos, and S. Falkow.** 1989. Coordinate regulation and sensory transduction in the control of bacterial virulence. *Science* **243:**916–922.

47. **Miller, V. L., B. B. Finlay, and S. Falkow.** 1988. Factors essential for the penetration of mammalian cells by *Yersinia. Curr. Top. Microbiol. Immunol.* **138:**15–39.

48. **Mintz, K. P., and P. M. Fives-Taylor.** 1993. Characterization of adhesion of *Actinobacillus actinomycetemcomitans* to epithelial cells. *J. Dent. Res.* **72:**325.

49. **Moulder, J. W.** 1985. Comparative biology of intracellular parasitism. *Microbiol. Rev.* **48:**298–337.

50. **Oldham, L. J., and F. G. Rodgers.** 1985. Adhesion, penetration and intracellular replication of *Legionella pneumophilia:* an in vitro model of pathogenesis. *J. Gen. Microbiol.* **131:**697–706.

51. **Ray, H. G., and B. Orban.** 1948. Deep necrotic foci in the gingiva. *J. Periodontol.* **19:**91–97.

52. **Rosan, R., J. Slots, R. J. Lamont, M. Listgarten, and G. M. Nelson.** 1988. *Actinobacillus actinomycetemcomitans* fimbriae. *Oral Microbiol. Immunol.* **3:**58–63.

53. **Rosenshine, I., and B. Brett Finlay.** 1993. Exploitation of host transduction pathways and cytoskeletal functions by invasive bacteria. *Bioessays* **15:**17–24.

54. **Rosenshine, I., V. Duronio, and B. B. Finlay.** 1992. Tyrosine protein kinase inhibitors block invasion-promoted bacterial uptake by epithelial cells. *Infect. Immun.* **60:**2211–2217.

55. **Rozengurt, E.** 1986. Early signals in the mitogenic response. *Science* **234:**161–166.

56. **Saglie, F. R.** 1977. A scanning electron microscopic study of the relationship between the most apically localized subgingival plaque and the epithelial attachment. *J. Periodontol.* **42:**105–115.

57. **Saglie, F. R., F. A. Carranza, Jr., M. G. Newman, L. Cheng, and K. J. Lewin.** 1982. Identification of tissue-invasive bacteria in human periodontal disease. *J. Periodontal Res.* **17:**452–455.

58. **Saglie, F. R., and J. J. Elbaz.** 1983. Bacterial penetration into the gingival tissue in periodontal disease. *J. West. Soc. Periodontal.* **31:**85–93.

59. **Saglie, F. R., C. T. Smith, M. G. Newman, F. A. Carranza, Jr., J. H. Pertuiset, L. Cheng, E. Auil, and R. J. Nisengard.** 1986. The presence of bacteria in the oral epithelium in periodontal disease. II. Immunohistochemical identification of bacteria. *J. Periodontol.* **57:**492–500.

60. **Saglie, R., T. Tollefsen, A. Marfeny, and J. Johansen.** 1988. *Actinobacillus actinomycetemcomitans* infection of a living gingival equivalent: an in vitro model of pathogenesis. *J. Dent. Res.* **67:**160.

61. **Sakai, T., C. Sasakawa, S. Makino, K. Kamata, and M. Yoshikawa.** 1986. Molecular cloning of a genetic determinant for Congo red binding ability which is essential for the virulence of *Shigella flexneri. Infect. Immun.* **51:**476–482.

62. **Sandros, J., P. A. Papapanou, and G. Dahlen.** 1993. *Porphyromonas gingivalis* invades oral epithelial cells in vitro. *J. Periodontol Res.* **28:**219–226.

63. **Sansonetti, P. J.** 1991. Genetic and molecular basis of epithelial cell invasion by *Shigella* species. *Rev. Infect. Dis.* **13:**(Suppl. 74):285–292.

64. **Shaw, J. H., and S. Falkow.** 1988. Model for the invasion of *Neisseria gonorrhoeae. Infect. Immun.* **56:**1625–1632.

65. **Slots, J., and R. J. Genco.** 1984. Black-pigmented *Bacteroides* species, *Capnocytophaga* species, and *Actinobacillus actinomycetemcomitans* in human periodontal disease: virulence factors in colonization, survival, and tissue destruction. *J. Dent. Res.* **63:**412–421.

66. **Small, P. L. C., R. R. Isberg, and S. Falkow.** 1987. Comparison of ability of enteroinvasive *Escherichia coli, Salmonella typhimurium, Yersinia pseudotuberculosis,* and *Yersinia enterocolitica* to enter and replicate within HEp-2 cells. *Infect. Immun.* **55:**1674–1679.

67. **Sreenivasan, P., and P. Fives-Taylor.** 1991. In vitro invasion model for *Actinobacillus actinomycetemcomitans. J. Dent. Res.* **68:**172.

68. **Sreenivasan, P. K., D. H. Meyer, and P. M. Fives-Taylor.** 1993. Requirements for invasion of epithelial cells by *Actinobacillus actinomycetemcomitans. Infect. Immun.* **61:**1239–1245.

69. **St. Geme, J. W., III, and S. Falkow.** 1990. *Haemophilus influenzae* adheres to and enters cultured human epithelial cells. *Infect. Immun.* **58:**4036–4044.

70. **Suzuki, J. B.** 1988. Diagnosis and classification of the periodontal diseases. *Dent. Clin. N. Am.* **32:**195–216.

71. **Sweet, S. P., T. W. MacFarlane, and L. P. Samaranayake.** 1989. An in vitro method to study the adherence of oral bacteria to HeLa cells. *Microbios* **60:**15–22.

72. **Takarada, H., M. Cattoni, A. Sujimoto, and G. G. Rose.** 1974. Ultrastructural studies of human gingiva. II. The lower part of the pocket epithelium in chronic periodontitis. *J. Periodontol.* **45:**155.

73. **Turner, J. G., and A. H. Drew.** 1918–1919. An experimental inquiry into the bacteriology of pyorrhea. *Proc. R. Soc. Med. Pt. 3 Odontol.* **12:**104–118.

74. **Une, T.** 1977. Studies on the pathogenicity of *Yersinia enterocolitica.* II. Interaction with cultured cells in vitro. *Microbiol. Immunol.* **21:**365–377.

75. **Wick, M. J., J. L. Madara, B. N. Fields, and S. J. Normark.** 1991. Molecular cross talk between epithelial cells and pathogenic microorganisms. *Cell* **67:**651–659.

76. **Winkler, J. R., S. R. John, R. H. Kramer, C. I. Hoover, and P. A. Murray.** 1987. Attachment of oral bacteria to a basement-membrane-like matrix and to purified matrix proteins. *Infect. Immun.* **55:**2721–2726.

77. **Zambon, J. J.** 1985. *Actinobacillus actinomycetemcomitans* in human periodontal disease: a review article. *J. Clin. Periodontol.* **12:**1–20.

Molecular Pathogenesis of Periodontal Disease
Edited by Robert Genco et al.
© 1994 American Society for Microbiology, Washington, DC 20005

Chapter 7

Molecular Biology of *Actinobacillus actinomycetemcomitans* Leukotoxin

Edward T. Lally and Irene R. Kieba

Actinobacillus is a member of the family *Pasteurellaceae,* a group of nonenteric, fermenting, gram-negative rods that are of considerable importance in both human and veterinary medicine. While various *Actinobacillus* species are found in animals, only *Actinobacillus actinomycetemcomitans* is routinely cultured from humans. This organism has been associated with a variety of infectious disease processes including endocarditis, brain abscesses, osteomyelitis, subcutaneous abscesses, and periodontal disease (7, 33, 48, 49). The mechanisms of *A. actinomycetemcomitans* pathogenesis are not well understood, but this organism produces a variety of potential virulence factors including leukotoxin (25, 28, 29), belonging to the RTX (repeats in toxin) family of bacterial cytolysins (41, 46).

    A. actinomycetemcomitans leukotoxin exhibits a unique cytolytic specificity in that it destroys human polymorphonuclear leukocytes and macrophages, whereas other types of cells, e.g., epithelial and endothelial cells, fibroblasts, erythrocytes, and platelets, are resistant to lysis (1, 42–44). The general mechanism of leukotoxin-mediated cell lysis involves the rapid formation of cation-selective pores that eventually leads to osmotic lysis of the target cell (23) (Fig. 1). Similar events occur in the lysis of mammalian erythrocytes and leukocytes by other RTX toxins derived from various gram-negative bacteria (4, 6, 30, 46). This suggests that the mechanisms of RTX toxin-mediated lysis may be similar despite the apparent differences in host cell specificity exhibited by the various toxins.

LEUKOTOXIN GENES

    RTX toxins such as *A. actinomycetemcomitans* leukotoxin are expressed by four-gene operons. We have designated the *A. actinomycetemcomitans* toxin genes *ltxC, ltxA, ltxB,* and *ltxD* (in transcriptional order) (Fig. 2) in order to distinguish these genes from a similar set of *lkt* genes that encode a related leuko-

Edward T. Lally and Irene R. Kieba • Department of Pathology, School of Dental Medicine, University of Pennsylvania, 4010 Locust Street, Philadelphia, Pennsylvania 19104-6002.

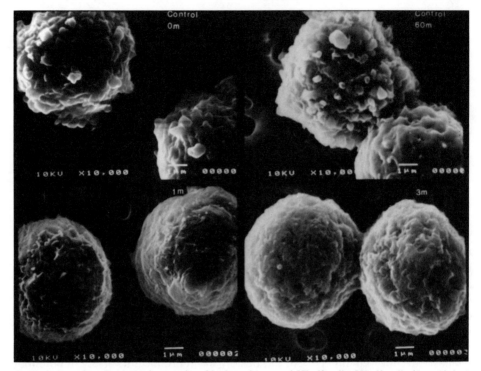

Figure 1. Scanning electron micrographs of leukotoxin-treated HL-60 cells. HL-60 cells (5 × 10⁵ in 1 ml) were incubated (1 h, 37°C) with a 95% lethal dose (22 ng/ml) of *A. actinomycetemcomitans* leukotoxin in Dulbecco modified Eagle medium. Aliquots of cells were then withdrawn at 1, 3, 10, and 60 min, fixed, and prepared for routine scanning electron microscopy.

toxin from *Pasteurella haemolytica*. The structural gene encoding the leukotoxin *(ltxA)* encodes a peptide of 116 kDa (25, 28) that exhibits between 40 and 50% identity with *Escherichia coli* α-hemolysin and the *P. haemolytica* leukotoxin (29).

The three remaining *ltx* genes (*ltxB, ltxC,* and *ltxD*) flanking *ltxA* are required for activating and transporting the leukotoxin. The first gene of the leukotoxin operon is the *ltxC* gene. The primary translation product of the structural toxin gene is "activated" by the *ltxC* gene product. Activated toxin is then transported to the cell surface by a signal peptide-independent mechanism requiring the products of the *ltxB* and *ltxD* genes. However, other genes may also be required in the transport process. For example, the RTX operons of *Erwinia chrysanthemi* and *Bordetella pertussis* contain genes analogous to *ltxB* and *ltxD*, but secretion of the RTX proteins expressed by these organisms requires another gene as well (*prtF* and *cyaE*, respectively). These genes are located immediately downstream of the corresponding *prtD* and *cyaD* genes and are transcribed as part of the RTX operon (17, 31). In contrast, the *tolC* gene of *E. coli* is not physically linked to the *hly* operon but has been shown to participate in the secretion of α-hemolysin (45).

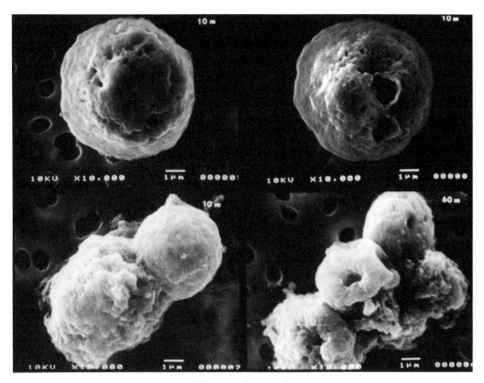

Figure 1. *Continued.*

ltxC Gene Product

Leukotoxin appears to be posttranslationally modified to its active form in the bacterial cytoplasm. In *E. coli,* the α-hemolysin is modified by fatty acid acylation reaction and requires the *hlyC* gene product and additional acyl carrier protein (19, 22). Although it has yet to be shown conclusively, it is likely that *ltxC* carries out a similar reaction in *A. actinomycetemcomitans.* The *ltxA* gene product expressed in the absence of the *ltxC* gene product will not kill target cells (unpublished results).

ltxA Gene Product

On the basis of our analysis (28), we divided the *ltxA* gene into four regions or domains: (i) the N-terminal domain, (ii) the central domain, (iii) the repeat domain, and (iv) the terminal domain (Fig. 3, segments a, b, c, and d, respectively). In the N-terminal domain (residues 1 to 408), all three proteins exhibit alternating hydrophobic and hydrophilic clusters, although the presence of a clearly delineated signal peptide is absent in all three proteins. A prediction of the secondary structure of this region suggested the presence of 22 α-helices containing 230 residues (49%). Helical wheel analysis (39) of all helices 10 residues or more long (seven cases)

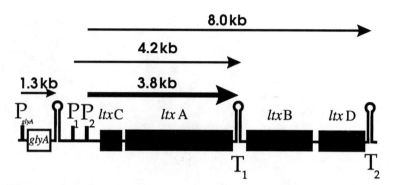

Figure 2. The leukotoxin operon. The *A. actinomycetemcomitans* serine hydroxymethyltransferase gene, its promoter, and its transcriptional terminator are located immediately upstream of multiple leukotoxin promoters (P_1 and P_2). Leukotoxin genes (*ltxC, ltxA, ltxB,* and *ltxD*) are represented in transcriptional order with two *rho*-independent transcriptional terminators situated between *ltxA* and *ltxB* (T_1) and after *ltxD* (T_2). A 4.2-kb message arises from P_1 while two (3.8- and 8.0-kb) messages are generated from P_2.

revealed five amphipathic helices, structures often associated with transmembrane segments and protein pores. This suggests that this region, which also contains two hydrophobic helices, may be responsible for toxin-target cell membrane interactions. Furthermore, mutations in this region resulted in loss of toxic activity (32; unpublished observation).

The central region (residues 409 to 730) contains largely hydrophilic residues, and although the three toxins present similar hydropathy profiles in this region, the conservation of structure appears less strong. The repeat domain (residues 730 to 900) consists of multiple copies of a tandemly repeated 9-amino-acid cassette having the consensus sequence GGXGXDXUX (where X is any amino acid, and U is L, I, V, W, Y, or F). Analysis of the "repeats" shows a novel "parallel β-roll" structure in which successive β strands (XUX) are wound in a right-handed spiral and Ca^{2+} ions are bound within the turns between the strands by a repeated GGXGXD (2). The exact function of the β roll remains to be determined, but it is possible that in the absence of Ca^{2+} ions the molecule would be unstable. Such instability could facilitate a membrane translocation. Once the molecule was secreted into the Ca^{2+}-rich extracellular milieu, the sites would quickly saturate and the toxin would condense into a tertiary "cytotoxic" shape.

The major structural differences between the *ltxA* gene product and its homologs lie after the repeats in the terminal domain (residues 900 to 1055). This region appears to be involved with the transportation of the molecule from the bacterial cytoplasm to the cell surface.

ltxB Gene Product

The translocation of the toxin from the bacterial cytoplasm to the cell membrane is remarkable in that it does not appear to require a canonical N-terminal signal sequence or some form of peptide cleavage/activation but rather is dependent on the *ltxB* and *ltxD* gene products.

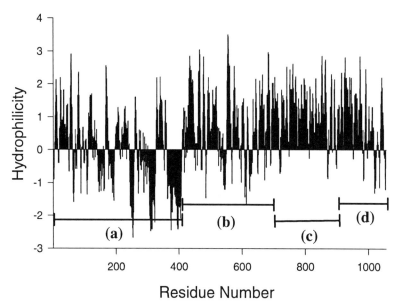

Figure 3. Hydropathy plot profile of *A. actinomycetemcomitans ltxA* gene. The method used is that of Kyte and Doolittle (26). Four regions have been identified: the N-terminal domain (a; residues 1 to 408), the central domain (b; residues 409 to 729), the repeat domain (c; residues 730 to 900), and the terminal domain (d; residues 901 to 1055).

The nucleotide and deduced amino acid sequences of *ltxB* (2,118 bp) exhibited strong homology to B genes found in other bacterial species, including the sequences of *E. coli* α-hemolysin *hylB* and *P. haemolytica lktB* (27). There was also significant homology between *ltxB* and mammalian genes (18) encoding proteins for multiple drug resistance (MDR) (9, 18) and the cystic fibrosis transmembrane conductance regulator (CFTR) (36), indicating some possible commonality of function. Furthermore, the carboxyl portion of the molecule was homologous to a much wider group of procaryotic and eucaryotic proteins that function in the transport of sugars (3, 16), vitamins (14), and oligopeptides (21) across cell membranes and to several genes that regulate cell division (15) and repair DNA (11), indicating that *ltxB* is part of a supergene family of transport proteins. The region of homology with these additional genes was centered on but not limited to two putative nucleotide-binding folds in the sequence with a short consensus nucleotide-binding sequence [-G-Xaa-G-(Xaa)$_2$-G-(Xaa)$_{16}$-R-] (9, 11, 18, 20, 39a) located at amino acid residues 601 to 623 of the *ltxB* gene product. It is also interesting that while proteins such as CFTR and MDR proteins also possess membrane-spanning hydrophobic domains that are similar in both number and relative location to those found in *ltxB* or other RTX cytolysins, they do not contain significant sequence similarity to *ltxB* or B genes of other RTX toxins. This finding suggests parallel evolution of the membrane-spanning domains of RTX and the eucaryotic transport proteins.

The deduced amino acid sequence of the *ltxB* gene product has been analyzed by prediction of secondary structure using a modification of the method of Chou and Fasman (10) and by hydropathy analysis by the method of Kyte and Doolittle (26). The *ltxB* gene product is predicted to consist of 35% helical residues and 42% β-sheet conformation residues with 28 β turns. This structure could presumably be folded into a highly compact structure, with the membrane-spanning helices forming a multiple loop structure through the membrane. The protein as a whole is neither net hydrophobic nor net hydrophilic, but analysis of charged residues suggests an isoelectric point of 10.1 for the *ltxB* gene product, assuming that all charged residues are available for titration. Three pairs of putative multiple membrane-spanning α-helices found near the amino terminus were homologous to similar regions in the *ltx*B-gene products of other bacteria.

ltxD Gene Product

Analysis of the DNA sequence of the fourth open reading frame (ORF) *ltxD* (1,431 bp) showed no evidence of homology with other procaryotic or eucaryotic genes in multiple database searches. Furthermore, the *ltxD* gene product contains no evidence of potential membrane-spanning elements and is predominantly hydrophilic in character. The most notable feature of the *ltxD* gene product is a large α helix (residues 102 to 175) in the central portion of the molecule. Portions of this helix show clustering of hydrophilic and hydrophobic amino acids, whereas other areas do not show this tendency (data not shown). The amphipathic nature of other helical segments of the *ltxD* gene product is demonstrated by hydrophobic moment analyses (12). Three predicted helices (residues 234 to 257, 262 to 289, and 377 to 390) show multiple frequency peaks for clusters of hydrophobic residues. In particular, the predicted helix near the C terminus (residues 377 to 390) demonstrates a strong peak at 100° that is characteristic of amphipathic helices. Since amphipathic helical structures are often associated with protein-protein interactions, these domains may be important for stabilizing and folding the *ltxA* protein (29) prior to secretion.

Effect of *ltxBD* Gene Expression on Cellular Localization of Leukotoxin

Supernatants from either *P. haemolytica* or *E. coli* lyse target cells, while culture supernatants from *A. actinomycetemcomitans* are not toxic. The reverse is true when whole bacteria are examined. Incubation of target cells with certain *A. actinomycetemcomitans* bacteria is potently leukotoxic, while neither *P. haemolytica* nor *E. coli* lyses target cells. These data may indicate that unlike the toxins of other RTX-secreting bacteria, *A. actinomycetemcomitans* leukotoxin remains associated with the bacterial cell outer membrane (5). To determine whether the presence of the *ltxBD* genes influences the cellular localization of recombinantly synthesized leukotoxin, we cloned *A. actinomycetemcomitans* leukotoxin in *E. coli* under two different conditions (27). In the first case, the cloned segment contained the *ltxCABD* genes, while in the second case, the segment contained only the *ltxC* and *ltxA* genes. Active toxin was synthesized in both instances; however, a definite difference in the cellular distribution of the toxins was demonstrated by immunohistochemical staining of the leukotoxin in *E. coli* containing *ltxACBD* genes or *ltxA* and *ltxC*. The leukotoxin was confined to the bacteria cytoplasm when only the *ltxA* and *ltxC* genes were present. In contrast, the leukotoxin was associated predominantly with the outer membrane in cells containing all four genes. The leukotoxin was not secreted into the bacterial culture medium in either case (data not shown). As yet, we have not directly demonstrated expression of the *ltxC*, *ltxB*, and *ltxD* genes, but the demonstration that translocation of biologically active toxin to the *E. coli* cell surface is dependent on the presence of the *ltxBD* genes strongly suggests that one or both of these genes are functionally expressed.

In addition to influencing the location of the structural gene product, the presence of *ltxB* and *ltxD* genes resulted in morphological changes in *E. coli* that were not observed in bacteria expressing only *ltxC* and *ltxA* or in uninduced controls. Expression of the *ltxB* and *ltxD* genes resulted in vacuolation of the bacterial cytoplasm and bleb formation in the cell membrane. Cultures induced for longer periods showed decreased optical density, indicative of cell lysis (data not shown). These experiments indicate that the membrane blebbing observed in certain strains of *A. actinomycetemcomitans* may be a direct result of *ltxB* and *ltxD* gene expression.

TRANSCRIPTIONAL REGULATION OF LEUKOTOXIN OPERON

Relatively little is known about the transcriptional regulation of *A. actinomycetemcomitans ltx* operon. At least two mRNAs are transcribed from the *ltxCABD* gene cluster. The predominant mRNA (3.8 kb) encodes *ltxC* and *ltxA* and presumably terminates at a *rho*-independent transcriptional terminator situated between *ltxA* and *ltxB* (24, 27). A less abundant mRNA of 8 kb, encoding *ltxCABD*, results when transcription fails to terminate at this *rho*-independent terminator and continues through *lktB* and *lktD*. Since overexpression of *ltxB* and *ltxD* appears to exert a detrimental effect on cell viability, expression of *ltxB* and *ltxD* gene products is relatively low and is controlled by the extent of transcriptional termination that occurs between *ltxA* and *ltxB*. The mechanism controlling the extent of termination at this site in *A. actinomycetemcomitans* remains to be determined. Leukotoxin expression also appears to vary among different strains of *A. actinomycetemcomitans*. Strains that express little or no detectable leukotoxic activity can readily be cultured from the human oral cavity (1). To further characterize the transcriptional regulation of the *ltx* operon in *A. actinomycetemcomitans*, we determined the nucleotide sequence, analyzed the promoter regions of *A. actinomycetemcomitans* JP2 and 652 (8), and found that the level of *ltx* expression is dependent on the structure of the *ltx* promoter region. A model illustrating the structure and activities of the *A. actinomycetemcomitans ltx* promoter region is shown in Fig. 2.

Analysis of *A. actinomycetemcomitans* JP2 *ltx* Promoter

Analysis of the nucleotide sequence of a 1.8-kbp region upstream of *ltxC* in strain JP2 identified a gene similar to *E. coli glyA* (33a). This gene terminates 582 bp upstream of *ltxC*. The intergenic region between *glyA* and *ltxC* contained several

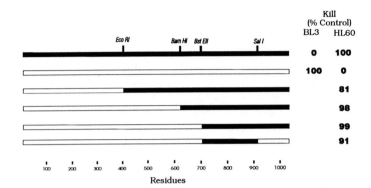

Figure 4. *ltxA-lktA* leukotoxin chimeras. Chimeric toxins were constructed by splicing increasingly larger fragments of *P. haemolytica* leukotoxin (▭) into the *A. actinomycetemcomitans* (■) *ltxA* gene. The construct was inserted into the pOTS-*Nco* plasmid, cloned into *E. coli,* and expressed under the control of the bacteriophage lambda promoter. Sonic extracts were tested for their abilities to kill HL-60 cells or BL3 cells. The results were expressed as percentage of control.

inverted repeats similar to those found downstream of the *E. coli glyA* gene. In addition, the intergenic region contained sequences resembling −10 and −35 elements approximately 350 bp upstream of *ltxC* that were suggested by Kraig et al. (25), on the basis of their similarity to the *E. coli* consensus −10 and −35 elements, to represent putative *ltx* promoter elements.

Primer extension of JP2 RNA from the oligonucleotide that anneals to sequence immediately upstream of *ltxC* start codon (residues −15 to +5) produced two major extension products of 40 and 340 nucleotides (8). The 340-nucleotide product is consistent with transcriptional initiation occurring at 335 and being driven by the −10 and −35 domains previously identified (25). In addition, Northern (RNA) blots identified a 4.2-kb *ltx* mRNA that hybridizes to probes encoding *ltxA* and the *ltxC* upstream region. The size of this mRNA is also consistent with transcription initiating at −335 and terminating at a putative *rho*-independent transcriptional terminator situated between *ltxA* and *ltxB* (25, 28). This mRNA had not been previously identified as a specific transcript of the JP2 *ltx* operon. Two additional *ltx* transcripts (3.8 and 8 kb) were identified in Northern blots; they represent the mRNAs previously identified (40). The 3.8-kb transcript, which is the predominant *ltx* mRNA of JP2, did not hybridize to the *ltxC* upstream probe, suggesting that its 5′ terminus differs from the 4.2-kb transcript. One explanation is that the 3.8-kb mRNA is generated by processing the 4.2-kb primary transcript. However, it is also possible that transcription initiates at multiple sites within the *ltx* upstream region. Either possibility could explain the generation of the 40-nucleotide primer extension product. The 8-kb mRNA was previously shown to encode *ltxCABD* (40) and, by analogy to the related RTX operons of *E. coli* and *P. haemolytica,* presumably arises from antitermination at the *rho*-independent terminator sequence between *ltxA* and *ltxB* (25, 27). It was previously suggested that this transcript encoded an additional secretion gene (40), but sequence of the region downstream of *ltxD* identified a second region resembling a *rho*-independent transcriptional terminator.

The promoter region of strain JP2 contains two promoters (P1 and P2) situated approximately 50 and 340 bp upstream from the *ltxC* translational initiation codon, and the sequence between P1 and P2 contains an ORF capable of encoding a peptide of ~10 kDa. The presence of this ORF appears to be unique to the *A. actinomycetemcomitans ltx* operon. The *E. coli hly* and *P. haemolytica lkt* operons are both transcribed from multiple promoters, and transcriptional initiation can occur several hundred nucleotides upstream of their respective *ltxC* genes. However, neither of these operons contains an ORF within this region (24, 41, 47). The ORF of *A. actinomycetemcomitans* JP2 is transcribed from P1, but we do not know whether it is translated in vivo.

RNA hybridization studies further substantiated the two-promoter concept. Total RNA was isolated from *A. actinomycetemcomitans* JP2 and probed with either a 1.5-kbp fragment from *ltxA* or a PCR-amplified fragment corresponding to the 470 bp upstream of *ltxC* and containing the ORF sequence hybridized to three mRNAs of 3.8, 4.2, and 8 kb. The upstream probe containing ORF sequences hybridized only with the 4.2-kb mRNA, a message not previously identified as a specific transcript of the *ltx* operon. These results indicate that the region situated upstream of *ltxC* is transcribed but is present in only one of the three *ltx*-specific mRNAs. This suggests that the *ltx* operon, like those of other RTX toxins, may be transcribed from multiple promoters or that a primary mRNA of 4.2 kb is processed to generate the 3.8-kb species.

In addition to primer extension experiments, truncations of the JP2 upstream region (8) were constructed and fused to a promoterless β-galactosidase gene. These constructs were introduced into *E. coli* DH5αF′, and β-galactosidase activity was determined by using the chromogenic substrate *o*-nitrophenyl-galactoside. The studies showed that both putative *ltx* promoters (P1 and P2) efficiently directed galactosidase expression.

Analysis of *A. actinomycetemcomitans* 652 *ltx* Promoter

As analyzed by Northern blotting, strain 652 expresses significantly lower levels of *ltx*-specific mRNA than JP2 does. The predominant RNA that hybridizes with the *ltxA* probe comigrates with the JP2 3.8-kb species (8). The nucleotide sequence of the 652 *ltx* promoter region was determined, and an additional 530 bp that was not present in the JP2 promoter was detected. Otherwise, the 652 promoter is virtually identical to the JP2 promoter from nucleotides 1 to 404 and from nucleotide 935 to the *ltxC* initiation codon (nucleotide 990). As a result, the P1 promoter is situated more than 800 bp upstream of *ltxC,* and Northern analysis failed to detect an mRNA consistent with *ltx* transcription being driven from this promoter. In addition, the structure of the upstream ORF differs in the two strains. The 652 ORF is larger (encoding 150 residues) and terminates 340 bp upstream of the *ltxC* start codon.

Relationship of *ltx* phenotype to promoter structure

The mechanism giving rise to the two different *ltx* promoters is not known. However, we have identified additional JP2-like strains (both serotype b) and found that they are genetically similar to other serotype b strains and distinct from organisms representing the remaining serotypes (34). This suggests that the JP2 promoter probably originated from the sequence present in 652, by deletion of the 530-bp domain. The strains which possess this deletion all expressed levels of toxicity similar to that of JP2. In contrast, 13 other 652-like strains exhibited 50% lethal doses 10- to 20-fold lower than those of JP2-like strains. Thus, the levels of toxicity expressed by these *A. actinomycetemcomitans* isolates appear to correlate with *ltx* promoter structure. However, only a limited number of JP2-like strains have been tested, and it is possible that other factors, e.g., *trans*-acting proteins, also contribute to the increased expression of *ltx* observed in these strains. Analysis of additional *A. actinomycetemcomitans* clinical isolates will be necessary to clarify the interesting relationship between the incidence of the JP2 promoter structure and its role in disease progression.

STRUCTURE/FUNCTION STUDIES OF THE *ltxA* GENE PRODUCT

The interrelationships between the structure of the leukotoxin and its biologic functions are of interest in our efforts to understand molecular aspects of the pathogenesis of periodontal disease. The cloning and sequencing of the leukotoxin gene(s) provided insight into the amino acid sequence of the leukotoxin gene. Furthermore, when analyzed and compared with the sequences of other RTX toxins, such as *E. coli* α-hemolysin and *P. haemolytica* leukotoxin, the deduced amino acid sequence of the *ltxA* gene product shows remarkable concordance in clustering of hydrophobic and hydrophilic residues. Delineation of the contributions of individual regions of the leukotoxin molecule to RTX intoxification will provide important baseline data in our preliminary study of the role *A. actinomycetem-comitans* plays in this disease process.

We have utilized both truncation mutants and chimeric toxins in attempts to understand the relationship of leukotoxin structure to its function. In these studies we used recombinantly produced leukotoxin as described previously (27). Briefly, *ltxC* and *ltxA* are expressed in tandem under Pλ control in pOTS-*Nco* expression vector (37, 38; a gift from Allan Shatzman and Martin Rosenberg, Smith, Kline Beckman Laboratories, Swedeland, Pa.). After ligation, the constructs are transformed into *E. coli* AR120. Protein synthesis was induced by adding nalidixic acid (60 μg/ml) to the culture medium for 4 h. In this system, nalidixic acid induces the endogenous *E. coli* SOS response, resulting in cleavage of the cI repressor by the *recA* protein and induction of expression. *ltxA* was modified by engineering a *Kpn*I site between residues 2 and 3. Placement of the *Kpn*I site required insertion of an additional glycine residue between residues 2 and 3, with no apparent effect on leukotoxicity. This minor modification permitted casetting of either truncation mutants or toxin chimeras of *ltxA* into a system for coexpression with *ltxC*.

3′ and 5′ Truncation Mutants of *ltxA*

The ability of leukotoxin truncation mutants to kill target cells was assessed by trypan blue exclusion assay. Washed cells (HL-60; 50 μl; 4 × 10⁶ cells per ml) were incubated (60 min, 37°C) with equal volumes of culture lysates from leukotoxin mutants. The ability of leukotoxin mutant to kill target cells was expressed as the ratio of percentage of target cells killed by the intact toxin molecule to percentage of target cells killed by the mutant toxin molecule.

Deletion of carboxyl-terminal residues 940 to 1055 of the leukotoxin molecule (1,055 residues total) had no obvious effect on toxic effects of the mutant molecule. However, further deletion (of residues 939 to 871) resulted in the loss of toxic activity. Removing N-terminal amino acids 3 to 80 did not affect toxicity, while a mutant with a deletion of residues 3 to 146 did not kill target cells.

The results show that leukotoxic activity resides in the structural and/or conformational characteristics imparted by the presence of residues 80 to 940.

ltxA-lktA Chimeric Toxins

While extensive amino acid homology exists among the various RTX, the cellular and species specificities remain unique for individual toxins. For example, *A. actinomycetemcomitans* leukotoxin will kill only human monomyelocytes, while the *P. haemolytica* leukotoxin kills bovine lymphoid cells. This provided a ready-made approach for determining the region of the leukotoxin molecule that is responsible for species specificity and indirectly for target cell recognition. In studies that involve truncation mutants, it is difficult to determine whether the observed loss of cytotoxicity is the result of deletion of a domain or simply of the mutant toxin's inability to fold properly. Chimeric toxins, on the other hand, would be expected to contain all of the functional domains of the native toxin, exhibit normal protein folding, and maintain their ability to kill either bovine or human target cells as long as the respective "species recognition" region was present. Furthermore, it is possible that the domain of the leukotoxin that is capable of distinguishing between a bovine and a human target cell is also the region that is responsible for target cell recognition.

In these experiments, PCR-generated fragments of *lktA* gene from *P. haemolytica* were spliced into naturally occurring restriction sites of the *A. actinomycetemcomitans ltxA* gene at *Eco*RI (1,217 bp), *Bam*HI (1,870 bp), *Bst*EII (2,061 bp), and *Sal*I (2,820 bp). After expression in tandem with the *ltxC* gene from the Pλ promoter, cells were sonicated and the abilities of the respective lysates to kill either HL-60 (human) or BL3 (bovine) cells was assessed by trypan blue exclusion. The results are shown in Fig. 4. Approximately 70% of the *ltxA* gene product can be replaced with sequences from *lktA* and have no effect on the organism's ability to kill the human cells. *Bst*EII encodes residues that are 43 amino acids from the first GGXGXDXUX repeat. Additionally, cloning a *Bst*EII-*Sal*I fragment from the *A. actinomycetemcomitans ltxA* gene into the *P. haemolytica* leukotoxin gene resulted in a switch in target cell specificity from bovine to human target cells. Initial attempts to produce mutants with splice sites between *Bst*EII and *Sal*I resulted in toxin that would kill neither cell, perhaps indicating that improper protein folding was occurring. Recently, the three-dimensional structure of a related protein with GGXGXDXUX repeat has been solved (2). Utilization of atomic coordinates from these data should permit construction of a model of the repeat region of the *A. actinomycetemcomitans* leukotoxin and subsequent development of toxin chimeras that further explore the mechanism of species specificity of the toxin.

Acknowledgments. We thank Ellis Golub, Robert Hooper, and Gerald Harrison for their assistance in the nucleotide sequence analyses and scanning electron microscopy.

This work was supported by National Institutes of Health grants DE09517 and DE07118.

REFERENCES

1. **Baehni, P. C., C. C. Tsai, W. P. McArthur, B. F. Hammond, B. J. Shenker, and N. S. Taichman.** 1981. Leukotoxic activity in different strains of the bacterium *Actinobacillus actinomycetemcomitans* isolated from juvenile periodontitis in man. *Arch. Oral Biol.* **26:**671–676.
2. **Baumann, U., S. Wu, K. M. Flaherty, and D. B. McKay.** 1993. Three-dimensional structure of the alkaline protease of *Pseudomonas aeruginosa:* a two-domain protein with a calcium binding parallel beta roll motif. *EMBO J.* **12:**3357–3364.
3. **Bell, A. W., S. D. Buckel, J. M. Groarke, J. N. Hope, D. H. Kingsley, and M. A. Hermodson.** 1986. The nucleotide sequences of the *rbs*D, *rbs*A, and *rbs*C genes of *Escherichia coli* K12. *J. Biol. Chem.* **261:**7652–7658.
4. **Benz, R., A. Schmid, W. Wagner, and W. Goebel.** 1989. Pore formation by the *Escherichia coli* α-hemolysin: evidence for an association-dissociation equilibrium of the pore-forming aggregates. *Infect. Immun.* **57:**887–895.
5. **Berthold, P. H., D. Forti, I. R. Kieba, J. Rosenbloom, N. S. Taichman, and E. T. Lally.** 1992. Electron immunocytochemical localization of *Actinobacillus actinomycetemcomitans* leukotoxin. *Oral Microbiol. Immunol.* **7:**24–27.
6. **Bhakdi, S., N. Mackman, J.-M. Nicaud, and I. B. Holland.** 1986. *Escherichia coli* α-hemolysin may damage target cell membranes by generating transmembrane pores. *Infect. Immun.* **52:**63–69.
7. **Block, P. J., A. C. Fox, C. Yoran, and A. J. Kaltman.** 1973. *Actinobacillus actinomycetemcomitans* endocarditis: report of a case and review of the literature. *Am. J. Med. Sci.* **266:**387–392.
8. **Brogan, J. M., E. T. Lally, and D. R. Demuth.** 1994. Regulation of *Actinobacillus actinomycetemcomitans* leukotoxin: analysis of the promoter regions of leukotoxic and minimally leukotoxic strains. *Infect. Immun.* **62:**501–508.
9. **Chen, C., J. E. Chin, K. Ueda, D. P. Clark, I. Pastan, M. M. Gottesman, and I. B. Roninson.** 1986. Internal duplication and homology with bacterial transport proteins in the mdr1 (P-glyco-protein) gene from multidrug-resistant human cells. *Cell* **47:**381–389.
10. **Chou, P. Y., and G. D. Fasman.** 1974. Conformational parameters for amino acids in helical, β-sheet and random cell regions calculated from proteins. *Biochemistry* **13:**211–222.
11. **Doolittle, R. F., M. S. Johnson, I. Husain, B. Van Houten, D. C. Thomas, and A. Sancar.** 1986. Domainal evolution of a procaryotic DNA repair protein and its relationship to active-transport proteins. *Nature* (London) **323:**451–453.
12. **Eisenberg, D., R. M. Weiss, and T. C. Terwilliger.** 1984. The hydrophobic moment detects periodicity in protein hydrophobicity. *Proc. Natl. Acad. Sci. USA* **81:**140–144.
13. **Felmlee, T., S. Pellett, and R. A. Welch.** 1985. Nucleotide sequence of the *Escherichia coli* chromosomal α-hemolysin. *J. Bacteriol.* **163:**94–105.
14. **Friedrich, M. J., L. C. de Veaux, and R. J. Kadner.** 1986. Nucleotide sequence of the *btuCED* genes involved in vitamin B_{12} transport in *Escherichia coli* and homology with components of periplasmic-binding-protein-dependent transport systems. *J. Bacteriol.* **167:**928–934.
15. **Gill, D. R., G. F. Hatfull, and G. P. Salmond.** 1986. A new cell division operon in *Escherichia coli. Mol. Gen. Genet.* **205:**134–145.

16. **Gilson, E., H. Nikaido, and M. Hofnung.** 1982. Sequence of the *mal*K gene in *E. coli* K12. *Nucleic Acids Res.* **10:**7449–7458.

17. **Glaser, P., H. Sakamoto, J. Bellalou, A. Ullman, and A. Danchin.** 1988. Secretion of cytolysin, the calmodulin sensitive adenylate cyclase-α-hemolysin bifunctional protein of *Bordetella pertussis*. *EMBO J.* **7:**3997–4004.

18. **Gros, P., J. Croop, and D. Houseman.** 1986. Mammalian multidrug resistance gene: complete cDNA sequence indicates strong homology to bacterial transport proteins. *Cell* **47:**371–380.

19. **Hardie, K. R., J.-P. Issartel, E. Koronakis, C. Hughes, and V. Koronakis.** 1991. In vitro activation of *Escherichia coli* prohaemolysin to the mature membrane targeted toxin requires *hlyC* and a low molecular weight cytosolic polypeptide. *Mol. Microbiol.* **5:**1669–1679.

20. **Hashimoto, E., E. Takio, and E. G. Krebs.** 1982. Amino acid sequence at a ATP binding site of cGMP-dependent protein kinase. *J. Biol. Chem.* **257:**727–733.

21. **Hiles, I. D., M. P. Gallagher, D. J. Jamieson, and C. F. Higgins.** 1987. Molecular characterization of the oligopeptide permease of *Salmonella typhimurium*. *J. Mol. Biol.* **195:**125–142.

22. **Issartel, J.-P., V. Koronakis, and C. Hughes.** 1991. Activation of *Escherichia coli* prohaemolysin to the mature toxin by acyl carrier protein dependent fatty acylation. *Nature* (London) **351:**759–761.

23. **Iwase, M., E. T. Lally, P. Berthold, H. M. Korchak, and N. S. Taichman.** 1990. Effects of cations and osmotic protectants on cytolytic activity of *Actinobacillus actinomycetemcomitans* leukotoxin. *Infect. Immun.* **58:**1782–1788.

24. **Koronakis, V., and C. Hughes.** 1988. Identification of the promoters directing in vivo expression of α-hemolysin genes in *Proteus vulgaris* and *Escherichia coli*. *Mol. Gen. Genet.* **213:**99–104.

25. **Kraig, E., T. Dailey, and D. Kolodrubetz.** 1990. Nucleotide sequence of the leukotoxin gene from *Actinobacillus actinomycetemcomitans*: homology to the α-hemolysin/leukotoxin gene family. *Infect. Immun.* **58:**920–929.

26. **Kyte, J., and R. F. Doolittle.** 1982. A simple method for displaying the hydrophobic character of a protein. *J. Mol. Biol.* **157:**105–132.

27. **Lally, E. T., E. E. Golub, I. R. Kieba, N. S. Taichman, S. Decker, P. Berthold, C. W. Gibson, D. R. Demuth, and J. Rosenbloom.** 1991. Structure and function of the *B* and *D* genes of the *Actinobacillus actinomycetemcomitans* leukotoxin complex. *Microb. Pathog.* **11:**111–121.

28. **Lally, E. T., E. E. Golub, I. R. Kieba, N. S. Taichman, J. Rosenbloom, J. C. Rosenbloom, C. W. Gibson, and D. R. Demuth.** 1989. Analysis of the *Actinobacillus actinomycetemcomitans* leukotoxin gene: delineation of unique features and comparison to homologous toxins. *J. Biol. Chem.* **264:**15451–15456.

29. **Lally, E. T., I. R. Kieba, D. R. Demuth, R. Rosenbloom, E. E. Golub, N. S. Taichman, and C. W. Gibson.** 1989. Identification and expression of the *Actinobacillus actinomycetemcomitans* leukotoxin gene. *Biochem. Biophys. Res. Commun.* **159:**256–262.

30. **Lalonde, G., T. V. McDonald, P. Gardner, and P. D. O'Hanley.** 1989. Identification of an α-hemolysin from *Actinobacillus pleuropneumoniae* and characterization of its channel properties in planar phospholipid bilayers. *J. Biol. Chem.* **264:**13559–13564.

31. **Letoffe, S., P. Delepelaire, and C. Wandersman.** 1990. Protease secretion by *Erwinia chrysanthemi*: the specific functions are analogous to those of *Escherichia coli* α-hemolysin. *EMBO J.* **9:**1375–1382.

32. **Ludwig, A., M. Vogel, and W. Goebel.** 1987. Mutations affecting activity and transport of haemolysin in *Escherichia coli*. *Mol. Gen. Genet.* **206:**238–245.

33. **Page, M. I., and E. O. King.** 1966. Infection due to *Actinobacillus actinomycetemcomitans* and *Haemophilius aphrophilus*. *N. Engl. J. Med.* **275:**181–188.

33a. **Plamann, M. D., L. T. Stauffer, M. L. Urbanowski, and G. W. Stauffer.** 1983. Complete nucleotide sequence of the *E. coli glyA* gene. *Nucleic Acids Res.* **11:**2065–2075.

34. **Poulsen, K., E. Theilade, E. T. Lally, D. R. Demuth, and M. Kilian.** Population structure of *Actinobacillus actinomycetemcomitans*: a framework for studies of disease-associated properties. *Microbiology,* in press.

35. **Rao, J. K. M., and P. Argos.** 1986. A conformational preference parameter to predict helices in integral membrane proteins. *Biochim. Biophys. Acta* **869:**197–214.

36. **Riordan, J. R., J. M. Rommens, B.-M. Kerem, N. Alon, R. Rozmahel, Z. Grzelczak, J. Zielenski,**

S. Lok, N. Plavsic, J.-L. Chou, M. L. Drumm, M. C. Iannuzzi, F. S. Collins, and L.-C. Tsui. 1989. Identification of the cystic fibrosis gene: cloning and characterization of complementary DNA. *Science* **245**:1066–1073.

37. **Rosenberg, M., and D. Court.** 1979. Regulatory sequences involved in the promotion and termination of RNA transcription. *Annu. Rev. Genet.* **13**:319–353.

38. **Rosenberg, M., Y.-S. Ho, and A. Shatzman.** 1983. The use of pKC30 and its derivatives for controlled expression of genes. *Methods Enzymol.* **101**:123–138.

39. **Ross, A. M., and E. E. Golub.** 1987. A computer program system for protein structure representation. *Nucleic Acids Res.* **16**:1801–1812.

39a. **Shoji, S., D. C. Parmlee, R. Wade, S. Kuma, L. Ericsson, K. Walsh, H. Neurath, G. Long, J. Demaille, E. Fischer, and K. Titani.** 1981. Complete amino acid sequence of the catalytic subunit of bovine cardiac muscle cyclic AMP-dependent protein kinase. *Proc. Natl. Acad. Sci. USA* **78**:848–851.

40. **Spitznagel, J., Jr., E. Kraig, and D. Kolodrubetz.** 1991. Regulation of leukotoxin in leukotoxic and nonleukotoxic strains of *Actinobacillus actinomycetemcomitans. Infect. Immun.* **59**:1394–1401.

41. **Strathdee, C. A., and R. Y. C. Lo.** 1989. Regulation of expression of the *Pasteurella haemolytica* leukotoxin determinant. *J. Bacteriol.* **171**:5955–5962.

42. **Taichman, N. S., R. T. Dean, and C. J. Sanderson.** 1980. Biochemical and morphological characterization of the killing of human monocytes by a leukotoxin drived from *Actinobacillus actinomycetemcomitans. Infect. Immun.* **28**:258–268.

43. **Taichman, N. S., B. J. Shenker, C. C. Tsai, L. T. Glickman, P. C. Baehni, R. C. Stevens, and B. F. Hammond.** 1984. Cytopathic effect of *Actinobacillus actinomycetemcomitans* leukotoxin on monkey blood leukocytes. *J. Periodontal Res.* **19**:133–145.

44. **Taichman, N. S., D. L. Simpson, S. Sakurada, M. Cranfield, J. DiRienzo, and J. Slots.** 1987. Comparative studies on the biology of *Actinobacillus actinomycetemcomitans* leukotoxin in primates. *Oral Microbiol. Immun.* **2**:97–104.

45. **Wandersman, C., and P. Delepelaire.** 1990. *TolC*, an *Escherichia coli* outer membrane protein required for α-hemolysin secretion. *Proc. Natl. Acad. Sci. USA* **87**:4776–4780.

46. **Welch, R. A.** 1991. Pore forming cytolysins of Gram negative bacteria. *Mol. Microbiol.* **5**:521–528.

47. **Welch, R. A., and S. Pellett.** 1988. Transcriptional organization of the *Escherichia coli* hemolysin genes. *J. Bacteriol.* **170**:1622–1630.

48. **Zambon, J. J.** 1985. *Actinobacillus actinomycetemcomitans* in human periodontal disease. *J. Clin. Periodontol.* **12**:1–20.

49. **Zambon, J. J., J. Slots, and R. J. Genco.** 1983. Serology of oral *Actinobacillus actinomycetemcomitans* and serotype distribution in human periodontal disease. *Infect. Immun.* **41**:19–27.

Molecular Pathogenesis of Periodontal Disease
Edited by Robert Genco et al.
© 1994 American Society for Microbiology, Washington, DC 20005

Chapter 8

Bacterial Polysaccharides as Microbial Virulence Factors

Robert E. Schifferle

ASSOCIATION OF BACTERIAL PS WITH VIRULENCE

Through the modulation of their cell surface components, bacterial pathogens can enhance their survival and suppress host defenses such as the complement system and phagocytosis (44). These bacterial components can include O-antigenic lipopolysaccharides (LPS), outer membrane proteins, and polysaccharide (PS) capsules.

Many bacterial species possess a PS covering external to their outer cell membrane. These exopolysaccharides may be present as a capsule attached to the bacterial cell or as a loose slime layer secreted by the bacteria (47). It is generally thought that the capsular PS are covalently attached to the outer membrane or anchored to the outer membrane via phospholipid moieties, with the attachment ranging from tightly adherent to loosely bound (24). The capsular domain can often be visualized with light microscopy by suspending bacterial cells in India ink. The capsule excludes the ink particles near the surface of the cell, resulting in a halo appearing around the cells (8). Alternatively, cells can be examined by electron microscopy (EM) employing a ruthenium red stain, which binds to the surface PS (5).

The PS capsule is often the first component to meet the host immune system on exposure of the host to foreign microorganisms. Encapsulation is frequently associated with virulence (4, 50). Bacterial species that cause meningitis and septicemia in humans are generally encapsulated and have anionic surface PS layers, thought to be critical to pathogenesis (13); these species include *Haemophilus influenzae, Escherichia coli, Neisseria meningitidis,* and the group B streptococci. The presence of an anionic PS has been demonstrated for *Porphyromonas gingivalis* (41).

Many bacterial PS have been identified, but not all are associated with disease. Six PS serotypes of *H. influenzae* are recognized, but only type b is associated with

Robert E. Schifferle • Department of Oral Biology, State University of New York at Buffalo, 318 Foster Hall, Buffalo, New York 14214-3092.

serious disease (26). Over 100 capsular PS (K antigens) are recognized for *E. coli*, but only a few K antigens are associated with pathogenic strains (34). Strains of *E. coli* that produce the K1 PS are observed in the majority of neonatal meningitis cases and are also highly represented in many cases of *E. coli* bacteremia and urinary tract infections (3). The PS of *N. meningitidis* B, a strain associated with adult meningitis, is chemically and structurally identical to the *E. coli* K1 PS (9).

Capsular PS often enhance bacterial virulence by increasing the microorganism's ability to inhibit phagocytosis, presumably through the prevention of opsonization (21). These antiphagocytic PS are generally acidic and may contain uronic acid, sialic acid, phosphate, or pyruvate residues. This anionic hydrophilic PS surface may physically reduce the possibility of hydrophobic interactions with phagocytic cells or may physically block cell surface structures that can activate complement. Alternatively, complement deposition may occur, but the capsule may block the binding of phagocytic receptors and/or inhibit binding of antibodies to bacterial cell surface structures.

PS composition may play a role in the activation of complement (20). Sialic acid plays a role in the inhibition of alternate pathway activation for both mammalian and bacterial cells. Treatment of group B streptococcal cells with neuraminidase to remove terminal sialic acid from the capsular PS converts the bacterial cell surface to a complement activator, resulting in rapid phagocytosis (10). In the group B streptococci, sialic acid appears to reduce activation of the alternate complement pathway through enhanced affinity for factor H, an inactivator of the alternate pathway. Anti-PS antibody can increase the efficiency of opsonization by two mechanisms. The binding of antibody to the PS can result in complement activation directly via the antibody Fc portion or indirectly by binding at and sterically blocking sites that contribute to down regulation.

Nonencapsulated mutants have been prepared for some microorganisms, and their virulence has been examined. An avirulent mutant of type III group B streptococci lacking the ability to express capsular PS was prepared through transposon-mediated mutagenesis (39). Although the nonencapsulated mutant was used at a 100-fold increase of the 50% lethal dose calculated for the encapsulated strain, it did not cause death in neonatal rats. Thus, the absence of a capsule in this strain greatly reduced its lethality in rats.

CHEMICAL AND STRUCTURAL CHARACTERIZATIONS

Bacteria can produce complex PS with a regular repeating sequence of one or more sugar residues (17, 43, 47). Homopolysaccharides are composed of only one type of sugar residue but may contain more than one type of glycosidic linkage. Heteropolysaccharides contain more than one sugar residue. Sugar residues that have been identified in bacterial PS include fucose, rhamnose, glucose, galactose, glucosamine, galactosamine, fucosamine, rhamnosamine, quinovosamine, sialic acid, dideoxysugars, uronic acids, and other monosaccharides. The PS are frequently anionic and may contain carboxylate, phosphate, pyruvate, acetyl, and phospholipid moieties.

Examples of homopolysaccharides include the glucose-containing glucans of *Streptococcus mutans* (28) and the *N*-acetylneuraminic acid polymers present as capsular antigens of *E. coli* (36). The K1 PS of *E. coli* is composed of α-(2-8)-linked NeuNAc residues (30), while the K92 PS is composed of both α-(2-8)- and α-(2-9)-linked NeuNAc (11). Examples of heteropolysaccharides include the group B streptococcal PS, composed of galactose, glucose, glucosamine, and *N*-acetyl-neuraminic acid (19), and the *Bacteroides fragilis* PS strain 23745, composed of fucose, galactose, glucose, quinovosamine, fucosamine, rhamnosamine, glucosamine, 3-amino-3,6-dideoxyhexose, and galacturonic acid (23).

Structures have been elucidated for many bacterial PS. Some PS are composed of linear sequences, while others have more complex structures. Examples of linear PS are the K1 PS of *E. coli*, composed of α-(2-8)-linked *N*-acetylneuraminic acid (30), and the K5 capsular PS of *Klebsiella*, composed of mannose, glucuronic acid, and glucose (15). Branched PS include the water-insoluble glucans produced by *S. mutans*, which contain both α-(1-3) and α-(1-6) linkages (28), and the type III PS of group B streptococci, which contains a backbone structure of repeating glucose, galactose, and glucosamine residues and a side chain composed of galactose and sialic acid linked to the backbone glucosamine residue (19). Nuclear magnetic resonance studies of this type III PS demonstrate conformational restraints imposed by the sialic acid via hydrogen bonding interactions with the glucosamine residue of the backbone. These conformational restraints contribute to the expression of specific immunological determinants. Thus, in addition to the primary structure, which is defined by the linked glycosidic residues, noncovalent interactions between the polar and/or charged residues present in the molecule may contribute to the presence of secondary structural characteristics.

IMMUNOLOGY OF PS

Encapsulated strains are implicated in serious bacterial infections such as meningitis and pneumonia, and despite antibiotic therapy, a high degree of mortality and morbidity remains. Antibiotic resistance makes therapy much more difficult (16). Thus, it is appropriate to try to prevent diseases associated with encapsulated bacteria through immunization.

PS of different bacterial species are currently being used commercially and experimentally as vaccines (18, 26). Greater safety can be achieved with purified components, such as bacterial PS, than with whole-cell vaccines. The current PS vaccines presently licensed for use in the United States are directed toward three different bacterial species: *N. meningitidis*, *Streptococcus pneumoniae*, and *H. influenzae* (26). The vaccines can be multivalent, such as the pneumococcal vaccine composed of 23 PS of the 83 recognized pneumococcal PS types, or they can be univalent, such as the *H. influenzae* vaccine, which contains only type b PS (26).

All recognized bacterial PS are antigenic and have various degrees of immunogenicity. The immunogenicity is related to molecular size, composition, and structure. PS vaccines may mimic carbohydrate sequences associated with host tissues and as such may not be recognized as foreign, thus acting to camouflage the bacterial surface. This reduces the ability of the host to mount an effective immune response, with poor immunogenicity likely when immunogens are similar to host molecules.

The immune response to PS is different from that to proteins (38). Proteins are T-dependent antigens whereby an interaction between T helper cells and B lymphocytes is required for a humoral immune response. In contrast, PS antigens are generally classified as T-independent antigens whereby the multiple repeating determinants are capable of interacting directly with the immunoglobulin receptors on the surface of B lymphocytes, resulting directly in a humoral immune response (6). T lymphocytes do, however, play a modifying role in the response to the T-independent antigens (1). The T-independent antigens generally induce a primary immunoglobulin M (IgM) response with limited IgA and IgG responses. This inability to produce a significant T-cell memory results in subsequent immunization failing to induce a secondary antibody response (14). PS antigens also demonstrate an age-related immunogenicity. The immune responses of infants and children to PS are generally poor (7). Unfortunately, this population is at greater risk of developing invasive disease by encapsulated bacteria (35).

One successful approach to increasing the immunogenicity of PS is to prepare a PS-protein conjugate (6). This conjugate can effect the transformation of a T-independent antigen to a T-dependent antigen, allowing occurrence of a secondary immune response and in turn production of a higher level of antibody. The antibodies produced by a conjugate are qualitatively similar to that of the purified PS but differ quantitatively in the level produced (35).

In the past, it was generally felt that the antigenicity of PS was defined by their primary structure, that is, their individual sugars and their linkages. Recent work, however, shows that conformationally dependent determinants also exist and that they may be the major immunodeterminants for some molecules (22). This immunodeterminant expression has been well studied for the group B streptococci. Although the PS serotypes contain the same sugar components, they are different serologically, with only minor cross-reactivity observed. The role of sialic acid in defining the immunodeterminants of the group B type III PS has been studied (19). The sialic acid residue contributes to the expression of the immunodeterminant, although the terminal NeuNAc-Gal sequence is not itself immunogenic or antigenic. Data from ^{13}C nuclear magnetic resonance studies demonstrated that hydrogen bonding between sialic acid and an *N*-acetylglucosamine residue contributed to the stabilization of a conformational immunodeterminant.

SYNTHESIS OF BACTERIAL PS

Bacterial PS are generally synthesized intracellularly via membrane-associated enzyme complexes prior to being transported to the bacterial surface. The structure of a specific PS is dictated by the specificity of the glycosyl transferases utilized by a given bacterium. Each linkage requires a specific enzyme to form a glycosidic linkage with a defined configuration. Sugar nucleotides function both as glycosyl donors and as precursors of other monosaccharides (17). PS chain assembly proceeds either by sequential addition of monosaccharide residues (monomeric mechanism) or by block addition of previously assembled repeating units (block mechanism). The final product is eventually translocated from the cytoplasmic membrane to the exterior of the cell.

The genes responsible for the production of PS have been characterized for some species. The gene cluster of *E. coli* K1 has received a fairly detailed analysis. The K1 capsular PS is a linear homopolysaccharide of α-(2-8)-linked *N*-acetylneuraminic acid. Characterization of the gene segment showed that it was composed of three spatially related gene clusters defined as regions 1, 2, and 3 (2). Region 2 appeared to code for the synthesis of enzymes responsible for the biosynthesis and activation of NeuNAc. Region 1 appeared to code for surface translocation of the PS. The third region appears to be involved in either PS biosynthesis or polymerization.

The cloned *E. coli* K1 PS biosynthesis gene was also employed in Southern blot experiments to evaluate the relationship of the biosynthetic pathways in the K1 PS and types K5 and K7 PS (37). K1 probes corresponding to regions 1 and 3 readily hybridized to the K5 and K7 genes. In contrast, a probe corresponding to region 2 did not demonstrate homology. Thus, for *E. coli*, all the genes responsible for PS biosynthesis appear to be similarly organized. Region 2 appears to be unique to the specific capsular type, while regions 1 and 3 appear to be associated with transport to the surface and possibly assembly into a functional capsule (2). It appears that a common transport is utilized for the different types of PS in *E. coli*.

CAPSULAR PS OF *P. GINGIVALIS*

The association of *P. gingivalis* and periodontal infection is well established. Numerous reports have described an association of *P. gingivalis* with periodontal infection in adults (12, 29, 45), in noninsulin-dependent diabetes mellitus patients presenting with periodontal disease (51), and in endodontic infections (49).

P. gingivalis and black-pigmented bacterial species have been shown by EM studies of ruthenium red-stained cells (27, 32) and by differential interference contrast microscopy (25) to have a capsule. In general, EM studies with ruthenium red staining have shown an electron-dense material external to the outer membrane that differs for different species. For some species, the ruthenium red-stained capsule appeared fibrous, while for other species, the capsule was compact and amorphous. One study noted no differences in the outer layers of *P. gingivalis* strains differing in serum sensitivity (32). A later study showed an intercellular matrix, which appeared to form a meshwork between individual cells of poorly phagocytosed strains (46).

Encapsulated strains of black-pigmented *Bacteroides* have shown greater resistance to phagocytosis than nonencapsulated strains (33, 46, 48). Van Steenbergen et al. (48) compared strains with various levels of virulence and noted that the more virulent strains were more resistant to the bactericidal effects of serum and to phagocytosis and had a thicker capsule, both by India ink exclusion and by EM with ruthenium red staining, than did the less virulent strains. Sundqvist et al. were able to divide strains of *P. gingivalis* into two distinct groups based on surface characteristics, noting that the group that was poorly phagocytosed appeared to possess a capsular structure when evaluated by EM (46).

PS has previously been isolated from *P. gingivalis* by phenol-water, alkali, acid, phenol, and EDTA extraction procedures, and the results have varied (27, 31, 32). PS isolated from *P. gingivalis* 382 by phenol-water extraction and Sepharose 4-B chromatography was composed of galactose, glucose, and glucosamine (27). Capsular material was isolated from *P. gingivalis* 381 by four different procedures and was shown to contain various ratios of rhamnose, glucose, galactose, mannose, and methylpentose (32). The PS from *P. gingivalis* A7A1-28 (ATCC 53977) was recently isolated and characterized (41). This preparation contained a higher proportion of amino sugars than that obtained in the other preparations and was distinctly different from the other reported strains, being composed of galactosamine, glucosamine, galactosaminuronic acid, and glucose. It did not contain galactose, a major component in the extracts from strains 381 and 382. The PS was chemically and antigenically distinct from the LPS of this strain.

The ability of the PS isolated from strain ATCC 53977 to activate serum complement was examined and compared with that of the LPS from strain ATCC 53977 and phenol-water extracts from four strains of *P. gingivalis*. Complement activation was assessed in an alternative-pathway-selective rabbit erythrocyte hemolytic assay. The phenol-water extracts and LPS readily activated the alternative pathway, but the PS fraction exhibited negligible activity (42). The PS from this strain does not activate the alternative pathway and may thus act to inhibit complement-mediated opsonophagocytosis.

The effect of immunization with a PS-protein conjugate containing PS from strain ATCC 53977 was evaluated in an experimental murine infection model. Serum antibody reactive to the PS, as determined by enzyme-linked immunosorbent assay, was elevated in the group of mice immunized with the conjugate but not in albumin-immunized control mice. Following microbial challenge, conjugate-immunized mice demonstrated less weight loss and smaller ulcerative lesions at secondary locations than the control group of mice. Thus, immunization of mice with *P. gingivalis* PS-containing conjugate could reduce the severity of but not prevent invasive infection (40).

Further studies are needed to better characterize the capsular PS from strains of *P. gingivalis* and to elucidate their role in the virulence of this bacterial species. Chemical and structural characterizations are needed to determine the roles of specific chemical and structural moieties in their interactions with host defenses. Goals that we should address in future studies include determination of the interaction of PS with the complement system, the role of PS in phagocytosis, and its effect on the host immunological response.

Acknowledgment. This study was supported by U.S. Public Health Service grant DE-09602.

REFERENCES

1. **Baker, P. J.** 1993. Suppressor T cells. *ASM News* **59:**123–128.
2. **Boulnois, G. J., and I. S. Roberts.** 1990. Genetics of capsular polysaccharide production in bacteria. *Curr. Top. Microbiol. Immunol.* **150:**1–18.
3. **Boulnois, G. J., I. S. Roberts, R. Hodge, K. R. Hardy, K. B. Jann, and K. N. Timmis.** 1987. Analysis of the K1 capsule biosynthesis genes of *Escherichia coli:* definition of three functional regions for capsule production. *Mol. Gen. Genet.* **208:**242–246.
4. **Brook, I.** 1987. Role of encapsulated anaerobic bacteria in synergistic infections. *Crit. Rev. Microbiol.* **14:**171–193.
5. **Costerton, J. W., R. T. Irvin, and K.-J. Cheng.** 1981. The bacterial glycocalyx in nature and disease. *Annu. Rev. Microbiol.* **35:**299–324.
6. **Dick, W. E., and M. Beurret.** 1989. Glycoconjugates of bacterial carbohydrate antigens. *Contrib. Microbiol. Immunol.* **10:**48–114.
7. **Douglas, R. M., J. C. Paton, S. J. Duncan, and D. J. Hansman.** 1983. Antibody response to pneumoccal vaccination in children younger than five years of age. *J. Infect. Dis.* **148:**131–137.
8. **Duguid, J. P.** 1951. The demonstration of bacterial capsules and slime. *J. Pathol. Bacteriol.* **63:**673–685.
9. **Echarti, C., B. Hirschel, G. J. Boulnois, J. M. Varley, F. Walsvogel, and K. N. Timmis.** 1983. Cloning and analysis of the K1 capsule biosynthesis genes of *Escherichia coli:* lack of homology with *Neisseria meningitidis* group B DNA sequences. *Infect. Immun.* **41:**54–60.
10. **Edwards, M. S., D. L. Kasper, H. J. Jennings, C. J. Baker, and A. Nicholson-Weller.** 1982. Capsular sialic acid prevents activation of the alternative complement pathway by type III, group B streptococci. *J. Immunol.* **128:**1278–1283.
11. **Egan, W., T.-Y. Liu, D. Dorow, J. S. Cohen, J. D. Robbins, E. C. Gottschlich, and J. B. Robbins.** 1977. Structural studies on the sialic acid polysaccharide antigen of *Escherichia coli. Biochemistry* **16:**3687–3692.
12. **Genco, R. J., J. J. Zambon, and L. A. Christersson.** 1988. The origins of periodontal infections. *Adv. Dent. Res.* **2:**245–259.
13. **Griffiss, J. M., M. A. Apicella, B. Greenwood, and H. P. Makela.** 1987. Vaccines against encapsulated bacteria: a global agenda. *Rev. Infect Dis.* **9:**176–188.
14. **Howard, J. G.** 1987. T cell independent responses to polysaccharides: their nature and delayed ontogeny, p. 221–231. *In* R. Bell and G. Torrigiani (ed.), *Towards Better Carbohydrate Vaccines.* John Wiley & Sons, Inc., New York.
15. **Isaac, D. H.** 1985. Bacterial polysaccharides, p. 141–184. *In* E. D. T. Atkins (ed.), *Polysaccharides: Topics in Structure and Morphology.* VCH, Deerfield Beach, Fla.
16. **Jacobs, M. R., H. J. Koornhof, and R. M. Robin-Browne.** 1978. Emergence of multiply resistant pneumococci. *N. Engl. J. Med.* **199:**735–740.
17. **Jann, B., and K. Jann.** 1990. Structure and biosynthesis of the capsular antigens of *Escherichia coli. Curr. Top. Microbiol. Immunol.* **150:**19–42.
18. **Jennings, H. J.** 1990. Capsular polysaccharides as vaccine candidates. *Curr. Top. Microbiol. Immunol.* **150:**97–127.
19. **Jennings, H. J., C. Lugowski, and D. L. Kasper.** 1981. Conformational aspects critical to the immunospecificity of the type III group B streptococcal polysaccharide. *Biochemistry* **20:**4511–4518.
20. **Jimenez-Lucho, V. E., K. A. Joiner, J. Foulds, M. M. Frank, and L. Leive.** 1987. C3b generation is affected by the structure of the O-antigen polysaccharide in lipopolysaccharide from *Salmonellae. J. Immunol.* **139:**1253–1259.
21. **Joiner, K. A.** 1985. Studies on the mechanism of bacterial resistance to complement-mediated killing and on the mechanism of action of bactericidal antibody. *Curr. Top. Microbiol. Immunol.* **121:**99–133.
22. **Kasper, D. L.** 1986. Bacterial capsule—old dogmas and new tricks. *J. Infect. Dis.* **153:**407–415.
23. **Kasper, D. L., A. Weintraub, A. A. Lindberg, and J. Lonngren.** 1983. Capsular polysaccharides

and lipopolysaccharides from two *Bacteroides fragilis* reference strains: chemical and immunochemical characterization. *J. Bacteriol.* **153:**991–997.

24. **Kuo, J. S.-C., V. W. Doelling, J. F. Graveline, and D. W. McCoy.** 1985. Evidence for covalent attachment of phospholipid to the capsular polysaccharide of *Haemophilus influenzae* type B. *J. Bacteriol.* **163:**769–773.

25. **Lambe, D. W., K. P. Ferguson, and D. A. Ferguson.** 1988. The *Bacteroides* glycocalyx as visualized by differential contrast microscopy. *Can. J. Microbiol.* **34:**1189–1195.

26. **Lee, C.-J.** 1987. Bacterial capsular polysaccharides—biochemistry, immunity and vaccine. *Mol. Immunol.* **24:**1005–1019.

27. **Mansheim, B. J., and D. L. Kasper.** 1977. Purification and immunochemical characterization of the outer membrane complex of *Bacteroides melaninogenicus* subspecies *asaccharolyticus*. *J. Infect. Dis.* **135:**787–799.

28. **Marsh, P., and M. Martin.** 1984. *Oral Microbiology,* 2nd ed. American Society for Microbiology, Washington, D.C.

29. **Mayrand, D., and S. C. Holt.** 1988. Biology of asaccharolytic black-pigmented *Bacteroides* species. *Microbiol. Rev.* **52:**134–152.

30. **McGuire, E. J., and S. B. Binckley.** 1964. The structure and chemistry of colominic acid. *Biochemistry* **3:**247–251.

31. **Millar, S. J., E. G. Goldstein, M. J. Levine, and E. Hausmann.** 1985. Modulation of bone metabolism by two chemically distinct lipopolysaccharide fractions from *Bacteroides gingivalis*. *Infect. Immun.* **51:**302–306.

32. **Okuda, K., Y. Fukumoto, I. Takazoe, J. Slots, and R. J. Genco.** 1987. Capsular structures of black-pigmented *Bacteroides* isolated from human. *Bull. Tokyo Dent. Coll.* **28:**1–11.

33. **Okuda, K., and I. Takazoe.** 1973. Antiphagocytic effects of the capsular structure of a pathogenic strain of *Bacteroides melaninogenicus*. *Bull. Tokyo Dent. Coll.* **49:**99–104.

34. **Ørskov, I., F. Ørskov, B. Jann, and K. Jann.** 1977. Serology, chemistry, and genetics of O and K antigens of *Escherichia coli*. *Bacteriol. Rev.* **41:**667–710.

35. **Robbins, J. B., R. Schneerson, S. C. Szu, A. Fattom, Y. Yang, T. Lagergard, C. Chu, and U. S. Sørensen.** 1989. Prevention of invasive bacterial diseases by immunization with polysaccharide-protein conjugates. *Curr. Top. Microbiol. Immunol.* **146:**169–180.

36. **Roberts, I., R. Mountford, N. High, D. Bitter-Suermann, K. Jann, K. Timmis, and G. Boulnois.** 1986. Molecular cloning and analysis of genes for production of K5, K7, K12, and K92 capsular polysaccharides in *Escherichia coli*. *J. Bacteriol.* **168:**1228–1233.

37. **Roberts, I. S., R. Mountford, R. Hodge, K. Jann, and G. J. Boulnois.** 1988. Common organization of gene clusters for production of different capsular polysaccharides (K antigens) in *Escherichia coli. J. Bacteriol.* **170:**1305–1310.

38. **Roitt, I. M.** 1988. *Essential Immunology,* 6th ed., p. 94–95. Blackwell Scientific Publications, Boston.

39. **Rubens, C. E., M. R. Wessels, L. M. Heggen, and D. L. Kasper.** 1987. Transposon mutagenesis of type III group B *Streptococcus:* correlation of capsule expression with virulence. *Proc. Natl. Acad. Sci. USA* **84:**7208–7212.

40. **Schifferle, R. E., P. B. Chen, L. B. Davern, A. Aguirre, R. J. Genco, and M. J. Levine.** 1993. Modification of experimental *Porphyromonas gingivalis* murine infection by immunization with a polysaccharide-protein conjugate. *Oral Microbiol. Immunol.* **8:**266–271.

41. **Schifferle, R. E., M. S. Reddy, J. J. Zambon, R. J. Genco, and M. J. Levine.** 1989. Characterization of a polysaccharide antigen from *Bacteroides gingivalis*. *J. Immunol.* **143:**3035–3042.

42. **Schifferle, R. E., M. E. Wilson, M. J. Levine, and R. J. Genco.** 1993. Activation of serum complement by polysaccharide-containing antigens of *Porphyromonas gingivalis*. *J. Periodontal Res.* **28:**248–254.

43. **Shibaev, V. N.** 1986. Biosynthesis of bacterial polysaccharide chains composed of repeating units. *Adv. Carbohydr. Chem. Biochem.* **44:**277–339.

44. **Slots, J., and R. J. Genco.** 1984. Black-pigmented *Bacteroides* species, *Capnocytophaga* species, and *Actinobacillus actinomycetemcomitans* in human periodontal disease: virulence factors in colonization, survival, and tissue destruction. *J. Dent. Res.* **63:**412–421.

45. **Slots, J., and M. A. Listgarten.** 1988. *Bacteroides gingivalis, Bacteroides intermedius* and *Acti-*

nobacillus actinomycetemcomitans in human periodontal diseases. *J. Clin. Periodontol.* **15:**85–93.

46. **Sundqvist, G., D. Figdor, L. Hänström, S. Sörlin, and G. Sandström.** 1991. Phagocytosis and virulence of different strains of *Porphyromonas gingivalis. Scand. J. Dent. Res.* **99:**117–129.

47. **Sutherland, I. W.** 1988. Bacterial surface polysaccharides: structure and function. *Int. Rev. Cytol.* **113:**187–231.

48. **van Steenbergen, T. J. M., F. G. A. Delemarre, F. Namavar, and J. de Graaff.** 1987. Differences in virulence within the species *Bacteroides gingivalis. Antonie van Leeuwenhoek* **53:**233–244.

49. **van Winkelhoff, A. J., A. W. Carlee, and J. de Graaff.** 1985. *Bacteroides endodontalis* and other black-pigmented *Bacteroides* species on odontogenic abscesses. *Infect. Immun.* **49:**494–497.

50. **Vermeulen, C., A. Cross, W. R. Byrne, and W. Zollinger.** 1988. Quantitative relationship between capsular content and killing of K1-encapsulated *Escherichia coli. Infect. Immun.* **56:**2723–2730.

51. **Zambon, J. J., H. Reynolds, J. G. Fisher, M. Shlossman, R. Dunford, and R. J. Genco.** 1988. Microbiological and immunological studies of adult periodontitis in patients with noninsulin-dependent diabetes mellitus. *J. Periodontol.* **59:**23–31

Molecular Pathogenesis of Periodontal Disease
Edited by Robert Genco et al.
© 1994 American Society for Microbiology, Washington, DC 20005

Chapter 9

Campylobacter rectus Toxins

M. Jane Gillespie and Joseph J. Zambon

Little is known of *Campylobacter rectus* pathogenic mechanisms despite the implication of this organism in oral and extraoral infections. This chapter will review two putative *C. rectus* virulence factors: the endotoxin, common to all gram-negative bacteria, and an extracellular cytotoxin. Preceding that review is a discussion of the investigations that identified *C. rectus* as a potential periodontal pathogen.

PATHOGENIC POTENTIAL OF *C. RECTUS*

Considering that Louis Pasteur and Robert Koch first began implicating bacteria in disease in the late 19th century, *C. rectus*, which was first described in 1979 (41), is very new among suspected bacterial pathogens. *C. rectus* was one of a group of gram-negative, asaccharolytic, agar-corroding bacteria isolated from lesions of humans with advanced destructive periodontitis. In subsequent studies, it was characterized as a serologically distinct (3), anaerobic, oxidase-positive, catalase-negative, rod-shaped bacterium (39). The rapid, darting motility characteristic of *C. rectus* was described, and in ultrastructural studies, a single polar flagellum and a distinctive array of hexagonally packed macromolecular subunits covering a typical gram-negative outer membrane were observed (22).

The guanine-plus-cytosine content (44 to 46 mol%) and the biochemical characteristics of this newly identified bacterium were similar to those of the rumen bacterium *Vibrio succinogenes;* however, its DNA was not homologous to *V. succinogenes* DNA. Therefore, a new genus, *Wolinella,* with two species, *Wolinella recta* and *Wolinella succinogenes,* was established (39). Later, a third bacterium isolated from a human alveolar abscess, *Wolinella curva,* was added to the genus *Wolinella* (42).

Wolinella spp. and many gram-negative, agar-corroding bacteria, including *Campylobacter sputorum, Campylobacter concisus, Bacteroides gracilis,* and *Bac-*

M. Jane Gillespie and Joseph J. Zambon • Departments of Oral Biology and Periodontology, State University of New York at Buffalo, Foster Hall, 3435 Main Street, Buffalo, New York 14214-3092.

teroides ureolyticus, required formate and fumarate to grow (39). As genetic techniques improved, it became apparent that the oral formate/fumarate-loving bacteria were more closely related to the campylobacters than to the rumen bacterium *W. succinogenes.* Recently, on the basis of 16S rRNA sequencing (31), rRNA-DNA hybridization, and immunotyping (44), *W. recta* and *W. curva* were transferred to the genus *Campylobacter* and given the species designations *C. rectus* and *Campylobacter curvus,* respectively.

Two 1985 reports reinforced interest in the pathogenic potential of *C. rectus.* Dzink et al. found significantly elevated levels of this organism in adult periodontitis lesions that were actively breaking down and in pockets deeper than 7 mm (8), and Moore et al., studying individuals with juvenile, moderate, and severe periodontitis, found *C. rectus* among those bacteria present at a higher proportion in diseased sites than in healthy sites (27). *C. rectus* has continued to be associated with adult forms of active destructive periodontal disease (23, 35, 40), and it has been noted at a high frequency in periodontitis patients with immune deficiencies related to diabetes mellitus (47) and AIDS (26, 28, 32, 46, 48). These reports have stimulated research on *C. rectus* virulence mechanisms. Two toxins, an extracellular cytotoxin and the endotoxin or lipopolysaccharide (LPS), are putative *C. rectus* virulence factors.

EVIDENCE OF A *C. RECTUS* CYTOTOXIN

In 1990, biological and genetic evidence that suggested that *C. rectus* could produce a cytotoxin emerged. Armitage and Holt (1) found that retinoic acid-induced cells of the promyelocytic HL-60 cell line were killed upon incubation with viable *C. rectus* cells for 1 to 4 days. Since retinoic acid induced HL-60 cells to differentiate into polymorphonuclear neutrophil (PMN)-like cells, this was considered evidence of leukotoxic activity.

At that time, the only periodontal pathogen known to produce a toxin was *Actinobacillus actinomycetemcomitans,* which elaborates an RTX leukotoxin (20). RTX refers to a large family of toxins with repeating subunits in their structural genes (24). The RTX toxins include the *Escherichia coli* hemolysin as well as a number of hemolysins and leukotoxins produced by veterinary pathogens of the *Pasteurella* and *Actinobacillus* spp. Sunday and coworkers (38) observed that synthetic oligonucleotide probes specific to conserved regions of the *E. coli* hemolysin and *Pasteurella haemolytica* leukotoxin genes hybridized to *C. rectus* genomic DNA. Others (16) used the PCR to demonstrate that 20-nucleotide-long primers prepared from conserved regions of the A genes of the *E. coli* hemolysin and the *P. haemolytica* and *A. actinomycetemcomitans* leukotoxins amplified a 560-bp PCR product from *C. rectus* genomic DNA that hybridized to the cloned *A. actinomycetemcomitans* leukotoxin A gene and to genomic DNA from *C. rectus* and *A. actinomycetemcomitans.* Thus, genetic studies suggested that *C. rectus* had an RTX-homologous gene.

LOCALIZATION OF THE *C. RECTUS* TOXIN

Most bacterial toxins are extracellular proteins; however, an important exception in periodontology is the cell-associated *A. actinomycetemcomitans* leukotoxin (43). Therefore, the origin of *C. rectus* toxicity was determined by testing both cell suspensions and medium supernatants of *C. rectus* cultures for toxicity against HL-60 cells and human PMNs. Viability assays of eucaryotic cells challenged with cell suspensions and medium supernatants for up to 4 h determined that cell suspensions of *C. rectus* did not kill these mammalian cells. However, ethanol and ammonium sulfate extracts of medium supernatants were toxic, killing 70% of 0.5×10^7 HL-60 cells in 2 h. This toxic activity was time and dose dependent as well as protease sensitive (11; Fig. 1).

Identification of the toxic component in the medium supernatants was accom-

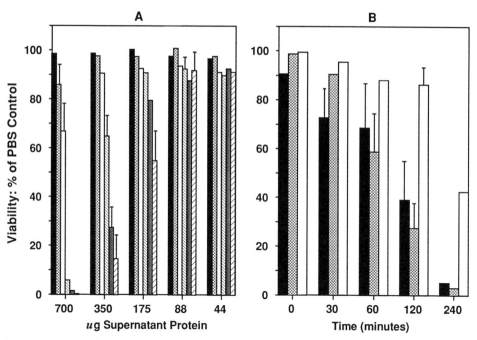

Figure 1. Time- and dose-dependent killing of HL-60 cells by a 60% $(NH_4)_2SO_4$ extract of medium supernatant from a *C. rectus* 33238 culture. Each assay was performed in duplicate and contained 0.5 ml of medium supernatant protein plus 0.5 ml of 10^7 HL-60 cells. At the indicated times, 50-μl aliquots were removed from each duplicate assay and diluted 1:10 and the number of viable cells was determined by trypan blue dye exclusion. The data are presented as the percentage of a phosphate-buffered saline control that contained no medium supernatant and are the average of five duplicate experiments. Error bars indicate the standard error of the mean (SE); no error is shown if the SE was less than 5. Symbols: ▨, 0 min; ▨, 30 min; ▭, 60 min; ▨, 120 min; ▨, 180 min; ▱, 240 min. (B) Protease-sensitive killing of HL-60 cells by medium supernatants from *C. rectus* 33238. The untreated sample contained 500 μg of supernatant protein per assay. The protease-treated sample was incubated overnight with 200 μg of proteinase K per ml. Symbols: ■, *A. actinomycetemcomitans* cell suspension; ▨, untreated medium supernatants; ▭, proteinase K-treated medium supernatants.

plished by growing *C. rectus* cultures that differed in toxicity. The experimental design was based on reports in the scientific literature that alterations in growth rates and iron availability affect toxin production in many bacteria, including *Escherichia, Pasteurella,* and *Actinobacillus* spp. (37) and *Pseudomonas* spp. (9). Both hemin and fumarate are known to affect *C. rectus* generation times (12), and hemin, although not required for *C. rectus* growth, is an important iron source for many oral bacteria. Therefore, *C. rectus* cultures were grown in media with and without hemin and in media with and without excess fumarate. Only medium supernatants from *C. rectus* cultures grown in excess fumarate were toxic (11). Coomassie blue-stained sodium dodecyl sulfate-polyacrylamide gel electrophoresis (SDS-PAGE) gels of *C. rectus* medium supernatants revealed that the toxic supernatants all contained a 104-kDa protein.

A common method for isolating proteins in solution was also useful in identifying the toxin. Ammonium sulfate precipitation of toxic medium supernatants concentrated the toxic activity into 60% ammonium sulfate precipitates (11). Coomassie blue staining of SDS-PAGE gels showed that the 104-kDa protein was also concentrated in these 60% ammonium sulfate preparations, while silver-stained SDS-PAGE gels of both toxic and nontoxic medium supernatants contained LPS (11). Immunoblot analysis of toxic medium supernatants verified that the protein and the LPS were specific to *C. rectus* (Fig. 2).

Medium supernatants from *C. rectus* clinical isolates also killed HL-60 cells and PMNs (11). The concentration of active toxin released by these clinical isolates varied. Specific activities were calculated to compare the number of an initial population of 0.5×10^7 eucaryotic cells killed per minute per microgram of supernatant protein. One microgram of supernatant protein from each of the three most toxic clinical isolates was sufficient to kill 100 to 300 HL-60 cells and 24 to

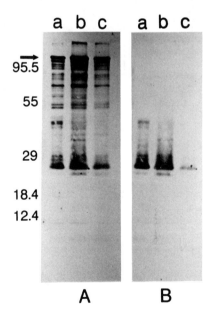

Figure 2. Western immunoblots of supernatants from *C. rectus* 33238 grown on Mycoplasma-formate-fumarate (12) with 50 mM fumarate. (A) Supernatants versus rabbit anti-*C. rectus* whole-cell antisera; (B) supernatants versus rabbit anti-*C. rectus* LPS. Lanes: a and b, 60% (NH₄)₂SO₄ precipitates at 60 and 120 μg of protein, respectively; c, 60 μg of protein from an ethanol extract. Arrow in panel A indicates the 104-kDa protein. Numbers at the left show the positions of the Diversified Biotech molecular size markers (in kilodaltons).

80 PMNs per min. Supernatants from the *C. rectus* clinical isolates also contained the 104-kDa protein (Fig. 3).

In summary, medium supernatants from the *C. rectus* American Type Culture Collection type strain and clinical isolates were toxic to HL-60 cells and PMNs. The toxic medium supernatants contained a 104-kDa protein and LPS, and the toxic activity was protease sensitive, implicating the 104-kDa protein (11).

These results concurred with those of Armitage and Holt (1). In their study, killing of the HL-60 cells required contact with live *C. rectus* for several days, indicating that *C. rectus* needed time to produce the toxic substance. In a related publication, Armitage and Holt (2) used electron microscopy to show that *C. rectus* resisted direct contact with PMNs and induced HL-60 cells, providing further evidence that the bacterium interacts through extracellular products.

The live *C. rectus* cells used by Armitage and Holt (1) killed only retinoic acid-induced HL-60 cells, whereas the toxic medium supernatants used by Gillespie et al. (11) killed uninduced HL-60 cells. This discrepancy was most likely due to receptor differences that may occur in various subpopulations of HL-60 cells. However, it is also possible that the *C. rectus* toxin displays binding sites for more than one HL-60 receptor, and these domains may have been more readily available in the high toxin concentrations used by Gillespie et al.

TOXIN IDENTIFICATION AND CHARACTERIZATION

The 104-kDa protein that was present in all toxic medium supernatants was identified as the *C. rectus* toxin in experiments demonstrating inhibition of supernatant toxicity by antibodies specific to the protein (14; Table 1). It is perhaps significant that many RTX toxins have molecular weights of 100 to 110 kDa (24).

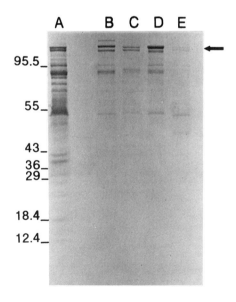

Figure 3. SDS-PAGE of concentrated medium supernatants from *C. rectus* clinical isolates on 10% acrylamide gels stained with 0.15% Coomassie blue. Isolates were 67759-5-#3 (lane A), WR 3A-2-#8 (lane B), 51.6 (lane C), WR 2A-2-#7 (lane D), and 45.5 (lane E). Each lane was loaded with approximately 60 μg of protein. Arrow indicates the position of the 104-kDa protein; numbers at the left show the positions of the Diversified Biotech molecular size markers (in kilodaltons).

Table 1. Lactate dehydrogenase assays for influence of inhibitors on killing of HL-60 cells by medium supernatant from 18-h *C. rectus* ATCC 33238 culture[a]

Activity	% of total LDH ± SD
Baseline toxic activity	
75.0 µg of protein (*n* = 5)..44.0 ± 15.6	
37.5 µg of protein (*n* = 4)..29.5 ± 13.8	
Effects of inhibitors[b]	
Antibody	
Control, no A104 .... 36.0	
A104, 37°C, 1 h 2.0	
Control, no ALPS.. 24.0	
ALPS, 37°C, 1 h 23.0	
PxB 32.0	
Control, no PxB.. 36.0	
438 µg of LPS 11.0	

[a] Viability was assayed by lactate dehydrogenase (LDH) release. LDH was calculated from Berger-Broida units (Sigma publication no. 500). Total cellular LDH was determined by incubating 10^7 HL-60 cells in 0.1% Triton X-100. Total cellular LDH was calculated as (supernatant LDH/total cellular LDH) − (phosphate-buffered saline control LDH/total cellular LDH). *n*, number of tests; A104, antibody to the 104-kDa protein; ALPS, antibody to *C. rectus* LPS; PxB, polymyxin B. Data are the averages of duplicate assays.
[b] Except for the LPS assay, 37.5 µg of supernatant protein was used in each.

The N-terminal amino acid sequence and the amino acid composition of the *C. rectus* toxin were determined on 104-kDa protein bands electroblotted from SDS-PAGE gels to Immobilon-P membranes. The sequence of the 16 amino-terminal amino acids of the *C. rectus* cytotoxin indicates a unique protein that has not been described previously (14). The toxin is not homologous to any bacterial sequences listed in the National Institutes of Health GenBank protein sequence database, nor is it homologous to published N-terminal sequences of three *C. rectus* outer membrane proteins (19; Fig. 4).

The amino acid composition of the *C. rectus* cytotoxin, including the characteristic lack of cysteine, is similar to those of the *A. actinomycetemcomitans* leukotoxin and other RTX toxins (14, 20). The differences between the *C. rectus* toxin and the RTX toxins could represent conservative amino acid substitutions in the *C. rectus* toxin. For example, the alanine, valine, and leucine residues in the *C. rectus* cytotoxin are more evenly distributed than those in the RTX toxins, and the ratio of serine to threonine is inverse to that of the RTX toxins. Comparisons between the *A. actinomycetemcomitans* leukotoxin and the *C. rectus* cytotoxin are

C. rectus **Cytotoxin:**

   ALTQTQVSQLYVTLFG

C. rectus **Outer Membrane Proteins:**

   43 kDa...ASAYNYELTPTIGGVHXEGNLGMNEQARIGLRVGTNDLN
   45 kDa...TPLEEAIKDVDFSGFAAYXYTGNKLXVNNK
   51 kDa...TPLEEAIKNVDLSGYAXYXYNNITVKKNPG

Figure 4. Comparisons of the N-terminal sequences of the *C. rectus* cytotoxin and three *C. rectus* outer membrane proteins. Data for *C. rectus* outer membrane proteins are from reference 19.

of interest because both are putative virulence factors of periodontal pathogens. The similar amino acid compositions of the two toxins indicate that they could share characteristics pertaining to their host ranges and mechanisms of action. Consistent with and possibly related to this is the observation that *A. actinomycetemcomitans* leukotoxin antiserum reacts with the *C. rectus* cytotoxin to indicate common epitopes on the two toxins (14).

The *C. rectus* LPS that was present in all toxic supernatants was not involved in killing, since purified LPS did not kill HL-60 cells and PMNs and since neither polymyxin B nor antibody to the LPS inhibited killing by the toxic medium supernatants (14; Table 1). Nevertheless, the presence of both the 104-kDa protein and the *C. rectus* LPS in medium supernatants indicates that these potential virulence factors may be released together in the oral cavity. Therefore, the release kinetics of the *C. rectus* toxin and LPS were studied in order to elucidate their roles and interaction in vivo (14). In individual bacteria, toxin release occurs through specific mechanisms. For example, clostridia release toxin by a cell autolytic mechanism whereby the protease that effects cell lysis also activates the toxin (18). In contrast, many toxins, including the *E. coli* hemolysin, *E. coli* colicin V, *Bordetella pertussis* adenylate cyclase, and *Vibrio cholerae* toxin, are actively secreted by specialized cell machinery (25).

Compared with those of the bacteria listed above, the secretory mechanisms leading to the release of the *A. actinomycetemcomitans* leukotoxin present an interesting paradox. Like the other RTX toxins, the *A. actinomycetemcomitans* leukotoxin is genetically programmed to produce secretory proteins; however, the toxin is cell associated in outer membrane blebs (4). The observation that the *C. rectus* toxin is found with LPS in medium supernatants suggests that it, too, may be released in outer membrane blebs, or, since a lytic cycle during *C. rectus* growth has been observed (12, 14, 15), the *C. rectus* toxin and LPS could be released upon cell lysis. However, studies of the release kinetics of *C. rectus* toxin and LPS do not support this possibility (14). The *C. rectus* toxin is released independently of cell lysis, beginning in early logarithmic phase. Its concentration in medium supernatants increases linearly throughout the *C. rectus* growth cycle and correlates with the time of growth. In contrast, LPS is released in two plateaus, the first occurring between 3 and 9 h and the second occurring between 15 and 24 h. These data indicate that the *C. rectus* toxin and LPS are released through different mechanisms and that the toxin is likely a secreted protein (14).

In secreted proteins, the amino acid that initiates translation, *n*-formylmethionine, is removed through posttranslational modification (29). The N terminus of the *C. rectus* cytotoxin is alanine (14; Fig 4). Although this does not prove that the toxin is an exported protein, it does prove that it is posttranslationally modified and supports the hypothesis, based on growth data, that it is secreted.

C. RECTUS ENDOTOXIN

In vitro biological assays have identified properties that make the *C. rectus* LPS capable of mediating the tissue destruction characteristic of periodontal disease. For example, the *C. rectus* LPS stimulates plasmin and plasminogen activator in cultured fibroblasts (30), its mitogenicity for murine lymphocytes and induction of the Shwartzman reaction is equivalent to that of enteric LPS (21), and it is a strong elicitor of prostaglandins and interleukin-1 (10, 13).

There is evidence that *C. rectus* LPS is superior to other LPS in inducing tissue-destructive immune modulators. In studies comparing prostaglandin E_2 (PGE) production in human peripheral monocytes exposed to LPS from a number of oral pathogens, including *C. rectus*, *Porphyromonas gingivalis*, *A. actinomycetemcomitans*, and *Prevotella intermedia*, *C. rectus* LPS was the most active, both in levels of PGE elicited and in maintenance of a maximum response over the 72-h duration of the experiment (10). This capacity of the *C. rectus* LPS to sustain a high PGE response over several days was also observed in experiments using mouse macrophages (13).

The remarkable biological activity of *C. rectus* LPS may be a consequence of an unusual structure. Preliminary chemical analysis by two laboratories indicates that *C. rectus* LPS likely has a conserved, enteric-type lipid A and an unusual polysaccharide (13, 21). The O antigen is composed of 88% rhamnose, and the core oligosaccharide has two molecular forms of heptose (13), which Kumada et al. identified as L-glycero- and D-glycero-D-mannoheptose (21). Gillespie et al. (13) also detected a dideoxyhexose in the *C. rectus* O antigen. Phytopathogens of *Pseudomonas* (36) and *Rhizobium* (33) spp. are the only bacteria with similar rhamnose and heptose compositions in their LPS polysaccharides.

The 2-keto-3-deoxyoctanoic acid (KDO) content of *C. rectus* LPS is of interest because many nonenteric LPS, including that of the oral pathogen *P. gingivalis* (5), either lack or have only minute amounts of this LPS-specific carbohydrate. Gillespie et al. (13) and Kumada et al. (21) estimated significantly different *C. rectus* KDO concentrations. Gillespie reported 0.6% KDO, which is typical of other oral LPS (13); Kumada reported 3.7% KDO, which is high even for enteric LPS (21).

Based on chemical analysis of the *C. rectus* lipid A, both Gillespie et al. (13) and Kumada et al. (21) predicted a typical diglucosamine backbone esterified to 24% (13) to 30% (21) β-hydroxymyristic acid and low concentrations of dodecanoic and tetradecanoic acids. They differed in the other fatty acids associated with the *C. rectus* lipid A. Gillespie et al. (13) detected seven fatty acids, including three (octadecenoate, octadecanoate, and hexadecanoate) not detected by Kumada's laboratory. Kumada et al. (21) reported six fatty acids; one, hexadecenoic acid, was not observed by Gillespie; another, β-hydroxyhexadecanoic acid, was present at four times the concentration detected by Gillespie. Phosphates are also important components of lipid A; Gillespie et al. reported 3.2% phosphate in the *C. rectus* LPS (13), and Kumada et al. reported only 0.8% (21).

These laboratories also differed in the proportion of lipid A reported per *C. rectus* LPS molecule. Gillespie reported 47% lipid A, while Kumada found only 16.5%. The SDS-PAGE results from these studies suggest that Gillespie isolated more core LPS. Each laboratory reported the same silver-stained *C. rectus* LPS SDS-PAGE pattern: a core band with an apparent molecular size of 13 kDa and two to four O-antigen bands at 26 kDa that comigrate with bands 5 through 8 of smooth *Salmonella* LPS. However, Gillespie's gels show a heavy core band equal to more than 33% of the total vertical width of the LPS (13), while Kumada's gels show a small, faint band with a vertical width equal to about 17% of the total LPS band width (21). Core LPS is equivalent to rough LPS and has a high proportion of lipid A (34), which could explain the high lipid A measurements obtained by Gillespie.

The disparities between these two laboratories concerning the total lipid A and the proportion of phosphates and long-chain fatty acids in the *C. rectus* LPS molecule may reflect differences in the methods used to extract and hydrolyze the LPS. Gillespie et al. used a cold $MgCl_2$-ethanol precipitation protocol that extracts total cellular LPS regardless of smooth or rough phenotype (7); hence, they detected more lipid A. This method also risks coextraction of bacterial phospholipids, which could explain the higher proportions of phosphates and long-chain fatty acids observed by these researchers. Kumada et al. extracted *C. rectus* LPS by the hot phenol-water technique (45), which favors isolation of smooth LPS (7) and explains their preferential isolation of the 26-kDa LPS with attached O antigen. The two laboratories also used different heat treatments. Kumada et al. (21) heated the bacterial cells at 121°C for 30 min prior to LPS extraction and hydrolyzed the LPS at 100°C in 5% acetic acid for 7.5 h, whereas Gillespie et al. (13) heated the cells at 100°C in 1% acetic acid for 1.5 h. Increased heating times and acid concentrations enhance nonspecific LPS hydrolysis, with consequent loss of phosphates and fatty acids as well as *trans*-esterification of fatty acids. These could account for the lower levels of phosphates and long-chain fatty acids detected by Kumada et al. (21).

It is also possible that *C. rectus*, like many bacteria, produces a heterogeneous LPS. If so, the growth conditions and extraction methods used by Gillespie and Kumada may have favored isolation of different molecular types. As discussed above, *C. rectus* spontaneously releases a cytotoxin-associated LPS (11). Analysis of this cell-free LPS compared to cell-bound LPS may clarify the complexity of the *C. rectus* LPS molecule. It may also elucidate an aggressive delivery system for this tissue-destructive *C. rectus* cell product.

VIRULENCE POTENTIAL OF *C. RECTUS* TOXINS

Immunoblotting has verified the in vivo expression of the *C. rectus* toxins, suggesting that they may be important antigens. Serum from non-insulin-dependent diabetics with significant bone loss and cultivable levels of *P. gingivalis* and *C. rectus* in their periodontitis lesions (47) contained antibody to purified *C. rectus* LPS (Fig. 5), and a patient demonstrating an oral *C. rectus* infection produced

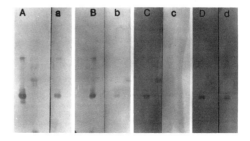

Figure 5. Western immunoblots of serum from non-insulin-dependent diabetics versus *C. rectus* LPS. The immunoglobulin M (IgM) response is shown in lanes A, B, C, and D; the IgG response is shown in lanes a, b, c, and d. Bands in unlabeled lanes are molecular weight markers. Lanes A through C and a through c are patient samples with cultivable levels of *P. gingivalis* and *C. rectus* and significant bone loss. The patient samples in lanes D and d had low titers to *P. gingivalis*, and patients showed no bone loss (47).

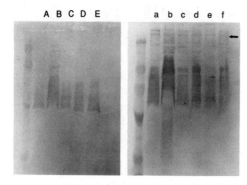

A B C D E   a b c d e f

Figure 6. Western immunoblots of medium supernatant protein and LPS from *C. rectus* clinical isolates versus serum from a patient with a *C. rectus* infection. The immunoglobulin (IgM) response is shown in lanes A through E; the IgG response is shown in lanes a through f. The arrow shows the position of the 104-kDa protein in *C. rectus* 33238. Clinical isolates are 45.5 (lanes A and a), 67759-5-#3 (lanes B and b), WR3A-2-#8 (lanes C and c), WR2A-2-#7 (lanes D and d), 51.6 (lanes E and e), and 3.11 (lane f).

antibodies to both the toxin and LPS (Fig. 6). These data, along with the reported release of the toxin and LPS throughout the *C. rectus* growth cycle and their close association in the growth medium (14), suggest that the cytotoxin and LPS are associated in vivo and may be a useful model for the study of interactions between bacterial virulence factors. Presumably, the *C. rectus* LPS would mediate tissue destruction and inflammation while the leukocidic toxin allowed evasion of host defense mechanisms. However, research with other bacterial toxins predicates a varied role for cytotoxins produced by periodontopathic bacteria. For example, sublethal doses of certain RTX toxins are known to exacerbate inflammation through release of tissue-oxidizing compounds (6) and PMN chemotactic factors (17), and the veterinary pathogen *Pasteurella multocida* produces an osteolytic toxin that stimulates osteoclastic bone resorption (24).

Acknowledgment. M. J. Gillespie's research is supported by grant DEO9661 from the National Institute of Dental Research.

REFERENCES

1. **Armitage, G. C., and S. C. Holt.** 1990. Effect of *Actinobacillus actinomycetemcomitans*, *Wolinella recta* and *Bacteroides gingivalis* on the viability of retinoic acid-induced and dimethyl sulfoxide-induced HL-60 cells. *Oral Microbiol. Immunol.* **5:**241–247.
2. **Armitage, G. C., and S. C. Holt.** 1990. Interaction of gram-negative periodontal pathogens with retinoic acid-induced and dimethyl sulfoxide-induced HL-60 cells. *Oral Microbiol. Immunol.* **5:**248–255.
3. **Badger, S. J., and A. C. R. Tanner.** 1981. Serological studies of *Bacteroides gracilis*, *Campylobacter concisus*, *Wolinella recta*, and *Eikenella corrodens*, all from humans with periodontal disease. *Int. J. Syst. Bacteriol.* **31:**446–451.
4. **Berthold, P., D. Forti, I. R. Kieba, J. Rosenbloom, N. S. Taichman, and E. T. Lally.** 1992. Electron immunocytochemical localization of *Actinobacillus actinomycetemcomitans* leukotoxin. *Oral Microbiol. Immunol.* **7:**24–27.
5. **Bramanti, T. E., G. G. Wong, S. T. Weintraub, and S. C. Holt.** 1989. Chemical characterization and biologic properties of lipopolysaccharide from *Bacteroides gingivalis* strains W50, W83 and ATCC 33277. *Oral Microbiol. Immunol.* **4:**183–192.
6. **Czuprynski, C. J., E. J. Noel, O. Ortiz-Carranza, and S. Srikumarman.** 1991. Activation of bovine neutrophils by partially purified *Pasteurella haemolytica* leukotoxin. *Infect. Immun.* **59:**3126–3133.
7. **Darveau, R. P., and R. E. W. Hancock.** 1983. Procedure for isolation of bacterial lipopolysac-

charides from both smooth and rough *Pseudomonas aeruginosa* and *Salmonella typhimurium* strains. *J. Bacteriol.* **155:**831–838.

8. **Dzink, J. L., A. C. R. Tanner, A. D. Haffajee, and S. S. Socransky.** 1985. Gram-negative species associated with active destructive periodontal lesions. *J. Clin. Periodontol.* **12:**648–659.

9. **Frank, D. W., D. G. Storey, M. S. Hindhal, and G. H. Iglewski.** 1989. Differential regulation by iron of *regA* and *toxA* transcript accumulation in *Pseudomonas aeruginosa*. *J. Bacteriol.* **171:**5304–5313.

10. **Garrison, S. W., S. C. Holt, and F. C. Nichols.** 1988. Lipopolysaccharide stimulated PGE$_2$ release from human monocytes: comparison of lipopolysaccharides prepared from suspected periodontal pathogens. *J. Periodontol.* **59:**684–687.

11. **Gillespie, J., E. DeNardin, S. Radel, J. Kuracina, J. Smutko, and J. J. Zambon.** 1992. Production of an extracellular toxin by the oral pathogen *Campylobacter rectus*. *Microb. Pathog.* **12:**69–77.

12. **Gillespie, J., and S. C. Holt.** 1987. Growth studies of *Wolinella recta*, a gram-negative periodontopathogen. *Oral Microbiol. Immunol.* **2:**105–111.

13. **Gillespie, J., S. T. Weintraub, G. G. Wong, and S. C. Holt.** 1988. Chemical and biological characterization of the lipopolysaccharide of the oral pathogen *Wolinella recta* ATCC 33238. *Infect. Immun.* **56:**2028–2035.

14. **Gillespie, M. J., J. Smutko, G. G. Haraszthy, and J. J. Zambon.** 1993. Isolation and partial characterization of the *Campylobacter rectus* cytotoxin. *Microb. Pathog.* **14:**203–215.

15. **Grenier, D., and D. Mayrand.** 1986. Nutritional relationships between oral bacteria. *Infect. Immun.* **53:**616–620.

16. **Haraszthy, V. I., M. J. Gillespie, E. T. Lally, and J. J. Zambon.** 1992. Molecular cloning of a *Campylobacter rectus* toxin, abstr. 933. *J. Dent. Res.* **71:**222.

17. **Henricks, P. A. J., G. J. Binkhorst, A. A. Drijver, and F. P. Nijkamp.** 1992. *Pasteurella haemolytica* leukotoxin enhances production of leukotriene B$_4$ and 5-hydroxyeicosatetraenoic acid by bovine polymorphonuclear leukocytes. *Infect. Immun.* **60:**3238–3243.

18. **Hill, G. B., O. Suydam, and H. P. Willett.** 1988. *Clostridium*, p. 549. *In* H. P. Joklik, H. P. Willett, B. D. Amos, and C. M. Wilfert (ed.), *Zinsser Microbiology*. Appleton & Lange, Norwalk, Conn.

19. **Kennel, W. L., and S. C. Holt.** 1991. Extraction, purification, and characterization of outer membrane proteins from *Wolinella recta* ATCC 33238. *Infect. Immun.* **59:**3740–3749.

20. **Kraig, E., T. Dailey, and D. Kolodrubetz.** 1990. Nucleotide sequence of the leukotoxin gene from *Actinobacillus actinomycetemcomitans*: homology to the alpha-hemolysin/leukotoxin gene family. *Infect. Immun.* **58:**920–929.

21. **Kumada, H., K. Watanabe, T. Umemoto, K. Kato, S. Kondo, and K. Hisatsune.** 1989. Chemical and biological properties of lipopolysaccharide, lipid A and degraded polysaccharide from *Wolinella recta* ATCC 33238. *J. Gen. Microbiol.* **135:**1017–1025.

22. **Lai, C.-H., M. A. Listgarten, A. C. R. Tanner, and S. S. Socransky.** 1981. Ultrastructures of *Bacteroides gracilis*, *Campylobacter concisus*, *Wolinella recta*, and *Eikenella corrodens*, all from humans with periodontal disease. *Int. J. Syst. Bacteriol.* **31:**465–475.

23. **Lai, C.-H., K. Oshima, J. Slots, and M. A. Listgarten.** 1992. *Wolinella recta* in adult gingivitis and periodontitis. *J. Periodontal Res.* **27:**8–14.

24. **Lo, R. Y. C.** 1990. Molecular characterization of cytotoxins produced by *Haemophilus*, *Actinobacillus*, *Pasteurella*. *Can. J. Vet. Res.* **54**(Suppl.):S33–S35.

25. **Lory, S.** 1992. Determinants of extracellular protein secretion in gram-negative bacteria. *J. Bacteriol.* **174:**3423–3428.

26. **Moore, L. V. H., W. E. C. Moore, C. Riley, C. N. Brooks, J. A. Burmeister, and R. M. Smibert.** 1993. Periodontal microflora of HIV positive subjects with gingivitis or adult periodontitis. *J. Periodontol.* **64:**48–56.

27. **Moore, W. E. C., L. V. Holdeman, E. P. Cato, R. M. Smibert, J. A. Burmeister, K. G. Palcanis, and R. R. Ranney.** 1985. Comparative bacteriology of juvenile periodontitis. *Infect. Immun.* **48:**507–519.

28. **Murray, P. A., J. R. Winkler, L. Sadkowski, K. S. Kornman, B. Steffensen, P. B. Robertson, and S. C. Holt.** 1988. Microbiology of HIV-associated gingivitis and periodontitis, p. 105–118. *In* P. B. Robertson and J. S. Greenspan (ed.), *Perspectives on Oral Manifestations of AIDS*. PSG Publishing, Littleton, Mass.

29. **Neidhardt, F. C., J. L. Ingraham, and M. Schaechter.** 1990. *Physiology of the Bacterial Cell*, p. 107. Sinauer Associates, Inc., Sunderland, Mass.

30. **Ogura, N., Y. Kamino, Y. Shibata, Y. Abiko, H. Izumi, and H. Takiguchi.** 1992. Effect of *Wolinella recta* LPS on plasmin activity by gingival fibroblasts, abstr. 1917. *J. Dent. Res.* **72:**755.

31. **Paster, B. J., and F. E. Dewhirst.** 1988. Phylogeny of campylobacters, wolinellas, *Bacteroides gracilis,* and *Bacteroides ureolyticus* by 16S ribosomal ribonucleic acid sequencing. *Int. J. Syst. Bacteriol.* **38:**56–62.

32. **Rams, T. E., M. Andriolo, Jr., D. Feik, S. N. Abel, T. M. McGivern, and J. Slots.** 1991. Microbiological study of HIV related periodontitis. *J. Periodontol.* **62:**74–81.

33. **Russa, R., and Z. Lorkiewica.** 1979. O-methylheptose in lipopolysaccharides of *Rhizobium trifolii* 24SM. *FEMS Microbiol. Lett.* **6:**71–74.

34. **Shands, J. W., Jr., and P. W. Chun.** 1980. The dispersion of gram-negative lipopolysaccharide by deoxycholate. *J. Biol. Chem.* **255:**1221–1226.

35. **Slots, J., L. J. Emrich, R. J. Genco, and B. G. Rosling.** 1985. Relationship between some subgingival bacteria and periodontal pocket depth and gain or loss of periodontal attachment after treatment of adult periodontitis. *J. Clin. Periodontol.* **12:**540–552.

36. **Smith, A. R. W., S. E. Zamze, and R. C. Hignett.** 1985. Composition of lipopolysaccharide from *Pseudomonas syringae* pv *morsprumorum* and its digestion by bacteriophage A7. *J. Gen. Microbiol.* **131:**963–974.

37. **Strathdee, C. A., and R. Y. C. Lo.** 1989. Regulation of expression of the *Pasteurella haemolytica* leukotoxin determinant. *J. Bacteriol.* **171:**5955–5962.

38. **Sunday, G. J., M. J. Gillespie, J. Kuracina, J. S. Smutko, E. DeNardin, S. Radel, and J. J. Zambon.** 1990. Abstr. 1468. *J. Dent. Res.* **69:**292.

39. **Tanner, A. C. R., S. Badger, C. H. Lai, M. Listgarten, R. A. Visconti, and S. S. Socransky.** 1981. *Wolinella* gen. nov., *Wolinella succinogenes* (*Vibrio succinogenes* Wolin et al.) comb. nov., and description of *Bacteroides gracilis* sp. nov., *Wolinella recta* sp. nov., *Campylobacter concisus* sp. nov., and *Eikenella corrodens* from humans with periodontal disease. *Int. J. Syst. Bacteriol.* **31:**432–445.

40. **Tanner, A. C. R., J. L. Dzink, J. L. Ebersole, and S. S. Socransky.** 1987. *Wolinella recta, Campylobacter concisus, Bacteroides gracilis,* and *Eikenella corrodens* from periodontal lesions. *J. Periodontal Res.* **22:**327–330.

41. **Tanner, A. C. R., C. Haffer, G. T. Bratthall, R. A. Visconti, and S. S. Socransky.** 1979. A study of the bacteria associated with advancing periodontitis in man. *J. Clin. Periodontol.* **6:**278–307.

42. **Tanner, A. C. R., M. A. Listgarten, and J. L. Ebersole.** 1984. *Wolinella curva* sp. nov.: "*Vibrio succinogenes*" of human origin. *Int. J. Syst. Bacteriol.* **34:**275–282.

43. **Tsai, C.-C., B. J. Shenker, J. M. DiRenzo, D. Malamud, and N. S. Taichman.** 1984. Extraction and isolation of a leukotoxin from *Actinobacillus actinomycetemcomitans* with polymyxin B. *Infect. Immun.* **43:**700–705.

44. **VanDamme, P., E. Falsen, R. Rossau, B. Hoste, P. Segers, R. Tytgat, and J. DeLey.** 1991. Revision of *Campylobacter, Helicobacter,* and *Wolinella* taxonomy: emendation of generic descriptions and proposal of *Arcobacter* gen. nov. *Int. J. Syst. Bacteriol.* **41:**88–103.

45. **Westphal, O., and K. Jann.** 1966. Bacterial lipopolysaccharides: extraction with phenol-water and further applications of the procedure. *Methods Carbohydr. Chem.* **5:**83–91.

46. **Zambon, J. J.** 1988. Overview of the microbiology of periodontal disease, p. 96–104. *In* P. B. Robertson and J. S. Greenspan (ed.), *Perspectives on Oral Manifestations of AIDS.* PSG Publishing, Littleton, Mass.

47. **Zambon, J. J., H. Reynolds, J. G. Fisher, M. Shlossman, R. Dunford, and R. J. Genco.** 1988. Microbiological and immunological studies of adult periodontitis in patients with noninsulin-dependent diabetes mellitus. *J. Periodontol.* **59:**23–31.

48. **Zambon, J. J., H. S. Reynolds, and R. J. Genco.** 1990. Studies of the subgingival microflora in patients with acquired immunodeficiency syndrome. *J. Periodontol.* **61:**699–704.

Molecular Pathogenesis of Periodontal Disease
Edited by Robert Genco et al.
© 1994 American Society for Microbiology, Washington, DC 20005

Chapter 10

Bacterial Endotoxic Substances and Their Effects on Host Cells

Shigeyuki Hamada, Taku Fujiwara, and Joji Mihara

Ample evidence has revealed that the initiation and progression of periodontal diseases are closely associated with quantitative and qualitative changes in the subgingival dental plaque flora. Following microbial plaque accumulation in the gingival crevice area, invasion of host tissues by microorganisms induces chronic inflammatory and immunopathologic reactions, which eventually lead to the development of gingivitis and periodontal diseases. Although a wide variety of bacterial species have been isolated from the gingival crevice, the emergence of certain gram-negative anaerobic rods has been associated with destructive periodontal disease. A limited number of species such as *Porphyromonas gingivalis*, *Prevotella intermedia*, and *Actinobacillus actinomycetemcomitans* have been implicated as important periodontopathogens (12).

Bacterial structural and extracellular components and metabolites may exhibit diverse effects on host cells, ultimately being capable of inducing inflammatory changes clinically observed in periodontal lesions. The outer membrane of a gram-negative bacterium contains different categories of substances that elicit modification of host responses from molecular to organ levels. Through a complicated process, however, various immunologic and biologic responses in the host, either enhanced or suppressed, have been induced upon stimulation with a number of macromolecules derived from gram-positive bacterial cells. For example, cell wall polysaccharide-peptidoglycan complex isolated from enzyme lysate of purified cell walls of *Streptococcus* and *Actinomyces* species induces immunologic and inflammatory responses (12, 47).

In this chapter, bacterial surface components exhibiting endotoxic activities against host cells will be described in relation to their etiologic implications in periodontal disease. Special attention will be paid to the lipopolysaccharides (LPS) of *P. gingivalis* and *P. intermedia,* i.e., former black-pigmented "*Bacteroides*" species.

Shigeyuki Hamada, Taku Fujiwara, and Joji Mihara • Department of Oral Microbiology, Faculty of Dentistry, Osaka University, Yamadaoka, Suita-Osaka, 565 Japan.

BACTERIAL COMPONENTS THAT EXHIBIT ENDOTOXIC EFFECTS

Endotoxins are generally defined as heat-stable LPS-protein complexes present in the cell envelopes of gram-negative bacteria, and therefore they are considered the somatic antigen of these organisms. LPS is a protein-free endotoxin that elicits a potent and pleiotropic stimulus for immune cells and has a direct role in the pathogenesis of gram-negative bacterial infection. LPS consists of three regions, i.e., lipid A, R core oligosaccharide, and O side chain heteropolysaccharide. Lipid A plays a central role in the induction by LPS of pathophysiologic effects in vitro and in vivo. Lipid A is the most conserved moiety of LPS molecules among different gram-negative species and is composed of hydrophilic D-glucosamine disaccharide-phosphate that is substituted with several hydrophobic fatty acid molecules. The core oligosaccharide is linked to the lipid A through a 3-deoxy-D-manno-2-octu-losonic acid (KDO) and comprises a set of sugar residues whose composition is species or genus specific. The O side chains contain repeating units composed of various sugars, which frequently determine the O-antigenic specificities of various microorganisms (39, 48).

It should be mentioned, however, that endotoxic substances other than lipid A may have profound pathologic and pharmacologic effects on mammalian cells. Experimental evidence indicates that the proteinaceous components of endotoxins have strong biologic activity independent of the LPS components, although LPS manifests a high affinity for outer membrane proteins. A low-molecular-weight lipoprotein that exists in a large number of copies per cell is a good example of an endotoxic protein exhibiting powerful stimulating activity against host cells, including immune and inflammatory cells. Thus, LPS can be released spontaneously or by a variety of mild procedures, accompanying the simultaneous release of outer membrane proteins. In fact, endotoxin substances may be isolated from cell-free culture supernatant of certain bacterial species (15, 34). Since both gram-negative and gram-positive bacterial organisms are heavily concentrated in the subgingival area where chronic localized inflammation occurs, the endotoxic components released from bacterial cells in dental plaque should have important roles in the pathogenic process in this area (7).

ENDOTOXIC SUBSTANCES AND PERIODONTAL DISEASES

The percentage of gram-negative bacteria in subgingival dental plaque increases markedly when healthy periodontal sites become periodontitis sites (12). Therefore, it is natural to assume that levels of endotoxic materials in periodontally affected sites correlate with the increased severity of local inflammation. This hypothesis was confirmed by several clinical studies that used the *Limulus* lysate test, a very sensitive assay used to detect and "quantitate" endotoxic substances in biologic samples. This assay was applied to dental plaque samples obtained from clinically healthy and diseased sites of subjects. The results of earlier studies are reviewed by Daly et al. (7). Although the *Limulus* test gave positive reactions with different kinds of microbial substances in addition to LPS or endotoxins, Wilson et al. (52) clearly showed through the use of polymyxin B-Sepharose 4B affinity chromatography that the major *Limulus*-reactive substance was LPS. They demonstrated that phenol-water extract of root surface material obtained from periodontally involved teeth markedly reduced *Limulus* activity after the extract was passed through an affinity column. It was noted that endotoxic LPS is loosely attached to the root surface or cementum in periodontally involved teeth (31).

More recently, a high degree of correlation between endotoxin levels and the ratio of gram-negative bacteria in both periodontally involved and healthy subgingival sites has been demonstrated (8). The consistency of experimental results has also been confirmed with multiple samplings of these sites over time, i.e., sampling on days 0 through 14. Furthermore, monoclonal antibodies (MAbs) recognizing specific sites of LPS of periodontopathogens, including *P. gingivalis*, *P. intermedia*, *A. actinomycetemcomitans*, and *Fusobacterium nucleatum,* have been used in bacterial or plaque concentration fluorescence immunoassays (43) to evaluate both live and dead cells in plaque. This method was 50-fold more sensitive than conventional cultural methods in detecting specific bacterial cells (53).

Periodontitis lesions are characterized by a local infiltration of lymphocytes, macrophages, and neutrophils. Among others, plasma cells become predominant by prolonged stimulation of plaque masses in the subgingival crevices of lesions. LPS present in plaque is a powerful antigenic substance as well as a nonspecific B-cell activator (51). In fact, plasma cells producing antibodies specific for *P. gingivalis* LPS were demonstrated by the enzyme-linked immunospot (ELISPOT) method, and higher frequencies of the LPS-specific antibody-producing cells were noted in the lesions of advanced cases of periodontitis than in those of moderate cases. However, no plasma cells producing antibodies specific for *Escherichia coli* LPS or cholera toxin were detected, which indicates that in stimulation of *P. gingivalis*, LPS should occur in the inflammatory lesions of periodontitis (35). It has recently been suggested that B-cell activation by LPS occurs in the milieu of interleukin-1 (IL-1) derived from LPS-stimulated adherent macrophages (5).

CHARACTERIZATION OF ENDOTOXINS OF PERIODONTAL PATHOGENS

Chemical Properties of LPS from *P. gingivalis*

In earlier studies, it was found that LPS from *Bacteroides fragilis* and other related *Bacteroides* species possess unique fatty acid profiles that are different from those of the LPS from members of the family *Entereobacteriaceae*. On the other hand, the LPS from most enterobacterial species contain even-numbered 3-hydroxy fatty acids, i.e., 3-hydroxymyristic acid (3-OH-14:0), which usually accounts for >50% of the total fatty acid content. On the other hand, 13-methyl-tetradecanoic (13-Me-13:0), 3-hydroxypentadecanoic (3-OH-15:0), 3-hydroxy hexadecanoic (3-OH-16:0), 3-hydroxy-15-methyl-hexadecanoic (3-OH-15-Me-16:0), and 3-hydroxy-heptadecanoic (3-OH-17:0) acids were demonstrated to be major components of fatty acids in *B. fragilis* LPS (39). The uniqueness of the LPS structure of *Bacteroides* species was supported by enzyme-linked immunosorbent assay (ELISA) inhibition of the lipid A and anti-lipid A reaction. The lipid A from *Bacteroides* species did not inhibit the anti-*Enterobacteriaceae* LPS reaction, while *Veillonella* and *Fusobacterium* lipid A did inhibit the reaction. Furthermore, the LPS from *Bacteroides* species were devoid of KDO and heptose, which are known as essential components of core polysaccharide in *Enterobacteriaceae* LPS, and the endotoxicity of these LPS preparations appeared very weak. Recently, oral black-pigmented *Bacteroides* species involved in the pathogenesis of periodontitis have been reclassified as *Porphyromonas* or *Prevotella* species; however, the basic properties of LPS obtained from these species are known to be very similar (15).

The properties of LPS of *P. gingivalis*, an important suspected periodontal pathogen, have been studied extensively in comparison with those of other black-pigmented *Porphyromonas* and *Prevotella* species. Like *B. fragilis* LPS, *P. gingivalis* 381 LPS was shown to have no heptose or KDO and to exhibit characteristic fatty acid profiles that are different from those of *E. coli* LPS (Fig. 1; 39, 48). A notable finding was that *P. gingivalis* LPS did stimulate B lymphocytes and macrophages from C3H/HeJ mice, an LPS low-responder strain of mice, while *E. coli* LPS did not (21). LPS molecules differed immunochemically among black-pigmented *Porphyromonas* and *Prevotella* species in immunodiffusion analysis. Antiserum to *P. gingivalis* 381 LPS reacted only with *P. gingivalis* LPS and not with those from other *Porphyromonas* and *Prevotella* species. Differences among these LPS were also demonstrated by the profile of sodium dodecyl sulfate-polyacrylamide gel electorophoresis (SDS-PAGE). It was further shown that strain variations exist within *P. gingivalis* (9, 10). The differences among various strains of *P. gingivalis* are similar to the differences in the immunochemical specificities of fimbriae from these strains (36). In this regard, excellent resolution of the LPS profiles of the

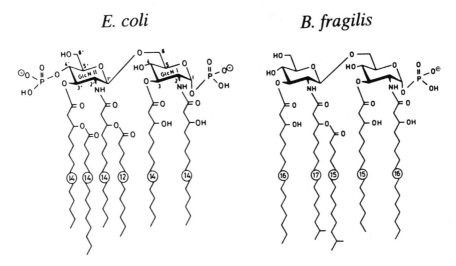

Figure 1. Structure of the predominant lipid A of LPS from *E. coli* and *B. fragilis*. Adapted from Rietschel et al. (39).

genus *Bacteroides* sensu strictu (e.g., nonoral, non-black-pigmented group of *Bacteroides* species) has been obtained by Maskell (27).

Early studies suggested that *P. gingivalis* LPS did not contain KDO and heptose (21). However, it was shown later that LPS of *P. gingivalis* as well as that of *B. fragilis* released atypical KDO, i.e., O-phosphorylated KDO, when the LPS was hydrolyzed in more severe conditions (2, 4, 10, 23). Heptose, on the other hand, has not been detected. It remains to be elucidated whether or not substituted heptose is present in the LPS of these bacterial species.

Fatty acids in *P. gingivalis* LPS have been examined by gas-liquid chromatography and mass spectrometry. *P. gingivalis* LPS preparations were found to contain branched 3-OH-C17:0, C16:0, branched 3-OH-C15:0, and 3-OH-C16:0 (10). Although some variation existed in analytical data, the fatty acid profile *P. gingivalis* LPS is basically similar to that of other related species, including *Prevotella* sp. and *B. fragilis* (4, 19, 39), It should be noted that genetic and some phenotypic heterogeneities exist among these bacterial species but that the presence of the fatty acids described above appears to be a common feature of the LPS from *Porphyromonas*, *Prevotella*, and *Bacteroides* species.

The polysaccharide content in the phenol-water extract of *P. gingivalis* varies among strains tested (10). Schifferle et al. (41) reported a polysaccharide antigen in which the immunologic specificity was different from that of LPS. This antigen was composed primarily of amino sugars, while the LPS was composed mainly of neutral sugars. The distribution of this type of polysaccharide antigen has not been elucidated.

Immunochemical Specificity of *P. gingivalis* LPS

Several investigators have used MAbs specific for *P. gingivalis* LPS to analyze the immunochemical properties of LPS of periodontal bacteria. The MAbs do not react with LPS from enterobacterial species or other black-pigmented anaerobic organisms except homologous species (30, 43). We have obtained a panel of MAbs against LPS from different strains of *P. gingivalis*. One MAb (i.e., 6/26-4) reacted with the LPS from all seven strains of *P. gingivalis* examined by ELISA and immunodiffusion, while another MAb (i.e., 381-3 or 6/26-2) reacted with only four of seven test strains, and the third MAb (i.e., 24D1) gave a positive reaction with the remaining three strains (Fig. 2). These results clearly indicate that *P. gingivalis* LPS possess species-specific epitopes and can simultaneously be classified into at least two serologic groups. *P. gingivalis* LPS serogroup I includes strains 381, ATCC 33277, BH18/10, and 6/26, while LPS serogroup II contains strains OMZ314, HW24D-1, and possibly OMZ409 (10).

LPS of Other Species of Periodontopathogens

LPS from other periodontal bacteria have been extensively studied. LPS from *A. actinomycetemcomitans*, *Campylobacter rectus* (formerly *Wolinella recta*), and *F. nucleatum* are similar to *E. coli* LPS, containing KDO, heptose, and β-hydroxymyristic acid (3-OH-C14:0) (11, 33, 37). LPS was isolated from a culture supernatant

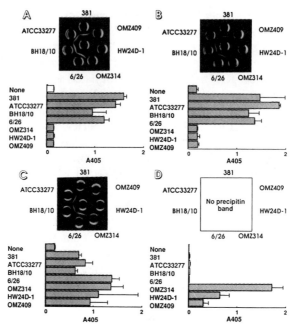

Figure 2. Reactivities of selected MAbs specific for *P. gingivalis* LPS with LPS preparations from seven strains of *P. gingivalis*. MAbs 381-3 (A), 6/26-2 (B), 6/26-4 (C), and 24D1 (D) were examined by immunodiffusion and ELISA for serologic specificities against LPS from *P. gingivalis* strains. No precipitin bands are seen between MAb 24D-1 and the LPS antigens examined (D). Modified from Fujiwara et al. (10).

of *A. actinomycetemcomitans*. The extracellular LPS was equivalent to that obtained by phenol-water extraction from whole cells, as revealed by SDS-PAGE (34). The polysaccharide portion of the LPS of this organism is immunodominant and is reactive with serum antibodies present in most patients with juvenile periodontitis (38). *Eikenella corrodens*, a conflicting periodontopathogen, has been reported to exhibit intraspecies heterogeneity among strains. There is no direct evidence indicating that certain LPS phenotypes are associated with periodontal disease (6).

BIOLOGICAL ACTIVITIES OF LPS

Endotoxic Properties

The general biologic aspects of bacterial LPS have been reviewed extensively, and the lipid A in LPS molecules has been recognized as an active center of biologic effects on the host (48). However, significant heterogeneity occurs in the lipid and polysaccharide portions of LPS from different bacterial species including periodontal pathogens, and inevitably, the endotoxicity of LPS exhibits variability among species.

With regard to *Limulus* lysate reaction, LPS preparations from different periodontal pathogens show strong activity comparable to that of *E. coli* LPS (9). In terms of Shwartzman activity considered to be specific for LPS or endotoxins, however, LPS from some periodontal pathogens exert very low endotoxicity (22). The LPS from *P. gingivalis* did not manifest local Shwartzman provoking activities in rabbits that had been prepared by *E. coli* LPS.

A unique, anaphylactoid reaction was induced by some selected LPS in mice primed with muramyl dipeptide. This resulted in convulsions, spreading, rolling, and finally death 5 to 20 min after the injection of LPS from *Porphyromonas, Prevotella,* or *Salmonella* species but not others. However, LPS from *P. gingivalis* and *P. intermedia* exhibited less toxicity in galactosamine-loaded mice than did *E. coli* LPS (46). Therefore, the lethal mechanism in the anaphylactoid reaction of the muramyl dipeptide-primed mice appears to be different from that involved in the usual endotoxic lethality.

Modification of Cellular Activity by LPS

It has been reported that *P. gingivalis* LPS stimulate lymphocytes and macrophages from C3H/HeJ as well as C3H/HeN mice. The former mouse strain has been known as a classic LPS nonresponder, while the latter is a classic LPS responder. LPS preparations from strains of *P. gingivalis* and other related black-pigmented species induced strong B-cell mitogenic responses and in vitro polyclonal B-cell activation in cultures of spleen cells of these two mouse strains (9, 10, 21). These LPS also enhanced production of IL-1, prostaglandin E, and tumor necrosis factor (TNF) by mouse macrophages (4, 9, 13). Furthermore, these LPS preparations activate mouse B lymphocytes polyclonally (51). More recently, Sveen and Skaug (45) reported that LPS from various oral or nonoral bacterial species induced significantly greater mitogenic responses and polyclonal B-cell activation in spleen cell cultures of BALB/c nude *(nu/nu)* mice than in those of normal BALB/c mice. In this study, *B. fragilis* LPS was a stronger mitogen than *P. intermedia* LPS.

LPS from oral anaerobic species such as *F. nucleatum, A. actinomycetemcomitans*, and *C. rectus* did induce mitogenic response and polyclonal B-cell activation in spleen cell cultures of C3H/HeN mice but not in those of C3H/HeJ mice (33, 37). Similarly, LPS of *A. actinomycetemcomitans, F. nucleatum,* and *C. rectus* did not stimulate C3H/HeJ macrophages to release IL-1 and/or prostaglandin E (11, 13). It should be added here that lipid A-associated protein prepared by butanol-water extraction of whole cells and extracellular proteinaceous endotoxic substance of *A. actinomycetemcomitans* were reported to be strong mitogens against lymphocytes of C3H/HeJ mice (33, 34).

CYTOKINES IN PERIODONTAL DISEASES

Implication of Cytokines in Periodontal Disease

Stimulation of monocytes and other cells by LPS from various sources resulted in the release of various cytokines. Some of these cytokines are widely known to be involved in inflammation and immune responses in the host (1). Recent evidence has clearly indicated that IL-1β is present in elevated amounts in gingival tissue from sites of active periodontitis lesions (16, 44). Jandinski et al. (18) have shown that larger numbers of IL-1β-containing cells occur in inflamed tissues than in clinically normal ones. Other investigators, however, claim that IL-1 activity found in crevicular fluid of patients with periodontitis appears to be due to IL-1α rather than IL-1β (20) or that both IL-1α and IL-1β were found in the diseased sites (26). Matsuki et al. (29) demonstrated expression of IL-1 mRNA in macrophages from inflamed gingival tissues of patients with periodontitis in which levels of expression of mRNA for IL-1α and IL-1β were similar. This type of phenomenon may be induced by LPS stimulation of macrophages. It is well known that endotoxic LPS activates cells of the monocytic system to synthesize cytokines (i.e., IL-1, IL-6, and TNF). However, repeated stimulation of monocytes by LPS causes diminished responses, i.e., desensitization, which results in a decrease in cytokine production and LPS-induced TNF mRNA. Stimulants other than LPS, i.e., *Staphylococcus aureus,* did not overcome desensitization (28, 54).

Inflammatory cytokines other than IL-1 have not been well studied. Levels of TNF-α in crevicular fluids were examined in relation to the activities of periodontitis, but no clear-cut relation was shown (40). Although IL-6 and IL-8 may participate in chronic inflammation in the gingival tissues, no evidence to support this hypothesis has been presented.

Bone Resorption Due to Host Cell Activation by Endotoxins

Both bacterial and host-derived factors have been implicated in the stimulation of bone loss from periodontitis. LPS is one of the most prominent bacterial components that induce bone resorption in in vitro organ culture systems. It was reported that *P. gingivalis* LPS was more active than LPS from other periodontal bacteria. *P. gingivalis* LPS requires the presence of osteoblasts to stimulate osteoclastic bone resorption. Fractions of *P. gingivalis* LPS inhibit collagen synthesis, as revealed by the increased release of ^{45}Ca from prelabeled fetal rat bones. A number of studies indicate that host-derived factors, such as IL-1, TNF-α, TNF-β, and prostaglandin E_2, can stimulate osteoclastic bone loss in vitro. *P. gingivalis* LPS has been shown to stimulate monocytes to release IL-1, which exerts strong osteoclast stimulation activity (3, 17), and to enhance synthesis of collagenase in articular chondrocyte cultures.

Interaction of Fibroblasts with LPS

LPS from *P. gingivalis*, *P. intermedia,* and *E. coli* cause a dose-dependent inhibition of cell growth of human gingival fibroblasts in vitro, although fairly large amounts of LPS are required to inhibit growth significantly (24). Serum lipoproteins known to interact with LPS did not affect the fibroblast growth inhibition (25). In a lower concentration of *P. intermedia* LPS, however, LPS promoted gingival fibroblast proliferation (14).

Fibroblasts cultured in vitro produce various cytokines in response to exogenous and endogenous stimulants. The latter may include cytokines released by other cells and fibroblasts (49). Among cytokines, IL-1, IL-6, and TNF-α are known to be major inflammatory cytokines (1). We have shown that LPS from various black-pigmented anaerobes such as *P. gingivalis* and *P. intermedia* induced cell-free IL-1β and IL-6 and cell-associated IL-1α in human gingival fibroblast cultures, while LPS from other species such as *E. coli* and *F. nucleatum* exhibited no significant cytokine-inducing activities (49).

Production of some cytokines by fibroblasts enhances synthesis of other cytokines exogenously provided. Fibroblasts pretreated with human beta interferon (IFN-β) or IFN-γ but not IFN-α synthesized enhanced levels of cell-associated IL-1α upon stimulation by *P. intermedia* LPS (Fig. 3). It was noted that no interferons exhibited direct IL-1-inducing activity (49). Other investigators have shown that human dermal fibroblast cultures secrete IL-8 like neutrophil chemotactic proteins when incubated with IL-1α or IL-1β but not with LPS (42). Tamura et al. (50), on the other hand, found that *P. intermedia* and *P. gingivalis* LPS stimulate human gingival fibroblasts to induce IL-8 mRNA. When fibroblasts were primed with IFN-γ or IFN-β, higher levels of IL-8 mRNA were expressed upon stimulation of *P. intermedia* LPS. It was also reported that IFN-γ-primed human monocytes released enhanced levels of prostaglandin through the stimulation of *P. intermedia* LPS but not significantly through *Salmonella typhimurium* LPS (32).

Thus, gingival fibroblasts as cytokine-producing cells as well as responders to a variety of cytokines may contribute to both inflammatory and immune responses in periodontal tissues.

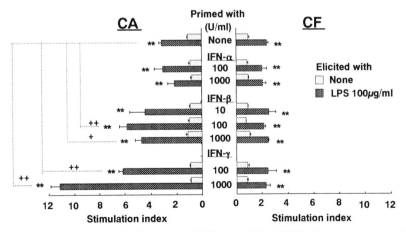

Figure 3. Enhanced production of cell-associated (CA) and cell-free (CF) IL-1α by human gingival fibroblast cultures primed with human IFN-β and IFN-γ but not IFN-α upon stimulation by *P. intermedia* LPS (100 μg/ml). IL-1α activity was quantitated by measuring the incorporation of [³H]thymidine in mouse thymocytes. Differences from the negative and positive (nonprimed and LPS-stimulated) controls were significant at $P < 0.01$ (** and + +) and $P < 0.05$ (+), respectively. Data are from Takada et al. (49).

SUMMARY

LPS, lipoproteins, peptidoglycans, and other surface components prepared from bacterial cells and their culture filtrates exhibit a wide variety of toxic and immunostimulating effects in the host. Since the majority of putative periodontopathic bacteria are gram-negative anaerobic organisms and the bacterial mass accumulated in the gingival crevice induces local inflammatory and immunopathologic responses, attention has been paid to LPS and related components present in the gram-negative anaerobes associated with periodontitis. Purified LPS preparations from *P. gingivalis* and other *Porphyromonas* and *Prevotella* species, i.e., the former black-pigmented *Bacteroides* group, exhibit low endotoxic activities such as local Shwartzman reaction, lethal toxicity, or pyrogenicity. However, they retain immunodulating activities comparable to that of the LPS of *Enterobacteriaceae* such as *E. coli*. A unique feature of *P. gingivalis* LPS is that it lacks the heptose that is widely found in classic LPS and has atypical KDO, i.e., phosphorylated KDO. Further, fatty acid compositions differ significantly between the LPS of *Porphyromonas* and *Prevotella* spp. and *E. coli*. These differences in chemical structure may be reflected in terms of stimulation of host cells. A clear difference was seen in stimulating B cells of "LPS high-responder" C3H/HeN and "LPS low-responder" C3H/HeJ mice. *Porphyromonas* and *Prevotella* LPS stimulate B cells of either C3H/HeN or C3H/HeJ mice, while *E. coli* LPS stimulate C3H/HeN B cells only. *P. gingivalis* LPS has also been reported to function as a local antigen and polyclonal B-cell activator in gingiva in the development of periodontitis; to stimulate fibroblasts to induce IL-1, IL-6, IL-8, and other cytokines; and to induce alveolar bone loss through induction of IL-1. These properties have been assumed to explain at least part of the virulence capabilities of the organism.

REFERENCES

1. **Akira, S., T. Hirano, T. Taga, and T. Kishimoto.** 1990. Biology of multifunctional cytokines: IL 6 and related molecules (IL 1 and TNF). *FASEB J.* **4:**2860–2867.
2. **Beckman, I., H. G. van Eijk, F. Meisel-Mikolajczyk, and H. C. S. Wallenburg.** 1989. Detection of 2-keto-3-deoxyoctonate in endotoxins isolated from six reference strains of the *Bacteroides fragilis* group. *Int. J. Biochem.* **21:**661–666.
3. **Bom-van Noorloos, A. A., J. W. M. van der Meer, J. S. van de Gevel, E. Schepens, T. J. M. van Steenbergen, and E. H. Burger.** 1990. *Bacteroides gingivalis* stimulates bone resorption via interleukin-1 production by mononuclear cells. *J. Clin. Periodontol.* **17:**409–413.
4. **Bramanti, T. E., G. G. Wong, S. T. Weintraub, and S. C. Holt.** 1989. Chemical characterization and biologic properties of lipopolysaccharide from *Bacteroides gingivalis* strains W50, W83, and ATCC 33277. *Oral Microbiol. Immunol.* **4:**183–192.
5. **Bucala, R.** 1992. Polyclonal activation of B lymphocytes by lipopolysaccharide requires macrophage-derived interleukin-1. *Immunology* **77:**477–482.
6. **Chen, C. K., and M. E. Wilson.** 1990. Outer membrane protein and lipopolysaccharide heterogeneity among *Eikenella corrodens* isolates. *J. Infect. Dis.* **162:**664–671.
7. **Daly, C. G., G. J. Seymour, and J. B. Kieser.** 1980. Bacterial endotoxin: a role in chronic inflammatory periodontal disease? *J. Oral Pathol.* **9:**1–15.
8. **Fine, D. H., C. Mendieta, M. L. Barnett, D. Furgang, A. Naini, and J. W. Vincent.** 1992. Endotoxin levels in periodontally healthy and diseased sites: correlation with levels of gram-negative bacteria. *J. Periodontol.* **63:**897–901.

9. **Fujiwara, T., T. Nishihara, T. Koga, and S. Hamada.** 1988. Serological properties and immunobiological activities of lipopolysaccharides from black-pigmented and related oral *Bacteroides* species. *J. Gen. Microbiol.* **134:**2867–2876.

10. **Fujiwara, T., T. Ogawa, S. Sobue, and S. Hamada.** 1990. Chemical, immunobiological and antigenic characterizations of lipopolysaccharides from *Bacteroides gingivalis* strains. *J. Gen. Microbiol.* **136:**319–326.

11. **Gillespie, J., S. T. Weintraub, G. G. Wong, and S. C. Holt.** 1988. Chemical and biological characterization of the lipopolysaccharide of the oral pathogen *Wolinella recta* ATCC 33238. *Infect. Immun.* **56:**2028–2035.

12. **Hamada, S., S. C. Holt, and J. R. McGhee (ed.).** 1991. *Periodontal Disease: Pathogens and Host Immune Responses.* Quintessence Publishing Co., Tokyo.

13. **Hamada, S., N. Okahashi, T. Fujiwara, T. Nishihara, and T. Koga.** 1988. Selective induction of prostaglandin E production in C3H/HeJ mouse macrophages by lipopolysaccharides from *Bacteroides gingivalis*. *Oral Microbiol. Immunol.* **3:**196–198.

14. **Hamada, S., H. Takada, J. Mihara, I. Nakagawa, and T. Fujiwara.** 1991. LPS of oral *Bacteroides* species: general properties and induction of cytokines in human gingival fibroblast cultures, p. 285–294. *In* A. Nowotny, J. J. Spitzer, and E. J. Ziegler (ed.), *Cellular and Molecular Aspects of Endotoxin Reactions.* Elsevier Science Publishers B. V., Amsterdam.

15. **Hamada, S., H. Takada, T. Ogawa, T. Fujiwara, and J. Mihara.** 1990. Lipopolysaccharides of oral anaerobes associated with chronic inflammation: chemical and immunomodulating properties. *Int. Rev. Immunol.* **6:**247–261.

16. **Hönig, J., C. Rordorf-Adam, C. Siegmund, W. Wiedemann, and F. Erard.** 1989. Increased interleukin-1 beta (IL-1β) concentration in gingival tissue from periodontitis patients. *J. Periodontal Res.* **24:**362–367.

17. **Hopps, R. M., and H. J. Sisney-Durrant.** 1991. Mechanisms of alveolar bone loss in periodontal disease, p. 307–320. *In* S. Hamada, S. C. Holt, and J. R. McGhee (ed.), *Periodontal Disease: Pathogens and Host Responses.* Quintessence Publishing Co., Tokyo.

18. **Jandinski, J. J., P. Stashenko, L. S. Feder, C. C. Leung, W. J. Peros, J. E. Rynar, and M. J. Deasy.** 1991. Localization of interleukin-1β in human periodontal tissue. *J. Periodontol.* **62:**36–43.

19. **Johne, B., I. Olsen, and K. Bryn.** 1988. Fatty acids and sugars in lipopolysaccharides from *Bacteroides intermedius*, *Bacteroides gingivalis* and *Bacteroides loescheii*. *Oral Microbiol. Immunol.* **3:**22–27.

20. **Kabashima, H., K. Maeda, Y. Iwamoto, T. Hirofuji, M. Yoneda, K. Yamashita, and M. Aono.** 1990. Partial characterization of an interleukin-1-like factor in human gingival crevicular fluid from patients with chronic inflammatory periodontal disease. *Infect. Immun.* **58:**2621–2627.

21. **Koga, T., T. Nishihara, T. Fujiwara, T. Nisizawa, N. Okahashi, T. Noguchi, and S. Hamada.** 1985. Biochemical and immunobiological properties of lipopolysaccharide (LPS) from *Bacteroides gingivalis* and comparison with LPS from *Escherichia coli*. *Infect. Immun.* **47:**638–647.

22. **Koga, T., C. Odaka, I. Moro, T. Fujiwara, T. Nishihara, N. Okahashi, and S. Hamada.** 1987. Local Shwartzman activity of lipopolysaccharides from several selected strains of suspected periodontopathic bacteria. *J. Periodontal Res.* **22:**103–107.

23. **Kumada, H., K. Watanabe, T. Umemoto, Y. Haishima, S. Kondo, and K. Hisatsune.** 1988. Occurrence of O-phosphorylated 2-keto-3-deoxyoctonate in the lipopolysaccharide of *Bacteroides gingivalis*. *FEMS Microbiol. Lett.* **51:**77–80.

24. **Larjava, H., V.-J. Uitto, E. Eerola, and M. Haapasalo.** 1987. Inhibition of gingival fibroblast growth by *Bacteroides gingivalis*. *Infect. Immun.* **55:**201–205.

25. **Layman, D. L., and L. A. Landreneau.** 1989. Suppression of fibroblast growth by *Bacteroides gingivalis* endotoxin is not reduced by serum lipoproteins. *J. Periodontol.* **60:**259–263.

26. **Masada, M. P., R. Persson, J. S. Kenney, S. W. Lee, R. C. Page, and A. C. Allison.** 1990. Measurement of interleukin-1α and -1β in gingival crevicular fluid: implications for the pathogenesis of periodontal disease. *J. Periodontal Res.* **25:**156–163.

27. **Maskell, J. P.** 1991. The resolution of bacteroides lipopolysaccharides by polyacrylamide gel electrophoresis. *J. Med. Microbiol.* **34:**253–257.

28. **Mathison, J. C., G. D. Virca, E. Wolfson, P. S. Tobias, K. Glaser, and R. J. Ulevitch.** 1990.

Adaptation to bacterial lipopolysaccharide controls lipopolysaccharide-induced tumor necrosis factor production in rabbit macrophages. *J. Clin. Invest.* **85:**1108–1118.

29. **Matsuki, Y., T. Yamamoto, and K. Hara.** 1991. Interleukin-1 mRNA-expressing macrophages in human chronically inflamed gingival tissues. *Am. J. Pathol.* **138:**1299–1305.

30. **Millar, S. J., P. B. Chen, and E. Hausmann.** 1989. Monoclonal antibody for identification of *Bacteroides gingivalis* lipopolysaccharide. *J. Clin. Microbiol.* **25:**2437–2439.

31. **Moore, J., M. Wilson, and J. B. Kieser.** 1986. The distribution of bacterial lipopolysaccharide (endotoxin) in relation to periodontally involved root surfaces. *J. Clin. Periodontol.* **13:**748–751.

32. **Nichols, F. C., J. F. Peluso, P. J. Tempro, S. W. Garrison, and J. B. Payne.** 1991. Prostaglandin E release from human monocytes with lipopolysaccharides isolated from *Bacteroides intermedius* and *Salmonella typhimurium:* potentiation by gamma interferon. *Infect. Immun.* **59:**398–406.

33. **Nishihara, T., T. Fujiwara, T. Koga, and S. Hamada.** 1986. Chemical composition and immunobiological properties of lipopolysaccharide and lipid-associated proteoglycan from *Actinobacillus actinomycetemcomitans. J. Periodontal Res.* **21:**521–530.

34. **Nishihara, T., T. Koga, and S. Hamada.** 1987. Extracellular proteinaceous substances from *Haemophilus actinomycetemcomitans* induce mitogenic responses in murine lymphocytes. *Oral Microbiol. Immunol.* **2:**48–52.

35. **Ogawa, T., M. L. McGhee, Z. Moldoveanu, S. Hamada, J. Mestecky, J. R. McGhee, and H. Kiyono.** 1989. *Bacteroides*-specific IgG and IgA subclass antibody-secreting cells isolated from chronically inflamed gingival tissues. *Clin. Exp. Immunol.* **76:**103–110.

36. **Ogawa, T., T. Mukai, K. Yasuda, H. Shimauchi, Y. Toda, and S. Hamada.** 1991. Distribution and immunochemical specificities of fimbriae of *Porphyromonas gingivalis* and related bacterial species. *Oral Microbiol. Immunol.* **6:**332–340.

37. **Okahashi, N., T. Koga, T. Nishihara, T. Fujiwara, and S. Hamada.** 1988. Immunobiological properties of lipopolysaccharides isolated from *Fusobacterium nucleatum* and *F. necrophorum. J. Gen. Microbiol.* **134:**1701–1715.

38. **Page, R. C., T. J. Sims, L. D. Engel, B. J. Moncla, B. Bainbridge, J. Stray, and R. P. Darveau.** 1991. The immunodominant outer membrane antigen of *Actinobacillus actinomycetemcomitans* is located in the serotype-specific high-molecular-mass carbohydrate moiety of lipopolysaccharide. *Infect. Immun.* **59:**3451–3462.

39. **Rietschel, E. T., L. Brade, O. Holst, V. A. Kulshin, B. Lindner, A. P. Moran, U. F. Schade, U. Zähringer, and H. Brade.** 1990. Molecular structure of bacterial endotoxin in relation to bioactivity, p. 15–32. *In* A. Nowotny, J. J. Spitzer, and E. J. Ziegler (ed.), *Cellular and Molecular Aspects of Endotoxin Reactions.* Elsevier Science Publishers B. V., Amsterdam.

40. **Rossomando, E. F., J. E. Kennedy, and J. Hadjimichael.** 1990. Tumour necrosis factor alpha in gingival crevicular fluid as a possible indicator of periodontal disease in humans. *Arch. Oral Biol.* **35:**431–434.

41. **Schifferle, R. E., M. S. Reddy, J. J. Zambon, R. J. Genco, and M. J. Levine.** 1989. Characterization of a polysaccharide antigen from *Bacteroides gingival. J. Immunol.* **143:**3035–3042.

42. **Schröder, J.-M., M. Sticherling, H. H. Henneicke, W. C. Preissner, and E. Christophers.** 1990. IL-1α or tumor necrosis factor-α stimulate release of three NAP-1/IL-8 related neutrophil chemotactic proteins in human dermal fibroblasts. *J. Immunol.* **144:**2223–2232.

43. **Shelburne, C. E., G. P. Sandberg, C. A. Binsfeld, L. F. Wolff, and R. A. Curry.** 1993. Monoclonal antibodies to lipopolysaccharide of four oral bacteria associated with periodontal disease. *J. Periodontal Res.* **28:**1–9.

44. **Stashenko, P., P. Fujiyoshi, M. S. Obernesser, L. Prostak, A. D. Haffajee, and S. S. Socransky.** 1991. Levels of interleukin 1β in tissue from sites of active periodontal disease. *J. Clin. Periodontol.* **18:**548–554.

45. **Sveen, K., and N. Skaug.** 1992. Comparative mitogenicity and polyclonal B cell activation capacity of eight oral or nonoral bacterial lipopolysaccharides in cultures of spleen cells from athymic (nu/nu-BALB/c) and thymic (BALB/c) mice. *Oral Microbiol. Immunol.* **7:**71–77.

46. **Takada, H., H. Hirai, T. Fujiwara, T. Koga, T. Ogawa, and S. Hamada.** 1990. *Bacteroides* lipopolysaccharides (LPS) induce anaphylactoid and lethal reactions in LPS-responsive and -nonresponsive mice primed with muramyl dipeptide. *J. Infect. Dis.* **162:**428–434.

47. **Takada, H., S. Kimura, and S. Hamada.** 1993. Induction of inflammatory cytokines by a soluble

moiety prepared from an enzyme lysate of *Actinomyces viscosus* cell walls. *J. Med. Microbiol.* **38:**395–400.

48. **Takada, H., and S. Kotani.** 1992. Structure-function relationships of lipid A, p. 107–134. *In* D. C. Morrison and J. L. Ryan (ed.), *Bacterial Endotoxic Lipopolysaccharides.* CRC Press, Inc., Boca Raton, Fla.

49. **Takada, H., J. Mihara, I. Morisaki, and S. Hamada.** 1991. Induction of interleukin-1 and -6 in human gingival fibroblast cultures stimulated with *Bacteroides* lipopolysaccharides. *Infect. Immun.* **59:**295–301.

50. **Tamura, M., M. Tokuda, S. Nagaoka, and H. Takada.** 1992. Lipopolysaccharides of *Bacteroides intermedius (Prevotella intermedia)* and *Bacteroides (Porphyromonas) gingivalis* induce interleukin-8 gene expression in human gingival fibroblast cultures. *Infect. Immun.* **60:**4932–4937.

51. **Tew, J., D. Engel, and D. Mangan.** 1989. Polyclonal B-cell activation in periodontitis. *J. Periodontal Res.* **24:**225–241.

52. **Wilson, M., J. Moore, and J. B. Kieser.** 1986. Identity of limulus amoebocyte lysate-active root surface materials from periodontally involved teeth. *J. Clin. Periodontol.* **13:**743–747.

53. **Wolff, L. F., L. Anderson, G. P. Sandberg, L. Reither, C. A. Binsfeld, G. Corinaldesi, and C. E. Shelburne.** 1992. Bacterial concentration fluorescence immunoassay (BCFIA) for the detection of periodontopathogens in plaque. *J. Periodontol.* **63:**1093–1101.

54. **Ziegler-Heitbrock, H. W. L., M. Blumenstein, E. Käfferlein, D. Kieper, I. Petersmann, S. Endres, W. A. Flegel, H. Northoff, G. Riethmüller, and J. G. Haas.** 1992. *In vitro* desensitization to lipopolysaccharide suppresses tumour necrosis factor, interleukin-1 and interleukin-6 gene expression in a similar fashion. *Immunology* **75:**264–268.

Molecular Pathogenesis of Periodontal Disease
Edited by Robert Genco et al.
© 1994 American Society for Microbiology, Washington, DC 20005

Chapter 11

CD14 and Lipopolysaccharide Binding Protein Control Host Responses to Bacterial Lipopolysaccharide

R. J. Ulevitch, J. C. Mathison, J.-D. Lee, J. Han, V. Kravchenko, J. Gegner, J. Pugin, S. Steinemann, D. Leturcq, A. Moriarty, T. Kirkland, and P. S. Tobias

The endotoxins (lipopolysaccharide [LPS]) of gram-negative bacteria stimulate phagocytic, endothelial, and epithelial cells to release cytokines and other inflammatory mediators (8, 9, 19). These responses contribute to host defense reactions that are important in elimination of the pathogen, but they can also initiate the cellular changes that result in the clinical syndrome known as septic shock (1). During the past decade, considerable progress has been made in defining the molecular properties and mechanism of action of the key mediators of endotoxin action. These mediators include products such as cytokines that are released from LPS-stimulated cells, lipid mediators derived from arachidonic acid or toxic oxygen radicals, and a variety of adhesive molecules expressed on cell surfaces.

Until recently, little was known about the cellular receptor for LPS that mediates cell activation. In this chapter, recent progress in understanding the molecular basis of cellular recognition of LPS will be reviewed (19). We will also briefly review what is known about the structural features of LPS that are essential for initiation of transmembrane signaling that leads to cell activation.

STRUCTURAL FEATURES OF LPS

Regardless of the type of gram-negative bacterium, the LPS consists of a carbohydrate-rich domain and a lipid-rich domain (12). The chemical structure of the

R. J. Ulevitch, J. C. Mathison, J.-D. Lee, J. Han, V. Kravchenko, J. Gegner, J. Pugin, S. Steinemann, and P. S. Tobias • Department of Immunology, The Scripps Research Institute, La Jolla, California 92037. *D. Leturcq and A. Moriarty* • The Robert Wood Johnson Pharmaceutical Research Institute, La Jolla, California 92121. *T. Kirkland* • Department of Pathology and Medicine, Veterans Administration Hospital Medical Center, The University of California, San Diego, California 92161.

carbohydrate domain varies considerably among different gram-negative organisms, while the lipid-rich domain, termed lipid A, exhibits a highly conserved structure (10). In general, lipid A consists of a diglucosamine backbone containing ester- and amide-linked long-chain fatty acids. These fatty acids are often hydroxylated and have additional fatty acids attached via the hydroxyl groups. Lipid A has been chemically synthesized, and synthetic lipid A induces biologic responses that are identical to those produced by naturally occurring LPS (10). While the sugars that are present in LPS modify the physicochemical properties of LPS, no evidence supports a role for these structures in LPS recognition at the level of the plasma membrane insofar as cell activation is involved. Thus, the membrane receptor for LPS is most likely a protein that contains a binding site for lipid A.

GENERAL ISSUES RELATING TO EXISTENCE OF MEMBRANE RECEPTORS FOR LPS

Two different types of mechanisms have been proposed to explain how LPS activates cells by lipid A-dependent mechanisms. (i) Activation is mediated by LPS binding to a specific membrane receptor for lipid A and receptor occupancy initiates transmembrane signaling. (ii) The "nonspecific" insertion of LPS into the lipid phase of the membrane results in a change in membrane structure that is responsible for production of a transmembrane signal.

The potency and specificity of cellular activation suggest that a specific membrane receptor is involved. LPS is active at picomolar to nanomolar concentrations, making its potency equal to or greater than those of protein agonists such as hormones or cytokines. Unfortunately, experiments using radiolabeled ligand to establish specific, saturable binding of lipid A have not yielded clear evidence for a specific lipid A receptor. The fact that the ligand used in these studies is always in a highly aggregated, high-molecular-weight form may contribute to the difficulties in establishing the ligand-binding characteristics of an LPS receptor that mediates signaling. Nevertheless, it seems clear that there is a specific LPS receptor that binds lipid A and initiates cell stimulation. In support of this are the observations that (i) partial structures of lipid A prepared by organic synthesis or enzymatic deacylation of natural lipid A act as LPS antagonists, suggesting that specific membrane structures, most likely proteins, recognize lipid A; (ii) application of chemical cross-linking strategies or radioactive lipid A binding to nitrocellulose-immobilized membrane proteins has led to the identification of proteins that are candidates for an LPS-lipid A receptor; (iii) macrophage (Mϕ) responses to LPS are regulated by adaptation or homologous desensitization, a mechanism that involves agonist-receptor interactions; and (iv) a specific protein of fully defined structure, CD14, present on the plasma membranes of phagocytic cells binds LPS, and engagement of CD14 leads to cell activation.

LBP AND ELUCIDATION OF A NOVEL, RECEPTOR-DEPENDENT PATHWAY OF CELL STIMULATION BY LPS

The identification of LPS binding protein (LBP) and characterization of its effects on LPS interactions with phagocytic cells revealed an unsuspected mechanism controlling LPS-induced cell activation. Experiments performed over the past several years by Ulevitch (19), Tobias and coworkers (14–18), and Wright et al. (20, 21) have demonstrated that LBP enhances cellular activation by LPS because complexes of LPS and LBP bind to a glycosylphosphatidylinositol (GPI)-anchored plasma membrane protein, CD14. Therefore, CD14 is an LPS receptor, and a large body of evidence now supports a crucial role for LBP and CD14 in cellular recognition of LPS.

The initial studies describing the discovery and characterization of LBP appeared in a series of publications by Tobias, Ulevitch, and coworkers (4–6, 9, 13, 14, 17–22). LBP is synthesized in hepatocytes as a single polypeptide, glycosylated, and released into blood as a 60-kDa glycoprotein (11). LBP concentrations are estimated to be <0.5 μg/ml in healthy rabbit serum and 5 to 10 μg/ml in healthy human serum. LBP is an acute-phase reactant, and LBP concentrations rise to 50 μg/ml 24 h after initiation of an acute-phase response; LBP concentrations in excess of 300 μg/ml have been detected in acute-phase human serum. A solid-phase competitive binding assay was used to show that LBP binds with high affinity to lipid A and binds similarly to LPS isolated from rough- or smooth-form gram-negative bacteria. Therefore, the polysaccharides of LPS do not significantly influence binding to lipid A. LBP does not bind to other charged polymers such as RNA, DNA, or heparin and binds weakly to high-molecular-weight dextran sulfate, lipoteichoic acid, or lipid X. Studies with Re595 LPS provided an estimate for a K_d in the nanomolar range and a binding stoichiometry of 1:1 (18).

Tobias et al. first reported that LBP and another LPS-lipid A binding protein, bactericidal permeability-increasing protein (BPI), share sequence homology (16). Protein sequence data base searches also revealed that LBP and BPI are related to the plasma protein cholesteryl ester transfer protein (3, 13). Comparative studies of LPS-induced cytokine production in the presence and absence of LBP with different Mφ sources expressing CD14 have been done (6, 7). Addition of LBP has several effects on cytokine production: the threshold stimulatory LPS dose is reduced as much as 1,000-fold, and the rate of cytokine release may be dramatically accelerated (15).

LBP also opsonizes particles containing LPS such as intact gram-negative bacteria or erythrocytes coated with LPS (21). A quantitative rosetting assay that takes advantage of the opsonic properties of LPB made possible the identification of CD14 as a receptor recognizing LPS-LBP complexes (20). CD14 is a 50-kDa membrane glycoprotein (mCD14) found on the surfaces of cells of monocytic origin (Mφ) and on polymorphonuclear leukocytes (22). mCD14 is a member of the family of proteins that are attached to the outer leaflet of the cell membrane via a GPI anchor (19, 22). A form of soluble CD14 lacking the GPI anchor is found in the plasma of humans and other animals (9, 22).

CD14 was first identified by reactivity with a group of monoclonal antibodies obtained from mice immunized with human monocytes. Cloning revealed the primary structure of CD14 and established that CD14 is a GPI-anchored protein (2). However, the function of CD14 was unknown until its identification as a receptor for LPS or LPS-LBP complexes (5, 19, 20). Numerous experimental studies provide evidence that CD14 plays a critical role in mediating the response of phagocytic cells to LPS under physiologic conditions (15, 19). However, there are many unanswered questions about how CD14 functions as a receptor for LPS.

We have developed a novel model system utilizing the LPS-responsive murine pre-B-cell line 70Z/3 to address questions related to how CD14 functions as an LPS receptor (5). LPS stimulates 70Z/3 cells to synthesize kappa light chains with consequent expression of membrane immunoglobulin M (IgM). 70Z/3 cells do not express CD14 and, in contrast to Mφ or polymorphonuclear leukocytes that are stimulated by picomolar or nanomolar LPS concentrations, require micromolar LPS doses to induce surface IgM expression (5). We speculated that the difference in sensitivity to LPS between 70Z/3 cells and Mφ might reside in the lack of expression of CD14 by the former cell line. To address this hypothesis, we established stably transfected lines of 70Z/3 cells that express human CD14. Transfection of 70Z/3 cells with DNA encoding human CD14 results in expression of GPI-anchored CD14 (70Z/3-hCD14 cells), and these cells displayed a phenotype different from that of the parental line with respect to LPS but not other agonists. For example, 70Z/3-hCD14 cells bind more LPS than do parental or 70Z/3 cells transfected with vector only (70Z/3-RSV). The enhanced LPS binding is blocked by pretreatment of cells with PI-PLC or anti-CD14 monoclonal antibodies. Most important, 70Z/3-hCD14 cells respond to 1,000-fold less LPS than do 70Z/3-RSV cells. Cellular responses measured by NF-κB activation (an early cell response to LPS) or surface IgM expression (a late cell response to LPS) reflect the increased sensitivity to LPS displayed by the 70Z/3-hCD14 cells.

These studies raised several important questions. (i) Is binding of LPS or LPS-LBP to CD14 sufficient to stimulate 70Z/3-CD14 cells? (ii) Because CD14-negative 70Z/3 cells respond to LPS, what is the relationship between CD14-dependent and -independent pathways? We have addressed these questions in two ways. First, we modified the DNA sequences encoding hCD14 so that hCD14 would no longer be expressed as a GPI-anchored protein but rather as a transmembrane protein. We achieved this by incorporating sequences encoding the transmembrane domains and cytoplasmic tails of two integral membrane proteins, tissue factor and murine class I. These constructs were used to transfect 70Z/3 cells, and after selection for stable transfectants, we characterized 70Z/3 cell lines expressing either GPI-anchored or integral membrane forms of CD14. We compared the responses of these transfected cell lines to that of LPS by measuring protein tyrosine phosphorylation, NF-κB activation, and surface expression of IgM. The responses of 70Z/3 cell lines expressing GPI-anchored or integral membrane CD14 are essentially identical, regardless of whether early activation events such as protein tyrosine phosphorylation and NF-κB activation or later activation events like surface IgM expression were measured. These findings show that conversion of CD14 to an integral membrane protein does not uncouple it from mediating LPS-induced cell activation. Because of these findings, we conclude that CD14 differs from other GPI-anchored receptors that require the presence of the GPI anchor to mediate cell activation. Moreover, these data have provided evidence to support a model in which CD14 functions as the ligand-binding subunit of a membrane receptor complex for LPS. The additional proteins that we suggest compose this functional receptor may also bind LPS, but binding is poor in the absence of CD14. This binding would represent the CD14-independent pathway observed with 70Z/3 cells but must play a key role in initiating transmembrane signaling. Obviously, an important task for future research will be to identify other membrane proteins that function together with CD14 to initiate a transmembrane signal after they are engaged by LPS.

SUMMARY

Progress during the past 5 years has provided evidence to support the contention that the interaction of LPS (lipid A) with cells is mediated by distinct membrane proteins. Some of these interactions are involved in removal and subsequent degradation of LPS, while others must play a crucial role in transmembrane signaling (19). Interactions that appear to be limited to LPS clearance involve a number of proteins, including the lipoprotein scavenger receptor or CD18, and LPS from rough-form bacteria, lipid A, or partial lipid A structures such as lipid IVa appear to be the preferred ligands. LPS from smooth-form organisms appears not to interact with these membrane proteins. Whether these interactions reflect events that occur in vivo remains to be unambiguously established. It is clear, however, that the scavenger receptor and CD18 do not have a role in mediating LPS-induced transmembrane signaling.

Photochemical cross-linking studies performed in several laboratories have revealed an LPS-binding membrane protein with an apparent molecular size of 70 to 80 kDa. This protein binds lipid A of LPS or the carbohydrate backbone of peptidoglycan. Studies with monoclonal antibodies to the 70- to 80-kDa protein show that the presence of antibody blocks LPS binding and indicate that engagement of this protein leads to transmembrane signaling. However, an unambiguous evaluation of the role of the 70- to 80-kDa protein in mediating LPS effects will require complete purification and/or gene cloning.

Perhaps the most important advance in our understanding of how LPS acts is derived from studies by Ulevitch, Tobias, and colleagues in which the LBP-CD14-dependent pathway of cell stimulation has been identified. Although this pathway has particular importance for LPS recognition and signaling by phagocytic cells that constitutively express CD14, recent studies have shown that complexes of LPS and soluble CD14 are potent agonists for some cell types that do not normally express membrane-bound CD14. The importance of the LBP-CD14-dependent pathway has been unambiguously demonstrated by experiments using immunologic, biochemical, and molecular biologic approaches. Available data are consistent with a model for a heteromeric membrane receptor complex for LPS that contains CD14 and an as-yet-unidentified additional component(s). Clearly, a major goal for future research will be to fully elucidate the additional proteins involved in the recognition of LPS.

REFERENCES

1. **Bone, R. C.** 1991. Gram-negative sepsis. Background, clinical features, and intervention. *Chest* **100**:802–808.
2. **Goyert, S. M., E. Ferrero, W. J. Rettig, A. K. Yenamandra, F. Obata, and M. M. LeBeau.** 1988. The CD14 monocyte differentiation antigen maps to a region encoding growth factors and receptors. *Science* **239**:497–500.
3. **Gray, P. W., G. Flaggs, S. R. Leong, R. J. Gumina, J. Weiss, C. E. Ooi, and P. Elsbach.** 1989. Cloning of the cDNA of a human neutrophil bactericidal protein. Structure and functional correlations. *J. Biol. Chem.* **264**:9505–9509.
4. **Kitchens, R. L., R. J. Ulevitch, and R. S. Munford.** 1992. Lipopolysaccharide (LPS) partial structures inhibit responses to LPS in a human macrophage cell line without inhibiting LPS uptake by a CD14-mediated pathway. *J. Exp. Med.* **176**:485–494.
5. **Lee, J. D., K. Kato, P. S. Tobias, T. N. Kirkland, and R. J. Ulevitch.** 1992. Transfection of CD14 into 70Z/3 cells dramatically enhances the sensitivity to complexes of lipopolysaccharide (LPS) and LPS binding protein. *J. Exp. Med.* **175**:1697–1705.
6. **Martin, T. R., J. C. Mathison, P. S. Tobias, R. J. Maunder, and R. J. Ulevitch.** 1992. Lipopolysaccharide binding protein enhances the responsiveness of alveolar macrophages to bacterial lipopolysaccharide: implications for cytokine production in normal and injured lungs. *J. Clin. Invest.* **90**:2209–2219.
7. **Mathison, J. C., P. S. Tobias, E. Wolfson, and R. J. Ulevitch.** 1992. Plasma lipopolysaccharide binding protein: a key component in macrophage recognition of gram-negative lipopolysaccharide (LPS). *J. Immunol.* **149**:200–206.
8. **Nathan, C. F.** 1987. Secretory products of macrophages. *J. Clin. Invest.* **79**:319–326.
9. **Pugin, J., C. C. Schurer-Maly, D. Leturcq, A. Moriarty, R. J. Ulevitch, and P. S. Tobias.** 1993. Lipopolysaccharide (LPS) activation of human endothelial and epithelial cells is mediated by LPS binding protein and soluble CD14. *Proc. Natl. Acad. Sci. USA* **90**:2744–2748.
10. **Raetz, C. R. H.** 1990. Biochemistry of endotoxins. *Annu. Rev. Biochem.* **59**:129–170.
11. **Ramadori, G., K.-H. Meyer zum Buschenfelde, P. S. Tobias, J. C. Mathison, and R. J. Ulevitch.** 1990. Biosynthesis of lipopolysaccharide binding protein in rabbit hepatocytes. *Pathobiology* **58**:89–94.
12. **Rietschel, E. T., and H. Brade.** 1992. Bacterial endotoxins. *Sci. Am.* **267**:54–61.
13. **Schumann, R. R., S. R. Leong, G. W. Flaggs, P. W. Gray, S. D. Wright, J. C. Mathison, P. S. Tobias, and R. J. Ulevitch.** 1990. Structure and function of lipopolysaccharide binding protein. *Science* **249**:1429–1433.
14. **Tobias, P., K. Soldau, and R. Ulevitch.** 1986. Isolation of a lipopolysaccharide-binding acute phase reactant from rabbit serum. *J. Exp. Med.* **164**:777–793.
15. **Tobias, P. S., J. Mathison, D. Mintz, J. D. Lee, V. Kravchenko, K. Kato, J. Pugin, and R. J. Ulevitch.** 1992. Participation of lipopolysaccharide binding protein in lipopolysaccharide dependent macrophage activation. *Am. J. Respir. Cell Mol. Biol.* **7**:239–245.
16. **Tobias, P. S., J. C. Mathison, and R. J. Ulevitch.** 1988. A family of lipopolysaccharide binding proteins involved in responses to gram-negative sepsis. *J. Biol. Chem.* **263**:13479–13481.
17. **Tobias, P. S., K. Soldau, L. Kline, J. D. Lee, K. Kato, T. P. Martin, and R. J. Ulevitch.** 1993. Crosslinking of lipopolysaccharide to CD14 on THP-1 cells mediated by lipopolysaccharide binding protein. *J. Immunol.* **150**:3011–3021.
18. **Tobias, P. S., K. Soldau, and R. J. Ulevitch.** 1989. Identification of a lipid A binding site in the acute phase reactant lipopolysaccharide binding protein. *J. Biol. Chem.* **264**:10867–10871.
19. **Ulevitch, R. J.** 1993. Recognition of bacterial endotoxins by receptor-dependent mechanisms. *Adv. Immunol.* **53**:267–289.

20. **Wright, S. D., R. A. Ramos, P. S. Tobias, R. J. Ulevitch, and J. C. Mathison.** 1990. CD14 serves as the cellular receptor for complexes of lipopolysaccharide with lipopolysaccharide binding protein. *Science* **249:**1431–1433.

21. **Wright, S. D., P. S. Tobias, R. J. Ulevitch, and R. A. Ramos.** 1989. Lipopolysaccharide (LPS) binding protein opsonizes LPS-bearing particles for recognition by a novel receptor on macrophages. *J. Exp. Med.* **170:**1231–1241.

22. **Ziegler-Heitbrock, H. W. L., and R. J. Ulevitch.** 1993. CD14: cell surface receptor and differentiation marker. *Immunol. Today* **14:**121–125.

Molecular Pathogenesis of Periodontal Disease
Edited by Robert Genco et al.
© 1994 American Society for Microbiology, Washington, DC 20005

Summary of Chapters 1 to 5

K. Okuda and R. J. Nisengard

Chapters 1 through 5 focused on salivary interactions and virulence factors of microorganisms, particularly *Actinobacillus actinomycetemcomitans* and *Porphyromonas gingivalis,* that may affect the ecology, colonization, and transmission of periodontal pathogens. Sophisticated methods including genetics and molecular biology further substantiated the role of these microorganisms in the pathogenesis of periodontitis.

In chapter 1, Zambon and coworkers presented studies on the transmission of *A. actinomycetemcomitans* and *P. gingivalis* in which an arbitrarily primed PCR (AP-PCR) with synthesized oligopeptides of random DNA sequences as primers was used. Patients are more likely to harbor a single clonal type than several types over extended periods of time, and where multiple clonal types occur, individual sites normally harbor a single clonal type. The authors further presented evidence that there is an intrafamilial transmission of these microorganisms and concluded that while infectivity may be low, transmission does occur in periodontal disease. The potential of AP-PCR as an epidemiological tool for studying transmission was also further reported on by Mouton and Ménard in chapter 4.

In chapter 2, Genco and colleagues reported studies on the fimbriae of *P. gingivalis,* which may enhance colonization by this organism. *P. gingivalis* contained a 37-kDA fimbrillin subunit and a lectin-like 12-kDa adhesin of fimbrillin. Each bacterial strain contained at least one copy of the *fimA* gene and had strain heterogeneity at this locus. With a mutant strain of *P. gingivalis* having an inactivated *fimA* gene, these authors demonstrated the importance of fimbriae in the colonization of the periodontal pocket by *P. gingivalis.* As a consequence, antifimbria antibodies induced by immunization may inhibit the colonization of *P. gingivalis.*

The proteases and collagenases of *P. gingivalis* were characterized in chapter 3 by Kuramitsu. He reported the completely sequenced protease genes *prtC,* with collagenase activity, and *tpr.* Evidence was presented that the ability of *P. gingivalis* strains to degrade type I collagen could play a direct role in tissue destruction and

K. Okuda • Department of Microbiology, Tokyo Dental College, 1-2-2 Masago, Chiba 261, Japan. *R. J. Nisengard* • Departments of Periodontology and Microbiology, Schools of Dental Medicine and Medicine, State University of New York at Buffalo, Buffalo, New York 14214.

aid in growth of the organism in periodontal pockets. In addition, the two protease genes also exhibited hemagglutinating activity, which is known to be a colonizing factor. Similarly, Genco et al. reported that *fimA*-negative mutants of *P. gingivalis* did not adhere to experimental pellicle but possessed hemagglutinating activity. The potential role of the hemagglutinating activity as a virulence factor for periodontopathogens requires further clarification.

In chapter 4, Mouton and Ménard reported on AP-PCR studies of *P. gingivalis* used as a means of discriminating pathogenic strains within the species. They demonstrated that DNA segments that amplify from one genomic DNA (random amplified polymorphic DNA) could be used as genetic markers for delineating subspecies clusters. They determined a genetic distance of 0.67 between human and animal biotypes of *P. gingivalis*. After identifying 29 patterns among *P. gingivalis* strains, they observed that common patterns occurred in spouses but not in unrelated individuals, suggesting transmission of strains.

Winston and Dyer reported in chapter 5 on the relationship between iron availability and *P. gingivalis* and *A. actinomycetemcomitans*. Many bacterial virulence determinants such as toxins and capsular polysaccharide production are regulated by iron availability. Microbial virulence determinants are known to be produced in significantly higher quantities when the microorganism is starved for iron. Their studies focused on the role of heme as an iron source for *P. gingivalis*. Their data suggest that porphyrins and transferrin, in addition to providing an iron source, may regulate the growth of *P. gingivalis*. They suggested that the iron-repressible outer membrane protein components of *P. gingivalis* and *A. actinomycetemcomitans* may be potential vaccine antigens.

Molecular Pathogenesis of Periodontal Disease
Edited by Robert Genco et al.
© 1994 American Society for Microbiology, Washington, DC 20005

Summary of Chapters 6 to 11

Stanley Holt and Robert E. Cohen

Chapters 6 to 11 discuss the mechanisms of bacterial invasion, functions of microbial toxins and receptors, and ability of a variety of bacterial components to act as host virulence factors. In chapter 6, Fives-Taylor et al. presented data that demonstrated the direct invasion of epithelial cells by some strains of *Actinobacillus actinomycetemcomitans. Haemophilus aphrophilus,* used as a negative control, was not invasive. Fives-Taylor et al. showed that epithelial-cell invasion in cell culture was associated with bacterial adherence, which was followed by receptor-mediated endocytosis. These events appeared to represent an active process that requires energy as well as induction of both bacterial and eucaryotic protein synthetic mechanisms. Further, invasion by *A. actinomycetemcomitans* appears to be associated with smooth colony morphology. In one strain that could be induced to switch from smooth to rough morphology, analysis by sodium dodecyl sulfate-polyacrylamide gel electrophoresis revealed differences in three peptides that also were associated with loss of invasiveness but increased the abilities of the peptides to adhere to epithelial cells. Preliminary findings from Fives-Taylor's laboratory have also suggested that a low-molecular-weight molecule induces the smooth-to-rough transition. Collectively, these data suggest that the requirements for *A. actinomycetemcomitans* invasion of eucaryotic cells are similar to the requirements of many but not all invasive bacteria.

The molecular biology of *A. actinomycetemcomitans* leukotoxin was reviewed in depth in chapter 7, by Lally and Kieba. *A. actinomycetemcomitans* leukotoxin appears to be produced by a contiguous four-gene operon similar in organization and DNA sequence to those for other toxins found in enteric and nonenteric gram-negative bacteria. An analysis of the deduced amino acid sequence of the *ltxA* gene product suggests that four separate domains may be associated with molecular function. These domains include hydrophobic, central, repeat, and C-terminal regions. Lally and Kieba utilized selective sequences of the *ltxA* gene to construct primers yielding precise, in-frame, 5' and 3' truncation mutants of the structural

Stanley Holt • Department of Periodontology, Health Science Center at San Antonio, University of Texas, San Antonio, Texas 78284.   *Robert E. Cohen* • Departments of Periodontology and Oral Biology, State University of New York at Buffalo, Buffalo, New York 14214-3092.

leukotoxin gene. Deletion of the carboxyl-terminal residues (residues 940 to 1055) of the leukotoxin molecule did not affect leukotoxin function, but further deletion of residues 939 to 871 did result in the loss of toxic activity. The results of further studies revealed that *A. actinomycetemcomitans* leukotoxic activity resides in the structural and/or conformational characteristics imparted by the presence of residues 80 to 140. These data provide additional information describing the interrelationships between the structure of the leukotoxin and its biological functions and may be of considerable utility in our efforts to understand the molecular aspects of the pathogenesis of periodontal diseases.

For many bacterial species, a polysaccharide outer capsule appears to defend against host defenses and confers an increased level of virulence. Although bacterial polysaccharide components are currently being used as vaccines, Schifferle reported in chapter 8 that problems may exist with such vaccines because of their T-lymphocyte independence. Such T-cell independence often results in a primary immune response characterized by a lack of any subsequent secondary response. Extensive characterization of *P. gingivalis* polysaccharide has resulted in new information detailing the structural features of this virulence factor. Further functional studies have suggested that the polysaccharide isolated from *P. gingivalis* ATCC 53977 cannot activate the alternative complement pathway, whereas lipopolysaccharide (LPS) is able to do so. In general, it appears that the polysaccharide component of *P. gingivalis* induces an immune response that is distinctly different from that obtained with LPS and that these differences may be important in periodontal diseases.

Campylobacter rectus is also considered an etiologic agent of periodontal disease. Accordingly, findings from Gillespie's laboratory (reported in chapter 9) have been directed toward the identification and characterization of a 104-kDa cytotoxin identified in culture media from *C. rectus* isolates. Gillespie and Zambon found that bacterial toxin release was independent of cell lysis, that the toxin could be inhibited by heat, but that it was not affected by polymyxin B or by antibodies specific for *C. rectus* LPS. The toxin also underwent N-terminal posttranslational modification, which is characteristic of secreted bacterial toxins. *C. rectus* toxin could be inhibited by an antibody specific to the 104-kDa protein and was reactive with antisera to *A. actinomycetemcomitans* leukotoxin when assessed by immunoblotting. Collectively, the data presented by Gillespie and Zambon indicated that *C. rectus* secretes a 104-kDa cytotoxin that may be a factor in the development and pathogenesis of periodontal diseases.

LPS, lipoproteins, cell wall peptidoglycans, and other bacterial cell components exhibit a variety of toxic and immunostimulating effects in the host. The role of such endotoxic substances, and their effects on host cells was discussed by Hamada et al. in chapter 10. Although purified LPS preparations from *P. gingivalis* and other *Porphyromonas* and *Prevotella* species generally exhibit low endotoxin activities, they retain immunomodulating activities comparable to that of the LPS of members of the family *Enterobacteriaceae*. However, differences in LPS fatty acid composition between *Porphyromonas* and *Prevotella* species and *Escherichia coli* may be reflected in corresponding differences in host cell stimulation. Such differences were detected after stimulation of B lymphocytes from LPS high-responder

(C3H/HeN) and LPS low-responder (C3H/HeJ) mouse strains. *Porphyromonas* and *Prevotella* LPS stimulates B lymphocytes either from C3H/HeN or C3H/HeJ mice, while *E. coli* LPS stimulates C3H/HeN B lymphocytes only. Evidence also suggests that *P. gingivalis* LPS may stimulate cellular signal transduction by activation of protein kinase C and may stimulate expression of cytokines in in vivo and in vitro systems. These properties may explain at least part of the virulence capabilities of these organisms.

Chapter 11, by Ulevitch et al., presents data describing endotoxin receptor and signaling pathways. CD14 is a glycosylphosphatidlinositol-anchored membrane protein found on leukocytes that is capable of binding complexes of LPS or LPS-binding protein. Binding of such complexes initiates cytokine production in leukocytes. Further studies from Ulevitch's laboratory have demonstrated that anti-CD14 antibody inhibits inflammatory cell activation, even after LPS-receptor interaction has occurred. The data presented indicated that CD14 localizes trace amounts of LPS and promotes subsequent interaction of LPS or the LPS-CD14 complex with additional membrane components involved in the signaling cascade. It also appeared that additional proteins that make up the receptor complex may include either a protein tyrosine kinase or a protein that interacts with LPS or LPS-CD14 complex and activates an intracellular tyrosine kinase. The proposed functional mechanism of CD14 may serve as a model for other members of the glycosylphosphatidlinositol-anchored family of proteins that serve as the principal ligand-binding units of membrane-bound receptor complexes.

SECTION II

HOST FACTORS: CYTOKINES AND OTHER EFFECTOR MOLECULES

Molecular Pathogenesis of Periodontal Disease
Edited by Robert Genco et al.
© 1994 American Society for Microbiology, Washington, DC 20005

Chapter 12

Cytokine Networks and Immunoglobulin Synthesis in Inflamed Gingival Tissues

Kohtaro Fujihashi, Jerry R. McGhee, Masafumi Yamamoto, Kenneth W. Beagley, and Hiroshi Kiyono

In the development of chronic inflammatory diseases, including periodontal disease (PD), rheumatoid arthritis, and inflammatory bowel disease, dysregulation of cytokine and production of immunoglobulin (Ig) at local disease sites have been considered major contributors to immunopathology. These diseases exhibit some common immunopathological features, i.e., localized inflammation associated with infiltration of T lymphocytes, macrophages (Mφ), neutrophils, and plasma cells (6, 22, 44). In localized chronic inflammation associated with gingival tissues (e.g., adult periodontitis [AP]), cellular and humoral immune responses are often upregulated (3, 23). For example, large numbers of plasma cells that produce IgG followed by IgA and essentially no IgM were seen in the inflamed gingivae of patients with advanced stages of AP (3, 18, 23, 27, 34, 35). These IgG antibodies possibly contribute to antibody-dependent cellular cytotoxicity, formation of immune complexes with tissue and/or microbial antigens, enhancement of opsonization, fixation of complement, and release of chemotactic factors. The inflammatory reactions triggered by the aberrant production of IgG at local disease sites likely contribute to the destruction of gingival tissues that occurs in patients with PD.

Among various cytokines, interleukin-2 (IL-2), IL-4, IL-5, and IL-6 are particularly important for regulation of B-cell responses, although these individual cytokines possess numerous other biological activities. In particular, IL-4 is involved in activation of resting B cells (36) by inducing quiescent B cells to enter the cell cycle and express major histocompatibility complex class II molecules (31, 37). Purified IL-5 isolated from patients with human T-cell leukemia can induce activated B cells to undergo cell division in humans (39). On the other hand, recombinant human IL-5, although active in the induction of eosinophil differentiation, fails to induce DNA replication in human B cells (5). In the murine system, IL-5 is a

Kohtaro Fujihashi, Masafumi Yamamoto, and Hiroshi Kiyono • Department of Oral Biology, University of Alabama at Birmingham, Birmingham, Alabama 35294. *Jerry R. McGhee and Kenneth W. Beagley* • Department of Microbiology, University of Alabama at Birmingham, Birmingham, Alabama 35294.

potent B-cell differentiation factor (26); however, its precise function in human Ig production remains to be examined.

IL-6 regulates the terminal differentiation of activated B cells to become high-rate Ig-secreting plasma cells (20). This cytokine induced high levels of IgM, IgG, and IgA synthesis in mitogen-stimulated human tonsillar B-cell cultures (30). Further, in situ-activated surface IgA1$^+$ and surface IgA2$^+$ B cells from human gut-associated lymphoreticular tissue, e.g., appendix, responded to recombinant human IL-6 and induced the differentiation of the surface IgA$^+$ cells into either IgA1- or IgA2-secreting plasma cells (13). Although IL-2 was originally described as a growth factor for T cells (43), this cytokine is capable of supporting both the proliferation and the differentiation of human B cells (17, 42). In this regard, IL-2 is an essential cytokine if *Staphylococcus aureus* Cowan I (SAC-1) mitogen-stimulated human B cells are to respond maximally to IL-6 (40). Taken together, these cytokines, including IL-2, IL-4, IL-5, and IL-6, form a cytokine network for B cells.

Dysregulation of this B-cell cytokine network at a local disease site could be considered a major contributor to the development of elevated B-cell responses in chronically inflamed gingival tissues. Thus far, however, no studies have directly addressed the exact profile of cytokine production by gingival mononuclear cells (GMC) isolated from diseased tissues.

In this chapter we summarize our characterization of IgG and IgA subclass antibody-secreting cells in inflamed gingival tissues and discuss the possible involvement of IgG and IgA subclass antibody production in the development of AP. Further, the exact profile of cytokine production by GMC is elucidated at both mRNA and protein levels in order to explain the contributions of various cytokines to the elevated B-cell responses found at local disease sites. The analysis of cytokine mRNA expression and production of IL-2, IL-4, IL-5, and IL-6 by GMC isolated from patients in the advanced stage of AP was performed by using cytokine-specific PCR, mRNA-cDNA dot blot hybridization, and biologic assays.

PROFILE OF IgG SUBCLASS ANTIBODY-PRODUCING CELLS IN GMC ISOLATED FROM PATIENTS WITH AP

By means of a novel enzymatic dissociation technique, functional mononuclear cells including viable plasma cells were isolated from chronically inflamed gingivae of patients with AP (27, 34). This allowed us to study the precise distribution of B-lineage cells including Ig-containing B-cell blasts and Ig-secreting plasma cells in single-cell preparations obtained from inflamed gingival tissues. When the numbers of spontaneous IgG subclass-specific antibody-producing cells were assessed in GMC isolated from AP patients, the IgG1 subclass generally predominated; however, significant numbers of IgG2-, IgG3-, and IgG4-producing cells were noted (33, 34) (Table 1). Further, IgG1 levels in gingival crevicular fluid from inflamed tissues were significantly increased over those in fluids from healthy sites (38). Interestingly, a similar IgG subclass distribution (IgG1>IgG2≥IgG3≥IgG4) was noted in mononuclear cells isolated from synovia of patients with rheumatoid arthritis (34). Thus, our findings that elevated numbers of IgG1-producing cells occur

Table 1. Immunological characteristics of inflamed gingival tissues in patients with AP

GMC are enriched for plasma cells (5–15%), lymphocytes (30–35%), and Mφ and monocytes (40–55%).

GMC contain a high frequency of Ig-producing cells. The ratios are as follows.

  (i) IgG > IgA >>> IgM

  (ii) IgG1 > IgG2 > IgG3 ≥ IgG4 (Occurrence of IgG3- and IgG4-producing cells is a unique characteristic.)

  (iii) IgA1 > IgA2 (increased numbers of IgA2 in the active stage of disease)

---

at the local gingival disease site provide strong support for an immunopathological role of IgG1 antibodies in the development of inflammatory disease, since this IgG subclass is most effective for induction of various immunological effector functions discussed above.

It was also interesting to note that elevated numbers of IgG4-producing cells occurred in GMC from inflamed tissues. Further, this subclass of antibody-secreting cells was absent in unstimulated peripheral blood mononuclear cells (PBMC) and spleen cells unless these populations of cells were stimulated with pokeweed mitogen (34). Also, low levels of IgG4 antibodies were present in healthy human serum. These IgG4 responses are most often associated with chronic exposure to protein antigens (1). This could be related to position in the *Igh* gene, where the $C_{\gamma 4}$ gene is the most 3′ among genes for the IgG subclasses and only the C_{ϵ} and $C_{\alpha 2}$ genes are more 3′ than this gene. Thus, constant antigenic exposure seems to induce IgG subclass switching to the IgG4 subclass in chronically inflamed gingival tissues, a process that is not associated with complement activation or opsonization. This tendency of constant antigenic exposure to induce a shift toward production of an anti-inflammatory type of IgG subclass suggests that elevated IgG4 antibodies may protect the host from tissue damage induced by IgG1 and IgG3 responses. However, this issue needs to be carefully examined in relation to PD.

IgA1 AND IgA2 SFC IN GINGIVAE OF PATIENTS WITH AP

IgA is considered more resistant to cleavage by proteolytic enzymes than the IgG isotype (4, 28). Even so, specific bacterial proteases produced by several pathogenic bacteria including PD-associated microorganisms selectively cleave IgA1 into Fab and Fc fragments; IgA2, however, is resistant to these proteases (4, 28). These differences in susceptibility of IgA subclasses to bacterial IgA1 proteases could profoundly influence the biological effector functions of IgA, especially during the course of inflammation caused by an infection. When the numbers of IgA subclass-specific spot-forming cells (SFC) in GMC were analyzed, IgA1-producing cells uniformly predominated over IgA2 cells (Table 1). However, the relative numbers of IgA2 plasma cells also increased in the severe stage of AP compared with numbers of GMC isolated during the moderate stage of disease (34). In this regard, the ratio of IgA2 to IgA1 was approximately 1:3 in the advanced stage and much higher (e.g., 1:7) in the moderate stage.

Changes in the distribution of IgA1 and IgA2 SFC responses during the progression of disease can be considered from both protective and destructive points of view. Certain strains of the subgingival bacterial microflora associated with PD are capable of secreting IgA1 proteases (9). The Fab fragments cleaved by IgA1 protease may obscure critical epitopes of the microorganisms, resulting in the inability of IgG or IgM antibodies to bind, and hence may protect the bacterium from subsequent complement-mediated killing and phagocytosis. Inasmuch as Ig Fc molecules possess an ability to polyclonally stimulate B cells in vitro, one may speculate that the local elaboration of Fc fragments of IgA induce polyclonal responses at disease sites and thus promote further inflammation (42). Thus, the occurrence of predominantly IgA1-producing cells at disease sites could be considered an immunopathological element.

Inasmuch as IgA1 antibody may become immunopathologically destructive to the host at the disease site, the immune system may attempt to overcome IgA1 antibodies by compensating with IgA2 responses. In this regard, unlike IgA1 antibodies, IgA2 antibodies are not susceptible to proteolytic enzymes secreted by bacterial species that occur in the oral region (4, 19). In addition, intact molecules of IgA2 bound to target organisms may mediate antibody-dependent cellular cytotoxicity via monocytes and lymphocytes, thus limiting bacterial multiplication without inducing activation of the complement pathway (19). Therefore, IgA2 responses at the local disease site could be considered anti-inflammatory reactions generated by the host immune system using this subclass of IgA.

DYSREGULATED CYTOKINE NETWORK FOR B-CELL RESPONSES IN INFLAMED GINGIVAE

As summarized above, inflamed tissues of patients in the advanced stage of AP can be considered hyper B-cell regions that contain large numbers of Ig-producing cells dominated by the IgG isotype followed by IgA (27, 34). Therefore, it was important to determine the exact profile of cytokine production by GMC in these inflamed tissues. To this end, production of several key cytokines (IL-4, IL-5, and IL-6) for B-cell responses by GMC was examined by using cytokine-specific enzyme-linked immunosorbent assay (ELISA) and biological assays. In these studies, GMC were incubated for 48 h in order to examine spontaneously produced cytokines. When 11 samples were assayed for IL-4 by using an IL-4-specific ELISA, none of these GMC culture supernatants contained detectable levels of IL-4 (11). In contrast, biologically active IL-5 was detected in GMC culture supernatants obtained from 16 different cases as determined by the BCL_1 assay (11). However, in two other cases, no IL-5 activity was found in GMC culture supernatants. Since the BCL_1 cells also respond to murine IL-4 and human granulocyte macrophage–colony-stimulating factor (GM-CSF) (32), the BCL_1 assay was performed in the presence of excess amounts of neutralizing goat anti-human GM-CSF or anti-human IL-4 antibody. In the presence of these antibodies, high levels of BCL_1 cell proliferation were still seen in cultures containing GMC culture supernatants. These findings showed that GMC isolated from inflamed gingivae produce the cytokine IL-5.

When IL-6 production was examined in GMC culture supernatants by using the IL-6-dependent KD 90 cell line, 10 of 10 samples contained this cytokine (11). This finding confirmed our previous observation of high levels of spontaneous IL-6 synthesis in GMC freshly isolated from surgical specimens (21). Taken together, our study has provided new evidence that GMC isolated from chronically inflamed tissues of AP patients spontaneously produce both IL-5 and IL-6 but not IL-4 (Table 2). Since IL-2 is also an important cytokine for B cells (17, 40, 42), the next studies were aimed to determine whether IL-2 was produced by GMC isolated from patients in the severe stages of AP. When GMC culture supernatants from 11 AP patients were examined for IL-2 activity by the HT-2 (or CTLL-2) assays either in the presence or the absence of goat anti-human IL-4 antibody, no detectable levels of IL-2 were found in 10 of 11 samples assayed (Table 2). In summary, an aberrant cytokine network for B cells in inflamed gingival tissues is represented by elevated IL-5 and IL-6 production with lack of IL-2 and IL-4 synthesis.

ANALYSIS OF IL-2-, IL-4-, IL-5-, AND IL-6-SPECIFIC mRNA LEVELS IN GMC

Because of the results obtained by cytokine analysis at the protein level, we next assessed IL-2-, IL-4-, IL-5-, and IL-6-specific mRNAs in RNA extracts prepared from freshly isolated GMC (11, 21). A dot blot hybridization analysis was initially performed by using the respective cytokine-specific cDNA probes. When mRNA was isolated from GMC and PBMC of the same patients and then hybridized with the IL-4-specific cDNA probe, no detectable levels of message (relative gray scale of 0 to 5) were seen in either GMC or PBMC (Table 3). Further, mRNA isolated from GMC and PBMC also did not have IL-2-specific message (relative gray scale of 0 to 2). GMC possessed high levels of mRNA for IL-5 (relative gray scale of 20 to 30), while PBMC from the same patient did not have detectable IL-5-specific mRNA. When the level of IL-6-specific message was also examined, the GMC were shown to contain mRNA for IL-6 (relative gray scale of 30 to 95), while PBMC from the same patients did not express detectable levels of IL-6 mRNA (11, 21). These results further showed that at the mRNA level, GMC possess high levels of message for IL-5 and IL-6 but not for IL-2 or IL-4. In addition, these findings suggest that AP is localized to diseased gingival tissues, since GMC from

Table 2. Biological analysis of IL-2, IL-4, IL-5, and IL-6 in culture supernatants of GMC isolated from patients with AP

Source of cells	Level of cytokine:			
	IL-2 (U/ml)	IL-4 (pg/ml)	IL-5 (SI)[a]	IL-6 (U/ml)
GMC	0–4	0–10	11–15	350–450
PBMC	2–8	17–23	2–3	5–10
PBMC + concanavalin A (2 μg/ml)	45–45	75–125	9–11	90–110

[a] SI, stimulation index (experiment/control).

Table 3. Examination of cytokine-specific mRNA expression in GMC isolated from inflamed periodontal tissues[a]

Cytokine	Level of hybridization with mRNA		
	GMC	PBMC	Appendix T cells (control)
IL-2	0–2	0–1	192
IL-4	0–5	0–4	15
IL-5	20–30	0–1	181
IL-6	30–95	0–1	27

[a] Cytokine-specific mRNA-cDNA dot blot hybridization with mRNAs from 2×10^5 cells was performed. Levels of hybridization were examined by scanning, and the actual numbers were expressed as a relative gray scale (0 to 200). The quality of mRNA was verified by use of a β-actin-specific cDNA probe.

inflamed sites express high levels of mRNA for selected cytokines, while PBMC from the same patients do not possess any message for IL-2, IL-4, IL-5, or IL-6 (11, 12, 21).

CYTOKINE-SPECIFIC RT-PCR ANALYSIS OF GMC ISOLATED FROM INFLAMED GINGIVAE

To formally prove that GMC isolated from inflamed tissues of the same patients produce IL-5 and IL-6 but not IL-2 or IL-4, cytokine-specific reverse transcriptase PCR (RT-PCR) was performed (11). In these studies, RNAs were obtained from GMC of six different patients in the severe stage of AP, reverse transcribed, and amplified by using the respective cytokine-specific 5' and 3' primers. After 35 cycles of amplification, distinct messages were observed for IL-6 in all cases. In the case of IL-5, three of six patients gave strong messages, while two others had weaker messages. When RT-PCR amplification was increased to 40 cycles, stronger messages were seen in these last two patients. However, one patient did not have any message for IL-5. In this same context, the message level for IL-6 was also lower in this patient than in other patients. The reason for this is not apparent. Nevertheless, one can conclude that B-cell regulatory cytokines such as IL-5 and IL-6 are actively produced by GMC at the disease site, since the pattern of cytokine mRNA expression analyzed by RT-PCR is in complete agreement with results obtained from IL-5- and IL-6-specific biological assays as well as cytokine-specific mRNA-cDNA dot blot hybridization. On the other hand, specific messages for IL-2 and IL-4 were not detected by RT-PCR using IL-2- and IL-4-specific 5' and 3' primers (Fig. 1). As positive controls, cDNA was reverse transcribed from RNA extracted from phytohemagglutin-stimulated tonsillar T cells, and these preparations showed specific messages for IL-2, IL-4, IL-5, and IL-6. These findings obtained by cytokine-specific RT-PCR further show that GMC isolated from patients with advanced stages of AP produce both IL-5 and IL-6 but not IL-2 or IL-4 (11) (Fig. 1 and 2).

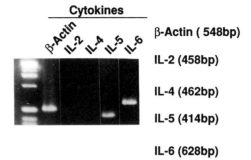

Figure 1. RT-PCR analysis of cytokine synthesis by GMC isolated from inflamed tissues. An RNA sample from the same patient was examined by RT-PCR using cytokine-specific primers. As internal control, β-actin-specific 5′ and 3′ primers were employed.

DISCUSSION

The GMC isolated from inflamed tissues spontaneously produce biologically active IL-5 and IL-6, which are considered key cytokines for the growth and proliferation of B cells and for their terminal differentiation to become Ig-producing plasma cells. Although the exact role of human IL-5 for B-cell responses still remains to be elucidated, it is clear that IL-5 possesses T-cell replacing factor and B-cell growth factor-II activities in murine B-cells (14, 41). Further, IL-5 can induce murine IgA-committed B cells to become IgA secreting cells (26). A recent study has provided evidence that human B cells isolated from PBMC and spleen express mRNA for the α and β chains of IL-5R (16). Further, recombinant human IL-5 enhanced Ig synthesis in *Moraxella catarrhalis*-stimulated human B-cell cultures. On the other hand, a standard human B-cell mitogen, SAC-I, did not provide an appropriate signal for human B cells to respond to IL-5. One possibility could be that the nature of the activation signal is important for the expression of IL-5R on human B cells. Since B cells that accumulate in inflamed gingival tissues are con-

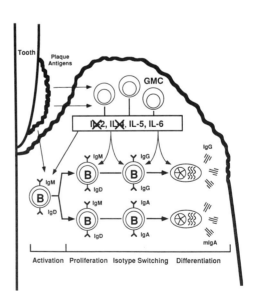

Figure 2. Cytokine profile of and Ig synthesis in inflamed periodontal tissues.

tinuously exposed to an array of oral antigens, it is possible that some PD-associated microorganisms provide an appropriate signal for human B cells to express IL-5R and to respond to IL-5 in disease sites.

In the case of IL-6, this cytokine has been shown to be an important terminal differentiation factor for postswitched, activated B cells to become high-rate Ig-secreting plasma cells of all isotypes in both human and murine systems (2, 13, 15, 30). Our previous work, as well as recent studies by others, shows that GMC isolated from inflamed tissues actively produce high levels of IL-6 (11, 12, 21). Further, IL-6 has significant biological activity, since the addition of culture supernatants containing IL-6 from the GMC of AP patients to PBMC cultures from the same subjects resulted in the induction of IgG and IgA subclass antibody-producing cells (21). The addition of anti-IL-6 antibody to these cultures inhibited the formation of antibody-producing cells. Finally, it was also shown that B cells isolated from inflamed gingival tissues expressed IL-6R (11, 12). Therefore, synthesis of IL-6 by GMC at the local disease site certainly contributes to the accumulation of plasma cells.

One surprising finding of these studies was that culture supernatants of GMC obtained from inflamed gingivae did not contain detectable levels of either IL-2 or IL-4. Our initial explanation for this finding was that high levels of IL-2 and IL-4 were also produced by GMC, but these cytokines may have been absorbed by activated, neighboring IL-2R$^+$ and IL-4R$^+$ cells (11). However, this was most likely not the case, since detectable levels of IL-2- and IL-4-specific message were not identified in GMC when both mRNA-cDNA dot blot hybridization and RT-PCR were used. This suggests that IL-2 and IL-4 are not produced in inflamed tissues. In support of this, decreased IL-2 production during chronic inflammation was also noted in patients with inflammatory bowel disease or rheumatoid arthritis (7). The other possibility could be that GMC isolated from inflamed tissues produce inhibitors for IL-2 and IL-4. In this regard, an inhibitor for IL-2 has been identified in synovial fluids of patients with rheumatoid arthritis (29). Since AP and these other two diseases possess immunologically similar features, the lack of IL-2 synthesis could be considered a common characteristic of chronic inflammation. The lack of IL-2 could also be considered a feedback pathway provided by the host to down-regulate chronically activated T-cell responses at the local disease site.

The absence of IL-4 synthesis by GMC could be an essential contributor to the persistence of increased numbers of Mφ and monocytes at local disease sites. To this end, high numbers of Mφ and monocytes were found in GMC isolated from inflamed tissues (27, 34). IL-4 has been shown to augment the process of DNA fragmentation (or apoptosis) in activated human monocytes but not in resting monocytes (24). Thus, addition of IL-4 to IL-1-, tumor necrosis factor alpha-, or lipopolysaccharide-stimulated monocyte cultures resulted in the reduction of apoptosis. Unavailability of IL-4 at the local disease site due to the lack of IL-4 production by GMC may provide an explanation for the persistence of Mφ and monocytes in the inflamed gingiva. Further, our most recent study has shown that high levels of production of gamma interferon, one of the potent inflammatory cytokines that possesses the ability to inhibit apoptosis (25), were produced by GMC (10). A unique profile of cytokine production at the local disease site (e.g., lack of IL-4 in the presence of gamma interferon) could provide the most suitable environment for Mø and monocytes to accumulate in the inflamed gingival tissues.

One important issue that needs to be addressed in our future work is the exact subset of mononuclear cells in GMC that produces IL-5 and IL-6. Since GMC enzymatically isolated from inflamed tissue contain large numbers of monocytes and Mø ($\sim 50\%$) in addition to lymphocytes (35 to 40%) and plasma cells (5 to 17%) (27), one likely source for excess amounts of IL-6 at local disease sites would be monocytes and Mø in GMC. However, it is possible that lymphocytes, including T and B cells, in GMC also contribute to the production of IL-6. In order to determine the exact subset of IL-5- and IL-6-producing cells in local disease sites, our current effort is focused on cytokine-specific mRNA levels in purified T cells, B cells, monocytes, and Mø by using respective cytokine-specific RT-PCR.

SUMMARY

Localized and chronically inflamed gingival tissues of patients with AP are characterized as hyper-B-cell regions in which increased numbers of plasma cells occur (Fig. 2). When GMC isolated from AP patients were examined for Ig production at the single-cell level, large numbers of IgG subclass antibody-secreting cells (e.g., IgG1 > IgG2 ≥ IgG3 ≥ IgG4) with significant numbers of IgA-producing cells, (e.g., IgA1 and IgA2) were noted. Since increased numbers of Ig-producing cells occurred in inflamed gingival tissues, it was important to characterize the cytokine profile of GMC, since these proteins are known modulators of B-cell responses. Thus, when GMC isolated from inflamed gingival tissues were examined at both the protein and the mRNA levels for IL-2, IL-4, IL-5, and IL-6 production, it was interesting that no IL-2 and IL-4 was detected, while high (and significant) levels of IL-5 and IL-6 were detected (Fig. 2). These results suggest that dysregulation of cytokine synthesis by GMC may contribute to the induction of hyper B-cell responses at local AP sites. Further, increased production of IL-5 and IL-6 by GMC could be an essential factor in the occurrence of large numbers of Ig-producing cells.

Acknowledgments. This work is supported by U.S. Public Health Service grants DE 08228, DE 09837, DE 04217, and AI 30366 and contract AI 15128. Hiroshi Kiyono is a recipient of NIH RCDA DE 00237.

We thank James E. Roberts, Alvin W. Stevens, and Donald Thompson for providing gingival tissues from their periodontal surgery patients. We thank Cindi Eastwood for preparation of the manuscript.

REFERENCES

1. **Aalberse, R. C., R. van der Gaag, and J. van Leeuwen.** 1983. Serologic aspects of IgG4 antibodies. I. Prolonged immunization results in an IgG4-restricted response. *J. Immunol.* **130:**722–726.
2. **Beagley, K. W., J. H. Eldridge, W. K. Aicher, J. Mestecky, S. DiFabio, H. Kiyono, and J. R. McGhee.** 1991. Peyer's patch B-cells with memory cell characteristics undergo terminal differentiation within 24 hours in response to interleukin-6. *Cytokine* **3:**107–116.
3. **Brandtzaeg, P., and F. W. Kraus.** 1965. Autoimmunity and periodontal disease. *Odontol Tidsskr.* **73:**281–393.

4. **Childers, N. K., M. G. Bruce, and J. R. McGhee.** 1989. Molecular mechanisms of immunoglobulin A defense. *Annu. Rev. Microbiol.* **43:**503–536.

5. **Clutterbuck, E., J. G. Shields, J. Gordon, S. H. Smith, A. Boyd, R. E. Callard, H. D. Campbell, I. G. Young, and C. J. Sanderson.** 1987. Recombinant human interleukin 5 is an eosinophil differentiation factor but has no activity in standard human B-cell growth factor assays. *Eur. J. Immunol.* **17:**1728–1750.

6. **Elson, C. O.** 1987. The immunology of inflammatory bowel disease, p. 97–164. *In* J. B. Kirsner and R. G. Shorter (ed.), *Inflammatory Bowel Disease,* 3rd ed. Lea and Febiger Publishers, Philadelphia.

7. **Fiocchi, C., M. L. Hilfiker, K. R. Youngman, N. C. Doerder, and J. H. Finke.** 1984. Interleukin-2 activity of human intestinal mucosal mononuclear cells. Decreased levels in inflammatory bowel disease. *Gastroenterology* **86:**734–742.

8. **Firestein, G. S., W. D. Xu, K. Townsend, D. Broide, J. Alvaro-Gracia, A. Glasebrook, and N. J. Zvaifler.** 1988. Cytokines in chronic inflammatory arthritis. I. Failure to detect T-cell lymphokines (interleukin 2 and interleukin 3) and presence of macrophage colony-stimulating factor (CSF-1) and a novel mast-cell growth factor in rheumatoid synovitis. *J. Exp. Med.* **168:**1573–1586.

9. **Frandseu, E. V., E. Theilade, B. Ellegaard, and M. Kilian.** 1986. Proportions and identity of IgA1-degrading bacteria in periodontal pockets from patients with juvenile and rapidly progressive periodontitis. *J. Periodontal Res.* **21:**613–623.

10. **Fujihashi, K., M. Yamamoto, J. R. McGhee, and H. Kiyono.** 1994. Type 1/type 2 cytokine production by CD4+ T cells in adult periodontitis (abstr. 818). *J. Dent. Res.* **73:**204.

11. **Fujihashi, K., K. W. Beagley, Y. Kono, W. A. Aicher, M. Yamamoto, S. DiFabio, J. Xu-Amano, J. R. McGhee, and H. Kiyono.** 1993. Gingival mononuclear cells from chronic inflammatory periodontal tissues produce interleukin (IL)-5 and IL-6 but not IL-2 and IL-4. *Am. J. Pathol.* **142:**1239–1250.

12. **Fujihashi, K., Y. Kono, K. W. Beagley, M. Yamamoto, J. R. McGhee, J. Mestecky, and H. Kiyono.** 1993. Cytokines and periodontal disease: immunopathological role of interleukins for B-cells responses in chronic inflamed gingival tissues. *J. Periodont.* **64:**400–406.

13. **Fujihashi, K., J. R. McGhee, C. Lue, K. W. Beagley, T. Taga, T. Hirano, T. Kishimoto, J. Mestecky, and H. Kiyono.** 1991. Human appendix B-cells naturally express receptors for and respond to interleukin 6 with selective IgA1 and IgA2 synthesis. *J. Clin. Invest.* **88:**248–252.

14. **Hara, Y., Y. Takahama, G. Murakami, S. Yamada, S. Ono, K. Takatsu, and T. Hamaoka.** 1985. B-cell growth and differentiation activity of a purified T-cell-replacing factor (TRF) molecule from B151-T-cell hybridoma. *Lymphokine Res.* **4:**243–249.

15. **Hirano, T., K. Yasukawa, H. Harada, T. Taga, Y. Watanabe, T. Matsuda, S. Kashiwamura, K. Nakajima, K. Koyama, A. Iwamatsu, S. Tsunasawa, F. Sakiyama, H. Matsui, Y. Takahara, T. Taniguchi, and T. Kishimoto.** 1986. Complementary DNA for a novel human interleukin (BSF-2) that induces B lymphocytes to produce immunoglobulin. *Nature* (London) **324:**73–76.

16. **Huston, D. P., J. P. Moore, and M. M. Huston.** 1993. Human B-cells express IL-5 receptor mRNA and respond to IL-5 with enhanced IgM production under selective mitogenic stimulation, abstr. 1034. *J. Immunol.* **150:**182A.

17. **Jelinek, D. F., J. B. Splawski, and P. E. Lipsky.** 1986 The roles of interleukin 2 and interferon-γ in human B-cell activation, growth and differentiation. *Eur. J. Immunol.* **16:**925–932.

18. **Kilian, M., B. Ellegaard, and J. Mestecky.** 1989. Distribution of immunoglobulin isotypes including IgA subclasses in adult, juvenile, and rapidly progressive periodontitis. *J. Clin. Periodontol.* **16:**179–184.

19. **Kilian, M., J. Mestecky, and M. W. Russell.** 1986. Defense mechanisms involving Fc-dependent functions of immunoglobulin A and their subversion by bacterial immunoglobulin A proteases. *Microbiol. Rev.* **52:**296–303.

20. **Kishimoto, T., and T. Hirano.** 1988. Molecular regulation of B lymphocyte response. *Annu. Rev. Immunol.* **6:**485–512.

21. **Kono, Y., K. W. Beagley, K. Fujihashi, J. R. McGhee, T. Taga, T. Hirano, T. Kishimoto, and H. Kiyono.** 1991. Cytokine regulation of localized inflammation. Induction of activated B-cells and IL-6 mediated polyclonal IgG and IgA synthesis in inflamed human gingiva. *J. Immunol.* **146:**1812–1821.

22. **Listgarten, M. A.** 1986. Pathogenesis of periodontitis. *J. Clin. Periodontol.* **13:**418–430.

23. **Mackler, B. F., K. B. Frostad, P. B. Robertson, and B. M. Levy.** 1977. Immunoglobulin bearing lymphocytes and plasma cells in human periodontal disease. *J. Periodontal Res.* **12:**37–45.

24. **Mangan, D. F., B. Robertson, and S. M. Wahl.** 1992. IL-4 enhances programmed cell death (apoptosis) in stimulated monocytes. *J. Immunol.* **148:**1812–1816.

25. **Mangan, D. F., and S. M. Wahl.** 1991. Differential regulation of human monocyte programmed cell death (apoptosis) by chemotactic factors and pro-inflammatory cytokines. *J. Immunol.* **147:**3408–3412.

26. **McGhee, J. R., J. Mestecky, C. O. Elson, and H. Kiyono.** 1989. Regulation of IgA synthesis and immune response by T-cells and interleukins. *J. Clin. Immunol.* **9:**175–199.

27. **McGhee, M. L., T. Ogawa, A. M. Pitts, Z. Moldoveanu, J. Mestecky, J. R. McGhee, and H. Kiyono.** 1989. Cellular analysis of functional mononuclear cells from chronically inflamed gingival tissue. *Reg. Immunol.* **2:**103–110.

28. **Mestecky, J., and M. W. Russell.** 1986. IgA subclasses. *Monogr. Allergy* **19:**277–301.

29. **Miossec, P., T. Kashiwado, and M. Ziff.** 1987. Inhibitor of interleukin-2 in rheumatoid synovial fluid. *Arthritis Rheum.* **30:**121–129.

30. **Muraguchi, A., T. Hirano, B. Tang, T. Matsuda, Y. Horii, K. Nakajima, and T. Kishimoto.** 1988. The essential role of B-cell stimulatory factor 2 (BSF-2/IL-6) for the terminal differentiation of B-cells. *J. Exp. Med.* **167:**332–344.

31. **Noelle, R., P. H. Krammer, J. Ohara, J. W. Uhr, and E. S. Vitetta.** 1984. Increased expression of Ia antigens on resting B-cells: an additional role for B-cell growth factor. *Proc. Natl. Acad. Sci. USA* **81:**6149–6153.

32. **O'Garra, A., D. Barbis, J. Wu, P. D. Hodgkin, J. Abrams, and M. Howard.** 1989. The BCL$_1$ B lymphoma responds to IL-4, IL-5 and GM-CSF. *Cell Immunol.* **123:**189–200.

33. **Ogawa, T., M. L. McGhee, Z. Moldoveanu, S. Hamada, J. Mestecky, J. R. McGhee, and H. Kiyono.** 1989. *Bacteroides*-specific IgG and IgA subclass antibody-secreting cells isolated from chronically inflamed gingival tissues. *Clin. Exp. Immunol.* **76:**103–110.

34. **Ogawa T., A. Tarkowski, M. L. McGhee, Z. Moldoveanu, J. Mestecky, H. Z. Hirsch, W. J. Koopman, S. Hamada, J. R. McGhee, and H. Kiyono.** 1989. Analysis of human IgG and IgA subclass antibody-secreting cells from localized chronic inflammatory tissue. *J. Immunol.* **142:**1150–1158.

35. **Okada, H., T. Kida, and H. Yamagami.** 1983. Identification and distribution of immunocompetent T-cells in inflamed gingiva of human chronic periodontitis. *Infect. Immun.* **41:**365–374.

36. **Paul, W. E.** 1987. Interleukin 4/B-cell stimulatory factor 1: one lymphokine, many functions. *FASEB J.* **1:**456–461.

37. **Rabin, E. M., J. Ohara, and W. E. Paul.** 1985. B-cell stimulatory factor 1 (BSF-1) activates resting B-cells. *Proc. Natl. Acad. Sci. USA* **82:**2935–2939.

38. **Reinhardt, R. A., T. L. McDonald, R. W. Bolton, L. M. DuBois, and W. B. Kaldahl.** 1989. IgG subclasses in gingival crevicular fluid from active versus stable periodontal sites. *J. Periodontol.* **60:**44–50.

39. **Shimizu, K., T. Hirano, K. Ishibashi, N. Nakano, T. Taga, K. Sugamura, Y. Yamamura, and T. Kishimoto.** 1985. Immortalization of BGDF (BCGF II)- and BCDF-producing T-cells by human T-cell leukemia virus (HTLV) and characterization of human BGDF (BCGF II). *J. Immunol.* **134:**1729–1733.

40. **Splawski, J. B., L. M. McAnally, and P. E. Lipsky.** 1990. IL-2 dependence of the promotion of human B-cell differentiation by IL-6 (BSF-2). *J. Immunol.* **144:**562–569.

41. **Swain, S. L., M. Howard, J. Kappler, P. Marrack, J. Watson, R. Booth, G. D. Wetzel, and R. W. Dutton.** 1983. Evidence for two distinct classes of murine B-cell growth factors with activities in different functional assays. *J. Exp. Med.* **158:**822–835.

42. **Teranishi, T., T. Hirano, B.-H. Lin, and K. Onoue.** 1984. Demonstration of the involvement of interleukin 2 in the differentiation of *Staphylococcus aureus* Cowan I-stimulated B-cells. *J. Immunol.* **133:**3062–3067.

43. **Watson, J., and D. Mochizuki.** 1980. Interleukin 2. A class of T-cell growth factors. *Immunol. Rev.* **51:**257–278.

44. **Ziff, M.** 1988. Emigration of lymphocytes in rheumatoid synovitis. *Adv. Inflammatory Res.* **12:**1–9.

Molecular Pathogenesis of Periodontal Disease
Edited by Robert Genco et al.
© 1994 American Society for Microbiology, Washington, DC 20005

Chapter 13

Modulatory Role of T Lymphocytes in Periodontal Inflammation

M. A. Taubman, J. W. Eastcott, H. Shimauchi, O. Takeichi, and D. J. Smith

LYMPHOCYTES IN PERIODONTAL DISEASES

T lymphocytes are prominent in all periodontal lesions (12, 22). It has been inferred and indirectly demonstrated that there are antigen-specific T lymphocytes in the gingival tissues. Investigation of peripheral blood T lymphocytes by limiting dilution analyses has also demonstrated the presence of lymphocytes specific for periodontopathic bacteria in peripheral blood (14). Interestingly, treatment (removal of plaque) appeared to reduce the frequency of detection of such cells from approximately 1/12,000 to approximately 1/28,000 T cells (15). T-cell lines and clones specific for periodontal bacteria have been isolated from the peripheral blood of periodontally diseased and healthy subjects (6, 23). The response of the T-cell clones was major histocompatibility complex class II restricted, and the clones provided help for B cells. In general, the clones demonstrated a high level of interleukin-4 (IL-4) activity, with two clones also exhibiting gamma interferon (IFN-γ) production and one other also producing IL-2 (6, 23).

COMPOSITION AND FUNCTION OF GINGIVAL T LYMPHOCYTES

Phenotypic and functional investigations of the composition of mononuclear cells from diseased human gingival tissues indicated that tissues were relatively enriched with CD8$^+$ lymphocytes, which exerted regulatory influences (29, 32). CD4$^+$ cells were also prominent, with the majority from adult periodontitis (AP) lesions expressing CD29 antigen and coexpressing CD45RA antigen (24). Others have shown that CD4$^+$ T cells from AP lesions also express the CD45RO antigen (5), suggesting a memory population.

M. A. Taubman, J. W. Eastcott, H. Shimauchi, O. Takeichi, and D. J. Smith • Department of Immunology, Forsyth Dental Center, 140 The Fenway, Boston, Massachusetts 02115.

Studies using two- and three-color immunofluorescence analyses indicated that a significant segment of the CD4[+] T-cell population obtained from periodontal disease lesions bore both CD45RA and CD29 cell markers (24). This finding in vivo of such activated populations, which had previously been described in cloned and isolated cell populations, indicates that most CD4[+] T cells in periodontal disease tissue are "memory"-type cells; only approximately 10% have the "naive" phenotype. These findings suggest that cells coexpressing both antigens represent restimulation of a memory population rather than stimulation of naive cells.

We have described the distribution of T-lymphocyte subsets in diseased gingival tissues and indicated an imbalance in which CD8[+] cell levels were relatively high (29, 32). A technique was devised to determine the functional potential of diseased gingival mononuclear cells (GMC) (28). Culture of these cells gave rise to production of immunoglobulin (Ig) and antibody. Ig production was suppressed after stimulation of GMC in vitro (9). This suppression was suggested to be related to CD8[+] cells producing a suppressive factor(s) and was found to be macrophage dependent (8). Studies of AP-diseased GMC indicated that depletion of the CD8[+] T-cell population resulted in increased Ig synthesis and also increased antibody synthesis. Further studies of GMC indicated that CD8[+] T lymphocytes produced a factor(s) that suppressed specific antibody production by peripheral blood cells from the same patient (33, 35).

Depletion of macrophages from GMC resulted in significant reduction of IgG synthesis, which could be restored by macrophage addition (33). The CD8 T-cell-mediated suppression of Ig production did not occur if macrophages were depleted (8). GMC produced more antibody in the presence of additives. These findings along with production of proinflammatory cytokines highlight the important role of macrophages in inflammatory periodontal diseases (8, 33). (For a presentation of a theoretical role for macrophages in periodontal diseases, see Fig. 3.)

Additional studies (31) indicated that IL-2R and HLA-DR expression on CD4[+] lymphocytes in inflamed gingival tissues is significantly greater than that on peripheral blood lymphocytes. Expression was elevated even more significantly on gingival crevice lymphocytes (31). These findings confirm that the gingival-tissue lymphocyte population is highly activated (mature) and that there might be a selective passage of immune T lymphocytes from gingival tissue into the periodontal pocket via the junctional epithelium (31).

ADOPTIVE TRANSFER OF T LYMPHOCYTES

We cloned rat T cells specific for *Actinobacillus actinomycetemcomitans* (2) and used these cloned T lymphocytes in periodontal disease experiments. The phenotypes and characteristics of a rat *A. actinomycetemcomitans*-specific Th2-type clone and a Th1 clone are presented in Table 1. Adoptive transfer of cloned *A. actinomycetemcomitans*-specific Th2 lymphocytes into rats gives systemic and local antibody formation, reduced disease (37), and infiltration of lymphocytes into the gingival tissues (Fig. 1). Higher numbers of total lymphocytes were recovered from

Table 1. Characteristics of *A. actinomycetemcomitans*-elicited Th1 and Th2 clone cells

Characteristic[a]	Clone G2	Clone A3
Origin	Spleen	Mesenteric lymph node
Phenotype		
W3/13 (CD43)	+	+
W3/25 (CD4)	−	+
OX8 (CD8)	−	−
OX22 (CD45RC)	−	−
OX19 (CD5)	+	+
R73 (αβ TCR)	−	+
Proliferation[b]		
A. actinomycetemcomitans Y4 (serotype b)	+	+ +
A. actinomycetemcomitans Y4 LPS	−	−
Streptococcus mutans	+	−
Other gram-negative bacteria[c]	+	−
Protein antigens[d]	−	−
Help for specific antibody	+	+ +
IL-2 production	+	−
DTH	?[e]	−
ConA response	+	+ +
T cell type designation	Th1	Th2

[a] LPS, lipopolysaccharide; DTH, delayed-type hypersensitivity; ConA, concanavalin A.
[b] Proliferation in vitro in the presence of antigen-presenting cells and antigen listed.
[c] *Capnocytophaga* spp., *Prevotella intermedia*, and *Porphyromonas gingivalis* 381.
[d] *Streptococcus sobrinus* glucosyltransferase and ovalbumin.
[e] ?, not tested.

the gingival tissues of the Plus A3 group, which received this T-cell clone, than from the No A3 group, (3.8±0.6 versus 1.4±0.3; $P < 0.05$), indicating that the gingivae were infiltrated with increased numbers of T cells. The numbers of gingival CD4$^+$ cells showed the greatest increase (3.2-fold), suggesting that cells of the clone phenotype might reside in the tissue. B-cell numbers were increased 1.7-fold. The nature of and requirements for T-lymphocyte migration to and retention in these gingival tissues were unknown and had not been previously investigated.

T-LYMPHOCYTE TRAFFIC TO GINGIVAL TISSUES

Naive and memory T lymphocytes demonstrate different trafficking patterns (11). The effect of antigenic stimulation is clonal expansion and differentiation of lymphocytes into effector and memory cells. Memory lymphocytes acquire the ability to migrate into tertiary lymphoid tissues. Once activated, lymphocytes are not relatively homogeneous like naive cells but show selectivity in homing behavior (10, 19). Some studies indicated that memory or activated lymphocytes preferentially traffic back to the tissue where they were first activated. Selective homing of

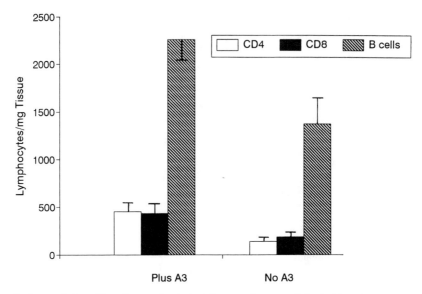

Figure 1. Accumulation of lymphocytes in gingival tissues and nature of gingival lymphocytes recovered from animals receiving clone A3 adoptively (Plus A3; $n = 10$) or no autologous cells (No A3; $n = 6$ to 15). All animals were superinfected with *A. actinomycetemcomitans*. Bars indicate mean recoveries of CD4$^+$, CD8$^+$, and B cells from gingival tissues. Error bars indicate standard errors of the means. (Data from reference 37.)

memory lymphocytes can be mediated by differential regulation of homing receptors, which are adhesion molecules needed to recognize and bind to endothelial cells within different tissue sites. These memory T lymphocytes can selectively home to nonlymphoid (or tertiary lymphoid) tissues (1, 18). Therefore, most T cells found in tissues such as gingivae are expected to be memory cells. Binding to endothelium (27) seems to be the first step in the transmigration of lymphocytes into tissues (7). This binding is regulated by the differential expression of surface molecules on the effector lymphocyte and also on the endothelium of the target tissue. Thus, lymphocyte homing is initiated by surface molecule expression.

After cells enter the tissue, other events that can be mediated by antigen and/ or cytokines occur. Recent evidence (20) suggests that the β-1 adhesion molecule CD29, a portion of both the adhesion receptor VLA-4 (CD49d/CD29) and VLA-1 (CD49a/CD29), may be significantly associated with the ability of lymphocytes to transmigrate through endothelial layers such as can be found in blood vessels. Thus, only CD29$^+$ T (either CD4$^+$ or CD8$^+$) cells transmigrated through monolayers of human umbilical cord endothelial cells prepared on a collagen gel matrix (20). Interestingly, human T lymphocytes binding to human gingival fibroblasts can be inhibited by monoclonal antibodies to VLA integrins (17). Also of considerable importance in this context is our finding that a majority of gingival CD4$^+$ T cells exhibit the β-1 adhesion molecule CD29 (24, 33). These molecules can play a major role in the traffic of T lymphocytes to gingival tissues.

HOMING AND RETENTION OF T HELPER (Th) CELL CLONES TO GINGIVAL TISSUES

The concept of homing of lymphocytes to periodontal tissue has only recently been addressed. Cells may enter tissues at random, specifically, or both specifically and randomly. Retention, however, can be based on antigen specificity. Therefore, bacterial-antigen-specific T cells should accumulate at a site of specific infection and would then have the potential to regulate immune responses in inflamed gingiva.

Th cells and clones can be divided into at least two subsets that, in general, regulate expression of antibody synthesis or delayed-type hypersensitivity (16). Classification as Th1 or Th2 cells is based on differences in the patterns of cytokine secretion and helper functions (13, 36). Also, a third type of Th cell (Th0) which produces both Th1 and Th2 type cytokines exists (4). Either Th clone type and/ or antigen-specificity may affect migration and T-cell-mediated immunoregulation in gingival tissues. We have reported that the adoptive transfer of *A. actinomycetemcomitans*-specific Th2 clone (A3) cells to *A. actinomycetemcomitans*-infected rats mediated protection with respect to periodontal bone loss (37). Recently, we have adoptively injected radioactively labeled cloned Th2 (A3) cells into infected rats. The cells appeared to be retained in the gingival lesions (25; see Fig. 2).

We performed a study to investigate homing of *A. actinomycetemcomitans*-specific cloned cells of the Th1 and Th2 types and to compare migration of these different clones into periodontal tissues of *A. actinomycetemcomitans*-infected and noninfected rats. The origins of the A3 and G2 clones are mesenteric lymph node and spleen, respectively (Table 1). The specificity of G2 is broader than that of clone A3, with proliferation to both gram-negative and gram-positive bacteria. This

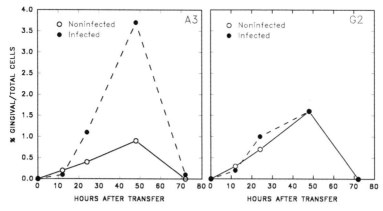

Figure 2. Migration of Th2 clone A3 (left) and Th1 clone G2 (right) to the gingival tissues of rats superinfected with *A. actinomycetemcomitans*. Shown are results of two representative experiments with two animals for each time point (12, 24, 48, and 72) after infusion of approximately 10⁷ *A. actinomycetemcomitans*-stimulated clone cells. Data show total numbers of radioactive lymphocytes in the gingival tissues expressed as a percentage of total radioactive lymphocytes recovered.

may indicate that specificity can be to a heat shock or stress protein, which can be found ubiquitously on both gram-negative and gram-positive microorganisms.

COMPARISON OF A3 AND G2 CLONE CELL MIGRATION AND RE-TENTION IN GINGIVAL TISSUES

We investigated the migration and retention of antigen-specific T-cell clones (A3; Th2 and G2; Th1) to *A. actinomycetemcomitans*-superinfected and nonsuperinfected gingivae and other tissues. Radioactive cells were traced in autoradiographs of the cell suspensions of the tissues. The total number of cells recovered from each of these tissues is expressed as a percentage of all the cells recovered. The recovery of radioactive cells from infected animals injected with clone A3 diminished from 40% at 12 h to 3% at 72 h, whereas the recovery of cells from G2-injected animals diminished from 23 to 6% over this time. Specific numeric differences in A3 clone cells seemed to indicate larger numbers of cells recovered from infected small intestine (also clone G2) and gingivae (Fig. 2).

Actual radioactive-cell recovery from the gingival tissues increased with time. Recovery of A3 was always higher in infected tissues, with significant differences occurring 48 h after cell infusion (Fig. 2, left). These data indicated that cells were retained in the gingival tissues and that retention was most pronounced in infected tissues.

As described in Table 1, the specificity of the Th1 G2 clone was less restricted than that of the A3 clone, and little or no difference was observed between superinfected and noninfected animals with respect to gingival cell recoveries (Fig. 2, right). However, the increasing numbers and increasing percentages of cells recovered from gingivae indicated that cells migrated to and were retained in the gingival tissues. Retention could be based on broad antigen specificity or on inflammation in general.

CYTOKINE PROFILES OF HUMAN GMC

Functional subsets of human T lymphocytes can be defined by cytokine profiles (13). Cytokines are potent regulators of immune effector mechanisms and correlate with secretory activities of mononuclear cells (21). Human mononuclear B cell progenitors migrate to inflamed gingivae, increase in number, and differentiate into Ig-secreting cells in a T-cell regulatory environment. To obtain information on the functional types of GMC from AP patients, we investigated the expression of RNA for cytokines by the reverse transcriptase PCR. We investigated mRNA expression of IL-1β, IL-2, IFN-γ, IL-5, and IL-6. β-Actin was also amplified as a control. Oligonucleotides for PCR were designed and synthesized using GenBank sequences. Collagenase digestion was used to isolate GMC from gingivae surgically removed from two sets of sites in each AP patient (28, 29, 32). RNA was extracted from the GMC before culture (day 0) and after culture for up to 16 days with or

without additives. After reverse transcription of the extracted RNA, PCR amplification products were electrophoresed on agarose gels. Only bands that were the
sizes predicted for the amplified cDNAs were considered. For analysis of PCR
amplification, Southern blots of hybridization with an internal probe were scanned
densitometrically. IFN-γ message was prominently expressed when cells were extracted before culture (day 0; Table 2). The RNA message (day 0, not cultured)
for cytokines from cells of 13 randomly obtained AP patients (10 females and 3
males, aged 45 to 64 years) was examined; 6 patients showed cytokine profiles
consistent with Th1-type cells (positive for IFN-γ and/or IL-2 message and negative
for IL-5 message). One sample presented a cytokine profile consistent with Th1
and Th2 (IL-2 and IL-5 positive) or possibly Th0. After culture and stimulation
(11 samples), 6 samples showed profiles consistent with Th1-type cells, and 5 showed
profiles consistent with Th1 and Th2 (or possibly Th0). Thus, the cytokine patterns
of these recovered mononuclear cells were consistent with the prominent presence
of Th1-type and/or CD8+-type T cells. Further in vitro growth or stimulation of
these cells gave rise to IL-5 and IL-6 message expression, possibly indicating the
presence of Th2-type cells but in relatively low numbers.

On the basis of the adoptive transfer experiments in rats and the type of cell
recoveries from diseased human tissues, we might speculate that Th2 cells can be
protective via the provision of help for antibody synthesis and that Th1 and CD8+
T lymphocytes might be destructive via the prominent production of IFN-γ and
the potential stimulation of macrophage secretion of IL-1 and subsequent bone-
destructive activity (26). Thus, T-cell features such as help for antibody, antigen
specificity, migration pattern, regulatory functions, surface marker expression, and

Table 2. Cumulative (before and after culture) cytokine RNA message expression
by GMC from AP patients

| Patient no. | Cell type[a] | | RNA message expression[b] | | | | |
	Day 0	After culture	IL-1β	IL-2	IFN-γ	IL-5	IL-6
1	?	Th1	+		+		−
2	Th1	Th1,Th2	+	+	+	+	+
3	?	Th1	+		+		−
4	Th1	Th1	+		+		+
5	?	Th1	+		+		−
6	Th1	Th1	−		+		−
7	?	Th1,Th2	−	−	+	+	+
8	?	Th1	−	−	+	−	+
9	Th1	Th1,Th2	+	+	+	+	+
10	Th1,Th2	(No culture)	+	−	+	−	−
11	Th1	Th1,Th2	+	+	+	+	−
12	?	Th1,Th2	−	+	+	+	−
13	Th1	(No culture)	+	+	−	−	−

[a] Cell type(s) was deduced from cytokine message expression by cells prior to culture or after culture with (usually)
no additives, pokeweed mitogen, concanavalin A, or (infrequently) recombinant IL-2. ?, insufficient information to
deduce cell type.
[b] +, particular cytokine message detected; −, no message detected. The absence of a symbol indicates that a particular
cytokine was not tested.

cytokine message expression are important modulators of periodontal inflammation (30) and therefore may be considered targets for disease intervention.

DESTRUCTIVE VERSUS PROTECTIVE ASPECTS OF T-LYMPHOCYTE MODULATION IN PERIODONTAL DISEASES

Adoptive transfer experiments in rodents with the Th2-type clone showed greatly increased levels of antibody (of all isotypes) and apparent abrogation of periodontal bone loss symptoms. T cells migrate to gingival and periodontal tissues. The mechanism of migration is currently not completely understood. T cells are retained in gingival and periodontal tissues. Retention can be related to the presence of antigen or to inflammation (which is less likely because inflammation exists in control rats that are not infected with *A. actinomycetemcomitans*). However, the *A. actinomycetemcomitans*-specific Th2 clone migrated to gingival tissues of superinfected animals and was clearly retained. Far less retention was observed in rats not superinfected. The Th1 clone, with broader specificity, migrated to the gingival tissues and was retained, but no differences between superinfected and nonsuperinfected animals were observed. Thus, the specific dynamics of entry into and retention of T cells in periodontal lesions can vary because of different subsets or specificities. This finding highlights the potential importance of the T lymphocyte in the genesis of periodontal disease and its potential role as a target for disease intervention.

In Fig. 3 we outline the hypothesis that we believe is consistent with the findings of the adoptive transfer (T-cell vaccine) experiments and the elucidation of the T-cell cytokine profiles in human diseased tissues. The hypothesis (which can be

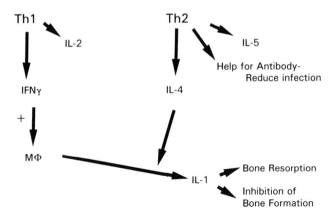

Figure 3. Potential mechanism to account for the protective and destructive aspects of Th2- and Th1-type lymphocytes, respectively, in periodontal bone loss. IFN-γ produced by Th1-type cells activates macrophages to produce IL-1, giving rise to increased bone resorption and decreased bone formation. Th2 lymphocytes produce IL-4, IL-5, and IL-6, which are necessary for B-cell production of antibody, which mediates protection against periodontal infection. Also of importance is IL-4, which down-regulates IL-1 production and up-regulates synthesis of IL-1 receptor antagonist. MΦ, macrophage.

subjected to further testing) suggests that Th2-type cells abrogate periodontal disease symptoms and that Th1 lymphocytes enhance disease. The mechanism (described in Fig. 3) indicates that IFN-γ produced by Th1 cells activates macrophages to produce IL-1, giving rise to increased bone resorption and decreased bone formation (26).

On the other hand, Th2 cells can produce IL-4, IL-5, and IL-6, which are essential for B-cell production of antibody. This production seems to be a key feature of protection against periodontal bone loss by Th2-cell vaccine (37). Presumably this function is implemented by antibody-mediated reduction of the pathogenic bacteria and their virulence (37). Furthermore (Fig. 3), IL-4 reduces steady-state IL-1 RNA levels by decreasing transcription and the half-lives of transcripts and also up-regulates synthesis of IL-1 receptor antagonist at the mRNA and protein levels (3, 34). It is important to note that Kiyono and his colleagues (see chapter 12 of this book) have demonstrated the presence of IL-4 message transcripts from cells in some but not all types of periodontal diseases. Also, IL-10 production by Th2 cells can inhibit murine Th1 cell function (4), but its role in the human system is not as clear. Thus, the nature of the host response to periodontal microflora can be modulated by different subsets of T cells (recognizable by the lymphokines secreted) to produce a destructive (Th1) or protective (Th2) immune response.

Acknowledgments. Portions of the work reported in this chapter have been supported by grants DE-03240, DE-04733, and DE-04881 from the National Institute of Dental Research.

We thank I. Saito and I. Moro for primers and probes, J. Haber for patient samples, T. Kawai for some adoptive transfer experiments, and Jan Schafer for excellent secretarial assistance.

REFERENCES

1. **Butcher, E. C.** 1986. The regulation of lymphocyte traffic. *Curr. Top. Microbiol. Immunol.* **128:**85–122.
2. **Eastcott, J. W., K. Yamashita, M. A. Taubman, and D. J. Smith.** 1990. Characterization of rat T cell clones with bacterial specificity. *Immunology* **71:**120–126.
3. **Fenton, M. J., and J. A. Burns.** 1992. IL-4 reciprocally regulates IL-1 and IL-1 receptor antagonist expression in human monocytes. *J. Immunol.* **149:**925–931.
4. **Fiorentino, D. F., M. W. Bond, and T. R. Mosmann.** 1989. Two types of mouse helper cells. IV. Th2 clones secrete a factor that inhibits cytokine production by Th1 clones. *J. Exp. Med.* **170:**2081–2095.
5. **Gemmell, E., B. Feldner, and G. J. Seymour.** 1992. CD45RA and CD45RO positive CD4 cells in human peripheral blood and periodontal disease tissue before and after stimulation with periodontopathic bacteria. *Oral Microbiol. Immunol.* **7:**84–88.
6. **Ishii, T., R. Mahanonda, and G. J. Seymour.** 1992. The establishment of human T cell lines reactive with specific periodontal bacteria. *Oral Microbiol. Immunol.* **7:**225–229.
7. **Janeway, C. A., and P. Golstein.** 1992. Lymphocyte activation and effector functions. Editorial overview. A return to intimacy. *Curr. Opin. Immunol.* **4:**241–345.
8. **Lundqvist, C. A., M. A. Taubman, E. D. Stoufi, J. Pappo, J. S. Sioson, and D. J. Smith.** Regulatory role of macrophages in immunoglobulin and antibody synthesis by human gingival cells. Submitted for publication.
9. **Lundqvist, C. A., M. A. Taubman, E. D. Stoufi, D. J. Smith, and X.-P. Liu.** 1992. Diminished immunoglobulin synthesis after stimulation of mononuclear cells from periodontal disease tissue. *Regional Immunol.* **4:**255–261.
10. **Mackay, C. R.** 1991. T-cell memory: the connection between function, phenotype and migration pathways. *Immunol. Today* **12:**189–192.

11. **Mackay, C. R., W. L. Marston, and L. Dudler.** 1990. Naive and memory T-cells show distinct pathways of lymphocyte recirculation. *J. Exp. Med.* **171**:801–817.

12. **Mackler, B. F., K. B. Frostad, P. B. Robertson, and B. M. Levy.** 1977. Immunoglobulin bearing lymphocytes and plasma cells in human periodontal disease. *J. Periodontal Res.* **12**:37–45.

13. **Maggi, E., P. Panonchi, R. Manetti, C. Simonelli, M.-P. Piccinni, F. S. Rugiu, M. DeCarli, M. Ricci, and S. Romanani.** 1992. Reciprocal regulatory effects of IFNγ and IL-4 on the *in vitro* development of human Th1 and Th2 clones. *J. Immunol.* **148**:2142–2147.

14. **Mahanonda, R., G. J. Seymour, L. W. Powell, M. F. Good, and J. W. Halliday.** 1989. Limit dilution analysis of peripheral blood T lymphocytes specific to periodontopathic bacteria. *Clin. Exp. Immunol.* **75**:245–251.

15. **Mahanonda, R., G. J. Seymour, L. W. Powell, M. F. Good, and J. W. Halliday.** 1991. Effect of initial treatment of chronic inflammatory periodontal disease on the frequency of peripheral blood T-lymphocytes specific to periodontopathic bacteria. *Oral Microbiol. Immunol.* **6**:221–227.

16. **Mosmann, T. R., and R. L. Coffman.** 1989. Th1 and Th2 cells: different patterns of lymphokine secretion lead to different functional properties. *Annu. Rev. Immunol.* **7**:145–173.

17. **Murakami, S., Y. Shimabuku, T. Saho, R. Isoda, K. Kameyama, K. Yamashita, and H. Okada.** 1993. Evidence for a role of VLA integrins in lymphocyte-human gingival fibroblast adherence. *J. Periodontal Res.* **28**:494–496.

18. **Picker, L. J., and E. C. Butcher.** 1992. Physiological and molecular mechanisms of lymphocyte homing. *Annu. Rev. Immunol.* **10**:561–591.

19. **Picker, L. J., L. W. M. M. Terstappen, L. S. Rott, P. R. Streeter, H. Stein, and E. C. Butcher.** 1990. Differential expression of homing associated adhesion molecules by T-cell subsets in man. *J. Immunol.* **145**:3247–3255.

20. **Pietschmann, P., J. J. Cush, H. Lipsky, and N. Oppenheimer-Marks.** 1992. Identification of subsets of human T-cells capable of enhanced transendothelial migration. *J. Immunol.* **149**:1170–1178.

21. **Scott, P., E. Pearce, A. W. Cheever, R. L. Coffman, and A. Sher.** 1992. Role of cytokines and CD4$^+$ T-cell subsets in the regulation of parasite immunity and disease. *Immunol. Rev.* **112**:161–182.

22. **Seymour, G. J., M. S. Crouch, R. N. Powell, D. Brooks, I. Beckman, H. Zola, J. Bradley, and G. F. Burns.** 1982. The identification of lymphoid cell subpopulations in sections of human lymphoid tissue and gingivitis in children using monoclonal antibodies. *J. Periodontal Res.* **17**:247–256.

23. **Seymour, G. J., E. Gemmell, R. A. Reinhardt, J. W. Eastcott, and M. A. Taubman.** 1993. Immunopathogenesis of chronic inflammatory periodontal disease: cellular and molecular mechanisms. *J. Periodontal Res.* **28**:478–486.

24. **Seymour, G. J., M. A. Taubman, J. W. Eastcott, E. Gemmell, and D. J. Smith.** Characterization of activated T-cells from human periodontal disease: coexpression of CD45RA and CD24 on CD4 positive cells. Submitted for publication.

25. **Shimauchi, H., M. A. Taubman, J. W. Eastcott, and D. J. Smith.** 1993. Migration of Th1 and Th2 *Aa*-specific clones into infected and noninfected gingival tissues. *J. Dent. Res.* **72**:243.

26. **Stashenko, P., F. E. Dewhirst, W. J. Peros, R. L. Kent, and J. M. Ago.** 1987. Synergistic interactions between interleukin 1, tumor necrosis factor and lymphotoxin in bone resorption. *J. Immunol.* **138**:1464–1468.

27. **Stoolman, L. J.** 1989. Adhesion molecules controlling lymphocyte migration. *Cell* **56**:907–910.

28. **Stoufi, E. D., M. A. Taubman, J. L. Ebersole, and D. J. Smith.** 1988. A method for studying immunoglobulin synthesis by gingival cells. *Oral Microbiol. Immunol.* **3**:108–112.

29. **Stoufi, E. D., M. A. Taubman, J. L. Ebersole, D. J. Smith, and P. P. Stashenko.** 1987. Phenotypic analyses of mononuclear cells recovered from healthy and diseased human periodontal tissues. *J. Clin. Immunol.* **7**:235–245.

30. **Takeichi, O., M. A. Taubman, D. J. Smith, J. Haber, and I. Moro.** 1993. Cytokine profiles of gingival mononuclear cells from periodontal disease patients. *J. Dent. Res.* **72**:159.

31. **Takeuchi, Y., H. Yoshie, and K. Hara.** 1991. Expression of interleukin-2 receptor and HLA-DR on lymphocyte subsets of gingival crevicular fluid in patients with periodontis. *J. Periodontal Res.* **26**:502–510.

32. **Taubman, M. A., E. D. Stoufi, J. L. Ebersole, and D. J. Smith.** 1984. Phenotypic studies of cells from periodontal disease tissues. *J. Periodontal Res.* **19**:587–590.

33. **Taubman, M. A., H.-Y. Wang, C. A. Lundqvist, G. J. Seymour, J. W. Eastcott, and D. J. Smith.** 1991. The cellular basis of host responses in periodontal diseases, p. 199–208. *In* S. Hamada, S. C. Holt, and J. R. McGhee (ed.), *Periodontal Disease: Pathogens and Host Immune Responses.* Quintessence Publishing Co., Tokyo.

34. **Wand, H. L., M. T. Lotze, L. M. Wahl, and S. M. Wahl.** 1992. Administration of recombinant IL-4 to humans regulates gene expression, phenotype and function in circulating monocytes. *J. Immunol.* **148:**2118–2125.

35. **Wang, H.-Y., M. A. Taubman, and D. J. Smith.** 1989. Effect of depletion of CD8+ cells on immunoglobulin synthesis by gingival lymphocytes. *J. Dent. Res.* **68:**221.

36. **Yamamura, M., X.-H. Wang, J. D. Ohmen, K. Uyemura, T. H. Rea, B. R. Bloom, and R. L. Modlin.** 1992. Cytokine patterns of immunologically mediated tissue damage. *J. Immunol.* **149:**1470–1475.

37. **Yamashita, K., J. W. Eastcott, M. A. Taubman, D. J. Smith, and D. S. Cox.** 1991. Effect of adoptive transfer of cloned *Actinobacillus actinomycetemcomitans*-specific T helper cells on periodontal disease. *Infect. Immun.* **59:**1529–1534.

Molecular Pathogenesis of Periodontal Disease
Edited by Robert Genco et al.
© 1994 American Society for Microbiology, Washington, DC 20005

Chapter 14

Human Immune Responses in Periodontal Disease: Analysis of Severe Combined Immunodeficient Mice Reconstituted with Lymphocytes from Patients with Localized Juvenile Periodontitis

Bruce J. Shenker, Sharon Wannberg, Caroline King, Irene Kieba, and Edward T. Lally

Despite recent advances in our understanding of periodontal disease, the role of the immune system in these inflammatory disorders remains poorly defined. There is little agreement as to whether the immune response acts to limit and slow the disease process or whether it contributes to the pathology of the disease. There are many reasons for this controversy; they result, in part, from individual investigations that argue for a static role (i.e., either protective or destructive) of the immune system. We believe, however, that the immune system plays a dynamic role in the disease process, one that changes with the course of the disease (12). In this regard, there is considerable evidence to suggest that inflammatory reactions initially serve a protective function in response to bacterial challenge. Therefore, the development of disease depends heavily on the outcome of the initial interaction or contact of pathogenic organisms with the host's defense system.

Recent studies suggest a role for altered host defenses as a possible underlying mechanism in the etiology and pathogenesis of periodontal disease (10–12, 14). Such a state of immunologic dysfunction would tend to favor colonization of the gingival crevice and the onset of disease. Clearly, this state of immunologic dysfunction must be transient, since most patients with various forms of periodontal disease eventually develop immunoreactivity to suspected pathogens. Therefore, a state of immunologic dysfunction in the earliest stages would be permissive to both infection and disease; this permissivity may be followed by a period of active

Bruce J. Shenker, Sharon Wannberg, Caroline King, Irene Kieba, and Edward T. Lally • Department of Pathology, School of Dental Medicine, University of Pennsylvania, Philadelphia, Pennsylvania 19104-6002.

immune reactivity that represents either a delayed or a depressed response. This delayed immune response may terminate the disease process in some individuals, while in others, it may contribute to tissue injury through immune reactions associated with hypersensitivity disease (9).

Such immunodynamics, whereby immunodeficiency is followed by periods of hypersensitivity, have been implicated in other infectious diseases (12). This model could explain the apparent contradictory finding of persistent infection in the face of high antibody titers in patients with periodontal disease. Such dynamics may also account for the cyclic or episodic nature of disease progression. If the immune response is viewed as an evolving process, then to accurately determine its relevant contribution to the pathogenesis of periodontal disease, we believe it is necessary to study the interrelationship between immunologic responsiveness, bacterial infection, and the onset of clinical symptoms. Furthermore, it is equally important to characterize both the qualitative and the quantitative aspects of the immune response to specific antigens and with respect to immunologic function. While clinical investigations can fulfill many of these goals, human studies are not always feasible for ethical and practical reasons.

USE OF ANIMAL MODELS TO STUDY PERIODONTAL DISEASE: POTENTIAL OF THE RECONSTITUTED SCID MOUSE

To better define the role of the host's immune system in the pathogenesis of periodontal disease, investigators have employed animal model systems. These animal studies have been and continue to be an important adjunct to both in vitro and clinical investigations. A wide range of animal models have been utilized to further our understanding of human disease in general and periodontal disease in particular. These models allow investigators to study selected microbiologic and immunologic parameters under controlled conditions (reviewed in references 5 and 13). Studies of experimental periodontitis have confirmed the pathogenicity of several suspected periodontal pathogens, including *Actinobacillus actinomycetemcomitans, Porphyromonas gingivalis, Capnocytophaga sputigena, Eikenella corrodens,* and *Fusobacterium nucleatum* among others (5, 10, 13). Specific immune reactions to these organisms, including both humoral and cell-mediated immune responses, have also been demonstrated (4, 5, 10, 15, 16). To evaluate the protective versus destructive nature of these responses, two general approaches have been employed: reconstitution of immunocompromised animals and selected immunization of otherwise healthy animals. For example, Taubman and his coworkers utilized such animal models to demonstrate that T cells can exert protective and/ or destructive effects in experimental periodontitis (13, 15, 16). Of particular relevance to our current investigation is the demonstration that athymic rats reconstituted with *A. actinomycetemcomitans*-specific T cells exhibited decreased bone loss when infected with this microorganism. Likewise, Chen et al. (4) showed that immunization against components of *P. gingivalis* improved the protective responses of rats to a subsequent challenge with this microorganism.

It is clear that animal models will continue to contribute to our understanding of the underlying mechanisms that are involved in the development of periodontal disease. However, it should also be noted that these systems have several short-comings, and the relevance of these models has often been questioned. Indeed, it is often difficult to extrapolate from in vivo animal studies to the human situation. For example, the animal response may be qualitatively different from that of humans; it is well known that the immune repertoire of animal species often differs from the human immune response. Likewise, the responses may differ quantitatively, resulting in dissimilar levels of immunoglobulin (Ig); other disparities may involve class and subclass distribution, which in turn may lead to functional discrepancies. Also, pathogenic mechanisms operative in the human form of the disease may be different from factors contributing to the development of lesions in animal systems.

In light of these concerns regarding the conventional animal paradigm, we initiated experiments to test the feasibility of using mice with severe combined immunodeficiency (SCID mice) to study the human immune response to periodontal pathogens. SCID mice have a defect in the recombinase system that impairs the rearrangement of antigen receptor genes in both T and B cells (2, 7). As a consequence, these animals lack a functional immune system and accept both allogenic and xenogenic grafts. These include grafts of malignant tumor cells, healthy human lymphoid tissue, and cells from individuals with autoimmune disease (3, 6, 8). In addition to surviving for extended periods in SCID mice, human blood lymphocytes spontaneously secrete Ig as well as specific antibody in response to antigenic challenge. Recently, Carlsson et al. (1) studied the basic aspects of immune function after immunization of SCID mice that had been repopulated with human lymphocytes. The results suggest that these animals exhibit several normal immune parameters of a human humoral response; these include high-dose tolerance, antigenic selection, and dose-dependent antigen-driven B-cell activation.

In this chapter we report on our initial observations regarding the reconstitution of SCID mice with lymphocytes from patients with localized juvenile periodontitis (SCID$_{LJP}$). We initiated these studies in order to test the efficacy of using SCID mice in the study of human immune responses to periodontal pathogens in general and to *A. actinomycetemcomitans* in particular. We felt that animals that permit in vivo growth of human lymphoid tissue would be highly desirable as a model for studying healthy immune function as well as pathogenic mechanisms operative in human disease. Such animal model systems allow manipulation of the human immune system in vivo that would not be feasible in human studies. The reconstitution of SCID mice with human lymphocytes allows the assessment of immunologic processes in which the kinetics, specificity, and function are indicative of the human response (i.e., the cell donor) and not of the animal per se. The results indicate that the reconstituted animals not only expressed human IgG, IgM, and IgA in their sera but also exhibited a range of antigenic recognition of *A. actinomycetemcomitans* similar to that of the sera of the patient cell donor. Additionally, the reconstituted lymphocytes were immunologically competent, as evidenced by their ability to respond to antigenic challenge.

RECONSTITUTION AND ASSESSMENT OF SCID$_{LJP}$ MICE

Human peripheral blood mononuclear cells were obtained from healthy blood donors (controls) with no history of periodontal disease and from three patients with LJP (17, 15, and 14 years of age; designated donors 1, 2, and 3, respectively). In addition to their clinical diagnosis, the LJP patients were selected on the basis of the documented presence of *A. actinomycetemcomitans* in diseased sites and the demonstration of serum anti-*A. actinomycetemcomitans* antibodies. Serum samples from the three LJP patients and the control subjects were screened for the presence of antibodies to *A. actinomycetemcomitans* by Western blot (immunoblot) analysis of sodium dodecyl sulfate-polyacrylamide gel electrophoresis (SDS-PAGE) gels of a crude extract of strain JP-2. These results, shown in Fig. 1, indicate that as expected, sera from healthy controls did not exhibit any immunoreactivity to the *A. actinomycetemcomitans* sonicate. In contrast, the LJP sera strongly reacted with a wide range of antigenic moieties. Similar but not identical reactions were seen for the three patients.

SCID mice (C.B.-17 *scid/scid*) were kindly supplied by John Cebra (University of Pennsylvania) and were maintained in microisolators under sterile conditions. The SCID mice (four or five animals per group) were reconstituted with 40×10^6 human peripheral blood mononuclear cells either from the periodontally healthy (control) individuals (SCID$_{CTL}$) or from LJP patients (SCID$_{LJP}$). In preliminary experiments, we observed that after intraperitoneal injection of cells, human IgG and IgM appeared in the mouse sera within 1 week, reached maximum levels between 2 and 3 weeks, and maintained these levels for at least 8 weeks. On the basis of these preliminary observations, we first bled the SCID$_{LJP}$ and SCID$_{CTL}$ mice 3 weeks following injection of the cells. Sera were analyzed by enzyme-linked immunosorbent assay (ELISA) for the presence of human immunoglobulin. As shown in Fig. 2, the mouse sera clearly contained detectable levels of human IgG, IgM, and, to a lesser degree, IgA; all four isotypes of IgG were present, with IgG$_1$ and IgG$_2$ representing approximately 70% of the total IgG. It should be noted that

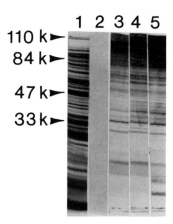

Figure 1. Presence of anti-*A. actinomycetemcomitans* antibodies in LJP patient sera. Crude soluble sonic extracts of *A. actinomycetemcomitans* JP-2 were resolved by SDS-PAGE and electrophoretically transferred to a nitrocellulose membrane. The membrane was blocked and cut into strips; individual strips were incubated with diluted sera obtained from each of the LJP patients and controls. The immunoblot was developed with an anti-human Ig antiserum conjugated to alkaline phosphatase and substrate. Lane 1, SDS-PAGE gel; lane 2, representative Western blot exposed to sera from control subjects; lanes 3 through 5, Western blots exposed to LJP donors 1, 2, and 3, respectively.

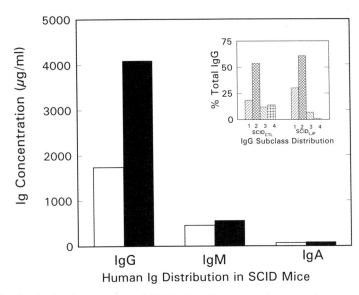

Figure 2. Ig distribution in reconstituted SCID mice. Serum samples were obtained 3 weeks after reconstitution of the SCID mice with human lymphoid cells. The sera were assayed by ELISA for total IgG, IgM, and IgA. Data represent the mean values for all mice in each group: 8 SCID$_{CTL}$ mice (□) and 14 SCID$_{LJP}$ mice (■). Additionally, IgG subclass distribution was assessed by ELISA; results are shown in the inset.

no murine immunoglobulin was detected in any of the animals, indicating that none of the mice had become "leaky." The SCID$_{CTL}$ sera contained 1.7 mg of IgG and 0.5 mg of IgM per ml. SCID$_{LJP}$ sera, on the other hand, contained 4.1 mg of IgG and 0.6 mg of IgM per ml. SCID$_{CTL}$ and SCID$_{LJP}$ mice exhibited approximately the same amounts of IgA, i.e., 13.8 and 47.1 μg/ml, respectively. In addition to assessing immunoglobulin class and subclass distribution, we also analyzed reconstituted SCID mouse sera for the presence of antibodies to *A. actinomycetemcomitans*. Figure 3 shows the immunologic profiles for the four animals reconstituted with cells from one LJP patient (LJP donor 1). All four animals exhibited a similar

Figure 3. Presence of human anti-*A. actinomycetemcomitans* antibodies in the sera of SCID$_{LJP}$ mice. Three weeks after reconstitution, SCID mice reconstituted with cells from either healthy controls or LJP donor 1 were bled. The sera were analyzed for anti-*A. actinomycetemcomitans* antibodies by Western blot of a crude sonic extract of strain JP-2. Lane 1, SDS-PAGE gel of the crude JP-2 extract; lane 2, representative immunoblot of sera from a SCID$_{CTL}$ animal; lanes 3 through 6, immunoblots of sera from SCID$_{LJP}$ mice that received cells from LJP donor 1; lane 7, immunoblot of the mouse serum shown in lane 3, but in this instance the blot was developed with goat anti-murine Ig antiserum conjugated to alkaline phosphatase instead of the goat anti-human Ig antiserum that was used for the other immunoblots.

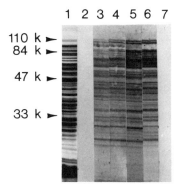

profile of immunoreactivity in a Western blot of *A. actinomycetemcomitans* strain JP-2 (Fig. 3, lanes 2 through 5). The antibodies were clearly derived from the transplanted human cells, as is evident in lane 6, where no immunoreactivity was observed when the immunoblot was developed with goat anti-mouse serum. It should be noted that none of the $SCID_{CTL}$ animals developed detectable immunologic reactivity to *A. actinomycetemcomitans* (Fig. 3, lane 2). We also compared the immunologic profile of each donor patient serum with that of the reconstituted mice by scanning densitometry analysis of representative immunoblots. As shown in Fig. 4, the $SCID_{LJP}$ mice developed an immunologic repertoire similar to that observed for their respective patient donors.

In addition to examining the total profile of antibodies to *A. actinomycetemcomitans,* we also determined whether the $SCID_{LJP}$ mouse sera contained specific human antibodies to the *A. actinomycetemcomitans* leukotoxin (LTX). First, we used immunoblot analysis of purified LTX to assay the sera for the presence of anti-LTX antibodies. As shown in Fig. 5, the sera of the $SCID_{LJP}$ mice reconstituted from all three LJP patients contained antibodies that recognized the LTX; as expected, the $SCID_{CTL}$ sera did not contain any anti-LTX antibody. It should be noted, however, that the $SCID_{LJP}$ mice reconstituted with cells from donor 3 had very low levels of antibody and required a much smaller dilution (1/10 versus 1/100) in order to detect immunoreactivity to the LTX. Likewise, the sera of all three LJP patients also contained anti-LTX antibodies, but patient 3 had approximately one-fifth of the titers of the other two patients (data not shown). Therefore, it appears that the reconstituted animals accurately reflect the immune potential of the human donor on both a qualitative and a quantitative basis.

In addition to being tested for immunoreactivity, the mouse sera were tested for their abilities to neutralize the cytotoxic activity of the LTX. As shown in Fig. 6, when the LTX was preexposed to the $SCID_{LJP}$ mouse serum (patient 1), its ability to kill HL-60 cells was greatly diminished. Similar results were observed for sera obtained from mice reconstituted with cells from LJP patient 2. By comparison, the sera from the $SCID_{CTL}$ mice were unable to alter the cytotoxic activity of the toxin.

ASSESSMENT OF IMMUNOCOMPETENCE OF $SCID_{LJP}$ MICE

Clearly, one of the benefits of employing animal models in general and the SCID system in particular is that they permit us to immunize and challenge the human immune response under relatively controlled and defined conditions. Therefore, we next wanted to determine whether the $SCID_{LJP}$ mice were immunologically competent. To achieve this, mice were challenged with either soluble antigen (LTX) or subcutaneous injections of live *A. actinomycetemcomitans*. In the first series of experiments, $SCID_{CTL}$ and $SCID_{LJP}$ (donors 1 and 3) mice were injected intravenously with 0.5 µg of LTX 3 weeks following reconstitution. The mice were bled prior to injection and again 7 days after antigenic challenge. The anti-LTX titer was determined in both pre- and postchallenge sera by ELISA. The results of these

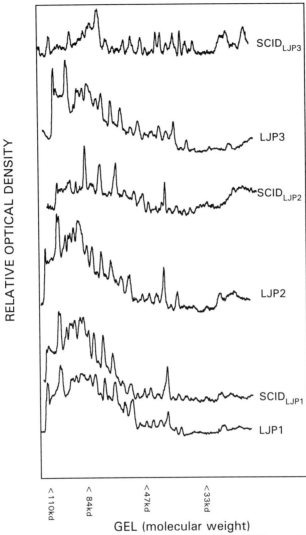

Figure 4. Comparison of anti-*A. actinomycetemcomitans* antibodies in LJP sera with those in reconstituted SCID mouse sera. Crude soluble sonic extracts of *A. actinomycetemcomitans* JP-2 were resolved by SDS-PAGE and electrophoretically transferred to a nitrocellulose membrane. The membrane was blocked and cut into strips; individual strips were incubated with diluted sera obtained from each of the LJP patients and a representative corresponding SCID$_{LJP}$ mouse. The immunoblot was developed with an anti-human Ig antiserum conjugated to alkaline phosphatase and substrate. The Western blots were then analyzed by laser scanning densitometry.

experiments are summarized in Table 1. The prechallenge anti-LTX titers in SCID$_{LJP2}$ and SCID$_{LJP3}$ were 962 and 0, respectively. It should be noted, however, that sera from SCID$_{LJP3}$ mice did have detectable immunoreactivity to the LTX on a Western blot (Fig. 5). Following challenge with LTX, the titers in both groups of SCID$_{LJP}$ mice increased significantly. While SCID$_{CTL}$ mice failed to respond to

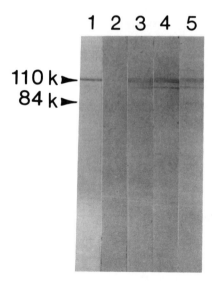

Figure 5. Detection of specific antibodies to *A. actinomycetemcomitans* LTX in the sera of SCID_LJP mice. Purified LTX was resolved by SDS-PAGE and electrophoretically transferred to nitrocellulose. The membrane was cut into strips and exposed to sera from representative SCID_LJP and SCID_CTL mice. Lane 1, SDS-PAGE gel of LTX; lane 2, SCID_CTL sera; lane 3, SCID_LPJ1 sera; lane 4, SCID_LJP2 sera; lane 5, SCID_LJP3 sera. The immunoblots were developed with goat anti-human Ig antiserum conjugated to alkaline phosphatase and substrate.

LTX, these mice also contained a functional immune system, as demonstrated by their abilities to respond to an antigenic challenge with tetanus toxoid (data not shown).

In a final series of experiments, we employed a modification of the murine abscess model to test the competency of the reconstituted immune systems of the SCID mice (4). In these experiments, SCID_CTL and SCID_LJP mice were challenged by subcutaneous injection with live or heat-killed *A. actinomycetemcomitans* JP-2. We then assessed and compared the immunocompetence of the mice to limit, retard, or prevent infection by monitoring lesion development at 24-h intervals. The results of these experiments are summarized in Table 2. All eight control animals developed

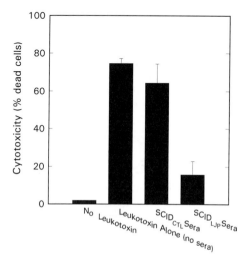

Figure 6. Neutralization of LTX by SCID mouse sera. Human promyelomonocytic HL-60 leukemic cells were used as the LTX-sensitive target cells. The LTX was incubated with and without mouse sera and then added to the HL-60 cells. Following a 60-min incubation, the viability of the cells was determined by trypan blue exclusion. Data are expressed as percent cytotoxicity (dead cells). Each bar represents the mean value of three SCID mouse sera assayed in duplicate.

Table 1. Effect of LTX challenge on anti-LTX titers

Mouse group	Anti-LTX titer[a]	
	Prechallenge	Postchallenge
SCID$_{CTL}$	0	0
SCID$_{LJP2}$	962 ± 573	$3,708 \pm 1,180$
SCID$_{LJP3}$	0^b	792 ± 479

[a] Anti-LTX titers were determined by ELISA. The titer was calculated by taking the reciprocal dilution of the serum that yielded an optical density of 0.2. Data represent the means of replicate samples for each of four mice per group.
[b] Although no titers were determined for these animals, they did contain antibodies to LTX, as demonstrated by immunoblot (Fig. 5).

relatively large lesions (44.3 mm^2) that exhibited severe ulceration and necrosis. As might be expected, these animals were not immunologically protected; the size and severity of the lesions were comparable to those observed in nonreconstituted animals (data not shown). No lesions developed when the animals were injected with heat-killed bacteria. The SCID$_{LJP2}$ mice, reconstituted with cells from LJP donor 2, exhibited a low incidence of cutaneous lesions (60%), and when lesions did develop, they were considerably smaller than those observed in the SCID$_{CTL}$ mice (19.5 versus 44.3 mm^2). In contrast, the SCID$_{LJP3}$ mice, reconstituted with cells from patient 3, exhibited lesions comparable to those of the controls. As demonstrated earlier, these mice (as well as the patient) had significantly lower levels of anti-LTX antibody than the other SCID$_{LJP}$ mice. To determine the relative abilities of anti-LTX antibodies to confer protection against *A. actinomycetemcomitans* infection (i.e., lesion development), some of the SCID$_{LJP3}$ mice

Table 2. Effect of reconstitution and immunization on SCID mouse resistance to live *A. actinomycetemcomitans* challenge[a]

Mouse group	Lesion incidence (%)[b]	Maximal lesion size (mm^2)[c]	Degree of ulceration and necrosis[d]	No. of days to maximum lesion size
SCID$_{CTL}$	100 (8/8)	44.3	3.2	3–4
SCID$_{LJP2}$	60 (3/5)	19.5	2.2	1
SCID$_{LJP3}$	100 (5/5)	44.0	4.0	3
SCID$_{LJP3-1}$	75 (3/4)	25.1	3.5	4

[a] SCID mice were challenged with 10^8 *A. actinomycetemcomitans*. Prior to challenge, sides and backs of mice were shaved; each animal received two injections of bacteria. Some animals also received injections of heat-killed bacteria. Mice were examined at 24-h intervals for the development of cutaneous lesions; results for both lesions on each animal were averaged.
[b] Numbers in parentheses indicate number of animals that developed lesions/total number of animals per group.
[c] Determined by obtaining two measurements at 90° to one another. These values were then averaged to determine the diameter of the lesion.
[d] Determined on a scale of 0 to 4 by estimating the percentage of the area of the lesion that exhibited these attributes: 0, no necrosis; 1, 25% necrosis; 2, 50% necrosis; 3, 75% necrosis; 4, 100% necrosis.

were immunized to the LTX prior to challenge with live *A. actinomycetemcomitans* (Table 1). As shown in Table 2, immunization to LTX not only resulted in higher titers of anti-LTX antibodies but also led to retardation of lesion development; these lesions were comparable in size to those of the more protected $SCID_{LJP2}$ animals.

SUMMARY AND CONCLUSIONS

In this initial study, we have demonstrated the feasibility of reconstituting SCID mice with an immune response derived from LJP patients. The immune repertoires of the reconstituted animals were similar to that of patient cell donors, as evidenced by the range of reactive human antibodies found in the mouse sera that recognized *A. actinomycetemcomitans* antigens. Furthermore, antibodies to specific antigens such as the *A. actinomycetemcomitans* LTX not only were immunoreactive but also exhibited the same function as human serum antibodies; the sera were able to neutralize the cytotoxic properties of the LTX. Our data also demonstrate that the transplanted cells were functional and the mice were immunologically competent when challenged by both soluble antigen and injection of live *A. actinomycetemcomitans*. Importantly, by combining the SCID system with the murine abscess model, we were able to discriminate between levels of immunologic competence in the different groups of mice; these dissimilarities presumably reflect variations in the levels of immune competence between the patients themselves. Finally, the most significant feature of this model was demonstrated by our ability to immunize the mice to a specific antigen, which in turn elevated the protective capacity of the host's immune response against *A. actinomycetemcomitans* infection.

In summary, our observations clearly demonstrate the feasibility of transplanting a functional immune system from patients with periodontal disease into SCID mice. While these findings must be considered preliminary, they do suggest that this animal model has great potential for future investigations. By manipulating such reconstituted mice under controlled conditions, questions regarding the nature of the immune system and its contribution to human periodontal disease can be studied and evaluated. This approach will be of particular significance as we begin to generate molecularly engineered peptides for relevant antigens in the hope of developing vaccines to periodontal pathogens.

Acknowledgments. This work was supported by U.S. Public Health Service grants DE06014 and DE07118.

REFERENCES

1. **Carlsson, R., C. Martensson, S. Kalliomake, M. Ohlin, and C. A. K. Borrebaeck.** 1992. Human peripheral blood lymphocytes transplanted into SCID mice constitute an *in vivo* culture system exhibiting several parameters found in a normal humoral immune response and are a source of immunocytes for the production of human monoclonal antibodies. *J. Immunol.* **148:**1065–1071.

2. **Carroll, A. M., and M. J. Bosma.** 1991. T-lymphocyte development in SCID mice is arrested shortly after the initiation of T-cell receptor and gene recombination. *Genes Dev.* **5:**1357–1366.

3. **Cesano, A., J. A. Hoxie, B. Lange, P. C. Nowell, J. Bishop, and D. Santoli.** 1992. The severe combined immunodeficient (SCID) mouse as a model for human myeloid leukemias. *Oncogene* **7:**827–836.

4. **Chen, P. B., L. B. Davern, R. Schifferle, and J. J. Zambon.** 1990. Protective immunization against experimental *Bacteroides* (*Porphyromonas*) *gingivalis* infection. *Infect. Immun.* **58:**3394–3400.

5. **Klausen, B.** 1991. Microbiological and immunological aspects of experimental periodontal disease in rats: a review article. *J. Periodontol.* **62:**59–73.

6. **Macht, L., N. Fukuma, K. Leader, D. Sarsero, C. A. S. Pegg, D. I. W. Phillips, P. Yates, S. M. McLachlan, C. Elson, and B. R. Smith.** 1991. Severe combined immunodeficient (SCID) mice: a model for investigating human thyroid autoantibody synthesis. *Clin. Exp. Immunol.* **84:**34–42.

7. **McCune, J. M., R. Namikawa, H. Kaneshima, L. D. Shultz, M. Lieberman, and I. L. Weissman.** 1988. The SCID-hu mouse: murine model for the analysis of human hematolymphoid differentiation and function. *Science* **241:**1632–1639.

8. **Mosier, D. E., R. J. Gulizia, S. M. Baird, and D. B. Wilson.** 1988. Transfer of a functional human immune system to mice with severe combined immunodeficiency. *Nature* (London) **335:**256–259.

9. **Ranney, R. R.** 1991. Immunologic mechanisms of pathogenesis in periodontal diseases: an assessment. *J. Periodontal Res.* **26:**243–254.

10. **Seymour, G. J.** 1987. Possible mechanisms involved in the immunoregulation of chronic inflammatory periodontal disease. *J. Dent. Res.* **66:**2–9.

11. **Seymour, G. J.** 1991. Importance of the host response in the periodontium. *J. Clin. Periodontol.* **18:**421–426.

12. **Shenker, B. J.** 1987. Immunologic dysfunction in the pathogenesis of periodontal diseases. *J. Clin. Periodontol.* **14:**489–498.

13. **Taubman, M. A., H. Yoshie, J. L. Ebersole, D. J. Smith, and C. L. Olson.** 1984. Host response in experimental periodontal disease. *J. Dent. Res.* **63:**455–460.

14. **Tenenbaum, H. C., D. Mock, and A. E. Simor.** 1991. Periodontitis as an early presentation of HIV infection. *Can. Med. Assoc. J.* **144:**1265–1269.

15. **Yamashita, K., J. W. Eastcott, M. A. Taubman, D. J. Smith, and D. S. Cox.** 1991. Effect of adoptive transfer of cloned *Actinobacillus actinomycetemcomitans*-specific T helper cells on periodontal disease. *Infect. Immun.* **59:**1529–1534.

16. **Yoshie, H., M. Taubman, C. L. Olson, J. L. Ebersole, and D. J. Smith.** 1987. Periodontal bone loss and immune characteristics after adoptive transfer of *Actinobacillus*-sensitized T cells to rats. *J. Periodontal Res.* **22:**499–505.

Molecular Pathogenesis of Periodontal Disease
Edited by Robert Genco et al.
© 1994 American Society for Microbiology, Washington, DC 20005

Chapter 15

Mechanisms of Regulation of Bone Formation by Proinflammatory Cytokines

Philip Stashenko, Lien Nguyen, and Yi-Ping Li

A proinflammatory cytokine cascade is induced by bacterial infections, including those that affect the periodontium. These cytokines include interleukin-1α (IL-1α), IL-1β, and tumor necrosis factor alpha (TNF-α) as primary constituents and IL-6, IL-8, and granulocyte macrophage–colony-stimulating factor (GM-CSF) as distal components (11). These mediators are highly pleiotropic and possess a spectrum of activities relevant to periodontal pathogenesis, including stimulation of prostaglandins and proteases, induction of T- and B-cell responses, and stimulation of bone resorption (8). The proinflammatory cytokines are induced in many different host cell types by periodontopathic bacteria and are present at biologically significant concentrations in diseased periodontal tissues and gingival crevicular fluid (24). In the case of IL-1β, levels in tissue have been correlated with recent episodes of tissue destruction (23).

In the past several years, studies from this laboratory have focused on yet another pathogenic property of proinflammatory cytokines, i.e., their ability to inhibit the biosynthetic program of osteoblasts and osteosarcoma cell lines and hence to down-regulate bone formation and repair. In this chapter we review recent work on the molecular mechanisms by which cytokines inhibit bone formation and indicate the relevance of these findings to the clinical treatment of periodontal diseases.

IL-1 AND TNF INHIBIT BONE FORMATION AND ACT AS UNCOUPLERS

Proinflammatory cytokines have been shown to down-regulate bone matrix protein synthesis and inhibit bone formation in vitro and in vivo. IL-1β, IL-1α, TNF-α, and TNF-β inhibit the synthesis of collagen, alkaline phosphatase, and osteocalcin by osteoblasts (4, 20, 22, 25). All of these mediators depress formation

Philip Stashenko, Lien Nguyen, and Yi-Ping Li • Department of Cytokine Biology, Forsyth Dental Center, 140 The Fenway, Boston, Massachusetts 02115.

of nodules of woven bone by disaggregated fetal rat osteoblasts (22). In contrast to resorption, inhibition of bone formation is prostaglandin independent. Osteoblast proliferation is not affected by cytokines, and inhibition is reversible if cytokines are removed prior to the matrix synthesis stage. These findings indicate that cytokines concomitantly stimulate resorption and inhibit bone formation and thereby possess the activities necessary to "uncouple" the bone resorption-formation linkage.

Evidence supporting this concept has been obtained by using in vivo systems in which normal coupling mechanisms are presumably operative. In our own studies, we have infused healthy adult rats with recombinant human IL-1β and observed the simultaneous effects on local and systemic bone resorption and bone formation (21). IL-1β stimulates clear increases in parameters of resorption, including elevated serum calcium and calcium/creatinine ratio in urine and increases in numbers of osteoclasts and resorption surface (Table 1). Bone formation is concomitantly inhibited, as determined by decreased serum osteocalcin levels, decreases in ash and dry weight of bones, and diminished apposition rate. Thus, IL-1β can be shown to disrupt normal coupling mechanisms, as suggested by in vitro studies.

Uncoupling is also seen in hypercalcemia of malignancy (26) and in multiple myeloma, in which IL-1β is produced in large amounts by myeloma cells (13). Although it has been reported that local infusions of IL-1 may under certain circumstances result in increased bone formation, suggesting a coupled response, this was observed only 3 to 4 days after cessation of IL-1 treatment (5). It is therefore likely that inhibition of bone formation and uncoupling in inflammatory and malignant diseases are dependent on the continued presence of inflammation and cytokine production by inflammatory and connective tissue cells.

TRANSCRIPTIONAL REGULATION OF OSTEOCALCIN

In recent studies, we have investigated the mechanisms by which cytokines inhibit the biosynthetic program of differentiated osteoblasts. Toward this end we have focused on the regulation of osteocalcin. Osteocalcin is a noncollagenous osteoblast-specific protein whose levels in serum closely correlate with whole-body bone formation rates and histomorphometric parameters of bone formation (6). Osteocalcin has also been implicated in control of bone resorption (28).

IL-1β and TNF-α both inhibited the production of endogenous osteocalcin protein and steady-state mRNA levels by ROS 17/2.8 osteosarcoma cells, whereas

Table 1. Effect of IL-1β on parameters of bone resorption and bone formation in infused rats[a]

Infusion	Bone resorption		Bone formation	
	Resorption surface (%)	No. of osteoclasts/mm²	Ash wt (g)	Apposition rate (μm/day)
Control (vehicle)	2.01 ± 0.39	0.6 ± 0.17	0.330 ± 0.007	5.68 ± 0.72
IL-1β	4.84 ± 0.70	1.76 ± 0.48	0.317 ± 0.005	2.75 ± 0.26

[a] From Nguyen et al. (21).

IL-6 had only a marginal effect at very high concentrations (Fig. 1) (16). Interestingly, TNF was more profoundly inhibitory than IL-1, with suppression in excess of 80%. The transcriptional regulation of osteocalcin expression by proinflammatory cytokines was then examined using an osteocalcin promoter-chloramphenicol acetyltransferase (CAT) fusion gene (PHOC-CAT). This construct consisted of a large portion of the 5' regulatory region plus the first exon and a portion of the first intron ($-1700/+299$). Following transfection of PHOC-CAT into ROS 17/2.8 osteosarcoma cells, reporter CAT activity was up-regulated in a dose-dependent fashion by 1,25-dihydroxyvitamin D3 (vitamin D) at concentrations above 10^{-12} M, as previously described for osteocalcin (7). In screening studies, TNF-α and IL-6 significantly inhibited vitamin D-stimulated osteocalcin transcription by 57% and 37%, respectively (Table 2). Surprisingly, IL-1α and IL-1β had no effect, despite

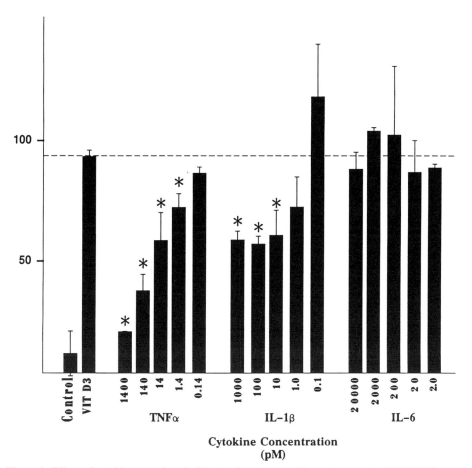

Figure 1. Effect of cytokines on vitamin D3-stimulated osteocalcin production by ROS 17/2.8 osteosarcoma cells. Cells were stimulated concomitantly with vitamin D (10^{-8} M) and cytokines, and 72-h supernatants were assayed for osteocalcin levels by radioimmunoassay. *, significant reduction ($P <$ 0.05) compared with vitamin D alone. From Li and Stashenko (16).

Table 2. Effect of cytokines on vitamin D-stimulated osteocalcin promoter activity[a]

Cytokine	Concn (nM)	No. of expt	% Control response[b]
IL-1α	1.0	1	106 ± 17
IL-1β	1.0	3	97 ± 4
IL-2	2.6	2	102 ± 16
IL-6	20.0	3	63 ± 7[c]
IL-7	0.6	2	111 ± 13
IL-8	12.5	4	107 ± 23
TNF-α	1.4	6	43 ± 12[c]
M-CSF	2.0	2	98 ± 8

[a] From Li and Stashenko (16).
[b] Control: CAT expression in presence of 10^{-8} M vitamin D.
[c] $P < 0.01$.

their clear inhibition of osteocalcin protein and mRNA. Other immune cytokines and growth factors, including IL-2, IL-3, IL-7, IL-8, and M-CSF, also failed to regulate osteocalcin transcription. TNF-α also inhibited basal transcription by approximately 50%, indicating a regulatory effect independent of the positive stimulus. Despite their lack of effect on transcription, IL-1α and IL-1β stimulated prostaglandin E_2 production by ROS 17/2.8, demonstrating that ROS 17/2.8 expresses IL-1 receptors and can respond to these mediators. In dose-response studies, down-regulation by TNF-α was significant at concentrations as low as 0.14 pM (0.1 U/ml). In contrast, IL-6 down-regulation required an approximately 10^4-fold-higher concentration (20 nM) to exert a significant effect. TNF-α-mediated down-regulation was unaffected by indomethacin, confirming our prior data that inhibition of bone formation is prostaglandin independent (29).

These data showed that of all cytokines studied, TNF-α alone potently down-regulates osteocalcin transcription. The lack of effect of IL-1 on the promoter suggests that this mediator acts by a different mechanism, perhaps by reducing mRNA stability. Although IL-6 had a slight inhibitory effect on transcription (but no effect on protein), it is probably not an important regulator of osteocalcin, particularly in light of the high concentrations necessary to observe transcriptional inhibition. Taken together, these studies demonstrate that activities of the proinflammatory cytokines are redundant but that they are also heterogeneous in terms of the mechanisms by which inhibition is mediated.

MECHANISM OF TRANSCRIPTIONAL INHIBITION BY TNF AND CHARACTERIZATION OF TNFRE

Mapping studies have subsequently been undertaken to characterize the element responsive to TNF (TNFRE) in the osteocalcin promoter (17). The long construct PHOC-CAT (−1700/+299) was digested with ExoIII, and the resultant 5' deletions were characterized by electrophoretic mobility and, in selected cases, by sequencing. Deletions were transfected into ROS 17/2.8 and tested for the effect of TNF on vitamin D-stimulated CAT activity. As shown in Fig. 2, deletion analysis localized the TNFRE to the -522 to -511 region, which was found to contain a 9-bp palindromic motif (AGGCTGCCT). A second homologous element was present at $-687/-677$ (TGGCTGTGCCA), the interruption of which also resulted in loss of TNF down-regulation. All deletions shorter than -511 also lost TNF down-regulation. A promoter segment containing the TNFRE sequence ($-580/+299$) down-regulated a heterologous simian virus 40 promoter in both the presence and the absence of vitamin D. In addition, this segment in reverse 3'-5' orientation also mediated inhibition, which is consistent with the activity of a classic repressor element. The TNFRE was then specifically modified by insertion of a HindIII restriction site (Fig. 3). The mutagenized TNFRE sequence lost the ability to inhibit vitamin D–stimulated CAT activity in the context of either the osteocalcin or the simian virus 40 promoters. These data provide convincing evidence that TNF regulates the human osteocalcin gene via a specific TNFRE.

A UNIQUE TRANSCRIPTION FACTOR BINDS TO TNFRE

Comparison of the TNFRE motif with other regulatory elements reveals some homology to the nuclear transcription factor NF-κB [consensus: GGGR(CAT) TYYCC] (15). In particular, the TNFRE is closely similar to the NF-κB2b site in

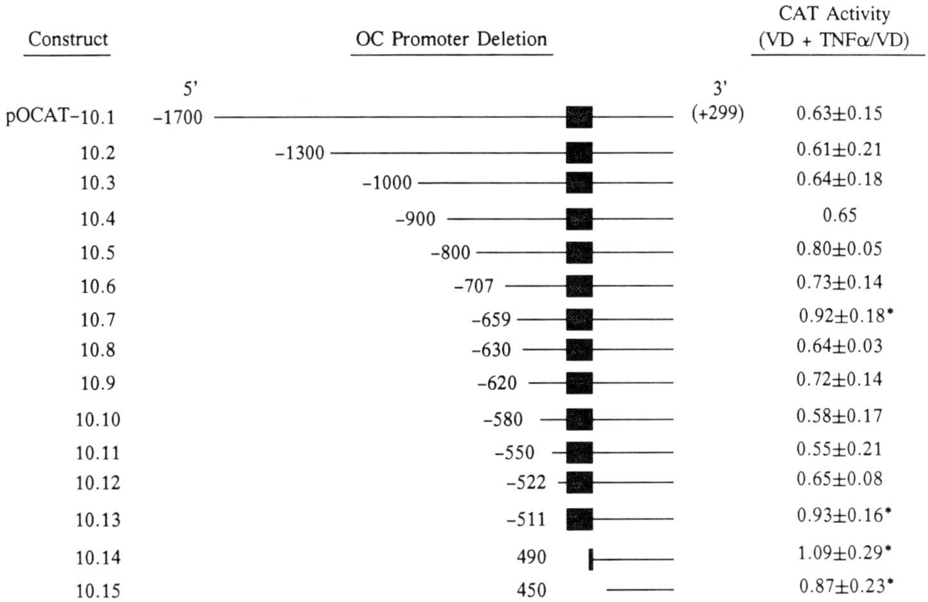

Construct	OC Promoter Deletion		CAT Activity (VD + TNFα/VD)
	5'	3'	
pOCAT–10.1	−1700 ————————————■————	(+299)	0.63±0.15
10.2	−1300 ————————■———		0.61±0.21
10.3	−1000 ————■———		0.64±0.18
10.4	−900 ———■———		0.65
10.5	−800 ———■———		0.80±0.05
10.6	−707 ——■———		0.73±0.14
10.7	−659 —■———		0.92±0.18*
10.8	−630 —■———		0.64±0.03
10.9	−620 —■———		0.72±0.14
10.10	−580 —■———		0.58±0.17
10.11	−550 ■———		0.55±0.21
10.12	−522 ■———		0.65±0.08
10.13	−511 ■———		0.93±0.16*
10.14	490 ⊢———		1.09±0.29*
10.15	450 ———		0.87±0.23*

Figure 2. Deletion analysis of the osteocalcin (OC) 5'-flanking region for localization of the TNF-α-responsive element. ■, Position of the VDRE ($-510/-487$) (21); *, constructs exhibiting loss of TNF-α down-regulation of the vitamin D response. Transfected cultures were tested for TNF-α-stimulated suppression of 1,25-dihydroxyvitamin D3 (vitamin D)-induced CAT activity. From Li and Stashenko (17).

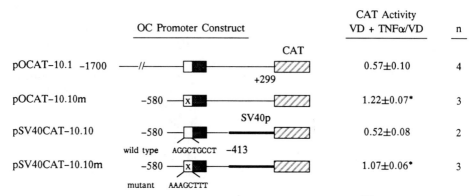

Figure 3. Response of site-directed mutants of the TNFRE. ▭, TNFRE; ⊠, *Hind*III site-specific mutant of TNFRE; ■, VDRE; *, deletions resulting in loss of TNF-α inhibition of the vitamin D response. An *Xho*I-*Pvu*II fragment ($-580/-413$) of the mutagenized TNFRE (pOCAT-10.10m) was also introduced into pSV40-CAT to test heterologous promoter regulation. The transfection, stimulation, and CAT assay protocol were carried out as described in the text. OC, osteocalcin; n, number of determinations. From Li and Stashenko (17).

the TNF promoter (GGGGCTGCCCC) (12). Mobility shift assays were performed to investigate this relationship and to determine the interactions of the TNFRE with DNA-binding regulatory proteins. Nuclear extracts were produced from TNF-stimulated and control ROS 17/2.8 cells. Extracts were reacted with a ^{32}P-labeled probe that contained the TNFRE ($-539/-505$). As shown in Fig. 4, unstimulated extracts contained a DNA-binding protein that retarded the mobility of the probe. The binding of this constitutively expressed protein was completely inhibited by unlabeled homologous probe as well as by an NF-κB consensus sequence but not by the consensus sequence for an unrelated motif (Fig. 4, Sp1, lane 6). Mutagenesis of the TNFRE by *Hind*III replacement decreased inhibition (lane 5). Interestingly, TNF stimulation induced a second TNFRE-binding protein that displaced the constitutive factor. The TNF-induced protein was inhibitable by homologous probe but was not inhibited by the NF-κB consensus sequence, suggesting the binding of a non-NF-κB factor.

Supershift assays were carried out in the presence of anti-NF-κB antisera to investigate the relationship between the constitutive and the TNF-induced proteins and NF-κB. The mobility of the constitutive factor was altered by an antiserum directed against the p50 subunit of NF-κB but not by antisera against the p65 subunit or *c-rel*. The constitutively expressed factor was therefore tentatively identified as the p50 homodimer of NF-κB, also termed KBF1 (3). In contrast, the TNF-induced protein was unreactive with all anti-NF-κB antisera, including anti-p50, anti-p65, and anti-*c-rel*. The TNF-induced TNFRE-binding protein may therefore represent a unique regulatory entity.

Finally, DNase footprinting was carried out to further define the region of interaction between these regulatory proteins and the promoter. Footprinting studies demonstrated that both factors protected the $-518/-497$ portion of the pro-

Figure 4. Electrophoretic mobility shift assay of DNA-binding proteins from TNF-stimulated or unstimulated ROS 17/2.8 cells. A 35-mer synthetic oligonucleotide containing the TNFRE ($-539/-505$) was end labeled with ^{32}P and utilized as a probe. Nuclear protein extracts were prepared from unstimulated or stimulated (10^{-9} M TNF) ROS 17/2.8 cells as described elsewhere (12). Binding reactions were carried out in the absence or presence of the following double-stranded DNA inhibitors: 35-mer homologous osteocalcin probe, 34-mer osteocalcin probe with TNFRE core replaced with a *Hind*III restriction site as in Fig. 3, 22-mer NF-κB consensus sequence (AGTTGAGGGGACTTTCCCAGGC) (22) (Promega, Madison, Wis.), and 22-mer Sp1 consensus sequence (ATTCGATCGGGGCGGGGC GAGC) (24) (Promega). Following the binding reaction, complexes were electrophoresed in a 4% polyacrylamide gel under low-salt conditions (28). Constitutive (C) and TNF-induced (I) DNA-binding proteins are indicated. Lanes 1 and 12, probe alone; lanes 2 through 6, uninduced extract (8 μg per lane); lanes 7 through 11, TNF-induced extract (8 μg); lanes 2 and 7, no inhibitor; lanes 3 and 8, homologous probe as inhibitor (100-fold excess); lanes 4 and 9, *Hind*III mutant (34-mer synthetic oligomer) (100-fold excess); lanes 5 and 10, NF-κB (100-fold excess); lanes 6 and 11, 100Sp1 (100-fold excess). From Li and Stashenko (17).

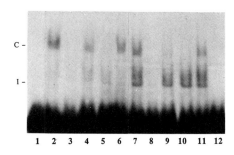

moter, which is consistent with mapping studies (Fig. 5). The protected region impinges on the vitamin D response element (VDRE) ($-510/-490$), suggesting that interactions occur between the TNFRE-binding proteins and the vitamin D receptor-VDRE complex.

Taken together, these data demonstrate the existence of a novel inhibitory pathway by which TNF suppresses osteocalcin expression (Fig. 6). The 5′ regulatory region of the rat osteocalcin gene also contains the TNFRE motif in similar proximity to the VDRE (18). Interestingly, a number of human genes known to be down-regulated by TNF, including other bone matrix proteins such as osteonectin, alkaline phosphatase, and collagen α1(I), also contain the TNFRE motif. None of these homologs has yet been tested for function. Nevertheless, it is tempting to speculate that the TNFRE represents a common mechanism by which TNF mediates inhibition of an entire array of bone matrix proteins.

SIGNIFICANCE OF CYTOKINE-MEDIATED INHIBITION FOR PERIODONTAL PATHOGENESIS AND REGENERATION

Periodontal diseases are bacterial infections characterized by loss of the bony support of the dentition. Most recent evidence has implicated as putative periodontal pathogens a relatively small group of microorganisms, including *Actinobacillus actinomycetemcomitans, Porphyromonas gingivalis, Prevotella intermedia, Bacteroides forsythus, Campylobacter rectus,* and others (9). These organisms are predominantly gram-negative anaerobes that produce lipopolysaccharides (LPS) and other cell wall components. These substances are potent stimulators of proinflammatory cytokine production by macrophages and other host cell types (8).

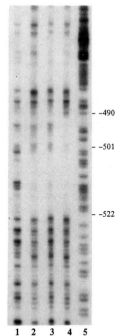

$- -490$

$- -501$

$- -522$

1 2 3 4 5

Figure 5. DNase I footprinting of TNFRE-binding proteins. The *Pvu*II-*Xho*I fragment ($-580/-413$) was labeled at the $3'$ end on the antisense strand. Lane 1, incubation with bovine serum albumin (12 μg); lanes 2 through 4, 12 μg of nuclear extracts from ROS 17/2.8; lane 2, uninduced; lane 3, TNF induced; lane 4, vitamin D (10^{-8} M) induced; lane 5, degradative sequence (C/T) residues. From Li and Stashenko (17).

Current periodontal therapy is directed toward elimination of these pathogens by mechanical and chemotherapeutic means. Clinical treatment success is usually defined as cessation of tissue destruction. However, newer procedures using guided tissue regeneration (10), alone or in combination with growth factor treatment (19),

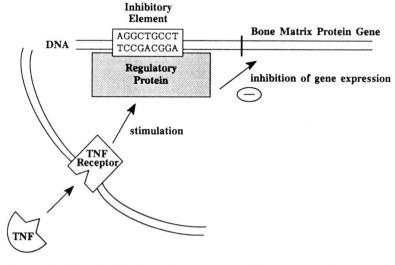

Figure 6. Model for TNF-stimulated down-regulation of bone matrix protein expression.

are attempting to replace lost supporting bone. These approaches have shown promise, yet they are often quite unpredictable in outcome. Given the role of proinflammatory cytokines in down-regulating osteoblastic activity, we suggest as a working hypothesis that the failure of natural regeneration (coupling) as well as therapeutic periodontal regenerative procedures may be due in part to continued cytokine production at sites of periodontal destruction. Although periodontal pathogens may be largely eliminated from the pocket following therapy, there is strong evidence that root dentin and cementum remain infected (14, 27). Infected radicular hard tissue may serve as a reservoir of reinfecting microorganisms as well as a source of LPS and other inducers of IL-1 and TNF (2). For example, Adriaens et al. (1) have demonstrated that bacteria invade dentinal tubules and can be cultured from root dentin in 87% of periodontally diseased caries-free teeth. Seventy-four percent of teeth had bacteria in the middle and inner dentinal zones, suggesting that even aggressive root-planing procedures are not capable of removing them. Many of these organisms were anaerobes that presumably produce endotoxin and activate cytokine expression.

In order to improve the prognosis for periodontal regeneration, a two-pronged approach is envisioned. First, more complete bacterial removal and "detoxification" of root surfaces must be accomplished. This may involve local treatment with antibacterial agents perhaps combined with new methods of sealing infected dentinal tubules from which bacteria and their products cannot be physically removed. Such treatments must of course result in a biologically compatible surface to which new periodontal structures can attach. Second, the inhibitory effects of cytokines may themselves be inhibited. Elucidation of the molecular mechanisms involved, such as the TNFRE and its binding protein discussed above, constitutes the first step in the design of targeted anti-inflammatory therapeutics that can block these inhibitory pathways. Such strategies used in combination with growth factors may yield more favorable regenerative outcomes in the future.

Acknowledgments. This work was supported by Public Health Service grant DE-07378 from the National Institute of Dental Research.

We thank W. Chen and P. Fujiyoshi for excellent technical assistance and Mary Ann Cugini for graphics and manuscript preparation.

REFERENCES

1. **Adriaens, P. A., J. A. De Boever, and W. J. Loesche.** 1988. Bacterial invasion in root cementum and radicular dentin of periodontally diseased teeth in humans. *J. Periodontol.* **59:**222–230.
2. **Aleo, J., F. De Renzis, P. Farber, and A. Varboncoeur.** 1974. The presence and biological activity of cementum-bound endotoxin. *J. Periodontol.* **45:**672–675.
3. **Baldwin, A. S., and P. A. Sharp.** 1988. Two transcription factors, NF-κB and H2Tf1, interact with a single regulatory sequence in the class I major histocompatibility complex promoter. *Proc. Natl. Acad. Sci. USA* **85:**723–727.
4. **Bertolini, D. R., G. E. Nedwin, T. S. Bringman, D. D. Smith, and G. R. Mundy.** 1986. Stimulation of bone resorption and inhibition of bone formation *in vitro* by human tumor necrosis factors. *Nature* (London) **319:**516–518.

5. **Boyce, B. F., T. B. Aufdemorte, I. R. Garrett, A. J. P. Yates, and G. R. Mundy.** 1989. Effects of interleukin-1 on bone turnover in normal mice. *Endocrinology* **125:**1142–1150.

6. **Delmas, P. D., L. Malaval, M. E. Arlot, and P. J. Meunier.** 1985. Serum bone Gla protein compared to bone histomorphometry in endocrine diseases. *Bone* **6:**339–345.

7. **Demay, M. B., D. A. Roth, and H. M. Kronenberg.** 1989. Regions of the rat osteocalcin gene which mediate the effect of 1,25-dilhydroxyvitamin D3 on gene transcription. *J. Biol. Chem.* **264:**2279–2282.

8. **Dinarello, C. A.** 1989. Interleukin-1 and its biologically related cytokines. *Adv. Immunol.* **44:**153–205.

9. **Dzink, J. L., A. C. R. Tanner, A. D. Haffajee, and S. S. Socransky.** 1985. Gram negative species associated with active destructive periodontal lesions. *J. Clin. Periodontol.* **12:**648–659.

10. **Hancock, E. B.** 1989. Regeneration procedures, p. 1–19. *Proceedings of the World Workshop in Clinical Periodontics,* vol. VI. American Academy of Periodontics, Chicago.

11. **Hesse, D. G., K. J. Tracey, Y. Fong. K. R. Manogue, M. A. Palladino, A. Cerami, G. T. Shires, and S. F. Lowry.** 1988. Cytokine appearance in human endotoxemia and primate bacteremia. *Surg. Gynecol. Obstet.* **166:**147–153.

12. **Jongeneel, C. V., A. N. Shakhov, S. A. Nedospasov, and J.-C. Cerottini.** 1989. Molecular control of tissue-specific expression at the mouse TNF locus. *Eur. J. Immunol.* **19:**549–552.

13. **Kawano, M., I. Yamamoto, K. Iwato, H. Tanaka, H. Aosuka, O. Tanabe, H. Ishikawa, et al.** 1989. Interleukin-1 beta rather than lymphotoxin as the major bone resorbing activity in human multiple myeloma. *Blood* **73:**1646–1649.

14. **Langeland, K., H. Rodrigues, and W. Dowden.** 1974. Periodontal disease, bacteria and pulpal histopathology. *Oral Surg.* **37:**257–270.

15. **Lenardo, M.J., and D. Baltimore.** 1989. NF-kappa B: a pleiotropic mediator of inducible and tissue-specific gene control. *Cell* **58:**227–229.

16. **Li, Y. P., and P. Stashenko.** 1992. Pro-inflammatory cytokines tumor necrosis factor-α and IL-6, but not IL-1, down-regulate the osteocalcin gene promoter. *J. Immunol.* **148:**788–794.

17. **Li, Y. P., and P. Stashenko.** 1993. Characterization of a tumor necrosis factor responsive element which down-regulates the human osteocalcin gene. *Mol. Cell. Biol.* **13:**3714–3721.

18. **Lian, J., C. Stewart, E. Puchacz, S. Mackowiak, V. Shalhoub, D. Collart, G. Zambetti, and G. Stein.** 1989. Structure of the rat osteocalcin gene and regulation of vitamin D-dependent expression. *Proc. Natl. Acad. Sci. USA* **86:**1143–1147.

19. **Lynch, S. E., R. C. Williams, and A. M. Polson.** 1989. A combination of platelet-derived and insulin-like growth factors enhances periodontal regeneration. *J. Clin. Periodontol.* **16:**545–548.

20. **Nanes, M. S., J. Rubin, G. Titus, G. Hendy, and B. D. Catherwood.** 1991. Tumor necrosis factor alpha inhibits 1,25-dihydroxyvitamin D3-stimulated bone Gla protein synthesis in rat osteosarcoma cells (ROS 17/2.8) by a pretranslational mechanism. *Endocrinology* **128:**2577–2584.

21. **Nguyen, L., F. E. Dewhirst, P. V. Hauschka, and P. Stashenko.** 1991. Interleukin-1β stimulates bone resorption and inhibits bone formation *in vivo. Lymphokine Cytokine Res.* **10:**15–21.

22. **Stashenko, P., F. E. Dewhirst, M. L. Rooney, L. A. DesJardins, and J. D. Heeley.** 1987. Interleukin 1β is a potent inhibitor of bone formation *in vitro. J. Bone Miner. Res.* **2:**559–565.

23. **Stashenko, P., P. Fujiyoshi, M. S. Obernesser, L. Prostak, A. D. Haffajee, and S. S. Socransky.** 1991. Levels of interleukin 1β in tissue from sites of active periodontal disease. *J. Clin. Periodontol.* **18:**548–554.

24. **Stashenko, P., J. J. Jandinski, P. Fujiyoshi, J. Rynar, and S. S. Socransky.** 1991. Tissue levels of bone resorptive cytokines in periodontal disease. *J. Periodontol.* **62:**504–509.

25. **Stashenko, P., M. S. Obernesser, and F. E. Dewhirst.** 1989. Effect of immune cytokines on bone. *Immunol. Invest.* **18:**239–249.

26. **Stewart, A. F., A. Vignery, A. Silverglate, N. Ravin, V. Livolsi, A. Broadus, and R. Baron.** 1989. Quantitative bone histomorphometry in humoral hypercalcemia of malignancy. *J. Clin. Endocrinol. Metab.* **55:**219–227.

27. **Waerhaug, J.** 1978. Healing of the dento-epithelial junction following subgingival plaque control. II. As observed on extracted teeth. *J. Periodontol.* **49:**119–134.

28. **Webber, D., P. Osdoby, P. Hauschka, and M. Krukowski.** 1990. Correlation of an osteoclast antigen and ruffled border on giant cells formed in response to resorbable substrates. *J. Bone Miner. Res.* **5:**401–410.
29. **Young, M. F., D. M. Findlay, P. Dominguez, P. D. Burbelo, C. McQuillan, J. B. Kopp, P. Gehron-Robey, and J. D. Termine.** 1988. Osteonectin promoter. *J. Biol. Chem.* **264:**450–456.

Molecular Pathogenesis of Periodontal Disease
Edited by Robert Genco et al.
© 1994 American Society for Microbiology, Washington, DC 20005

Chapter 16

Regulation of Chronic Inflammation

Sharon M. Wahl, Keith L. Hines, Tomozumi Imamichi, Hongsheng Tian, Sharnn Shepheard, and Nancy L. McCartney-Francis

The organized accumulation of microorganisms at the tooth surface (14) triggers humoral and cellular mechanisms of host defense (32). Microbial release of chemotactic factors, endotoxins, cell wall fragments, toxins, and other products (5, 20, 27, 32) (Fig. 1) engages leukocyte populations that emigrate from the vasculature, become activated, and attempt to clear the offending pathogens (Table 1). While this series of events is characteristic of the host response to injury, bacteria, or antigen deposition, the inability of the host to eliminate or sequester persistently accumulating numbers of pathogens that occur in the oral cavity results in prolonged inflammation. This progression to chronic periodontal disease may be associated with the destruction of soft tissues and alveolar bone (32), and similarly, chronic inflammation in other tissues such as lung, liver, and synovium may also result in tissue destruction (30). Understanding the sequelae of events progressing to pathologic tissue injury is central to the identification of targets for modulating chronic inflammatory processes as an adjunct to antimicrobial therapy.

LEUKOCYTE RECRUITMENT

Recruitment of leukocytes to a site of infection is dependent on the initial adhesion of the cells to the vascular wall and extracellular matrix components (29). This selective adhesion is a multistep process involving initial rolling of leukocytes along the vessel wall followed by a firmer attachment that enables the cells to localize to the target site and emigrate into the tissues. The considerable interest focused on these early events in the evolution of an inflammatory response has led to the identification of adhesion molecule receptor families and their independent but interrelated contributions to leukocyte recruitment (Table 2). The initial slowing down of the leukocytes at an inflammatory site appears to be mediated by selectins, which are carbohydrate-binding lectins capable of low-affinity adherence interac-

Sharon M. Wahl, Keith L. Hines, Tomozumi Imamichi, Hongsheng Tian, Sharnn Shepheard, and Nancy L. McCartney-Francis • Laboratory of Immunology, National Institute of Dental Research, National Institutes of Health, Bethesda, Maryland 20892-0001.

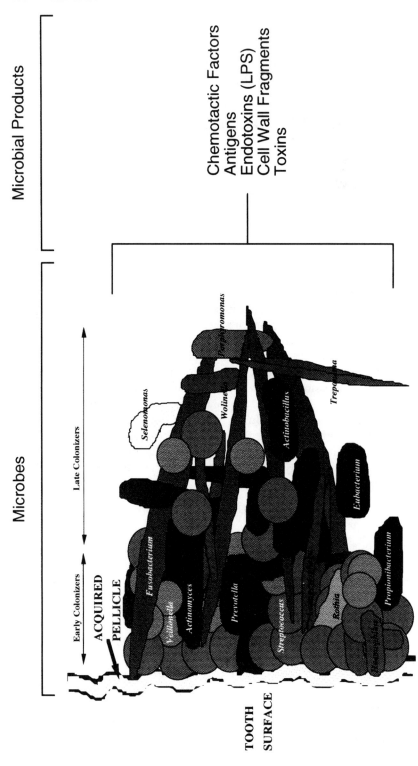

Figure 1. Microbial products recruit and stimulate inflammatory cells. Because of their adhesive properties, bacteria sequentially colonize the tooth surface (14), where they release a plethora of products capable of triggering inflammatory cell extravasation, recruitment, and activation. Activation of the infiltrating leukocytes results in the release of a cascade of inflammatory mediators that are involved in host defense but may become pathologic if persistently produced. Modified from Kolenbrander et al. (14) with permission. LPS, lipopolysaccharide.

Table 1. Inflammatory events triggered by trauma, infection, or foreign materials

Vascular changes
Cell adhesion
Leukocyte migration
Leukocyte activation
Phagocytosis
Enzymatic activity
Mediator production
Elimination of foreign material
Resolution of inflammation
Tissue repair

tions between endothelial cells and leukocytes (3, 15). Engagement of these receptor-ligand pairs is responsible for targeted leukocyte rolling (3). Subsequent interactions involve the α, β-heterodimer integrin receptors, including the β_2 leukocyte integrins, leukocyte function antigen (LFA-1), and MAC-1, which mediate attachment to intracellular adhesion molecules (ICAM-1 and ICAM-2) (29). Additionally, leukocyte interactions with vascular cell adhesion molecules (VCAM-1) (4, 24) and extracellular matrix, promoted by β_1 integrins (12, 25), all combine to selectively target circulating leukocytes to the appropriate site of infection and inflammation (Table 2).

INTERRUPTION OF LEUKOCYTE RECRUITMENT

Because this complex adhesion process is mandatory in both physiologic and pathologic inflammation, studies have focused on mechanisms to block adhesion with the potential for modulating the inflammatory response and any associated pathology. While emphasis has focused on blocking leukocyte β_2 interactions with endothelial cells (13) and, recently, on interrupting selectin-mediated adhesion (22), our studies have explored the potential of β_1 integrins as a target for blocking leukocyte adhesion and recruitment. Several of the β_1 integrins ($\alpha_5\beta_1$, $\alpha_4\beta_1$, $\alpha_3\beta_1$) bind to various domains of the extracellular matrix glycoprotein fibronectin. Fibronectin is found in plasma, cell matrices, and basal laminae and also on cell

Table 2. Leukocyte adhesion molecule interactions at sites of inflammation[a]

Leukocyte	Endothelial cell(s)	ECM
L-Selectin	Sialyl Lew[a]	
GlyCAM-1	E and P selectins	
LFA-1 ($\alpha_L\beta_2$)	ICAM-1, ICAM-2	
MAC-1 ($\alpha_M\beta_2$)	ICAM-1, ICAM-2	
VLA-3 ($\alpha_3\beta_1$)		Fibronectin, laminin, collagen
VLA-4 ($\alpha_4\beta_1$)	VCAM-1	Fibronectin (CS-1, heparin)
VLA-5 ($\alpha_5\beta_1$)		Fibronectin (RGD)

[a] LFA-1, leukocyte function antigen; MAC-1, CD11/CD18; VLA, very late antigen; ICAM-1 and -2, intracellular adhesion molecules 1 and 2; VCAM-1, vascular cell adhesion molecule; ECM, extracellular matrix.

surfaces (1), providing ready targets for cells bearing the appropriate receptors. Interference with leukocyte-fibronectin interactions may prevent leukocyte adhesion and migration at sites of inflammation. In this regard, exogenous addition of synthetic fibronectin peptides (8 to 25 amino acids) derived from the Arg-Gly-Asp (RGD) cell-binding domain as well as from the heparin-binding region of this glycoprotein inhibits leukocyte binding to fibronectin substrates in vitro (10, 31). Moreover, when administered systemically in an animal model in which animals are rendered arthritic by the injection of bacterial cell wall peptidoglycan-polysaccharide complexes, peptides containing the RGD motif have a dramatic effect in suppressing the development of acute and chronic phases of the arthritic lesions (10, 31). Intravenous injection of RGD-containing peptides inhibited leukocyte recruitment into the synovium, likely by interfering with the binding of $\alpha_5\beta_1$, $\alpha_3\beta_1$, or some other receptor to fibronectin matrix proteins. Peptides incorporating the CS-1 domain were also inhibitory by virtue of their selective binding to $\alpha_4\beta_1$ receptors, which may negate cell-cell and cell-matrix adhesion.

In additional studies, three peptides (FNI, II, and V) derived from the 33-kDa heparin-binding carboxyl-terminal region of the fibronectin A chain that could block inflammation by apparent RGD-independent pathways were identified. Whether coupled to a carrier protein (ovalbumin) or administered as free peptides, microgram quantities of these peptides administered sequentially for only 5 days reduced the clinical parameters of arthritis >80% compared to those in untreated animals during the chronic stages of the disease, as reflected by the reduced articular index (31). Recent evidence suggests that the heparin-binding domain of fibronectin interacts with leukocyte populations via the $\alpha_4\beta_1$ integrin (26) and that this receptor is up-regulated during the course of inflammatory disease in the bacterium-induced arthritis animal model (12a). These data lend credence to the concept that the initial adherence and recruitment of leukocytes are important targets for therapeutically manipulating an aberrant immune or inflammatory response.

LEUKOCYTE ACTIVATION

As the leukocytes enter the tissues at a site of inflammation, they become exposed to a barrage of stimuli that may be derived from serum, bacteria, platelets, other inflammatory cells, and stromal cells. In response to this plethora of stimulatory agents, the leukocytes are induced to differentiate, become activated, and produce a host of inflammatory mediators, including lipid metabolites, proteins, enzymes, and reactive oxygen species. While these leukocyte-generated products cooperate to kill and/or sequester foreign materials and microbial pathogens, failure to eliminate the microorganisms and their products may result in prolonged production and secretion of potentially toxic inflammatory molecules.

Among the molecules that when delivered in modest amounts are protective but in excess become destructive are reactive oxygen species. Stimulated macrophages release reactive oxygen intermediates (ROI) including O_2^- and H_2O_2, which have microbistatic and microbicidal properties. In the development of chronic inflammatory lesions, excess ROI have been linked to tissue pathology (28). Antagonists of ROI, including superoxide dismutase and catalase, effectively diminish inflammatory processes and thereby inhibit the tissue destruction associated with persistent immune activation (28). Another toxic free radical involved in host defense is nitric oxide (NO). As a mediator associated with killing of leishmania, toxoplasma, and other organisms (23), NO appears to have important microbicidal activity and is up-regulated following macrophage activation (9, 23). Stimulated macrophages express inducible nitric oxide synthase, which catalyzes the generation of NO from arginine (9, 35). Rapidly metabolized to stable and quantifiable nitrites and nitrates, NO has also been identified in tissues undergoing chronic inflammatory responses (18, 19), implicating excess levels of this toxic molecule as potentially detrimental.

SUPPRESSION OF NO

In order to define a role for NO in the dysregulation of normal tissue architecture associated with certain types of inflammatory lesions, inhibitors of NO synthesis have been administered exogenously and monitored for their therapeutic potential in chronic inflammatory disease (18). For example, when N^G-monomethyl-L-arginine, an inhibitor of nitric oxide synthase, is delivered intravenously in the bacterial-cell-wall-induced model of arthritis, it significantly suppresses the evolution of tissue-destructive synovial lesions (18).

While the production of NO by macrophages is less well defined in humans (8), inflamed periodontal tissues have recently been shown to generate NO, as reflected by increased nitrite levels compared to those of uninflamed gingival tissue specimens (19). In the context of the oral cavity, continued leukocyte recruitment and activation by periodontopathic microorganisms and their products may lead to an overzealous attempt on the part of the leukocytes to destroy these pathogens. In so doing, the leukocytes may release into the milieu excess levels of toxic products that target host tissues in addition to the pathogens. Clearly, the evidence in experimental systems suggests that inhibition of the release of such molecules may be of significant benefit in alleviating the associated tissue pathology. Thus, antibiotic therapy coupled with inhibitors of toxic molecules such as NO may provide an important approach to alleviating the severe injury accompanying microbe-induced chronic periodontitis.

CYTOKINES WITH ANTI-INFLAMMATORY POTENTIAL

While selected pharmacologic inhibitors of O_2^- and NO may provide some protection under conditions in which inflammation becomes chronic and destructive, recent evidence indicates that endogenous cytokines may provide alternative methods of anti-inflammatory therapy. In particular, products of the type 2 T helper (Th2) subpopulation of lymphocytes (6) appear to have unique anti-inflammatory activities. While promoting humoral immune pathways, Th2-derived interleukin-4 (IL-4) and IL-10 also suppress many of the actions associated with Th1 (helper)

Table 3. Suppression of macrophage inflammatory mediators by IL-4 and IL-10[a]

Macrophage product	IL-4	IL-10
IL-1	↓	↓
IL-1ra	↑	↑
TNF	↓	↓
O_2^-	↓	↓
NO	↓	↓
Collagenase	↓	↓

[a] IL-1ra, IL-1 receptor antagonist; TNF, tumor necrosis factor; ↓, suppression; ↑, up-regulation.

lymphocytes. Especially relevant are the abilities of IL-4 and IL-10 to target macrophages and suppress their release of cytokines, ROI, and NO (Table 3), thereby markedly diminishing cell-mediated immune pathways (11, 33, 34). In addition to inhibiting the production of inflammatory mediators, IL-4 induces programmed cell death, or apoptosis, to reduce the number of locally infiltrating inflammatory macrophages (16, 17) and up-regulates the production of IL-1 receptor antagonist, a potent inhibitor of inflammatory pathways (33). Moreover, Th2 products appear to be lacking in chronic inflammatory lesions (7, 21), suggesting that a dysregulation in the production of IL-4 and/or IL-10 may contribute to the inability of an inflammatory response to resolve and to the evolution of a chronic inflammatory lesion. In this regard, IL-4 is deficient in chronic periodontal tissues (7) as well as in rheumatoid synovium (21), which favors the notion that exogenous application of IL-4 might foster resolution of such lesions. Indeed, systemic administration of IL-4 has a beneficial impact on the course of chronic disease initiated by bacterial peptidoglycan-polysaccharide complexes (2). Thus, IL-4 appears to provide a network of intersecting mechanisms responsible for inhibiting pathologic inflammation, implicating dysregulation of immunomodulatory molecules in disease etiology as well as elucidating targets for intervention.

SUMMARY

As our understanding of the basic cellular, molecular, and genetic mechanisms of immune and inflammatory events expands, a clearer view of the dysregulation of these events that lead to aberrancy and disease emerges. Identification of aberrant leukocyte adherence provided the impetus for the use of antiadhesive peptides, and the demonstration of excess NO synthesis implicated the use of NO synthase inhibitors to ameliorate tissue pathology. Continued pursuit of these and other pathways and their targeted disruption may provide novel therapeutic approaches.

REFERENCES

1. **Akiyama, S. K., and K. M. Yamada.** 1987. Fibronectin. *Adv. Enzymol.* **57:**1–57.
2. **Allen, J. B., H. L. Wong, G. L. Costa, M. Bienkowski, and S. M. Wahl.** 1993. Suppression of

monocyte function and differential regulation of IL-1 and IL-1ra by IL-4 contribute to resolution of experimental arthritis. *J. Immunol.* **151:**4344–4351.

3. **Bevilacqua, M. P., and R. M. Nelson.** 1993. Selectins. *J. Clin. Invest.* **91:**379–387.

4. **Elices, M. J., L. Osborn, Y. Takada, C. Crouse, S. Luhowskyj, M. J. Hemler, and R. Lobb.** 1990. VCAM-1 on activated endothelial cells interacts with the leukocyte integrin VLA-4 at a site distinct from the VLA-4/fibronectin binding site. *Cell* **60:**577–584.

5. **Finlay, B. B., and S. Falkow.** 1989. Common themes in microbial pathogenicity. *Microbiol. Rev.* **53:**210–230.

6. **Fiorentino, D. F., M. W. Bond, and T. R. Mosmann.** 1989. Two types of mouse helper T cell. IV. Th2 clones secrete a factor that inhibits cytokine production by Th1 clones. *J. Exp. Med.* **170:**2081–2095.

7. **Fujihasi, K., J. Yoshiharu, K. W. Beagley, M. Yamamoto, J. R. McGhee, J. Mestecky, and H. Kiyono.** 1993. Cytokines and periodontal disease: immunopathological role of interleukins in chronic inflamed gingival tissues. *J. Periodontol.* **64:**400–406.

8. **Geller, D. A., C. J. Lowenstein, R. A. Shapiro, A. K. Nussler, M. Di Silvio, S. C. Wang, D. K. Nakayama, R. L. Simmons, S. H. Snyder, and T. R. Billiar.** 1993. Molecular cloning and expression of inducible nitric oxide synthase from human hepatocytes. *Proc. Natl. Acad. Sci. USA* **90:**3491–3495.

9. **Hibbs, J. B., Jr., R. R. Taintor, and Z. Vavrin.** 1987. Macrophage cytotoxicity: role for L-arginine deiminase and imino nitrogen oxidation to nitrite. *Science* **235:**473–476.

10. **Hines, K. L., T. Imamichi, J. B. McCarthy, L. T. Furcht, and S. M. Wahl.** 1993. Fibronectin peptides inhibit inflammatory cell infiltration and arthritis. *J. Immunol.* **150:**139A.

11. **Howard, M., A. O'Garra, H. Ishida, R. DeWaal Malefyt, and J. De Vries.** 1992. Biological properties of interleukin 10. *J. Clin. Immunol.* **12:**239–247.

12. **Hynes, R. O.** 1987. Integrins. A family of cell surface receptors. *Cell* **48:**549–554.

12a. **Imamichi, T.** Unpublished observations.

13. **Isobe, M., H. Yagita, K. Okumura, and A. Ihara.** 1992. Specific acceptance of cardiac allograft after treatment with antibodies to ICAM-1 and LFA-1. *Science* **255:**1125–1128.

14. **Kolenbrander, P., N. Ganeshkumar, F. J. Cassels, and C. V. Hughes.** 1993. Coaggregation: specific adherence among human oral plaque bacteria. *FASEB J.* **7:**406–413.

15. **Lasky, L. A.** 1992. Selectins: interpreters of cell-specific carbohydrate information during inflammation. *Science* **258:**964–969.

16. **Mangan, D. F., S. E. Mergenhagen, and S. M. Wahl.** 1993. Apoptosis in human monocytes: possible role in chronic inflammatory diseases. *J. Periodontol.* **64:**461–466.

17. **Mangan, D. F., B. Robertson, and S. M. Wahl.** 1992. IL-4 enhances programmed cell death (apoptosis) in stimulated human monocytes. *J. Immunol.* **148:**1812–1816.

18. **McCartney-Francis, N., J. B. Allen, D. E. Mizel, J. E. Albina, Q. Xie, C. F. Nathan, and S. M. Wahl.** 1993. Suppression of arthritis by an inhibitor of nitric oxide synthase. *J. Exp. Med.* **178:**749–754.

19. **McCartney-Francis, N., S. Shepheard, K. L. Hines, R. Bingham, G. Henley, and S. M. Wahl.** Unpublished data.

20. **Mims, C. A.** 1987. *The Pathogenesis of Infectious Disease.* Academic Press, Ltd., London.

21. **Moissec, P., J. Briolay, J. Dechaniet, J. Wijdenes, H. Martinez-Valdez, and J. Banchereau.** 1992. Inhibition of the production of proinflammatory cytokines and immunoglobulins by interleukin-4 in an ex vivo model of rheumatoid synovitis. *Arthritis Rheum.* **35:**874.

22. **Mulligan, M. S., J. Varani, M. K. Dame, C. L. Lane, C. W. Smith, D. C. Anderson, and P. A. Ward.** 1991. Role of ELAM-1 in neutrophil-mediated lung injury in rats. *J. Clin. Invest.* **88:**1396–1406.

23. **Nathan, C.** 1992. Nitric oxide as a secretory product of mammalian cells. *FASEB J.* **6:**3051.

24. **Osborn, L., C. Hession, R. Tizard, C. Vassallo, S. Luhowskyj, M. J. Hemler, and R. Lobb.** 1989. Direct expression cloning of vascular cell adhesion molecule 1, a cytokine-induced endothelial protein that binds to lymphocytes. *Cell* **59:**1203.

25. **Ruoslahti, E.** 1991. Integrins. *J. Clin. Invest.* **87:**1–5.

26. **Sanchez-Aparicio, P., O. C. Ferreira, Jr., and A. Garcia-Pardo.** 1993. α4β1 recognition of the Hep

II domain of fibronectin is constitutive on some hemopoietic cells but requires activation on others. *J. Immunol.* **150**:3506–3514.

27. **Schiffmann, E., B. A. Corcoran, and S. M. Wahl.** 1975. N formylmethionyl peptides as chemoattractants for leucocytes. *Proc. Natl. Acad. Sci. USA* **72**:1059–1063.

28. **Skaleric, U., J. B. Allen, P. D. Smith, S. E. Mergenhagen, and S. M. Wahl.** 1991. Inhibitors of reactive oxygen intermediates suppress bacterial cell wall-induced arthritis. *J. Immunol.* **147**:2559–2564.

29. **Springer, T. A.** 1990. Adhesion receptors of the immune system. *Nature* (London) **346**:425–434.

30. **Wahl, S. M.** 1991. Cellular and molecular interactions in the induction of inflammation in rheumatic diseases, p. 101–132. *In* T. F. Kresina (ed.), *Monoclonal Antibodies, Cytokines and Arthritis: Mediators of Inflammation and Therapy.* Marcel Dekker, Inc., New York.

31. **Wahl, S. M., J. B. Allen, K. L. Hines, T. Imamichi, A. M. Wahl, L. T. Furcht, and J. B. McCarthy.** Synthetic fibronectin peptides suppress arthritis by interrupting leukocyte adhesion and recruitment. Submitted for publication.

32. **Williams, R. C.** 1990. Periodontal disease. *N. Engl. J. Med.* **322**:373–376.

33. **Wong. H. L., G. L. Costa, M. T. Lotze, and S. M. Wahl.** 1993. Interleukin-4 differentially regulates monocyte IL-1 family gene expression and synthesis in *vitro* and *in vivo*. *J. Exp. Med.* **177**:775–781.

34. **Wong, H. L., M. T. Lotze, L. M. Wahl, and S. M. Wahl.** 1992. Administration of recombinant IL-4 to humans regulates gene expression, phenotype and function in circulating monocytes. *J. Immunol.* **148**:2118–2125.

35. **Xie, Q.-W, H. J. Cho, J. Calaycay, R. A. Mumford, K. M. Swiderek, T. D. Lee, A. Ding, T. Troso, and C. Nathan.** 1992. Cloning and characterization of inducible nitric oxide synthase from mouse macrophages. *Science* **256**:225–228.

Molecular Pathogenesis of Periodontal Disease
Edited by Robert Genco et al.
© 1994 American Society for Microbiology, Washington, DC 20005

Chapter 17

Host-Mediated Extracellular Matrix Destruction by Metalloproteinases

Henning Birkedal-Hansen

A body of evidence suggests that microbe-induced degradation of the extracellular matrix is mediated primarily by host cells. Microorganisms may cause tissue destruction without intervention by host cells by release of proteolytic enzymes that directly attack stromal structures (3, 50), but this pathway appears to be the exception rather than the rule. In most cases, microbial products (toxins, lipopolysaccharide, enzymes) trigger resident and immigrant cell populations for expression of degradative enzymes or provoke an immune response that results in release of proinflammatory cytokines from mononuclear cell infiltrates. Each of these events culminates in activation of one or more endogenous degradative pathway(s) (4, 23, 38, 66). Most of the cell types that populate the periodontal tissues are capable of responding to cytokines and other proinflammatory mediators by expression of a degradative phenotype. Such is the case not only for immigrant polymorphonuclear leukocytes (PMN) and macrophages but also for indigenous fibroblasts, keratinocytes, endothelial cells, and possibly even the osteoblasts that line the osseous surfaces of the alveolar bone. The extracellular matrix is composed of perhaps as many as 50 different gene products, but the structural basis for the cohesion and mechanical stability of interstitial connective tissues resides almost exclusively with fibrils of type I and III collagen. Since attachment loss is directly linked to degradation of Sharpey's fibers and alveolar bone, it is reasonable to assume that the enzymatic and cellular mechanisms by which collagen fibrils are dissolved or disintegrated play an important role in periodontal disease progression. The complexity of these mechanisms is becoming increasingly apparent. Bone and "soft" connective tissues appear to be degraded by two distinct pathways that differ markedly in terms of enzymatic, cellular, and molecular reactions. The constituents of unmineralized connective tissues are degraded primarily by an extracellular matrix metalloproteinase (MMP)-dependent pathway. By contrast, mineralized tissues are degraded by an osteoclastic pathway that relies on dissolution of mineral crystals by organic acids and degradation of the collagenous matrix by acidic lysosomal

Henning Birkedal-Hansen • National Institute of Dental Research, National Institutes of Health, Building 30, Room 132, Bethesda, Maryland 20892

proteinases. For more detailed information on each of these mechanisms, the reader is referred to recent reviews (2, 70, 71, 84). This chapter will focus primarily on the metalloproteinase pathway.

OSTEOCLASTIC BONE RESORPTION

Dissolution of the collagenous matrix of bone apparently is mediated by lysosomal cathepsins released to an acidified closed microenvironment between the osteoclast membrane and the bone surface (5). Bone matrix collagen (fibrillar type I collagen) is quite resistant to proteolysis at neutral pH, but studies by Etherington and coworkers (14, 15, 56) have shown that bone matrix can be degraded by lysosomal cathepsins at pH^2 2 to 4, particularly at high Ca^{2+} concentrations. It is likely that osteoclastic bone resorption is preceded by removal of the osteoid seam by osteoblasts via a collagenase-dependent process (71), but the precise role of MMPs in bone resorption remains to be fully elucidated (5, 49, 71).

MATRIX METALLOPROTEINASES

The role of individual MMPs in specific developmental or pathological processes, including human periodontal diseases, is still incompletely understood, and the supportive evidence remains largely circumstantial and indirect. MMPs have broad and overlapping substrate specificities that all but preclude an accurate definition of their natural substrates. Another complicating factor is that most induced cells at any one time express a fairly extensive complement of MMPs. The common finding that cells and interstitial fluids of periodontal tissues often harbor MMP or their transcripts in vivo, however, supports the notion that these enzymes are involved in remodeling of the gingival stroma. The MMP family consists of nine or more metalloendopeptidases that collectively are capable of degrading all of the constituents of the extracellular matrix (Fig. 1). The enzymes share extensive sequence homology but differ somewhat in terms of substrate specificity and transcriptional regulation. Structurally, all MMPs are derived from a collagenase/stromelysin (SL)-like prototype formed by either addition or deletion of regulatory domains as shown in Fig. 1. The catalytic activity resides in an $\approx M_r$ 18,000 domain that harbors the active-site Zn^+-binding sequence (6, 72). The catalytic domain is preceded by a propeptide that endows the enzyme with catalytic latency at the time of secretion and is followed by a proline-rich hinge region that marks the transition to the largest domain, a pexin-like COOH-terminal sequence that plays a role in determining substrate specificity. The two gelatinases, in addition, contain a gelatin-binding insert (fibronectin type II-like repeats) in the catalytic domain (11, 19). The pexin-like domain is absent from the smallest MMP, matrilysin. The MMP may be divided into three functional groups.

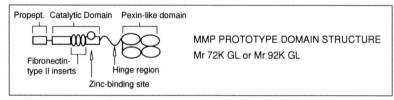

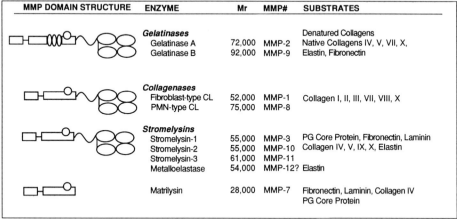

Figure 1. MMP family. CL, collagenase; PG, proteoglycan.

Type I Collagenases

Only two of the nine members of the MMP family cleave native type I collagen fibrils (fibroblast-type [FIB-CL] and PMN-type [PMN-CL] collagenases). While FIB-CL is expressed after appropriate stimulation by a variety of cell types (fibroblasts [20, 64], keratinocytes [34, 48], endothelial cells [25, 40], macrophages [9, 78], osteoblasts [45, 49], and chondrocytes [32]), PMN-CL is expressed only by PMN. PMN-CL is synthesized and packaged in specific storage granules before the cells emigrate from the bloodstream and is rapidly released when the cells are triggered. By contrast, FIB-CL is synthesized on demand in response to growth factor/cytokine activation of gene transcription. The fully functional protein emerges from the cell after a 6- to 12-h lag period required for activation of the transcriptional and translational machinery.

Gelatinases/Type IV Collagenases

The two gelatinases/type IV collagenases cleave types IV, V, VII, and X collagens; elastin; fibronectin; and perhaps other matrix substrates as well (16, 54, 55, 80). M_r 72,000 gelatinase/type IV collagenase (M_r 72K GL) is expressed by most cell types (fibroblasts [53], keratinocytes [52], endothelial cells [28], macrophages [18], osteoblasts [46], and chondrocytes [33]) but characteristically not by PMN. M_r 92K GL is produced by PMN (26, 41) but also by a number of other cell types, including keratinocytes (52, 82), macrophages (35), and occasionally fibroblasts. It is interesting that PMN express a unique and distinct interstitial collagenase gene (for PMN-CL) but utilize the same type IV collagenase gene (for M_r 92K GL) as other cells, although in a manner that yields a storable rather than a secreted enzyme.

Stromelysin Group

SL-1 and SL-2 cleave a wide range of extracellular matrix proteins: proteoglycan core protein, type IV and V collagens, fibronectin, and laminin (43). Induced stromal cells, macrophages, and endothelial cells express the SL-1 gene, whereas keratinocyte express the homologous SL-2 gene. SL-3 transcripts are expressed by stromal cells of human mammary carcinomas and by embryonic fibroblasts. The deduced amino acid sequence suggests that the gene for SL-3 encodes a fully functional metalloproteinase, and recent studies have confirmed that the recombinant enzyme possesses activity against some extracellular matrix substrates. Matrilysin cleaves a wide range of substrates, including fibronectin, laminin, and gelatin. It has been isolated from the involuting rat uterus and from rectal carcinoma cells (39, 85). Macrophage metalloelastase shares with several other members of the MMP family the ability to cleave elastin, but little is known yet of its pattern of expression.

ACTIVATION OF MMP

The latency of MMP precursors is maintained at least in part by a coordination bond that links an unpaired Cys residue in the propeptide to the active-site Zn^{2+} (61, 73, 83). Disruption of the $Cys-Zn^{2+}$ bond may be achieved (i) by modification of the Cys residue by organomercurial agents, metal ions, thiol reagents, and oxidants; (ii) by conformational change of the polypeptide backbone induced by certain chaotropic agents (KI, NaSCN) and detergents (sodium dodecyl sulfate); and (iii) by excision of a portion of the propepeptide by proteolytic enzymes. The enzyme ultimately catalyzes one or more autolytic cleavages to generate the fully processed form (42, 65). Although we have gained considerable knowledge of the chemistry of the latent and active forms and have identified many means for conversion of the precursor to the fully catalytically competent forms, the actual biologic activation mechanisms are still poorly understood. Among the most likely candidate mechanisms are (i) limited proteolysis by plasmin, (ii) oxidation by HOCl, and (iii) activation by as yet unidentified plasma membrane components.

TRANSCRIPTIONAL REGULATION OF MMP

Induction or repression of expression of most MMP genes (FIB-CL, SL-1, SL-2, SL-3, and M_l 92K GL) is controlled by growth factors and cytokines (Table 1).

Table 1. Regulation of MMP expression[a]

Type of regulation	
Induction	IL-1α, IL-1β, TNF-α, TGF-α, EGF, PDGF, bFGF, NGF, TGF-β, PGE$_2$, 1,25-di(OH)D$_3$
Repression	IFN-γ, IL-4, TGF-β, glucocorticoids, retinoic acid

[a] PDGF, platelet-derived growth factor; bFGF, basic fibroblast growth factor; NGF, nerve growth factor; PGE$_2$, prostaglandin E$_2$.

Stimulation of MMP transcription characteristically results in up to a 50-fold change in mRNA and protein levels. Transcription of responsive MMP genes is induced by a variety of cytokines (interleukin-1β [IL-1β] and tumor necrosis factor alpha [TNF-α]) and growth factors (platelet-derived growth factor, transforming growth factor alpha [TGF-α], epidermal growth factor [EGF], basic fibroblast growth factor, and nerve growth factor) and with a few exceptions repressed by TGF-β, gamma interferon, and IL-4 (for a review, see reference 2). TGF-β plays an interesting role in that it possesses both anabolic and catabolic effects. TGF-β ablates transcription of FIB-CL and SL-1 in fibroblasts yet up-regulates M_r 72K GL by two- to fourfold in the same cells. Moreover, TGF-β up-regulates M_r 92K GL by up to eightfold in keratinocytes (47, 52). Alteration of cellular behavior to shift the balance between anabolic or catabolic functions relies on the coordinate up- and down-regulation of a multitude of genes. Induction or repression of MMP genes represents only one element required for a shift of this balance. It is interesting in this context that the most important anabolic and catabolic cytokines regulate both the metalloproteinase and the osteoclastic pathways. Growth factors of the TGF-β superfamily (TGF-βs and bone morphogenetic proteins) promote deposition of stroma and down-regulate both MMP expression and osteoclastic bone resorption, while proinflammatory cytokines (IL-1 and TNF-α) activate MMP transcription and induce bone resorption (1, 22, 24).

Some mediators regulate transcription of MMP genes in a cell-specific or cell-selective manner. Expression of FIB-CL is induced in osteoblasts by parathyroid hormone and 1,25-di(OH)D$_3$ (49) and in macrophages by lipopolysaccharide (75–79). Prostaglandins, particularly prostaglandin E$_2$, have profound catabolic, proinflammatory effects that are mediated predominantly through macrophages and bone cells, for instance, by up-regulation of MMP gene transcription via a cyclic-AMP-dependent signaling pathway (37, 76). Prostaglandins primarily target degradative processes mediated by cells of macrophage lineage (MMP expression by macrophages, development of osteoclasts) and seem to have less dramatic influences on stromal-cell populations. Glucocorticoids are perhaps the most potent biologic repressors of MMP gene expression, and it is likely that their anti-inflammatory effect is mediated in large measure by down-regulation of MMP expression. Retinoids down-regulate MMP expression in stromal cells (8, 44), but the effect of retinoids on tissue destruction is somewhat ambiguous, as they stimulate rather than suppress bone and cartilage resorption and keratinocyte-mediated collagen degradation (81).

MMP INHIBITORS

The activity of MMP in the tissues is tightly controlled by plasma-derived proteinase inhibitors (α-macroglobulins) or locally produced tissue inhibitors of metalloproteinases (TIMPs) (7, 58, 63). TIMPs are synthesized by a variety of cell types (fibroblasts, keratinocytes, macrophages, and endothelial cells). These inhibitors form classic biomolecular, noncovalent complexes with the active forms of MMP and in some instances also with latent MMP zymogens. Three genetically distinct members of the TIMP family have been identified, but others may exist (10, 12, 13, 27, 62). TIMP-1 and -2 are functionally and structurally homologous but display some level of preference among members of the MMP family. TIMP-2 is more effective than TIMP-1 against the two type IV collagenases, whereas TIMP-1 more effectively inhibits FIB-CL. The TIMP-1 and TIMP-2 genes are differently regulated. Expression of TIMP-1 is stimulated by growth factors (EGF, TNF-α, IL-1, and TGF-β), phorbol esters, retinoids, and glucocorticoids, whereas TIMP-2 is down-regulated by TGF-β and fails to respond to phorbol esters. Active MMPs are also captured by α-macroglobulins, which are broad-spectrum plasma proteinase inhibitors (59, 60). Cleavage of the inhibitor "bait region" by the attacking proteinase results in entrapment or covalent cross-linking of the proteinase. The complex is eliminated either in the liver or locally in the tissues after receptor-mediated endocytosis. α-Macroglobulins are primarily vascular inhibitors, but they also exist in the extravascular compartment, particularly in areas of inflammation, and are found in gingival crevicular fluid in substantial concentrations.

MMPs AND CYTOKINES IN HUMAN GINGIVITIS AND PERIODONTITIS

Recent studies have confirmed that proinflammatory cytokines and MMPs are indeed expressed in inflamed human gingivae. Both IL-1 and TNF-α are present at physiologically meaningful concentrations in gingival crevicular fluids from inflamed sites (36, 51). IL-1 appears to be by far the most potent cytokine, with maximal activity at concentrations as low as 10^{-10} M, whereas TNF-α requires ≈ 10-fold-higher concentrations to elicit a similar response. Most cell types of the human periodontal tissues express a fairly extensive complement of MMP in vitro either constitutively or after appropriate induction (Table 2), but it is not immediately apparent that these enzymes are also expressed in vivo. Human FIB-CL has been identified by immunfluorescence staining in inflamed human gingivae, particularly in stromal cells close to the epithelial/mesenchymal interphase (86). Studies from our laboratory (67) have shown that both FIB-CL and M_r 72K GL are often expressed by stromal cells near inflammatory infiltrates, by cells in the subepithelial stroma, and by macrophages. These enzymes are also occasionally expressed by endothelial cells in capillary vessel walls. The metalloproteinase inhibitor TIMP-1 is primarily found in macrophage-like cells scattered throughout the tissue and occasionally in stromal cells as well. The epithelium, on the other hand, except for

Table 2. Expression of MMPs by cell types of human periodontal tissues

PMN	Fibroblast	Keratinocyte	Macrophage	Endothelial cell
PMN-CL	FIB-CL	FIB-CL	FIB-CL	FIB-CL
	SL-1		SL-1	SL-1
		SL-2		
	SL-3			
	M_r 72K GL	M_r 72K GL	M_r 72K GL	M_r 72K GL
M_r 92K GL	M_r 92K GL	M_r 92K GL	M_r 92K GL	M_r 92K GL
	Matrilysin	Matrilysin		
			Metalloelastase	

Langerhans cells, which express both M_r 72K GL collagenase and TIMP-1, is usually devoid of metalloproteinases and their inhibitors. Analyses of gingival crevicular fluids have shown that collagenase, probably of the PMN type, is fairly abundant at inflamed sites (17, 21, 29–31, 57, 69, 74). Gangbar et al. (17) and Teng et al. (68) also found M_r 92K GL but not the homologous M_r 72K GL enzyme. These findings taken together suggest that both the M_r 92K GL and the interstitial collagenase are PMN products released by degranulation. M_r 72K GL and SL-1, two MMPs that might represent "spillover" from stromal cells, have not yet been identified.

The data summarized so far permit us to develop a model for microbe-induced, host-mediated destruction of periodontal tissues. Signals emitted by the microorganisms either in the form of antigens or LPS give rise to a mononuclear cell infiltrate in the underlying connective tissue. Macrophages and lymphocytes of the infiltrate release proinflammatory cytokines and growth factors such as IL-1β, TNF-α, and perhaps TGF-α that activate transcription of MMP genes in fibroblast and keratinocyte populations and induce resorption of the alveolar bone. Fibroblasts, keratinocytes, macrophages, and endothelial cells all respond in similar fashion to catabolic growth factors and cytokines, namely, by up-regulation of MMP expression. MMPs released from the cells are activated extracellularly and collectively cleave the constituents of the extracellular matrix, while osteoclasts formed or attracted in response to the same signals resorb the adjacent bone. While this model is probably correct in its essential elements, many important details, not all of them trivial, remain to be elucidated. The rapid advancements in this area, however, will soon enable us to describe in considerable detail the expression pattern in time and locale of both mediators and effector enzymes. This "road map" will unquestionably provide important clues to the resolution of the remaining mechanistic questions.

Acknowledgments. This work was supported by U.S. Public Health Service grants DE08228 and DE06028.

REFERENCES

1. **Bertolini, D. R., G. E. Nedwin, T. S. Bringman, D. D. Smith, and G. R. Mundy.** 1986. Stimulation of bone resorption and inhibition of bone formation *in vitro* by human tumour necrosis factors. *Nature* (London) **319:**516–518.

2. **Birkedal-Hansen, H., W. G. I. Moore, M. K. Bodden, L. J. Windsor, B. Birkedal-Hansen, A. DeCarlo, and J. A. Engler.** 1993. Matrix metalloproteinases: a review. *Crit. Rev. Oral Biol. Med.* **4:**197–250.

3. **Birkedal-Hansen, H., R. E. Taylor, J. J. Zambon, P. K. Barwa, and M. Neiders.** 1988. Characterization of collagenolytic activity from strains of *Bacteroides gingivalis. J. Periodontal Res.* **23:**258–264.

4. **Birkedal-Hansen, H., B. R. Wells, H.-Y. Lin, P. W. Caufield, and R. E. Taylor.** 1984. Activation of keratinocyte-mediated collagen (type I) breakdown by suspected human periodontopathogen: evidence of a novel mechanism of connective tissue breakdown. *J. Periodontal Res.* **19:**645–650.

5. **Blair, H. C., A. J. Kahn, E. C. Crouch, J. J. Jeffrey, and S. L. Teitelbaum.** 1986. Isolated osteoclasts resorb the organic and inorganic components of bone. *J. Cell. Biol.* **102:**1164–1172.

6. **Bode, W., F. X. Gomis-Ruth, R. Huber, R. Zwilling, and W. Stöcker.** 1992. Structure of astacin and implication of astacins and zinc-ligation of collagenases. *Nature* (London) **358:**164–166.

7. **Boone, T. C., M. J. Johnson, Y. A. DeClerck, and K. E. Langley.** 1990. cDNA cloning and expression of a metalloproteinase inhibitor related to tissue inhibitor of metalloproteinases. *Proc. Natl. Acad. Sci. USA* **87:**2800–2804.

8. **Brinckerhoff, C. E., J. W. Coffey, and A. C. Sullivan.** 1983. Inflammation and collagenase production in rats with adjuvant arthritis reduced with 13-cis-retinoic acid. *Science* **221:**756–758.

9. **Campbell, E. J., J. D. Cury, C. J. Lazarus, and H. G. Welgus.** 1987. Monocyte procollagenase and tissue inhibitor of metalloproteinases. *J. Biol. Chem.* **262:**15862–15868.

10. **Carmichael, D. F., A. Sommer, R. C. Thompson, D. C. Anderson, C. G. Smith, H. G. Welgus, and G. P. Stricklin.** 1986. Primary structure and cDNA cloning of human fibroblast collagenase inhibitor. *Proc. Natl. Acad. Sci. USA* **83:**2407–2411.

11. **Collier, I. E., P. A. Krasnov, A. Y. Strongin. H. Birkedal-Hansen, and G. I. Goldberg.** 1992. Alanine scanning mutagenesis and functional analysis of the fibronectin-like collagen binding domain from human 92kDa type IV collagenase. *J. Biol. Chem.* **267:**6776–6781.

12. **DeClerck, Y. A., N. Perez, H. Shimada, T. C. Boone, K. E. Langley, and S. M. Taylor.** 1992. Inhibition of invasion and metastasis in cells transfected with an inhibitor of metalloproteinases. *Cancer Res.* **52:**701–708.

13. **Docherty, A. J. P., A. Lyons, B. J. Smith, E. M. Wright, P. E. Stephens, T. J. R. Harris, G. Murphy, and J. J. Reynolds.** 1985. Sequence of human tissue inhibitor of metalloproteinases and its identity to erythroid-potentiating activity. *Nature* (London) **318:**66–69.

14. **Etherington, D. J.** 1977. The dissolution of insoluble bovine collagens by cathepsin B1, collagenolytic cathepsin and pepsin; the influence of collagen type, age and chemical purity on susceptibility. *Connect. Tissue Res.* **5:**135–145.

15. **Etherington, D. J., and H. Birkedal-Hansen.** 1987. The influence of dissolved calcium salts on the degradation of hard-tissue collagens by lysosomal cathepsins. *Collagen Relat. Res.* **7:**185–199.

16. **Gadher, S. J., T. M. Schmid, L. W. Heck, and D. E. Woolley.** 1989. Cleavage of collagen type X by human synovial collagenase and neutrophil elastase. *Matrix* **9:**109–115.

17. **Gangbar, S., C. M. Overall, A. G. McCulloch, and J. Sodek.** 1990. Identification of polymorphonuclear leukocyte collagenase and gelatinase activities in mouthrinse samples: correlation with periodontal disease activity in adult and juvenile periodontitis. *J. Periodontal Res.* **25:**257–267.

18. **Garbisa, S., M. Ballin, D. Daga-Gordini, G. Gastelli, M. Naturale, A. Negro, G. Semenzato, and L. A. Liotta.** 1986. Transient expression of type IV collagenolytic metalloproteinase by human mononuclear phagocytes. *J. Biol. Chem.* **261:**2369–2375.

19. **Goldberg, G. I., B. L. Marmer, G. A. Grant, A. Z. Eisen, S. Wilhelm, and C. He.** 1989. Human 72-kilodalton type IV collagenase forms a complex with a tissue inhibitor of metalloproteases. designated TIMP-2. *Proc. Natl. Acad. Sci. USA* **86:**8207–8211.

20. **Goldberg, G. I., S. M. Wilhelm, A. Kronberger, E. A. Bauer, G. E. Grant, and A. Z. Eisen.** 1986. Human fibroblast collagenase. Complete primary structure and homology to an oncogene transformation-induced rat protein. *J. Biol. Chem.* **261:**6600–6605.

21. **Golub, L. M., M. Wolfe, H. M. Lee, T. F. McNamara, N. S. Ramamurthy, J. Zambon, and S. Ciancio.** 1985. Further evidence that tetracyclines inhibit collagenase activity in human crevicular fluid and other mammalian sources. *J. Periodontal Res.* **20:**12–23.

22. **Gowen, M., and G. R. Mundy.** 1986. Actions of recombinant interleukin 1, interleukin 2, and interferon-γ on bone resorption in vitro. *J. Immunol.* **136:**2478–2482.

23. **Heath, J. K., S. J. Atkinson, R. M. Hembry, J. J. Reynolds, and M. C. Meikle.** 1987. Bacterial antigens induce collagenase and prostaglandin E_2 synthesis in human gingival fibroblasts through a primary effect on circulating mononuclear cells. *Infect. Immun.* **55:**2148–2154.

24. **Heath, J. K., J. Saklatvala, M. C. Meikle, S. J. Atkinson, and J. J. Reynolds.** 1985. Pig interleukin-1 (catabolin) is a potent stimulator of bone resorption *in vitro. Calcif. Tissue Int.* **37:**95–97.

25. **Herron, G. S., M. J. Banda, E. J. Clark, J. Gavrilovic, and Z. Werb.** 1986. Secretion of metalloproteinases by stimulated capillary endothelial cells. I. Production of procollagenase and prostromelysin exceeds expression of proteolytic activity. *J. Biol. Chem.* **261:**2810–2813.

26. **Hibbs, M. S., K. A. Hasty, J. M. Seyer, A. H. Kang, and C. M. Mainardi.** 1985. Biochemical and immunological characterization of the secreted forms of human neutrophil gelatinase. *J. Biol. Chem.* **260:**2493–2500.

27. **Howard, E. W., E. C. Bullen, and M. J. Banda.** 1991. Regulation of the autoactivation of human 72-kDa progelatinase by tissue inhibitor of metalloproteinases-2. *J. Biol. Chem.* **266:**13064–13069.

28. **Kalebic, T., S. Garbisa, B. Glaser, and L. A. Liotta.** 1983. Basement membrane collagen: degradation by migrating endothelial cells. *Science* **221:**281–283.

29. **Kryshtalskyj, E., and J. Sodek.** 1987. Nature of collagenolytic enzyme and inhibitor in crevicular fluid from healthy and inflamed periodontal tissues of beagle dogs. *J. Periodontal Res.* **22:**264–269.

30. **Kryshtalskyj, E., J. Sodek, and J. M. Ferrier.** 1986. Correlation of collagenolytic enzymes and inhibitors in gingival crevicular fluid with clinical and microscopic changes in experimental periodontitis. *Arch. Oral Biol.* **3:**21–31.

31. **Larivee, J., J. Sodek, and J. M. Ferrier.** 1986. Collagenase and collagenase inhibitor activities in crevicular fluid of patients receiving treatment for localized juvenile periodontitis. *J. Periodontal Res.* **21:**702–715.

32. **Lefebvre, V., C. Peeters-Joris, and G. Vaes.** 1990. Modulation by interleukin-1 and tumor necrosis factor α of production of collagenase, tissue inhibitor of metalloproteinases and collagen types in differentiated and dedifferentiated articular chondrocytes. *Biochim. Biophys. Acta* **1052:**366–378.

33. **Lefebvre, V., C. Peeters-Joris, and G. Vaes.** 1991. Production of gelatin-degrading matrix metalloproteinases ("type IV collagenases") and inhibitors by articular chondrocytes during their dedifferentiation by serial subcultures and under stimulation by interleukin-1 and tumor necrosis factor α. *Biochim. Biophys. Acta* **1094:**8–18.

34. **Lin, H. Y., B. R. Wells, R. E. Taylor, and H. Birkedal-Hansen.** 1987. Degradation of type I collagen by rat mucosal keratinocytes. *J. Biol. Chem.* **262:**6823–6831.

35. **Mainardi, C. L., M. S. Hibbs, K. A. Hasty, and J. M. Seyer.** 1984. Purification of a type IV collagen degrading metalloproteinase from rabbit alveolar macrophages. *Collagen Rel. Res.* **4:**479–492.

36. **Masada, M. P., R. Persson, J. S. Kenney, S. W. Lee, R. C. Page, and A. C. Allison.** 1990. Measurement of interleukin-1α and -1β in gingival crevicular fluid: implications for the pathogenesis or periodontal disease. *J. Periodontal Res.* **25:**156–163.

37. **McCarthy, J. B., S. M. Wahl, J. Rees, C. E. Olsen, A. L. Sandberg, and L. M. Wahl.** 1980. Mediation of macrophage collagenase production by 3'-5' cyclic adenosine monophosphate. *J. Immunol.* **124:**2405–2409.

38. **Meikle, M. C., S. J. Atkinson, R. V. Ward, G. Murphy, and J. J. Reynolds.** 1989. Gingival fibroblasts degrade type I collagen films when stimulated with tumor necrosis factor and interleukin-1: evidence that breakdown is mediated by metalloproteinases. *J. Periodontal Res.* **24:**207–213.

39. **Miyazaki, K., Y. Hattori, F. Umenishi, H. Yasumitsu, and M. Umeda.** 1990. Purification and characterization of extracellular matrix-degrading metalloproteinase, matrin (Pump-1) secreted from human rectal carcinoma cell line. *Cancer Res.* **50:**7758–7764.

40. **Moscatelli, D., E. Jaffe, and D. B. Rifkin.** 1980. Tetradecanoyl phorbol acetate stimulates latent collagenase production by cultured human endothelial cells. *Cell* **20:**343–351.

41. **Murphy, G., R. Ward, R. M. Hembry, J. J. Reynolds, K. Kuhn, and K. Tryggvason.** 1989. Characterization of gelatinase from pig polymorphonuclear leukocytes. *Biochem. J.* **258:**463–472.

42. **Nagase, H., J. J. Enghild, K. Suzuki, and G. Salvesen.** 1990. Stepwise activation of the precursor

of matrix metalloproteinase 3 (stromelysin) by proteinases and (4-aminophenyl) mercuric acetate. *Biochemistry* **29:**5783–5789.

43. **Nicholson, R., G. Murphy, and R. Breathnach.** 1989. Human and rat malignant-tumor-associated mRNAs encode stromelysin-like metalloproteinases. *Biochemistry* **28:**5195–5203.

44. **Nicholson, R. C., S. Mader, S. Nagpal, M. Leid, C. Rochette-Egly, and P. Chambon.** 1990. Negative regulation of the rat stromelysin gene promoter by retinoic acid is mediated by an AP-1 binding site. *EMBO J.* **9:**4443–4454.

45. **Otsuka, K., J. Sodek, and H. Limeback.** 1984. Synthesis of collagenase and collagenase inhibitors by osteoblast-like cells in culture. *Eur. J. Biochem.* **145:**123–129.

46. **Overall, C. M., and J. Sodek.** 1987. Initial characterization of a neutral metalloproteinase, active on native 3/4-collagen fragments, synthesized by ROA 17/2.8 osteoblastic cells, peridontal fibroblasts, and identified in gingival crevicular fluid. *J. Dent. Res.* **66:**1271–1282.

47. **Overall, C. M., J. L. Wrana, and J. Sodek.** 1991. Transcriptional and post-transcriptional regulation of 72-kDa gelatinase/type IV collagenase by transforming growth factor-β1 in human fibroblasts. *J. Biol. Chem.* **266:**14064–14071.

48. **Petersen, M. J., D. T. Woodley, G. P. Stricklin, and E. J. O'Keefe.** 1987. Production of procollagenase by cultured human keratinocytes. *J. Biol. Chem.* **262:**835–840.

49. **Quinn, C. O., D. K. Scott, C. E. Brinckerhoff, L. M. Matrisian, J. J. Jeffrey, and N. C. Partridge.** 1990. Rat collagenase. Cloning, amino acid sequence comparison, and parathyroid hormone regulation in osteoblastic cells. *J. Biol. Chem.* **265:**22342–22347.

50. **Robertson, P. B., M. Lantz, P. T. Marucha, K. S. Kornman, C. L. Trummel, and S. C. Holt.** 1982. Collagenolytic activity associated with bacteroides species and Actinobacillus actinomycetemcomitans. *J. Periodontal Res.* **17:**275–283.

51. **Rossomando, E. F., and L. B. White.** 1993. A novel method for the detection of TNF-alpha in gingival crevicular fluid. *J. Periodontol.* **64**(Suppl.)**:**445–449.

52. **Salo, T., J. G. Lyons, F. Rahemtulla, H. Birkedal-Hansen, and H. Larjava.** 1991. Transforming growth factor-β1 up-regulates type IV collagenase expression in cultured human keratinocytes. *J. Biol. Chem.* **266:**11436–11441.

53. **Seltzer, J. L., S. A. Adams, G. A. Grant, and A. Z. Eisen.** 1981. Purification and properties of a gelatin-specific neutral protease from human skin. *J. Biol. Chem.* **256:**4662–4668.

54. **Seltzer, J. L., A. Z. Eisen, E. A. Bauer, N. P. Morris, R. W. Glanville, and R. E. Burgeson.** 1989. Cleavage of type VII collagen by interstitial collagenase and type IV collagenase (gelatinase) derived from human skin. *J. Biol. Chem.* **264:**3822–3826.

55. **Senior, R., G. Griffin, C. J. Fliszar, S. D. Shapiro, G. I. Goldberg, and H. G. Welgus.** 1991. Human 92- and 72-kilodalton type IV collagenases are elastases. *J. Biol. Chem.* **266:**7870–7875.

56. **Silver, I. A., R. J. Murrills, and D. J. Etherington.** 1988. Microelectrode studies on the acid microenvironment beneath adherent macrophages and osteoclasts. *Exp. Cell Res.* **175:**266–276.

57. **Sorsa, T., V.-J. Uitto, M. Suomolainen, M. Vauhkonen, and S. Lindy.** 1988. Comparison of interstitial collagenases from human gingiva, sulcular fluid and polymorphonuclear leukocytes. *J. Periodontal Res.* **23:**386–393.

58. **Sottrup-Jensen, L.** 1989. Structure, shape, and mechanism of proteinase complex formation. *J. Biol. Chem.* **264:**11539–11542.

59. **Sottrup-Jensen, L., H. F. Hansen, and U. Christensen.** 1983. Generation and reactivity of "nascent" α2-macroglobulin: localization of cross-links in α2-macroglobulin-trypsin complex. *Ann. N.Y. Acad. Sci.* **421:**188–208.

60. **Sottrup-Jensen, L., O. Sand, L. Kristensen, and G. H. Fey.** 1989. The α-macroglobulin bait region. Sequence diversity and localization of cleavage sites for proteinases in five mammalian α-macroglobulins. *J. Biol. Chem.* **264:**15781–15789.

61. **Springman, E. B., E. L. Angleton, H. Birkedal-Hansen, and H. E. Van Wart.** 1990. Multiple modes of activation of latent human fibroblast collagenase: evidence for the role of a Cys[73] active-site zinc complex in latency and a "cysteine switch" mechanism for activation. *Proc. Natl. Acad. Sci. USA* **87:**364–368.

62. **Stetler-Stevenson, W. G., P. D. Brown, M. Onisto, A. T. Levy, and L. A. Liotta.** 1990. Tissue inhibitor of metalloproteinases-2 (TIMP-2) mRNA expression in tumor cell lines and human tumor tissues. *J. Biol. Chem.* **265:**13933–13938.

63. **Stetler-Stevenson, W. G., H. C. Krutzsch, and L. A. Liotta.** 1989. Tissue inhibitor of metalloproteinase (TIMP-2). A new member of the metalloproteinase inhibitor family. *J. Biol. Chem.* **264:**17374–17378.

64. **Stricklin, G. P., E. A. Bauer, J. J. Jeffrey, and A. Z. Eisen.** 1977. Human skin collagenase: isolation of precursor and active forms from both fibroblast and organ cultures. *Biochemistry* **16:**1607–1615.

65. **Suzuki, K., J. J. Enghild, T. Morodomi, G. Salvesen, and H. Nagase.** 1990. The activation of tissue procollagenase by matrix metalloproteinase 3 (stromelysin). *Biochemistry* **29:**10261–10270.

66. **Takada, H., J. Mihara, I. Morisaki, and S. Hamada.** 1991. Induction of interleukin-1 and -6 human gingival fibroblast cultures stimulated with *Bacteroides* lipopolysaccharides. *Infect. Immun.* **59:**295–301.

67. **Takeyama, H., B. Birkedal-Hansen, W. G. I. Moore, M. K. Bodden, and H. Birkedal-Hansen.** 1993. Immunolocalization of fibroblast-type collagenase, Mr 72K gelatinase and TIMP-1 in inflamed human gingiva. *J. Dent. Res.* **72:**437 (abstr.).

68. **Teng. Y. T., J. Sodek, and C. A. G. McCulloch.** 1992. Gingival crevicular fluid gelatinase and its relationship to periodontal disease in human subjects. *J. Periodontal Res.* **27:**457–552.

69. **Uitto, V.-J., K. Suomolainen, and T. Sorsa.** 1990. Salivary collagenase. Origin, characteristics and relationship to periodontal health. *J. Periodontal Res.* **25:**135–142.

70. **Vaes, G.** 1988. Cellular biology and biochemical mechanism of bone resorption. *Clin. Orthop. Relat. Res.* **231:**239–271.

71. **Vaes, G., J.-M. Delaisse, and Y. Eeckhout.** 1992. Relative roles of collagenase and lysosomal cysteine-proteinases in bone resorption, p. 383–388. *In* H. Birkedal-Hansen, Z. Werb, H. G. Welgus, and H. E. Van Wart (ed.), *Matrix Metalloproteinases and Inhibitors.* Gustav Fischer Verlag, Stuttgart, Germany.

72. **Vallee, B. L., and D. S. Auld.** 1992. Active zinc binding sites of zinc metalloenzymes, p. 5–19. *In* H. Birkedal-Hansen, Z. Werb, H. G. Welgus, and H. E. Van Wart (ed.) *Matrix Metalloproteinases and Inhibitors.* Gustav Fischer Verlag, Stuttgart, Germany.

73. **Van Wart, H. E., and H. Birkedal-Hansen.** 1990. The cysteine switch: a principle of regulation of metalloproteinase activity with potential applicability to the entire matrix metalloproteinase gene family. *Proc. Natl. Acad. Sci. USA* **87:**5578–5582.

74. **Villela, B., R. B. Cogen, A. A. Bartolucci, and H. Birkedal-Hansen.** 1987. Collagenolytic activity in crevicular fluid from patients with chronic adult periodontitis, localized juvenile periodontitis and gingivitis, and from healthy control subjects. *J. Periodontal Res.* **22:**381–389.

75. **Wahl, L., S. M. Wahl, S. E. Mergenhagen, and G. R. Martin.** 1974. Collagenase production by endotoxin-activated macrophages. *Proc. Natl. Acad. Sci. USA* **71:**3598–3601.

76. **Wahl, S. M., and L. M. Wahl.** 1985. Regulation of macrophage collagenase, prostaglandin, and fibroblast-activating-factor production by anti-inflammatory agents: different regulatory mechanisms for tissue injury and repair. *Cell Immunol.* **92:**302–312.

77. **Wahl, S. M., L. M. Wahl, J. B. McCarthy, L. Chedid, and S. E. Mergenhagen.** 1979. Macrophage activation by mycobacterial water soluble compounds and synthetic muramyl dipeptide. *J. Immunol.* **122:**2226–2231.

78. **Welgus, H. G., E. J. Campbell, Z. Bar-Shavit, R. M. Senior, and S. L. Teitelbaum.** 1985. Human alveolar macrophages produce a fibroblast-like collagenase and collagenase inhibitor. *J. Clin. Invest.* **76:**219–224.

79. **Welgus, H. G., E. J. Campbell, J. D. Cury, A. Z. Eisen, R. M. Senior, S. M. Wilhelm, and G. I. Goldberg.** 1990. Neutral metalloproteinases produced by human mononuclear phagocytes. Enzyme profile, regulation, and expression during cellular development. *J. Clin. Invest.* **86:**1496–1502.

80. **Welgus, H. G., C. J. Fliszar, J. L. Seltzer, T. M. Schmid, and J. J. Jeffrey.** 1990. Differential susceptibility of type X collagen to cleavage by two mammalian interstitial collagenases and 72kDa type IV collagenase. *J. Biol. Chem.* **265:**13521–13527.

81. **Wells, B. R., and H. Birkedal-Hansen.** 1985. Retinoic acid stimulates degradation of interstitial collagen fibrils by rat mucosal keratinocytes *in vitro. J. Dent. Res.* **64:**1186–1190.

82. **Wilhelm, S. M., I. E. Collier, B. L. Marmer, A. Z. Eisen, G. A. Grant, and G. I. Goldberg.** 1989. SV40-transformed human lung fibroblasts secrete a 92kDa type IV collagenase which is identical to that secreted by normal human macrophages. *J. Biol. Chem.* **264:**17213–17221.

83. **Windsor, L. J., H. Birkedal-Hansen, B. Birkedal-Hansen, and J. A. Engler.** 1991. An internal cys-

teine plays a role in the maintenance of the latency of human fibroblast collagenase. *Biochemistry* **30:**641–647.

84. **Woessner, J. F., Jr.** 1991. Matrix metalloproteinases and their inhibitors in connective tissue remodeling. *FASEB J.* **5:**2145–2154.

85. **Woessner, J. F., and C. J. Taplin.** 1988. Purification and properties of a small latent matrix metalloproteinase of the rat uterus. *J. Biol. Chem.* **263:**16918–16925.

86. **Woolley, D. E., and R. M. Davies.** 1981. Immunolocalization of collagenase in periodontal tissues. *J. Periodontal Res.* **16:**292–297.

Molecular Pathogenesis of Periodontal Disease
Edited by Robert Genco et al.
© 1994 American Society for Microbiology, Washington, DC 20005

Chapter 18

Role of Prostaglandins in High-Risk Periodontitis Patients

Steven Offenbacher, John G. Collins,
Behnaz Yalda, and Geoffrey Haradon

The role of prostaglandins in the pathogenesis of periodontal disease has been the subject of several recent comprehensive reviews (17, 18, 21). The current data summarizing the relationship between periodontal disease severity, activity, and prostaglandin E_2 (PGE_2) levels have been discussed in detail elsewhere (18). We recently reviewed the therapeutic application of nonsteroidal anti-inflammatory drugs (18, 19) and the use of analyses of crevicular fluid levels of PGE_2 as a diagnostic tool (17, 21). In addition, a hypothetical model that uses the PGE_2 level within the periodontal tissues as a key determinant of pathogenesis (17, 18, 21) and emphasizes the monocyte as playing a central role in regulating the local PGE_2 level has been proposed (17, 18). The extensive nature of these reviews bears witness to the considerable evidence that indicates that PGE_2 is an important regulatory molecule in pathogenesis. However, it is highly unlikely that any single mediator of inflammation, irrespective of whether it is a pluripotent mediator of destruction or not, can ever serve as the hallmark of the dominant pathway of molecular pathogenesis. There is far too much redundancy and feedback in the inflammatory processes for any mediator to assume an omnipotent pivotal position. The host response is multifactorial, involving both genetic and environmental contributions to the components of the response, including inflammation, phagocytic cell function, antibody secretion, healing capacity, and many other attributes. Nonetheless, despite this complexity, some aspects of the inflammatory response are clearly distinct components of the host response. The intent of this chapter is to place in perspective our current understanding of the regulatory mechanisms that influence the inflammatory response, focusing on prostaglandin synthesis as an important element of that inflammatory process. The objective is to provide insight into the possible molecular basis of host susceptibility to disease.

The expression of periodontal disease severity represents a balance between

Steven Offenbacher, John G. Collins, Behnaz Yalda, and Geoffrey Haradon • School of Dentistry, CB 7455, Dental Research Center, University of North Carolina at Chapel Hill, Chapel Hill, North Carolina 27599-7455.

two major systems, the microbial flora and the host response. Both are complex homeostatic systems that are stabilized by interdependent pathways characterized by redundancy and feedback control. Both systems probably have genetically defined or genetically limited response characteristics that can be highly modified or "titrated" depending on the magnitude and quality of environmental influences. Increasing data suggest that there are two major models of disease. In the "plaque and local factor" model of periodontitis, the cumulative plaque and debris serve as the principal determinant of disease severity. In this model, specific pathogens within the microbial flora dominate over the host response in controlling disease expression. The microbial numbers are high, placing an excessive burden on the normal host defenses. This plaque-dependent form of disease probably represents the majority of the disease in the population, but it also responds well to debridement therapy. In modern American society, the popularity of the toothbrush, regular oral hygiene measures, and the preventive orientation of dentistry have all been offered as plausible explanations for the observed decrease in this type of periodontal disease in the United States (22).

The other major category of periodontal disease is one in which the "compromised host" plays the dominant role. This type of disease is less prevalent and includes early-onset, refractory, and diabetes-associated periodontitis as examples. In contrast to the plaque and local factor model, treatment with debridement therapy in these patients provides less predictable results. The severity and rate of progression are often advanced and do not correlate well with local factors such as the amount of plaque or calculus. For these reasons, deficiencies in the host response have often been cited as possible explanatory variables for this "high-risk" state. However, many subcomponents of the host response may be involved, including the innate defense system and phagocyte function as well as the quantity and quality of the antibody, inflammatory mediator, and wound-healing responses. Neutrophil impairment has been convincingly demonstrated to predispose an individual to severe forms of periodontal disease (see reference 12 for a review). Beyond that observation, however, abnormalities in other host defense mechanisms have not been fully characterized as possible risk factors. In this discussion, we present early findings that suggest that abnormal PGE_2 secretion, as part of an exaggerated inflammatory response, may be considered a host response risk marker for susceptibility in these high-risk patients. It is our opinion that the previous association between disease severity and elevated crevicular-fluid PGE_2 levels is a reflection of an underlying abnormal, host-dependent inflammatory response in certain patients. In other words, the hyperinflammatory trait does not appear to be limited to the local sites of periodontal infection but may also be evident at healthy periodontal sites and perhaps even systemically. The data suggest that this type of inflammatory response abnormality is common in patients with early-onset, refractory, and insulin-dependent diabetes mellitus (IDDM)-associated periodontitis. At this point, the data for a systemic hyperinflammatory trait are neither overwhelming nor convincing. This chapter should therefore be considered as further developing the concept and describing the current supportive data for this conceptual framework.

PGE$_2$ AND PERIODONTAL DISEASE ACTIVITY

PGE$_2$ is a "well-behaved," robust marker for periodontal disease severity and activity. In the crevicular fluid, it is low or nondetectable in health and elevated in disease. In diseased patients, only four periodontal sites need to be sampled to provide an estimate of the underlying patient whole-mouth mean PGE$_2$ level of gingival crevicular fluid (GCF) at an error rate of $\alpha = 0.05$ and $\beta = 0.20$. Thus, four sites per patient are usually sufficient to provide a "reliable" estimate of the patient's mean GCF PGE$_2$ response (21). There is an elevation in GCF PGE$_2$ level during the induction of experimental gingivitis in humans from 20.5 to 53.5 ng/ml at 4 weeks (8). Although there is a steady rise in inflammation during the first 3 weeks of this model, GCF PGE$_2$ levels do not rise until week 4, when there is a dramatic increase in bleeding scores (17, 18). This increase in GCF PGE$_2$ level at 4 weeks is equivalent to that seen cross-sectionally in adult periodontitis patients (59.5 ± 6.3 ng/ml) (18, 21). Thus, it appears that GCF PGE$_2$ levels, when measured at many periodontal sites and expressed as a mean patient value, do not discriminate between late-stage gingivitis and adult periodontitis (AP) patients examined cross-sectionally (see reference 17 for further discussion). In contrast, the levels at actively progressing sites are significantly elevated at 305.6 ± 56.5 ng/ml (20). Although the mean GCF PGE$_2$ level for a patient can be used to assess the patient risk for future attachment loss with an overall predictive value of 0.92 to 0.95, the use of PGE$_2$ as a site-specific indicator for adult periodontitis is more complex (21). Previous studies that demonstrated that the GCF PGE$_2$ level can be used as a predictive risk factor were based on the longitudinal monitoring of AP patients who had already received full-mouth scaling and root planing (18, 20, 22). Our recent data suggest that GCF PGE$_2$ levels are substantially higher in certain high-risk patients such as refractory, early-onset periodontitis (EOP), or diabetic patients. Furthermore, the GCF PGE$_2$ levels at healthy and diseased sites within these special patient groups are quite different from those levels associated with clinically similar sites in patients with the more prevalent form of disease, AP. Thus, the interpretation of GCF PGE$_2$ levels in these three patient groups needs to be considered in more detail.

PGE$_2$ IN HIGH-RISK PATIENTS

The data in Table 1 demonstrate the relationship between pocket depth and GCF PGE$_2$ levels in healthy, AP, and EOP patients. Healthy individuals have very low GCF PGE$_2$ levels (20.5 ± 7.6 ng/ml) compared with patients with AP (59.5 ± 6.3 ng/ml) or EOP (139.3 ± 15.3 ng/ml). These differences are significant at $P < 0.005$. However, if one looks at the shallow sites (0 to 3 mm deep) in the AP patients, the GCF PGE$_2$ levels are significantly elevated compared to those in healthy patients, i.e., 49.0 ± 5.2 versus 20.5 ± 7.6 ng/ml ($P = 0.006$). This elevation of GCF PGE$_2$ at shallow sites (0 to 3 mm) in AP patients is present even if sites with redness and bleeding are excluded (data not shown). The elevation in GCF

Table 1. Relationship between pocket depth and GCF PGE$_2$ level

Periodontal status	No. of patients	Pocket depth (mm)	No. of sites	Mean GCF PGE$_2$ ± SE (ng/ml)
Health	21	0–3	168	20.5 ± 7.6
AP	16	0–3	192	49.0 ± 5.2
		4–11	176	71.0 ± 11.8
		0–11	368	59.5 ± 6.3
EOP	12	0–3	131	119.1 ± 20.7
		4–11	152	156.8 ± 22.1
		0–11	283	139.3 ± 15.3

PGE$_2$ at shallow AP sites is not as great as that in patients with experimental gingivitis (53.5 ng/ml), but it is quite close. Thus, clinically healthy sites in diseased patients appear to have higher basal levels of GCF PGE$_2$ than clinically healthy sites in healthy patients. This baseline elevation may reflect the response to changes in the flora in AP versus gingivitis patients. Alternatively, it may be a reflection of the increased numbers of inflammatory cells present within the adjacent periodontium. As expected, there remains a significant increase in GCF PGE$_2$ levels at deeper sites in AP patients relative to levels at shallow sites, i.e., 71.0 versus 49.0 ng/ml ($P = 0.02$). The same trend, although more dramatic, is apparent in the EOP patients. In these individuals, the shallow sites have GCF PGE$_2$ levels almost sixfold higher than those seen in healthy controls and more than twofold higher than the GCF PGE$_2$ levels at shallow sites in AP patients ($P = 7.8 \times 10^{-6}$ and 0.001, respectively). Deeper pockets (4 to 11 mm) also demonstrate a significant increase in EOP patients compared to comparable pockets in AP patients, i.e., 156.8 versus 71.0 ng/ml ($P = 0.001$). These data suggest that in diseased patients, even sites that are clinically healthy have elevated GCF PGE$_2$ levels. The magnitude of this elevation in AP patients is not particularly large and remains below levels in gingivitis patients. However, in EOP patients, the elevation in GCF PGE$_2$ that is observed in shallow sites is quite remarkable (119.1 ng/ml). These levels are above the 66.2-ng/ml level that was used to define high risk in an AP patient population (18, 20, 21) and suggest that these sites are at risk. Certainly shallow sites are at risk for attachment loss in diseased patients, as was recently demonstrated by epidemiological incidence data from Beck et al. (1). On a patient level, there are usually more shallow sites than deep sites; thus, the risk of attachment loss occurring at a shallow site is greater than the risk at a deep site (1). However, on a site level, deeper sites have greater risk for future loss of attachment (1). Thus, the GCF PGE$_2$ data are consistent with these findings in that a shallow site in an EOP patient may be at greater risk than a shallow or perhaps even a deep site in an AP patient. What is remarkable, however, is the dramatic range of GCF PGE$_2$ concentrations that appears at sites that are clinically similar in different periodontal disease conditions. This observation rhetorically confirms previous data that demonstrate that the local flora and/or host response associated with EOP is vastly different from that associated with AP. However, it further suggests that the GCF PGE$_2$ level reflects the collective response of the patient's periodontium, that is, a response not as a collection of sites that function independently but rather as an organ that reflects the patient's systemic response to a local infection. Previous data have demonstrated that GCF PGE_2 levels increase in synchrony at many periodontal sites simultaneously during periods of active attachment loss, even at sites that are stable and relatively healthy (20). This finding—that the inflammatory response is synchronized at multiple sites, including healthy sites—suggests that an underlying systemic response rather than a local site response is driving the inflammatory process. The elevated PGE_2 response at healthy sites in EOP patients suggests that an underlying systemically based hyperinflammatory response may be present in EOP patients but not in healthy or AP patients. If the cause of the increased GCF PGE_2 level in EOP patients is local infection, one might not expect the infected sites to be clinically healthy. It is possible, however, that the unique flora of the EOP patient induces a systemic up-regulation of the inflammatory response that results in elevated GCF PGE_2 levels even at healthy sites but that the destructive capacity of PGE_2 is held in abeyance by some compensatory component of the host response.

The elevation of GCF PGE_2 levels seen in EOP patients is also observed in refractory cases and in IDDM patients. In a group of 15 refractory patients, we observed a mean GCF PGE_2 level of 90.9 ± 15.8 ng/ml (unpublished data), which is greater than that of AP patients ($P < 0.01$) but less than that of EOP patients (Table 1). Similarly, for a group of 39 IDDM patients with mild to severe periodontitis, we demonstrated a mean GCF PGE_2 level of 283.2 ± 9.7 ng/ml (31). This GCF level in IDDM patients is approximately double that of EOP patients and triple that of AP patients. Thus, these three patient populations (EOP, refractory, and IDDM) share an exaggerated GCF PGE_2 inflammatory response compared to that of AP patients, and this response may be a reflection of an underlying system-based abnormal host response.

This concept is further supported when one examines the responses of these patients to therapy. Figure 1 shows the GCF PGE_2 levels for AP, EOP, and refractory patients before and after treatment. As can be seen, the treatment of 13 AP patients with scaling and root planing resulted in a significant drop in the GCF PGE_2 level to 30.7 ± 7.2 ng/ml. This reduction brings the basal GCF PGE_2 level down to that present in healthy individuals (Table 1), even though pocketing and attachment loss persist at this 6-week post-initial-therapy evaluation. This suggests that the systemic or local inflammatory response to the bacterial challenge seen in AP patients can be readily reversed by reducing the bacterial burden. In contrast, patients with refractory or EOP forms of disease do not respond with a dramatic drop in PGE_2 levels. The refractory patients (22) were treated with scaling and root planing plus augmentin (see reference 3 for details). The drop in GCF PGE_2 level from a pretreatment level of 136.9 ± 34.5 ng/ml to a posttreatment level of 104.3 ± 30.9 ng/ml was not significant. The seven EOP patients had pretreatment GCF PGE_2 levels of 133.4 ± 39.2 ng/ml, and these levels decreased only slightly, to 106.9 ± 28.0 ng/ml, following treatment with scaling and root planing plus tetracycline (1 g/day for 2 weeks). Thus, in contrast to what happened with AP patients, a reduction in the bacterial burden in these EOP patients did not attenuate

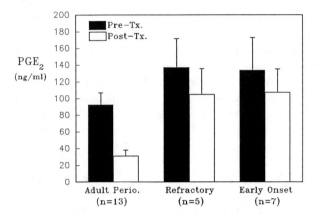

Figure 1. Effects of scaling and root planing on levels of GCF PGE$_2$. Crevicular fluid was collected and assayed for PGE$_2$ at baseline and at 6 weeks posttherapy in a group of 13 AP patients (advanced AP types II and III), 5 refractory periodontitis patients, and 7 early-onset cases. All patients received full-mouth scaling and root planing. The refractory and EOP groups were given 2 weeks of antibiotic therapy with augmentin or tetracycline, respectively. Further details regarding the patient population and treatment provided are given elsewhere (3). The treatment resulted in a significant reduction ($P < 0.005$) in the GCF PGE$_2$ level in AP patients.

the local GCF PGE$_2$ response. Although the patient numbers are modest in this preliminary data set, the trial provides a clear example that these individuals maintain an exaggerated PGE$_2$ inflammatory response even with a diminution in the bacterial burden.

MONOCYTIC PGE$_2$ RELEASE IN HIGH-RISK PATIENTS

Data obtained using peripheral blood monocytes suggest that the elevated GCF PGE$_2$ levels seen in patients (EOP, refractory, and IDDM) is also systemically associated with a hyperresponsive monocytic PGE$_2$ trait. Adherent monocytes obtained from the patients were maintained in culture for 18 to 24 h in the presence of [^{3}H]-arachidonic acid to measure the basal rate of PGE$_2$ synthesis. The supernatants were harvested, extracted with organic solvents, and prepared for high-pressure liquid chromatography (HPLC) by reverse-phase preparative chromatography (25). [^{3}H]-PGE$_2$ synthesis was quantitated by HPLC as described previously (21). PGE$_2$ counts per milliliter of medium for 16 patients are shown in Fig. 2. As can be seen from these preliminary data, basal secretion of PGE$_2$ in the four juvenile periodontitis and four refractory patients was greater than that seen in healthy control or AP patients. These data are certainly preliminary, but they are consistent with and confirm the findings of Shapira et al. (26). In a study of 28 EOP patients, Shapira et al. observed higher basal levels of PGE$_2$ secretion in the EOP patients than in 14 control patients. Furthermore, the amount of PGE$_2$ secreted in response to various levels of lipopolysaccharide (LPS) challenge was consistently three- to fivefold greater in the EOP patients than in the healthy

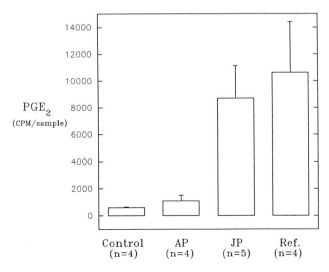

Figure 2. Monocyte prostaglandin E_2 production. Adherent monocytes were cultured in RPMI medium supplemented with 10% human AB serum and 1% penicillin-streptomycin at 37°C in 5% CO_2 for 18 to 24 h in the presence of tritiated arachidonic acid. Supernatants were collected, extracted, and prepared for HPLC according to methods described by Powell (25). Radiolabeled PGE_2 was quantitated using a Beta Flow in-line scintillation counter following straight-phase HPLC (25). Basal levels of radiolabeled PGE_2 released into monocyte culture medium are shown as patient mean values ± standard errors of the means, expressed as counts per minute (CPM) per sample. Representative samples of adherent cells were assayed for total DNA (2) to ensure equal final plating densities. JP, juvenile periodontitis form of EOP.

controls at each LPS concentration (26). Thus, the basal responsiveness of monocytes from EOP patients appears to be up-regulated with regard to resting cyclooxygenase levels, and the response to LPS results in greater PGE_2 synthesis than in control patients.

The data for a systemic monocytic hyperresponsiveness are even more compelling in IDDM patients (see reference 31 for details). Figure 3 illustrates the PGE_2 response of adherent monocytes to a single dosage of *Porphyromonas gingivalis* LPS (30 ng/ml). The amount of PGE_2 secreted by 21 AP patients was 60.4 ± 9.4 pg/ml (mean ± standard error of the mean) at this LPS concentration. At this same LPS challenge, the monocytes from 39 IDDM patients secreted significantly more PGE_2 than did those from the AP group (253.6 ± 36.4 pg/ml; $P = 0.0003$). This PGE_2 hypersecretion effect is seen throughout the LPS dose-PGE_2 response curve (31). Thus, as a group, the IDDM patients all share the hyper-PGE_2 secretory response as expressed in both GCF PGE_2 level and monocyte secretion in response to a standardized LPS challenge. Furthermore, within the diabetic patients, those with more severe periodontal disease (advanced AP type III or IV) have significantly elevated levels of monocytic PGE_2 secretion (289.0 ± 45.0 pg/ml) compared with IDDM patients with milder (advanced AP type I or II) forms of disease (151.0 ± 44.9 pg/ml; $P < 0.05$). This same trend is paralleled by the differences in the GCF PGE_2 levels when those two groups of IDDM patients

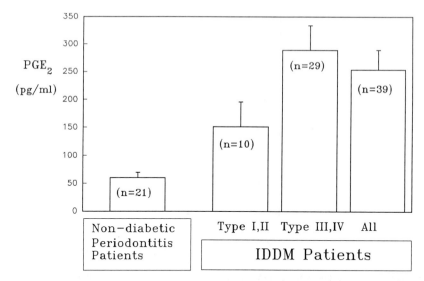

Figure 3. Monocytic PGE$_2$ release for 21 nondiabetic and 39 IDDM AP patients. The monocytes were cultured in the presence of 30 ng of *P. gingivalis* LPS per ml, and the amount of PGE$_2$ released into the culture supernatant was determined by radioimmunoassay. The 39 IDDM patients are further subdivided into those with advanced AP type I or II and those with type III or IV periodontal disease.

with periodontal disease are compared (31). These differences in monocytic and GCF PGE$_2$ responses are independent of the level of metabolic control, as reflected in the glycosylated hemoglobin levels, which are not different in these two subgroups of IDDM patients. Overall, the monocytic PGE$_2$ secretion is three- to sixfold greater in these IDDM patients than in AP nondiabetic patients. When considered together, these data suggest that in many of the EOP, refractory, and IDDM patients, GCF PGE$_2$ level is higher than that in AP patients and that this elevated GCF PGE$_2$ level is associated with an exaggerated systemic monocytic secretion of PGE$_2$.

POTENTIAL MECHANISMS OF MONOCYTIC
HYPERRESPONSIVENESS

Although the underlying mechanism for the hypersecretory monocyte and its relationship to the GCF PGE$_2$ seen in these high-risk periodontitis patients are unknown, there is considerable evidence regarding monocytic hyperresponsiveness in other disease conditions, including periodontitis (see Whicher and Evans for a review [29]). The concept of a monocytic hypersecretory state (Mϕ^+) was initially described for "periodontal-disease-susceptible" individuals by Garrison et al.(6). A Mϕ^+ trait has been described for IDDM patients by Mølvig (13, 14) and Pociot et al. (23, 24) and for rheumatoid arthritis patients by Goto et al. (7), Shore et al. (27), and Fujii et al. (5). Although more data are needed, it is our impression that "plaque-associated" AP, even severe type IV disease, may not be associated with a systemic Mϕ^+ trait. Some transient elevations of systemic Mϕ^+ responses may conceivably coincide with episodes of disease progression in these individuals, but the magnitude of the monocyte up-regulation in AP patients relative to that in healthy controls (Fig. 2) is negligible in comparison to that in EOP, refractory, and IDDM patients.

There are several possible explanations for the apparent PGE_2-Mϕ^+ trait. In EOP patients, the reported neutrophil chemotactic abnormality is a good model for the potential controversy surrounding this concept. Periodontal infection may cause the systemic Mϕ^+ trait. Chronic infection and LPS exposure might lead to systemic elevations of tumor necrosis factor alpha (TNF-α), interleukin-1β (IL-1β), and granulocyte macrophage–colony-stimulating factor, all of which are capable of up-regulating monocytic PGE_2 synthesis. T-cell activation is also known to influence monocytic PGE_2 secretion. Th1 responses elicit higher PGE_2 secretion via gamma interferon signaling, and Th2 activation induces IL-4 release, which attenuates monocytic PGE_2 release. There are also extensive data establishing the genetic basis of the monocytic secretory response in mice and humans. In humans, many of the genes that regulate monocytic cytokine response have been mapped to the *HLA-DR* region of chromosome 5 in the area of the TNF-β genes (see reference 4 for a review). Perhaps there is a genetic basis for the observed Mϕ^+ trait. This is an attractive hypothesis in light of the known genetic heritability of both EOP and IDDM. Recently, the regulatory upstream sequences of genes for various cytokines, including TNF-α genes, from different individuals have been compared (11, 23, 24). There appear to be sequence differences in the 5' promoter region of the TNF-α gene. In certain patients, specific response element sequences have been deleted (11, 23, 24). Thus, by analogy, a differential presence or absence of a promoter region of the cyclooxygenase-2 (COX-2) gene may be one possible mechanism for a genetically determined hyperresponsiveness trait.

Two cyclooxygenase genes produce two isoenzymes, COX-1 and COX-2, with approximately 64% overall identity (9, 15, 16, 28). These isoenzymes share enzymatic activity and important structural features, including a conserved hydrophobic region, a heme-binding domain, and glycosylation sites. COX-1 is expressed constitutively, whereas COX-2 is the inflammatory cyclooxygenase that is apparently expressed only upon ligand-coupling stimulation. Stimulation of monocytes with IL-1β or LPS, for example, results in de novo transcription and synthesis of COX-2 without induction of COX-1 (4, 15, 30). The promoter region of the COX-2 gene contains a TATA box and a variety of enhancer elements, including a serum response element, an AP-1, a C-*mos*, an NF-κB, and several SP-1 and AP-2 sites (30). Thus, there are several potential sites for environmental influence on the transcriptional regulation of COX-2. Most of the data indicate that monocytic PGE_2 synthesis in response to stimulus is principally dependent upon de novo transcription of COX-2 and is not controlled at a posttranscriptional or posttranslational level (4). Rather, COX-2 is a member of the immediate or early-response gene family that includes stress proteins, growth peptides, and other cytokines (4, 10). If COX-2 regulatory sequences differ among different patients, then the resting basal levels of COX-2 and the response to challenge may also differ among patients. In the multifactorial model of periodontal diseases, the genetic determination of COX-2 gene regulatory sequences may define the PGE_2 response capacity of an individual.

The particular combination of response elements within that regulatory region may define the PGE_2 response to specific environmental influences.

Although this discussion is clearly an extrapolation beyond the existing data, it does provide a reasonable model for further examining the contribution that PGE_2 synthesis plays in the molecular pathogenesis of periodontal diseases. Furthermore, the association of abnormal monocytic and GCF PGE_2 responses with these high-risk periodontal conditions provides further insight into the molecular basis of periodontal disease susceptibility.

Acknowledgment. This work was supported by NIH grant HD26652.

REFERENCES

1. **Beck, J. D., G. C. Koch, and S. Offenbacher.** Attachment loss trends over three years in community-dwelling blacks and whites. *J. Periodontal Res.,* in press.
2. **Brunk, C. F., K. C. Jones, and T. James.** 1979. Assay for nanogram quantities of NDA in cellular homogenates. *Anal. Biochem.* **92:**497–500.
3. **Collins, J. G., S. Offenbacher, and R. R. Arnold.** 1993. Effects of a combination therapy to eliminate *P. gingivalis* in refractory periodontitis. *J. Periodontol.* **64:**998–1007.
4. **Fu, J.-Y., J. L. Masferrer, K. Seibert, A. Raz, and P. Needleman.** 1990. The induction and suppression of prostaglandin H_2 synthase (cyclooxygenase) in human monocytes. *J. Biol. Chem.* **265:**16737–16740.
5. **Fujii, I., M. Shingu, and M. Nobunaga.** 1990. Monocyte activation in early onset rheumatoid arthritis. *Ann. Rheum. Dis.* **49:**497–503.
6. **Garrison, S. W., S. C. Holt, and F. C. Nichols.** 1988. Lipopolysaccharide-stimulated PGE_2 release from human monocytes. *J. Periodontol.* **59:**684–687.
7. **Gogo, M., M. Sasano, T. Miyamoto, K. Nishioka, K. Nakamura, S. Aotsuka, and R. Yokohari.** 1989. Lack of association of HHLA-DR4 with interleukin 1β secretion from peripheral blood monocytes in patients with rheumatoid arthritis. *J. Rheumatol.* **16:**1025–1028.
8. **Heasman, P. A., J. G. Collins, and S. Offenbacher.** 1993. Changes in crevicular fluid levels of IL-1β, LTB_4, PGE_2, TxB_2 and TNFα in experimental gingivitis in humans. *J. Periodontal Res.* **28:**241–247.
9. **Kujubu, D. A., B. S. Fletcher, B. C. Varnum, R. W. Lim, and H. R. Herschman.** 1991. TIS10 a phorbol ester tumor promoter-inducible mRNA from Swiss 3T3 cells, encodes a novel prostaglandin synthetase cyclooxygenase homologue. *J. Biol. Chem.* **266:**12866–12872.
10. **Maier, J. A. M., T. Hla, and T. Maciag.** 1990. Cyclooxygenase is an immediate-early gene induced by interleukin-1 in human endothelial cells. *J. Biol. Chem.* **265:**10805–10808.
11. **Messer, G., U. Spengler, M. C. Jung, G. Honold, K. Blömer, G. R. Pape, Riethmüller, and E. H. Weiss.** 1991. Polymorphic structure of the tumor necrosis factor (TNF) locus: an NcoI polymorphism in the first intron of the human TNF-β gene correlates with a variant amino acid in position 26 and a reduced level of TNF-β production. *J. Exp. Med.* **173:**209–219.
12. **Miyasaki, K. T.** 1991. The neutrophil mechanisms of controlling periodontal bacteria. *J. Periodontol.* **62:**761–774.
13. **Mølvig, J.** 1992. A model of pathogenesis of insulin-dependent diabetes mellitus. *Dan. Med. Bull.* **39**(6):509–541.
14. **Mølvig, J., L. Pociot, L. Bæk, H. Worsaae, L. Dall Wogensen, P. Christensen, L. Staub-Nielsen, T. Mandrup-Poulsen, K. Manogue, and J. Nerup.** 1990. Monocyte function in IDDM patients and healthy individuals. *Scand. J. Immunol.* **32:**297–311.
15. **O'Banion, M. K., H. B. Sadowski, V. Winn, and D. A. Young.** 1991. A serum- and glucocorticoid-regulated 4-kilobase mRNA encodes a cyclooxygenase-related protein. *J. Biol. Chem.* **266:**23261–23267.
16. **O'Banion, M. K., V. D. Winn, and D. A. Young.** 1992. cDNA cloning and functional activity of a glucocorticoid-regulated inflammatory cyclooxygenase. *Proc. Natl. Acad. Sci. USA* **89:**4888–4892.

17. **Offenbacher, S., J. G. Collins, and R. R. Arnold.** 1993. New clinical diagnostic strategies based on pathogenesis of disease. *J. Periodontal Res.* **28:**523–535.
18. **Offenbacher, S., P. A. Heasman, and J. G. Collins.** 1993. Modulation of host PGE_2 secretion as a determinant of periodontal disease expression. *J. Periodontol.* **64:**432–444.
19. **Offenbacher, S., B. M. Odle, M. D. Green, C. S. Mayambala, M. A. Smith, M. E. Fritz, T. E. Van Dyke, K. C. Yeh, and F. J. Sena.** 1990. Inhibition of human periodontal prostaglandin E_2 synthesis with selected agents. *Agents Actions* **29:**232–238.
20. **Offenbacher, S., B. M. Odle, and T. E. Van Dyke.** 1986. The use of crevicular fluid prostaglandin E_2 levels as a predictor of periodontal attachment loss. *J. Periodontal Res.* **21:**101–112.
21. **Offenbacher, S., A. Soskolne, and J. G. Collins.** 1991. Markers of periodontal disease susceptibility and activity present within the gingival crevicular fluid, p. 313–337. *In* N. W. Johnson (ed.), *Risk Markers for Oral Diseases*, vol. 3. Cambridge University Press, Cambridge.
22. **Oliver, R. C., L. J. Brown, and H. Löe.** 1991. Variations in the prevalence and extent of periodontitis. *J. Am. Dent. Assoc.* **122:**43–47.
23. **Pociot, F., J. Mølvig, L. Wogensen, H. Worsaae, H. Dalbøge, L. Bæk, and J. Nerup.** 1991. A tumour necrosis factor beta gene polymorphism in relation to monokine secretion and insulin-dependent diabetes mellitus. *Scand. J. Immunol.* **33:**37–49.
24. **Pociot, F., J. Mølvig, L. Wogensen, H. Worsaae, and J. Nerup.** 1991. A Taq1 polymorphism in the human interleukin-1β (IL-1β) gene correlates with IL-1β secretion *in vitro. Eur. J. Clin. Invest.* **22:**396–402.
25. **Powell, W. S.** 1987. High-pressure liquid chromatography in the analysis of arachidonic acid metabolites, p. 75–98. *In* C. Benedetto, R. G. McDonald-Gibson, S. Nigam, and T. F. Slater (ed.), *Prostaglandins and Related Substances: a Practical Approach.* IRL Press Ltd., Oxford.
26. **Shapira, L., W. A. Soskolne, M. N. Sela, S. Offenbacher, and V. Barak.** 1994. The secretion of PGE_2, IL-1β, IL-6 and TNFα by monocytes from early onset periodontitis patients. *J. Periodontol.* **65:**139–146.
27. **Shore, A., S. Jaglal, and E. C. Keystone.** 1986. Enhanced interleukin 1 generation by monocytes in vitro is temporally linked to an early event in the onset or exacerbation of rheumatoid arthritis. *Clin. Exp. Immunol.* **65:**295–302.
28. **Smith, W. L., T. E. Eling, R. J. Kulmacz, L. J. Marnett, and A. L. Tsai.** 1992. Tyrosyl radicals and their role in hydroperoxide-dependent activation and inactivation of prostaglandin endoperoxide synthase. *Biochemistry* **31:**3–7.
29. **Whicher, Y. T., and S. W. Evans.** 1990. Cytokines in disease. *J. Clin. Chem.* **36:**1269–1281.
30. **Xie, W., J. R. Merrill, W. S. Bradshaw, and D. L. Simmons.** 1993. Structural determination and promoter analysis of the chicken mitogen-inducible prostaglandin G/H synthase gene and genetic mapping of the murine homolog. *Arch. Biochem. Biophys.* **300:**247–252.
31. **Yalda, B., J. G. Collins, R. R. Arnold, and S. Offenbacher.** 1994. Monocytic and crevicular fluid PGE_2 and IL-1β in diabetic patients. *J. Dent. Res.* **73:**2337.

Molecular Pathogenesis of Periodontal Disease
Edited by Robert Genco et al.
© 1994 American Society for Microbiology, Washington, DC 20005

Chapter 19

Leukocyte Adhesion Molecules CD11/CD18 and Their Role in Periodontal Diseases

Yoji Murayama, Kyoko Katsuragi, Shogo Takashiba, and Hidemi Kurihara

The normal response of the immune system to infection of the skin or mucosal tissue requires peripheral blood leukocytes to mobilize to the site of infection. The molecular mechanisms that enable leukocytes to migrate across blood vessels have been elucidated in part by the characterization of a recently defined disease called human leukocyte adhesion deficiency (LAD). Patients with this heritable disease suffer from severe recurrent bacterial infections of soft tissue. In the early 1980s, several laboratories demonstrated that this disease is due to a cell surface deficiency of the leukocyte adhesion molecules (LFA-1, Mac-1, and pl50,95). More recently, it has been shown that LAD is due to heterogeneous mutation in the β subunit (CD18) common to all three of the leukocyte glycoproteins. Although LAD is a rare disease, the molecular analyses of periodontal patients with manifestations of altered β2 integrin will greatly increase our understanding of pathogenesis in other periodontal diseases.

In order to assess the role of β2 integrin in the pathogenesis of periodontal disease, we first give a summary account of β2 integrin with reference to the excellent review articles of Anderson and Springer (7), Hynes (35), Haynes et al. (29), Arnaout (8), Springer (67), Figdor et al. (23), and Pardi et al. (58). Subsequently, we discuss the biological features and molecular study of the individuals with decreased expression of β2 integrin on leukocyte cell surfaces, including two sibling patients with generalized prepubertal periodontitis (G-PP) caused by LAD and one patient with localized juvenile periodontitis and no remarkable systemic diseases.

LEUKOCYTE ADHESION MOLECULES CD11/CD18

Subclass and Distribution of CD11/CD18

Table 1 summarizes some properties of the integrin family of adhesion molecules. With the exception of the complex isolated from chicken cells, all integrin

Yoji Murayama, Kyoko Katsuragi, Shogo Takashiba, and Hidemi Kurihara • Department of Periodontology and Endodontology, Okayama University Dental School, 2-5-1 Shikata-cho, Okayama 700, Japan.

Table 1. Members of integrin receptor family[a]

Subfamily and members	Structure	Ligand	Cell distribution	Function
β_1 integrins				Cell-cell and cell-matrix adhesion; leukocyte homing
VLA-1	$\alpha_1\beta_1$	CL, LM	F, BM, B', T'	
VLA-2	$\alpha_C\beta_1$	CL	P, F, EN, EP, T	
VLA-3	$\alpha_3\beta_1$	FN, LM, CL	EP, F	
VLA-4	$\alpha_4\beta_1$	VCAM-1, FN, PP HEV	NC, F, B, T, M, LGL	
VLA-5	$\alpha_F\beta_1$	FN	F, EP, EN, P, Th, T	
VLA-6	$\alpha_6\beta_1$	LM	P, T	
β_2 integrins				
CD11a/CD18	$\alpha_L\beta_2$	ICAM-1, ICAM-2	B, T, M, G	
CD11b/CD18	$\alpha_M\beta_2$	iC3b, factor X, FB, LPS	M, G	
CD11c/CD18	$\alpha_X\beta_2$	iC3b	M, G	Leukocyte adhesion
β_3 integrins				
gpIIb/IIIa	$\alpha_{IIb}\beta_3$	FN, VN, FB, vWF, TSP?		
Vitronectin receptor	$\alpha_V\beta_3$	VN, FB, vWF	P, EN	Cell-cell and cell-matrix adhesion
β_4 integrins	$\alpha_E\beta_4$		EP	Cell matrix adhesion
β_5 integrins	$\alpha_V\beta_X$	FN, VN	EP	Cell matrix adhesion
β_6 integrins	$\alpha_4\beta_P$		T, B	Peyer's patches-lymphocyte adhesion

[a] Abbreviations: CL, collagen; FN, fibronectin; FB, fibrinogen; LM, laminin; ICAM, intercellular adhesion molecule; iC3b, proteolytic fragment of complement protein C3; factor X, coagulation factor X; vWF, von Willebrand's factor; TSP, thrombospondin; PP HEV, Peyer's patch high endothelial venule; EN, endothelial cells; EP, epithelial cells; F, fibroblasts or other connective tissue; NC, neural crest melanocytes; P, platelets; BM, basement membrane associated; B, B lymphocytes; T, T lymphocytes; prime ('), activated lymphocytes only; Th, thymocytes; M, monocytes; G, granulocytes; LGL, large granular lymphocytes; VCAM, vascular cell adhesion molecules; LPS, lipopolysaccharide; VN, vitronectin.

molecules consist of two noncovalently linked subunits, a β subunit (90- to 130-kDa protein) and an α subunit (140- to 200-kDa protein) (16, 32, 33, 37). Six mammalian subfamilies (β_1 through β_6) of integrins have been described and classified according to the types of their highly homologous (40 to 48%) β subunits (20, 24, 41, 46, 61, 64, 68). The β2-integrin leukocyte adhesion molecules (Leu-CAM, CD11/CD18) consist of three surface membrane heterodimeric glycoproteins named LFA-1 (CD11a/CD18), Mac-1 (CD11b/CD18), and p150,95 (CD11c/CD18) (47, 70, 71). The α subunits (C11a, CD11b, and CD11c) are distinct, with apparent molecular masses of 180, 170, and 150 kDa, respectively, each associating with a common β subunit (CD18) of 95 kDa. Monoclonal antibodies against the β chain are referred to as CD18, and antibodies against the α chain are CD11a, CD11b, and CD11c. CD11/CD18 is expressed only in leukocytes. CD11a/CD18 is present on all leukocytes, and CD11b/CD18 and CD11c/CD18 are restrictively expressed on monocytes, macrophages, neutrophils, and natural killer cells. CD11c/CD18 is also expressed on some B-cell lines such as hairy cell leukemia cells and certain cloned cytotoxic T lymphocytes (62). The relative abundance of CD11/CD18 varies depending on the cell type and the state of cell activation and differentiation. T and B lymphocytes normally express only CD11a/CD18. Activated granulocytes express far more CD11b/CD18 than the other two antigens. In resting monocytes, the relative abundance of these antigens is CD11a/CD18 > CD11b/CD18 > CD11c/CD18. In tissue macrophages, the relative amounts expressed are CD11c/CD18 > CD11a/CD18 > CD11b/CD18 (25, 51).

Structure of CD11/CD18

The major structural features of the α and β subunits of CD11/CD18 are outlined in Fig. 1 (8). The CD11 a, b, and c subunits are characterized by a cytoplasmic region, a transmembrane region, seven homologous tandem repeats, and an additional domain in the extracellular portion. The cytoplasmic regions of the CD11 a, b, and c subunits are 53, 19, and 29 amino acids long, respectively, and include the distinctive amino acids serving as potential targets of phosphorylation (Fig. 2) (15). The mechanisms operating in the integrin-mediated cell adhesion pathway from intracellular signal transduction to inside-out signaling were recently discussed in association with the role of cytoplasmic domains of integrins, particularly the CD11a/CD18 β subunit (23, 58, 67). Binding of antigen to T-cell receptor–CD3 complex stimulates phospholipase C, which catalyzes the hydrolysis of phosphatidyl inositol *bis*-phosphate, thus producing inositol triphosphate, which mobilizes intracellular calcium and diacylglycerol. Protein kinase C (PKC), which is activated by diacylglycerol, subsequently phosphorylates the intracellular portion of the adhesion molecule. The CD11a/CD18 heterodimer appears to be itself a substrate for PKC-dependent phosphorylation upon antigen receptor triggering in T cells (59). In consequence of serving as a substrate for PKC-mediated phosphorylation, CD11a/CD18 becomes physically associated with the actin-based cytoskeleton upon antigen receptor cross-linking in T cells. Many subcortical proteins are thought to be part of a multimolecular hinge connecting the plasma membrane to the cytoskeleton in activated cells (44, 56). In other words, phosphorylated CD11a/

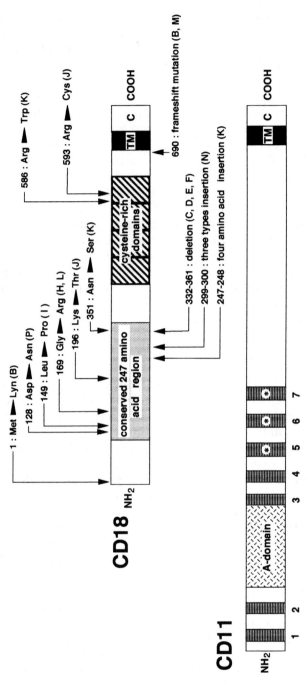

Figure 1. Schematic diagram of the structure of CD11/CD18 and the positions of mutations in CD18. The main structural features of CD11 and CD18 are outlined. The hatched region indicates the four cysteine-rich repeats that are characteristic of CD18 and other β subunits of integrins. The stippled area indicates the conserved 247-amino-acid region. The black area represents the putative transmembrane site (TM), and C represents the cytoplasmic end. The position of the seven tandem repeats in the extracellular portion of CD11 is outlined, as is the additional domain (A domain). Repeats 5 through 7 each contain a putative metal-binding site (*). Upward and downward arrows and numbers show positions of amino acid substitutions, deletions, insertions, or frameshift mutations. Letters in parentheses correspond to the patients listed in Table 4.

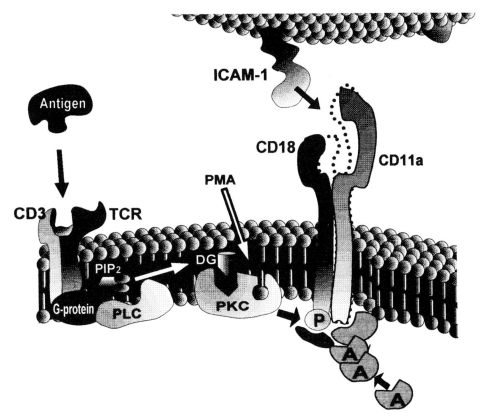

Figure 2. Proposed model for conformational change of CD11a/CD18 (LFA-1). Binding of antigen to TCR-CD3 stimulates phospholipase C (PLC), which catalyzes the hydrolysis of phosphatidyl inositol *bis*-phosphate (PIP2), producing inositol triphosphate, which mobilizes intracellular calcium and diacylglycerol (DG). PKC activated by DG subsequently phosphorylates the intracellular portion of CD18. This phosphorylation results in the formation of a core of actin (A) polymerization. The actin-based cytoskeleton or biochemical modifications of cytoplasmic domains induce conformational changes of the extracellular domain of CD11a/CD18 that lead to high-affinity binding to ICAM-1. Phorbol myristate acetate (PMA) activates CD11a/CD18 by direct activation of PKC.

CD18 and/or cytoskeletal proteins undergo an allosteric transition resulting in the association of one or more subcortical proteins that are directly or indirectly involved in binding to actin with membrane-bound CD11a/CD18. This creates a core of actin polymerization. Finally, the activated cytoskeleton-associated isoforms of CD11/CD18 appear to be conformationally altered at the extracellular domain level, as demonstrated by the acquisition of neoepitopes (3, 73). In addition, the CD11a/CD18 β-subunit cytoplasmic domain is required for functional activity in binding to ICAM-1, as shown by transfection of patient cell lines that genetically lack the CD18 with intact and truncated CD18 cDNA (12, 31, 76). On the basis of these findings, the cytoplasmic domain of the CD18 appears to play a fundamental role in the functioning of the mature molecular complex, especially in the transient acquisition of the high-avidity state (30). The transmembrane region is highly conserved among CD11 a, b, and c subunits (88% amino acid identity), suggesting a critical functional role in the stabilization of αβ associations, in signal transduction, or in interaction with membrane lipids. The extracellular regions in the CD11 a, b, and c subunits contain seven homologous tandem repeats (each approximately 60 amino acids), which could have arisen by a series of duplication events and subsequent diversion, leaving the most important structural components conserved. Repeats 5 through 7 each contain the metal-binding consensus nanopeptide DXXDXGXXD (D is aspartic acid, G is glycine, and X is any amino acid). A large insertion of 187 amino acids between the second and third homologous repeats is not present in other integrins. This additional domain (A domain) is 64% identical in CD11b and CD11c and only 36% homologous between these two receptors and CD11a, suggesting that it may be responsible in part for common functions mediated by CD11b/CD18 and CD11c/CD18, such as iC3b-binding.

The structure of the CD18 subunit constitutes the characteristic feature of four highly conserved regions (Fig. 1). The primary structure is a transmembrane glycoprotein with a short cytoplasmic region (46 amino acids) containing distinctive amino acid residues that account for stimulus-induced phosphorylation of CD18 (Fig. 2) (15). The second domain is the transmembrane region, which may mediate functions similar to those proposed above for the transmembrane region in the α subunits. The third is a highly conserved cysteine-rich region existing in the extracellular region of CD18 and consisting of four tandem repeats of an 8-cysteine motif. Some mutations in this region prevent cell surface expression of CD18, as discussed below (9, 53). The fourth highly conserved domain is a 247-amino-acid region N terminal to the cysteine-rich domain that contains four segments in which identity with other β subunits is 60 to 89%. Conservation of these segments among β integrins suggests that they serve some common and essential functions such as ligand binding and/or heterodimer formation. In β1, β3, and β5, the first segment in this region contains a binding site for RGD (Arg-Gly-Asp) sequences that play a role in adhesion to ligands such as fibronectin and vitronectin (22, 66). Although there are no RGD sequences in the 247-amino-acid region of the CD11/CD18 molecule, the region is somehow related to adhesion functions, as shown by the fact that mutations within the 247-amino-acid region in CD18 impair cell surface expression of CD18, as described below (9, 18, 38, 42, 48, 53, 75).

DISEASES ASSOCIATED WITH ABERRANT CD11/CD18 EXPRESSION

Diseases with Decreased CD11/CD18 Expression

Deficiency of CD11/CD18 is related to a rare congenital disease, LAD, characterized by the occurrence of recurrent bacterial infections of soft tissue, delayed umbilical cord separation, impaired pus formation, and persistent leukocytosis (4, 19). About 60 patients have been described worldwide.

The level of CD11/CD18 expression is heterogeneous among LAD patients and correlates with the severity of the clinical symptoms and the loss of leukocyte adhesive functions. For this reason, two LAD phenotypes, severe and moderate, depending on whether the CD11/CD18 is undetectable or ranges between 3 and 30% of normal levels, have been defined (6). These clinical symptoms are all secondary to defective leukocyte integrin-mediated functions such as neutrophil and monocyte mobilization into inflammatory sites (e.g., chemotaxis, aggregation, phagocytosis, adherence to endothelial cell, and complement iC3b binding) and antibody-dependent cellular toxicity (4, 5, 11, 19). Reduced surface expression of CD11/CD18 is also observed in specific granule deficiency and in neonatal granulocytes and is due to an absence or a relative deficiency, respectively, of the major intracellular storage sites for CD11b,c/CD18 (4, 55). Deficient CD11a/CD18 expression was also observed in certain lymphomas and was cited as a likely contributory factor in leukemic transition or escape from immunosurveillance (17). These deficiencies also contribute to defective cell adhesion and the increased incidence of bacterial infections in these patients.

Diseases with Increased CD11/CD18 Expression

Increased expression of CD11/CD18 on circulating leukocytes occurs in several inflammatory disorders associated with neutrophil activation: in patients with burns (52), systemic lupus erythematosis (13), or diabetes mellitus (63) or on hemodialysis (10). Increased expression is often mediated by inflammatory mediators generated in vivo and may contribute, through enhanced cell adhesion, to neutropenia and/or the microvascular injury observed in these diseases. In Down's syndrome, a persistent small increase in expression of CD11a/CD18 on lymphocytes is observed (69). This is associated with enhanced lymphocyte adhesion and may contribute to some of the impaired immune responses and increased incidence of infections seen in these patients.

Periodontal Patients with Decreased CD11/CD18 Expression

It is thought that individuals having the early-onset aggressive forms of periodontitis are predisposed to bacterial infections by an underlying abnormality in the host defense systems. In fact, decreased expression of CD11/CD18 on circulating leukocytes seems to occur in some patients with early-onset periodontitis, but descriptions of the periodontal manifestation of the CD11/CD18 deficiency syndrome have been limited to only a few reports. Page et al. (57) studied the expression of CD11/CD18 on leukocytes from 29 patients with various forms of periodontitis, including one family in which the father and two children had G-PP, the mother had suggestive periodontal manifestations of postjuvenile periodontitis, and two other children were periodontally healthy. The expression levels of CD11/CD18 from five of the 29 patients (four children and one adult) with G-PP were abnormally low. Then Waldrop et al. (74) reported additional studies of members of the family just mentioned. The three family members with G-PP manifested recurrent necrotic soft tissue infections, impaired pus formation, delayed wound healing, granulocytosis, severe abnormalities of adhesion-dependent granulocyte functions, and a profound deficiency (3 to 6% of healthy levels) of CD11/CD18 on cell surfaces. The others in this family demonstrated approximately half of healthy CD11/CD18 expression but showed normal adhesion-dependent leukocyte functions and no susceptibility to systemic infections. However, their periodontal manifestations were divided into two forms; i.e., two children demonstrated no periodontal manifestation, but the mother exhibited clinical and radiographic features of typical postjuvenile periodontitis. In the following three individuals with periodontal disease of early-onset type, the biological functions of CD11/CD18 and the structural aspects of CD18 mRNA were examined (26, 38, 43, 45, 48, 54).

Patients N and O

Patients N (a 5-year-old girl) and O (a 3-year-old boy) were Japanese siblings with a severe form of LAD (26, 43). Both of them lacked three membrane glycoproteins of CD11/CD18. Their leukocyte counts were extremely high. Chemotactic and phagocytic activities of their neutrophils (Table 2) were extremely low, but bactericidal capacities were normal. Neutrophil adhesion was also low. These patients demonstrated severe involvement of the gingiva at the time of eruption of the primary dentition and acute generalized inflammation and proliferation of gingival tissues thereafter, despite intermittent administration of antibiotics. Both patients developed severe periodontal symptoms, e.g., periodontal abscess, tooth mobility, gingival bleeding, and periodontal pocket formation, and the clinical diagnosis for both siblings was G-PP (38). Incidentally, their parents, nonconsanguineous, had reduced levels of CD11/CD18, but they were relatively healthy and had had no remarkable inflammatory diseases including periodontal diseases, suggesting an autosomal recessive mode of inheritance. The mother's neutrophils showed a slight impairment of chemotaxis and normal phagocytosis; the father's neutrophils revealed normal chemotaxis and mildly impaired phagocytosis (43, 48).

Patient Q

Patient Q was a 16-year-old Japanese girl with typical localized juvenile periodontitis. She demonstrated extensive depression of neutrophil functions such as chemotaxis, random migration, phagocytic capacity, and superoxide production but normal levels of bacterial attachment (Table 2) (58). Kurihara et al. (45) studied the expression of CD11/CD18 on the neutrophil surface membranes of patient Q. Western blotting (immunoblotting) analysis with CD18 monoclonal antibody revealed that the amount of protein corresponding to CD18 molecule of peripheral neutrophils from patient Q was lower than amounts in control subjects, including healthy subjects and patients with various types of periodontal diseases. The expression of CD11/CD18 on the cell surface was examined before and after activation with N-formyl-methionyl-leucyl phenylalanine (fMLP) by flow cytometry. Although patient Q expressed the same levels of CD18 molecule on the intact cells as the control subjects did before neutrophil activation with fMLP, there was a wide difference between patient Q and control subjects after activation; i.e., levels in patient Q were not raised as much as levels in the control subjects. These findings indicate that the CD11/CD18 in patient Q was likely originally stored in a small

Table 2. Neutrophil functions of patients N, O, and Q[a]

Patient	Migration (%)		Phagocytic capacity (%)				Bactericidal capacity (%)[b]	Adhesion (%)[c]
	Chemotactic	Random	Attached bacteria		Incorporated bacteria			
			Per 100 neutrophils	Per functional neutrophils	Per 100 neutrophils	Per functional neutrophils		
N (39)	41.1	47.4	ND	ND	30.1	ND	150	7.1
O (24)	52.4	55.5	ND	ND	19.4	ND	ND	14.9
Q (40)	22.2	26.8	98.3	110	34.4	63.5	23.2	ND

[a] Values represent percentages of healthy control subject levels. ND, not determined.
[b] Superoxide production.
[c] Anchoring, i.e., rate of cells spread out and flattened onto a plastic surface.

quantity in the intracellular granules. As a result, she was likely to manifest a partial deficiency of CD11/CD18 and depression of neutrophil functions. However, bactericidal activity was relatively normal in patients N and O, as seen in most LAD patients, while that in patient Q was impaired. Incidentally, patient Q's mother revealed no deficiency of CD11/CD18 but showed depression of neutrophil functions similar to that of patient Q. The mother had a rapidly progressive periodontitis supposedly based on a juvenile periodontitis (54). In addition, the functional properties of neutrophils crossed between patients N and O and patient Q (Table 2). These facts may reflect differences in the pathogenic mechanisms causing periodontal disease in patients N and O and in patient Q.

BIOLOGICAL FUNCTIONS ASSOCIATED WITH QUANTITATIVE AND QUALITATIVE MODIFICATIONS IN CD11/CD18

Intercellular adhesion is now recognized as a fundamental process in leukocyte physiology (21, 60, 67). It has become clear that CD11/CD18 acts as a general cellular adhesion molecule that mediates a variety of heterotypic and homotypic cell-cell interactions (28, 34, 39, 49, 50). Homotypic adhesion of leukocytes is dependent on CD11/CD18 and ICAM-1 and -2 but is not necessarily accompanied by an increase in the expression of CD11/CD18 or ICAM-1 and -2. Adhesion of leukocytes to one another can be measured as cluster formation, i.e., aggregation (27). (Cluster formation is also induced by treatment with phorbol ester.)

The appearance of homotypic adhesion by Epstein-Barr virus (EBV)-transformed B lymphocytes from patients N, O, and Q was examined in the absence of phorbol ester (phorbol myristate acetate [PMA]) and in its presence by quantitative and qualitative aggregation assays (Table 3) (38). Cells from patients N and O did not aggregate and form clusters in either the absence or the presence of PMA, while cells from patient Q and control subjects, including other patients with periodontitis and periodontally healthy subjects, demonstrated 65 and 90% aggregation in the absence and presence of PMA, respectively. The clusters formed exhibited some differences between patient Q and the control subjects in the presence of PMA. The clusters of patient Q were smaller and more numerous, and each cluster comprised a small number of cells. Every cluster formed in the absence of PMA was medium sized and loosely aggregated. In addition, the homotypic cell aggregate formation was inhibited by anti-CD11a/CD18 monoclonal antibody. Cells from patients N and O failed to aggregate, indicating a functional disorder of the cells caused by a CD11a/CD18 deficiency on the cell surface. Cells from patient Q showed a characteristic PMA-stimulated lymphoid cell adhesiveness, suggesting that quantitative or qualitative changes occurred in CD11a/CD18 after cell activation, although there is no direct experimental evidence to substantiate this possibility. The functional mechanism of cell adhesion mainly correlates with conformational rather than quantitative changes in CD11a/CD18 (1, 2, 67, 77). In particular, additional or alternative modifications may be required for regulating the various adhesion interactions mediated by adhesion molecules (7). The main regulatory factor is CD11a/CD18 phosphorylation (Fig. 2) (15, 72). All CD11 subunits are intrinsically

Table 3. Homotypic adhesion by EBV-transformed B lymphocytes from patients N, O, and Q

Subject	Quantitative aggregation[a]			Qualitative aggregation					
				−PMA			+PMA		
	−PMA	+PMA	+PMA+MAb	Shape and size of clusters	No. of clusters	No. of constitutional cells (10^3)	Shape and size of clusters	No. of clusters	No. of constitutional cells (10^3)
Patient N (39)	−	−	−	None	0	0	None	0	0
Patient O (24)	−	−	−	None	0	0	None	0	0
Patient Q (40)	+	++	−	Loose, medium	32	2.1	Dense, small	51	1.8
Controls with periodontal diseases	+	++	−	Loose, medium	33	2.1	Dense, large	27	3.4
Healthy people	+	++	−	Loose, medium	32	2.2	Dense, large	28	3.2

[a] −PMA, cells were not treated with PMA; +PMA, cells were treated with PMA; +PMA+MAb, anti-CD18 monoclonal antibody; −, little if any cell aggregation; +, cells aggregated; ++, cells aggregated strongly.

phosphorylated in isolated human mononuclear cells, while the common CD18 subunit is detected with little or no phosphorylation (14, 15). Because phosphorylation of the CD18 subunit is increased after stimulation with PMA, Pardi et al. (58) recently proposed a model of activation-dependent transition for the ligand (Fig. 2). This appears to result from direct or indirect activation of PKC (15, 17, 58). Consequently, phosphorylation of the CD18 subunit may induce a conformational change in the CD11a/CD18 molecule and predispose individual host defense systems to periodontitis.

MOLECULAR BASIS FOR DEFICIENT EXPRESSION OF CD11/CD18

Molecular Basis of LAD

LAD is inherited on an autosomal recessive basis in the majority of cases. The underlying cause for deficient expression of CD11/CD18 in patients is heterogeneous mutations that alter the expression and/or structure of the common CD18 subunit, so knowing the molecular basis for this heterogeneity is invaluable in developing a detailed structure-function map of the CD18 molecule. LAD patients have previously been categorized as severely deficient (<1% of healthy levels of expression) or moderately deficient (3 to 10% of healthy levels) (Table 4) (6). The severity of the clinical complications is reflected in the extent of the deficiency. Severely deficient patients tend to succumb to overwhelming microbial infections early in childhood. Moderately deficient patients suffer from severe recurrent infections but can survive to adulthood with medical care. However, within the severe and moderate phenotypes, there are variations in clinical features that suggest further heterogeneity in the mutations causing LAD. In fact, five distinct classes of mutations have been defined on the basis of sizes and levels of expression of the CD18 mRNA and the CD18 precursor protein (40).

The molecular basis of LAD has so far been described for 16 patients (Table 4). Although one patient categorized as severe phenotype (patient A in Table 4) had no detectable CD18 subunit mRNA expression and precursor synthesis (43) and the phenotype of another patient (patient M) has not been determined (65), the other patients with the severe phenotype had detectable mRNA expression and/or protein precursor. The half of the patients with detectable mRNA showed low levels of CD18 mRNA expression and/or precursor synthesis without distinction of phenotype, and the other half showed healthy levels of both mRNA expression and precursor synthesis. Of the eight patients who revealed aberrant mRNA expression or precursor synthesis, four belonged to one family (6, 42), two were siblings (26, 43), and two were unrelated (6, 36). Four related patients (patients C, D, E, and F) who synthesized an aberrantly small CD18 precursor had an unusual RNA splicing defect that resulted in an in-frame deletion of 30 amino acids from an extracellular region that is highly conserved among other integrin β subunits (42). One of the siblings (patient N) with low mRNA expression had aberrant mRNA that was 1.2 kb longer than usual because of a G-to-A substitution at the splice donor site of a 1.2-kb intron (48). The molecular study of LAD in six unrelated patients who synthesized a normal-sized CD18 precursor revealed point mutations

Table 4. Patients with aberrant expression of CD11/CD18 and mutations in CD18 causing aberrant expression

Class of mutation	Patient(s)	Phenotype	Size of β precursor	Amt and size of β mRNA	Mutation(s)[a]	Reference(s)
I	A	Severe	Not detected	Not detected	Not reported	7
II	B	Moderate	Trace amount	Low	T → A (74 nt) Single-base deletion (2,142 nt)	36, 65
III	C, D, E, F (related)	Moderate	Aberrantly small	Normal	G → C (5′ site of intron)	6, 42
IV	G	Severe	Aberrantly large	Normal	Not reported	6
V	H	Severe	Normal size	Normal	G → A (577 nt)	6, 75
	I	Moderate	Normal size	Normal	T → C (517 nt)	6, 75
	J	Moderate	Normal size	Normal	A → C (587 nt), C → T (1,777 nt)	9
	K	Moderate	Normal size	Normal	C → A (3′ site of intron), A → G (1,052 nt), C → T (1,756 nt)	53
Unclassified	L	Severe	Normal size	Normal	G → A (577 nt)	18
	M	Severe	Not reported	Not reported	Single-base deletion (2,142 nt)	65
	N	Severe	Not detected	Low, large	G → A (5′ site of intron)	43, 48
	O	Severe	Not reported	Low, large	Not reported	26, 38
	P	Severe	Not reported	Normal	G → A (454 nt)	48
	Q	Periodontal disease	Not reported	Normal	Normal	26, 45

[a] nt, nucleotide.

in the CD18 molecule irrespective of the clinical penotype, but there was hetero-geneity in the point mutation among the patients (9, 18, 48, 53, 75). Molecular cloning of six different patients' CD18 gene revealed that four patients (patients H, I, L, and P) had one mutant allele, but two patients (patients J and K) were double heterozygous, with two mutation alleles, both of which drastically reduce CD18 expression in a transient mammalian expression system (53). Further, cDNA clones from patient K contained a 12-bp insertion resulting in an in-frame addition of 4 amino acids and generating an aberrant splice acceptor site (53). Eight kinds of point mutations were sequenced. For the substitution Asp for Asn at position 128 (Asp-128 → Asn) (patient P), Leu-149 → Pro (patient I), Gly-169 → Arg (patients H and L), Lys-196 → Thr (patient J), and Asn-351 → Ser (patient K), the mutation regions are located in the highly conserved 247-amino-acid region that is required for association of CD18 with CD11. For Arg-586 → Trp (patient K) and Arg-593 → Cys (patient J), the mutation regions are located in the cysteine-rich region that is also important in formation of the stable heterodimer CD11/CD18, and for Met-1 → Lys (patient B), the mutation region is in the initiation codon (Fig. 1 and Table 4). In the cases of double mutations, it is uncertain whether one or both alleles contributed to the occurrence of LAD, but it is of interest that one of the identified mutations invariably is located in the highly conserved 247-amino-acid region. Besides, Sligh et al. (65) recently reported two cases of LAD (in patients B and M) caused by CD18 gene mutations that varied from those discussed above. The mutation in the first patient (patient B), who had a moderate phenotype and expressed only 9% CD11/CD18 on leukocytes, was characteristic of an ATG-to-AAG alteration in the initiation codon arising from a paternal defect and a frameshift mutation with the deletion of a single T in the aspartic acid codon arising from a maternal defect. The second patient (patient M), who had the severe phenotype (<1% of CD11/CD18), also demonstrated the frameshift mutation.

Molecular Study of Patients with Periodontal Diseases

Katsuragi et al. (38) recently did a molecular study of patients N, O, and Q with aberrant CD11/CD18 expression on their leukocytes. The quantity and quality of CD18 mRNA synthesized in their EBV-transformed lymphoblastoid cells were analyzed by Northern (RNA) blotting, reverse transcription PCR, and RNase protection assay. Northern blot analysis revealed low levels of large CD18 mRNA in patients N and O and healthy levels of normal-sized CD18 mRNA in patient Q, indicating that the CD18 gene of patient Q was normally and actively transcribed in the leukocytes. To locate the genetic defect within the CD18 mRNA, they tried to amplify by reverse transcription PCR two regions of CD18 cDNA that together covered the whole coding region. The whole coding region of CD18 cDNA was obtained from neither patient N nor patient O but was amplified from patient Q, suggesting an aberrantly large mRNA or base mismatches on the reverse primer in patients N and O. RNase protection assay of the patient's total RNA hybridized with antisense RNA probes, which together covered the whole coding region of CD18 cDNA, revealed excessive aberrant fragments in patients N and O but not in patient Q. All of the fragments together from patients N and O were estimated to be longer than the fully protected probe. On the basis of inserts in the mRNA between nucleotides 965 and 1450, patients N and O had the same mutation, and patient Q had no structural anomaly in the CD18 mRNA coding region.

Consequently, of the 14 mutations so far reported, 7 are missense mutations, 1 is an initiation codon mutation, 3 are splicing mutations, 2 are single-base deletions, and 1 is a frameshift mutation identified as causing LAD. All the LAD patients with one exception (patient B) show at least one mutation region within the highly conserved 247-amino-acid region on the CD18 subunits that bound directly to the corresponding CD11 subunits. However, the cysteine-rich region may also impair heterodimer formation by preventing to various degrees the spatial orientation needed to permit an optimal fit with the corresponding CD11 subunits.

CONCLUSION

Through detailed analyses of LAD, significant progress has been made in elucidating the critical role of CD11/CD18 leukocyte adhesion molecules in host defense. The abnormality of CD11/CD18 can account for the defect in cell adherence, the suppressed neutrophil functions, and the enhanced susceptibility to infections. However, studies of the mechanism underlying this abnormality are limited to individuals with LAD except for one patient with localized juvenile periodontitis (38, 45), and the underlying mechanisms were based on heterogeneous molecular mutations in CD18. The exceptional patient (patient Q) showed anomalies of neutrophil functions but without remarkable systemic diseases except severe periodontal manifestations. The exact molecular mechanisms underlying the allosteric transition responsible for the biological dysfunction in patient Q have not yet been defined, since a series of intracellular events have not been thoroughly examined in the corresponding study (38). However, certain conformational changes of CD11/CD18 molecules induced by phosphorylation of the β subunit are suspect as mechanisms underlying this periodontitis. The mechanism is likely to correlate with certain conformational changes of CD11/CD18 molecules induced by phosphorylation of the β subunit of CD11/CD18, although those in LAD patients were based on heterogeneous molecular mutations on CD18. A few reports have described the periodontal features demonstrated in some LAD patients, suggesting that LAD is susceptible to a particular form of periodontitis, i.e., G-PP. Thus, information derived from these studies should provide invaluable insights that could lead to understanding the role of CD11/CD18 in the onset and progression of certain periodontal diseases. However, the exact mechanism by which CD11/CD18-mediated cell adhesion is regulated in association with periodontal diseases has remained obscure. The molecular basis of proteins related to the regulation of CD11/CD18-mediated adhesion should be studied in periodontal patients with structurally and functionally aberrant expression of CD11/CD18.

Acknowledgments. This work was supported by a Grant-in-Aid for Scientific Research (03454441) from the Ministry of Education, Science and Culture in Japan.
We thank Midori Hayakawa for typing the manuscript.

REFERENCES

1. **Altiere, D. D., R. Bader, P. M. Mannucci, and T. S. Edgington.** 1988. Oligospecificity of the cellular adhesion receptor Mac-1 encompasses an inducible recognition specificity for fibrinogen. *J. Cell Biol.* **107:**1893–1900.

2. **Altieri, D. C., and T. S. Edgington.** 1988. The saturable high affinity association of factor X to ADP-stimulated monocytes defines a novel function of the Mac-1 receptor. *J. Biol. Chem.* **263:**7007–7015.

3. **Altieri, D. C., and T. S. Edgington.** 1988. A monoclonal antibody reacting with distinct adhesion molecules defines a transition in the functional state of the receptor CD11b/CD18 (Mac-1). *J. Immunol.* **141:**2656–2660.

4. **Anderson, D. C., K. L. Becker-Freeman, B. Heerdt, B. J. Hughes, R. M. Jack, and C. W. Smith.** 1987. Abnormal stimulated adherence of neonatal granulocytes: impaired induction of surface Mac-1 by chemotactic factors or secretagogues. *Blood* **70:**740–750.

5. **Anderson, D. C., L. J. Miller, F. C. Schmalsteig, R. Rothlein, and T. A. Springer.** 1986. Contributions of the MAC-1 glycoprotein family to adherence-dependent granulocyte functions: structure-function assessments employing subunit-specific monoclonal antibodies. *J. Immunol.* **137:**15–27.

6. **Anderson, D. C., F. C. Schmalsteig, M. J. Finegold, B. J. Hughes, R. Rothlein, L. J. Miller, S. Kohl, M. F. Tosi, R. L. Jacobs, T. C. Waldrop, A. S. Goldman, W. T. Shearer, and T. A. Springer.** 1985. The severe and moderate phenotypes of heritable Mac-1, LFA-1 deficiency: their quantitative definition and relation to leukocyte dysfunction and clinical features. *J. Infect. Dis.* **152:**668–689.

7. **Anderson, D. C., and T. A. Springer.** 1987. Leukocyte adhesion deficiency: an inherited defect in the Mac-1, LFA-1, and p150,95 glycoproteins. *Annu. Rev. Med.* **38:**175–194.

8. **Arnaout, M. A.** 1990. Structure and function of the leukocyte adhesion molecules CD11/CD18. *Blood* **75:**1037–1050.

9. **Arnaout, M. A., N. Dana, S. K. Gupta, D. G. Tenen, and D. M. Fathallah.** 1990. Point mutations impairing cell surface expression of the common β subunit (CD18) in a patient with leukocyte adhesion molecule (Leu-CAM) deficiency. *J. Clin. Invest.* **85:**977–981.

10. **Arnaout, M. A., R. M. Hakim, R. F. Todd III, N. Dana, and H. R. Colten.** 1985. Increased expression of an adhesion-promoting surface glycoprotein in the granulocytopenia of hemodialysis. *N. Engl. J. Med.* **312:**457–462.

11. **Arnaout, M. A., L. L. Lanier, and D. F. Faller.** 1988. Relative contribution of the leukocyte molecules Mol, LFA-1 and p150,95 (Leu M5) in adhesion of granulocytes and monocytes to vascular endothelium is tissue- and stimulus-specific. *J. Cell. Physiol.* **137:**305–309.

12. **Back, A. L., W. W. Kwok, M. Adam, S. J. Collins, and D. D. Hickstein.** 1990. Retroviral-mediated gene transfer of the leukocyte integrin CD18 subunit. *Biochem. Biophys. Res. Commun.* **171:**787–795.

13. **Buyon, J. P., N. Shadick, R. Berkman, P. Hopkins, J. Dalton, G. Weissmann, R. Winchester, and S. B. Abramson.** 1988. Surface expression of Gp165/95, the complement receptor CR3, as a marker of disease activity in systemic *Lupus erythematosis. Clin. Immunol. Immunopathol.* **46:**141–149.

14. **Buyon, J. P., S. G. Slade, J. Reibman, S. B. Abramson, M. R. Philips, G. Weissmann, and R. Winchester.** 1990. Constitutive and induced phosphorylation of the α- and β-chains of the CD11/CD18 leukocyte integrin family. Relationship to adhesion-dependent functions. *J. Immunol.* **144:**191–197.

15. **Chatila, T., R. S. Geha, and M. A. Arnaout.** 1989. Constitutive and stimulus-induced phosphorylation of CD11/CD18 leukocyte adhesion. *J. Cell Biol.* **109:**3435–3444.

16. **Cheresh, D. A., J. W. Smith, H. M. Cooper, and V. Quaranta.** 1989. A novel vitronectin receptor integrin ($\alpha_v\beta_x$) is responsible for distinct adhesive properties of carcinoma cells. *Cell* **57:**59–69.

17. **Clayberger, C., A. Wright, L. J. Medeiros, T. D. Koller, M. P. Link, S. D. Smith, R. A. Warnke, and A. M. Krensky.** 1987. Absence of cell surface LFA-1 as a mechanism of escape from immunosurveillance. *Lancet* **ii:**533–536.

18. **Corbi, A. L., A. Vara, A. Ursa, M. C. G. Rodriguez, G. Fontan, and F. Sanchez-Madrid.** 1992. Molecular basis for a severe case of leukocyte adhesion deficiency. *Eur. J. Immunol.* **22:**1877–1881.

19. **Danna, N., and M. A. Arnaout.** 1988. Leukocyte adhesion molecules (CD11/CD18) deficiency. *Baillieres Clin. Immunol. Allergy* **2:**453–476.

20. **DeSimone, D. W., and R. O. Hynes.** 1988. Xenopus leavis integrins. Structural conservation and evolutionary divergence of integrin beta subunits. *J. Biol. Chem.* **263:**5333–5340.

21. **Dransfield, I., A.-M. Buckle, and N. Hogg.** 1990. Early events of the immune response mediated by leukocyte integrins. *Immunol. Rev.* **114:**29–44.

22. **D'Souza, S. E., M. H. Ginsberg, T. A. Burke, S. C.-T. Lam, and E. F. Plow.** 1988. Localization of an Arg-Gly-Asp recognition site within an integrin adhesion receptor. *Science* **242:**91–93.

23. **Figdor, C. G., Y. van Kooyk, and G. D. Keizer.** 1990. On the mode of action of LFA-1. *Immunol. Today* **11:**277–280.

24. **Fitzgerald, L. A., B. Steiner, S. C. Rall, Jr., S. S. Lo, and D. R. Philips.** 1987. Protein sequence of endothelial glycoprotein IIIa derived from a cDNA clone: identity with platelet glycoprotein IIIa and similarly to integrin. *J. Biol. Chem.* **262:**3936–3939.

25. **Freyer, D. R., M. L. Morganroth, C. E. Rogers, M. A. Arnaout, and R. F. Todd III.** 1988. Regulation of surface glycoproteins CD11/CD18 (Mol, LFA-1, p150/95) by human mononuclear phagocytes. *Clin. Immunol. Immunopathol.* **46:**272–283.

26. **Fujita, K., K. Kobayashi, and T. Kajii.** 1985. Impaired neutrophil adhesion: a new patient in a previously reported family. *Acta Pediatr. Jpn.* **27:**527–534.

27. **Hamann, A. D., Jablonski-Westrich, A. Raedler, and H. G. Thiele.** 1984. Lymphocytes express specific antigen-independent contact interaction sites upon activation. *Cell. Immunol.* **86:**14–32.

28. **Haskard, D., D. Cavender, P. Beatty, T. Springer, and M. Ziff.** 1986. T lymphocyte adhesion to endothelial cells: mechanisms demonstrated by anti-LFA-1 monoclonal antibodies. *J. Immunol.* **137:**2901–2906.

29. **Haynes, B. F., L. P. Hale, S. M. Denning, P. T. Le, and K. H. Singer.** 1989. The role of leukocyte adhesion molecules in cellular interactions: implications for the pathogenesis of inflammatory synovitis. *Springer Semin. Immunopathol.* **11:**163–185.

30. **Hibbs, M. L., S. Jakes, S. A. Stacker, R. W. Wallace, and T. A. Springer.** 1991. The cytoplasmic domain of the integrin lymphocyte function-associated antigen 1 β subunit: sites required for binding to intercellular adhesion molecule 1 and the phorbol ester-stimulated phosphorylation site. *J. Exp. Med.* **174:**1227–1238.

31. **Hibbs, M. L., A. J. Wardlaw, S. A. Stacker, D. C. Anderson, A. Lee, T. M. Roberts, and T. A. Springer.** 1990. Transfection of cells from patients with leukocyte adhesion deficiency with an integrin beta subunit (CD18) restores LFA-1 expression and function. *J. Clin. Invest.* **85:**674–681.

32. **Holzmann, B., B. W. McIntyre, and I. L. Weissman.** 1989. Identification of a murine Peyer's patch-specific lymphocyte homing receptor as an integrin molecule with an α chain homologous to human VLA-4α. *Cell* **56:**37–46.

33. **Holzmann, B., and I. L. Weissman.** 1989. Peyer's patch-specific lymphocyte homing receptors consist of a VLA-4-like α chain associated with either of two integrin β chains, one of which is novel. *EMBO J.* **8:**1735–1741.

34. **Howard, D. R., A. C. Eaves, and F. Takei.** 1986. Lymphocyte function-associated antigen (LFA-1) is involved in B cell activation. *J. Immunol.* **136:**4013–4018.

35. **Hynes, R. O.** 1987. Integrins: a family of cell surface receptors. *Cell* **48:**549–554.

36. **Issekutz, A. C., K. Y. Lee, and W. D. Biggar.** 1979. Combined abnormality of neutrophil chemotaxis and bactericidal activity in a child with chronic skin infections. *Clin. Immunol. Immunopathol.* **14:**1–10.

37. **Kajiji, S., R. N. Tamura, and V. Quaranta.** 1989. A novel integrin (αEβ4) from human epithelial cells suggests a fourth family of integrin adhesion receptors. *EMBO J.* **8:**673–680.

38. **Katsuragi, K., S. Takashiba, H. Kurihara, and Y. Murayama.** Molecular basis of leukocyte adhesion molecules in early-onset periodontitis patients with decreased CD11/CD18 expression on leukocytes. *J. Periodontol.,* in press.

39. **Keizer, G. D., J. Borst, C. G. Figdor, H. Spits, F. Miedema, C. Terhorst, and J. E. de Vries.** 1985. Biochemical and functional characteristics of the human leukocyte membrane antigen family LFA-1, Mo-1 and p150,95. *Eur. J. Immunol.* **15:**1142–1148.

40. **Kishimoto, T. K., N. Hollnder, T. M. Roberts, D. C. Anderson, and T. A. Springer.** 1987. Het-

erogeneous mutations in the β subunit common to the LFA-1, Mac-1, and p150,95 glycoproteins cause leukocyte adhesion deficiency. *Cell* **50:**193–202.

41. **Kishimoto, T. K., K. O'Connor, A. Lee, T. M. Roberts, and T. A. Springer.** 1987. Cloning of the β subunit of the leukocyte adhesion proteins: homology to an extracellular matrix receptor defines a novel supergene family. *Cell* **48:**681–690.

42. **Kishimoto, T. A., K. O'Connor, and T. A. Springer.** 1989. Leukocyte adhesion deficiency: aberrant splicing of a conserved integrin sequence causes a moderate deficiency phenotype. *J. Biol. Chem.* **264:**3588–3595.

43. **Kobayashi, K., K. Fujita, F. Okino, and T. Kajii.** 1984. An abnormality of neutrophil adhesion: autosomal recessive inheritance associated with missing neutrophil glycoproteins. *Pediatrics* **73:**606–610.

44. **Kupfer, A., and S. J. Singer.** 1989. Cell biology of cytotoxic and helper T cell functions: immunofluorescence microscopic studies of single cells and cell couples. *Annu. Rev. Immunol.* **7:**309–337.

45. **Kurihara, H., M. Kobayashi, T. Chihara, H. Hongyo, N. Shimizu, S. Takashiba, T. Yoshino, and Y. Murayama.** The deficient expression of CD18 on the surface of neutrophil from patients with early-onset periodontitis. Submitted for publication.

46. **Law, S. K. A., J. Gagnon, J. E. K. Hildreth, C. E. Wells, A. C. Wills, and A. J. Wong.** 1987. The primary structure of the β subunit of the cell surface adhesion glycoproteins LFA-1, CR3 and p150,95 and its relationship to the fibronectin receptor. *EMBO J.* **6:**915–919.

47. **LeBien, T. W., and J. H. Kersey.** 1980. A monoclonal antibody (TA-1) reactive with human lymphocytes and monocytes. *J. Immunol.* **125:**2208–2214.

48. **Matsuura, S., F. Kishi, M. Tsukahara, H. Nunoi, I. Matsuda, K. Kobayashi, and T. Kajii.** 1992. Leukocyte adhesion deficiency: identification of novel mutations in two Japanese patients with a severe form. *Biochem. Biophys. Res. Commun.* **184:**1460–1467.

49. **Mentzer, S. J., D. V. Faller, and S. J. Burakoff.** 1986. Interferon-gamma of LFA-1 mediated homotypic adhesion of human monocytes. *J. Immunol.* **137:**108–113.

50. **Mentzer, S. J., S. H. Gromkowski, A. M. Krensky, S. J. Burakoff, and E. Martz.** 1985. LFA-1 membrane molecule in the regulation of homotypic adhesion of human B lymphocytes. *J. Immunol.* **135:**9–11.

51. **Miller, J. M., R. Schwarting, and T. A. Springer.** 1986. Regulated expression of the Mac-1, LFA-1, p150,95 glycoprotein family during leukocyte differentiation. *J. Immunol.* **137:**2891–2900.

52. **Nelson, R. D., S. R. Hasslen, D. H. Ahrenholz, E. Haus, and L. D. Solem.** 1986. Influence of minor thermal injury on expression of complement receptor CR3 on human neutrophils. *Am. J. Pathol.* **125:**563–570.

53. **Nelson, C., H. Rabb, and M. A. Arnaout.** 1992. Genetic cause of leukocyte adhesion molecule deficiency: abnormal splicing and a missense mutation in a conserved region of CD18 impair cell surface expression of β2 integrins. *J. Biol. Chem.* **267:**3351–3357.

54. **Nishimura, F., A. Nagai, K. Kurimoto, O. Isoshima, S. Takashiba, M. Kobayashi, I. Akutsu, H. Kurihara, Y. Nomura, Y. Murayama, H. Ohta, and K. Kato.** 1990. A family study of a mother and daughter with increased susceptibility to early-onset periodontitis: microbiological, immunological, host defensive, and genetic analyses. *J. Periodontol.* **61:**755–765.

55. **O'Shea, J. J., E. J. Brown, B. E. Seligman, J. A. Metcalf, M. M. Frank, and J. I. Gallin.** 1985. Evidence of distinct intracellular pools of receptors for C3b and C3bi in human neutrophils. *J. Immunol.* **134:**2580–2587.

56. **Otey, C. A., F. M. Pavalko, and K. Burridge.** 1990. An interaction between α-actinin and β1 integrin subunit in vitro. *J. Cell Biol.* **111:**721–729.

57. **Page, R. C., P. Beatty, and T. C. Waldrop.** 1987. Molecular basis for the functional abnormality in neutrophils from patients with generalized prepubertal periodontitis. *J. Periodontal Res.* **22:**182–183.

58. **Pardi, R., L. Inverardi, and J. R. Bender.** 1992. Regulatory mechanisms in leukocyte adhesion: flexible receptors for sophisticated travelers. *Immunol. Today* **13:**224–230.

59. **Pardi, R., L. Inverardi, C. Rugardi, and J. R. Bender.** 1992. Antigen-receptor complex stimulation triggers protein kinase C-dependent CD11a/CD18-cytoskeleton association in T lymphocytes. *J. Cell Biol.* **116:**1211–1220.

60. **Patarroyo, M., J. Brieto, J. Rincon, T. Timonen, C. Lundberg, L. Lindbom, B. Asjo, and C. G. Gahmberg.** 1990. Leukocyte-cell adhesion: a molecular process fundamental in leukocyte physiology. *Immunol. Rev.* **114:**67–108.

61. **Ramaswamy, H., and M. E. Hemler.** 1990. Cloning, primary structure and properties of a novel human integrin β subunit. *EMBO J.* **9:**1561–1568.

62. **Sanchez-Madrid, F., J. Nagy, E. Robbins, P. Simon, and T. A. Springer.** 1983. A human leukocyte differentiation antigen family with distinct α subunits and a common β subunit: the lymphocyte function-associated antigen (LFA-1), the C3bi complement receptor (OKM1/Mac-1), and the p150,95 molecule. *J. Exp. Med.* **158:**1785–1803.

63. **Setiadi, H., J. L. Wautier, A. Courillon-Mallet, P. Passa, and J. Caen.** 1987. Increased adhesion to fibronectin and Mo-1 expression by diabetic monocytes. *J. Immunol.* **138:**3230–3234.

64. **Sheppard, D., C. Rozzo, L. Starr, V. Quaranta, D. J. Erle, and R. Pytela.** 1990. Complete amino acid sequence of a novel integrin β subunit (β6) identified in epithelial cells using the polymerase chain reaction. *J. Biol. Chem.* **265:**11502–11507.

65. **Sligh, J. E., Jr., M. Y. Hurwitz, C. Zhu, D. C. Anderson, and A. L. Beaudet.** 1992. An initiation codon mutation in CD18 in association with the moderate phenotype of leukocyte adhesion deficiency. *J. Biol. Chem.* **267:**714–718.

66. **Smith, C. W., S. D. Marlin, R. Rothlein, C. Toman, and D. C. Anderson.** 1989. Cooperative interactions of LFA-1 and Mac-1 with ICAM-1 molecule in facilitating adherence and transendothelial migration of human neutrophils in vitro. *J. Clin. Invest.* **83:**2008–2017.

67. **Springer, T. A.** 1990. Adhesion receptors of the immune system. *Nature* (London) **346:**425–435.

68. **Suzuki, S., and Y. Naitoh.** 1990. Amino acid sequence of a novel integrin beta 4 subunit and primary expression of the mRNA in epithelial cells. *EMBO J.* **9:**757–763.

69. **Taylor, G. M., H. Haigh, A. Williams, S. W. D'Souza, and R. Harris.** 1988. Down's syndrome lymphoid cell lines exhibit increased adhesion due to the over-expression of lymphocyte function-associated antigen (LFA-1). *Immunology* **64:**451–456.

70. **Todd, R. F., III, L. M. Nadler, and S. F. Schlossmann.** 1981. Antigens on human monocytes identified with monoclonal antibodies. *J. Immunol.* **126:**1435–1442.

71. **Trowbridge, I. S., and M. B. Omary.** 1981. Molecular complexity of surface glycoproteins related to the macrophage differentiation antigen Mac-1. *J. Exp. Med.* **154:**1517–1524.

72. **Valmu, L., M. Autero, P. Siljander, M. Patarroyo, and C. G. Gahmberg.** 1991. Phosphorylation of the β-subunit of CD11/CD18 integrins by protein kinase C correlates with leukocyte adhesion. *Eur. J. Immunol.* **21:**2857–2862.

73. **van Kooyk, Y., P. Weder, F. Hogervorst, A. J. Verhoeven, G. van Seventer, A. A. te Velde, J. Borst, G. D. Keizer, and C. G. Figdor.** 1991. Activation of LFA-1 through a Ca^{2+}-dependent epitope stimulates lymphocyte adhesion. *J. Cell Biol.* **112:**345–354.

74. **Waldrop, T. C., D. C. Anderson, W. W. Hallmon, F. C. Schmalstieg, and R. L. Jacobs.** 1987. Periodontal manifestations of the heritable Mac-1, LFA-1, deficiency syndrome: clinical, histopathologic and molecular characteristics. *J. Periodontol.* **58:**400–416.

75. **Wardlaw, A. J., M. L. Hibbs, S. A. Stacker, and T. A. Springer.** 1990. Distinct mutations in two patients with leukocyte adhesion deficiency and their functional correlates. *J. Exp. Med.* **172:**335–345.

76. **Wilson, J. M., A. J. Ping, J. C. Krauss, L. Mayo-Bond, C. E. Rogers, D. C. Anderson, and R. F. Todd III.** 1990. Correction of CD18-deficient lymphocytes by retrovirus-mediated gene transfer. *Science* **248:**1413–1416.

77. **Wright, S. D., P. A. Reddy, M. T. C. Jong, and B. W. Erickson.** 1987. C3bi receptor (complement receptor type 3) recognizes a region of complement protein C3 containing the sequence Arg-Gly-Asp. *Proc. Natl. Acad. Sci. USA* **84:**1965–1973.

Molecular Pathogenesis of Periodontal Disease
Edited by Robert Genco et al.
© 1994 American Society for Microbiology, Washington, DC 20005

Chapter 20

Role of the Extracellular Matrix in Inflammation

Majed M. Hamawy, Reuben P. Siraganian, and Stephan E. Mergenhagen

Inflammation is the normal response of the body to infection and tissue injury. The inflammatory response is characterized by dilation of the local blood vessels, an increase in the permeability of the capillaries, and the accumulation of neutrophils, monocytes, eosinophils, basophils, and mast cells at these sites (7, 22, 26, 41). The increase in numbers of cells is the result of chemotactic factors released by tissue injury or infection (10, 17, 18, 22, 28, 48, 55, 76, 78, 80, 82). Not only are these factors chemotactic, but they also activate cells and lead to an increase in the number and/or affinity of adherence receptors on the surfaces of both endothelial and inflammatory cells. These adhesion receptors play an important role in the migration of inflammatory cells out of the circulation into the sites of injury. Thus, to reach sites of inflammation, these cells have to adhere to and pass through the vascular endothelium and then migrate through the surrounding basement membrane and the connective tissue stroma. Once in the tissue, inflammatory cells actively eliminate the causative agent of injury and participate with resident tissue cells in wound healing and tissue remodeling by secreting various components of the extracellular matrix (ECM) (2, 72). Thus, the recruitment of inflammatory cells into areas of injury is crucial for a successful inflammatory response.

The ECM consists of glycoproteins and proteoglycans that are secreted by cells and assembled locally into an organized network (35, 66). The components of the ECM vary between tissues but generally include one or more types of collagen and proteoglycans. Other molecules frequently incorporated into the ECM are fibronectin, fibrin(ogen), entactin, tenascin, thrombospondin, von Willebrand factor, vitronectin, and laminin. The ECM has important functions: it maintains the integrity and strength of tissues; it provides the structural components of tissues such as bone, cartilage, and tendon; and it is crucial for many cellular and developmental processes, including cell anchorage, shape, migration, proliferation, and differentiation.

Majed M. Hamawy, Reuben P. Siraganian, and Stephan E. Mergenhagen • Laboratory of Immunology, National Institute of Dental Research, National Institutes of Health, Bethesda, Maryland 20892.

The concentration of ECM glycoproteins increases significantly at sites of inflammation (1, 14, 49). These molecules are chemotactic for inflammatory cells both in vitro (55, 71, 75) and in vivo (19). Inflammatory cells also adhere to the different ECM proteins (4, 6, 15, 33, 46, 53, 57, 71, 74). This adherence to ECM molecules affects the function of the inflammatory cells (see below). Thus, ECM glycoproteins play an important role in the inflammatory reaction.

This review will focus mainly on the effect of adherence to the ECM on the function of inflammatory cells.

INTEGRINS: RECEPTORS FOR THE ECM

Integrins are the best-characterized membrane receptors that mediate cell binding to the ECM (36, 61). Each integrin is a noncovalently linked heterodimer of α and β subunits. There are at least 8 different β subunits and 14 α subunits. The integrins are divided into groups based on the β subunit. Integrins are expressed on many cells, and most cells express several integrins. A further level of complexity is that there is alternative splicing of the cytoplasmic domains of several of the β and α subunits (5). This variation may play a role in the different interactions of the cytoplasmic domains of integrins with the actin-based cytoskeleton (12, 13, 56).

Some integrins bind to several ECM proteins, and more than one integrin binds to the same ECM glycoprotein. Short peptide sequences in the ECM are some of the ligands for integrin binding. Members of the β_1, β_3, β_5, and β_6 subfamilies recognize, in addition to other sequences, the Arg-Gly-Asp (RGD) sequence that is present on many ECM glycoproteins (62, 84). Unlike other β_1 integrins, $\alpha_4\beta_1$ (also called VLA-4) binds the Leu-Asp-Val (EILDV) sequence in the alternatively spliced CS1 region of fibronectin (42, 84). The β_2 integrin subfamily (CD18, CD11a, CD11b, CD11c) is leukocyte specific and has three members: $\alpha_L\beta_2$ (LFA-1), $\alpha_M\beta_2$ (Mac-1), and $\alpha_X\beta_2$ (p150,95). Table 1 summarizes the expression of integrins on inflammatory cells and the ECM glycoproteins that they bind.

Until recently, the effect of the ECM on cells was believed to be due mainly to the capacity of the ECM components to induce changes in cell shape and in the

Table 1. Integrins on inflammatory cells[a]

Integrin	Cell(s)	Natural ligand(s)	References(s)
$\alpha_2\beta_1$	MC, Bas	LM, CO	68
$\alpha_4\beta_1$	MC, Bas, M, E	FN, VCAM-1	8, 20, 29, 68, 86
$\alpha_5\beta_1$	MC, Bas, M	FN	11, 29, 68
$\alpha_6\beta_1$	MC, M	LM	64, 79
$\alpha_L\beta_2$	Bas, E, N	ICAM-1, ICAM-2	8, 68
$\alpha_M\beta_2$	Bas, N, M	C3bi, FB, FN, ?VN	3, 23, 28, 38, 58, 68
$\alpha_X\beta_2$	Bas, MC, N	FB	44, 68
$\alpha_V\beta_3$	M, MC	VN	6, 29, 43, 68
$\alpha_4\beta_7$	MC	FN, VCAM-1	20, 30

[a] Abbreviations: Bas, basophils; CO, collagen; E, eosinophil; FB, fibrinogen; FN, fibronectin; ICAM-1 and -2, intercellular cell adhesion molecule-1 and -2; LM, laminin; M, monocytes; MC, mast cell; N, neutrophils; VCAM-1, vascular cell adhesion molecule-1.

organization of the cytoskeleton. However, more recent studies indicate that adhesion receptors such as integrins not only act as a physical link between the ECM and the cell but also, when aggregated, transduce signals into the cell that are independent of the changes in cell shape and cytoskeleton reorganization (39). The intracellular signals induced by adherence include protein tyrosine phosphorylation, phosphatidylinositol hydrolysis, and changes in intracellular pH and calcium concentrations. Adherence also induces the expression of several genes. Thus, in addition to inducing cell spreading, the ECM transduces signals and modulates cell function by activating integrins.

REGULATION OF INFLAMMATORY CELL ADHERENCE TO THE ECM

Modulation of the specificity and affinity of integrin receptors occurs. Cells can spontaneously adhere to ECM glycoproteins, but cell activation usually enhances such binding. The activation of the cells may be the result of factors released at inflammation sites (e.g., lipopolysaccharide) or of adherence-induced stimulation of the cells. The enhanced binding of the cells is a consequence of an increase in the binding affinity, an increase in the number of receptors, or the expression of new types of receptors. Some changes in adherence are regulated by the cytoplasmic domain of the integrins; e.g., the tyrosine phosphorylation of the cytoplasmic domain of integrins may modulate cell adherence to the ECM (64, 70).

There are numerous examples of the modulation of cell adherence by different stimuli. Neutrophils activated with phorbol myristic acetate or the chemotactic factor formylmethionyl-Leu-Phe bind to laminin better than nonactivated cells do (85). Tissue macrophages adhere spontaneously to fibronectin but require activation by gamma interferon or lipopolysaccharide to bind to the basement membrane glycoproteins laminin and collagen (63), an effect that may be due to phosphorylation of the cytoplasmic domain of the integrin $\alpha_6\beta_1$ (64). Granulocyte macrophage–colony-stimulating factor (GM-CSF) and tumor necrosis factor (TNF) induce expression of the adhesion receptor CD11/CD18 integrin in neutrophils and monocytes, whereas interleukin-3 (IL-3) increases the expression of CD11/CD18 in monocytes. Although monocytes spontaneously adhere to fibronectin (11), lipopolysaccharide and the cytokines gamma interferon, GM-CSF, TNF, but not IL-1 enhance the binding of normal human monocytes to fibronectin (57, 60). Changes in cell adhesion occur in inflammatory diseases; e.g., monocytes from patients with bronchiectasis exhibit enhanced adhesion to fibronectin in vitro compared with monocytes from healthy subjects (57). This adhesion of monocytes to fibronectin is mediated by CD11/CD18 integrins via both RGD-dependent and RGD-independent mechanisms. Bone marrow-derived cultured mast cells adhere spontaneously to vitronectin through integrins (6); however, binding to fibronectin (15) and laminin (73, 74) is markedly enhanced by cell activation, for example, by immunoglobulin E-receptor aggregation, by stimulation of protein kinase C by phorbol myristic acetate, or by increase in intracellular calcium with calcium ionophores. Although rat basophilic leukemia (RBL-2H3) cells adhere spontaneously to fibronectin (33), activation enhances their binding to surfaces coated with suboptimal concentrations of fibronectin (unpublished data). The precise mechanisms by which cell activation modulates cell adherence are not fully understood but may involve the redistribution of surface integrins, the enhanced expression of surface integrins, and/or the modulation of integrin affinity to their ligands.

REGULATION OF INFLAMMATORY CELL FUNCTIONS BY THE ECM

Monocytes/Macrophages

Monocytes are in the circulation, but when activated, they migrate into tissues and differentiate into tissue macrophages. The intracytoplasmic lysosomes of these cells contain several acid hydrolases and peroxidase. The functions of monocytes/macrophages include phagocytosis and killing of pathogenic microorganisms and secretion of numerous cytokines, such as interferon, IL, CSF, and TNF. The $\alpha_4\beta_1$ integrins and the β_2 integrins CD11/CD18 mediate monocyte adherence to the ECM (3, 44, 57, 86).

In vitro, human monocytes in the presence of serum differentiate into macrophages (50). Similarly, monocytes cultured on collagen rather than glass phenotypically resemble resident tissue macrophages (40). The adherence of monocytes to the ECM increases concentrations of mRNAs for c-*fos* and c-*jun*, early growth response genes that may be important for the transcription of other genes (65). Thus, adherence to the ECM may initiate and facilitate monocyte differentiation into tissue macrophages.

Adherence of monocytes to the ECM results in the induction of inflammatory genes, including those for IL-1β, TNF, CSF, and several monocyte adherence-derived inflammatory genes (24, 69). Remarkably, monocyte adhesion to different ECM components selectively regulates the levels and types of the genes induced (21, 34). Not only adherence but also the aggregation of β_1 integrins by monoclonal antibodies (MAbs) results in the expression of these genes (86). Thus, adherence-induced gene expression is mediated by the integrins. Adherence by itself is insufficient to cause efficient translation and secretion of certain cytokines; it requires activation by a second signal such as bacterial endotoxin (21, 34). Secretion induced with MAb to β_1 integrins is dependent on the anti-integrin MAb used (86). Thus, some antibodies require a second costimulatory signal, while for others, the binding of MAb alone induces the production of these proteins. The binding of human monocytes/macrophages to collagen or to collagen fragments induces the release of prostaglandin E_2 and IL-1/mononuclear cell factor (IL-1/MCF) (16). By stimulating the production of collagenase, IL-1/MCF plays a significant role in the pathogenesis of tissue destruction in arthritic diseases.

Adherence to the ECM also modulates the function of monocytes. The rate of phagocytosis and the number of opsonized bacteria phagocytized by collagen-adherent monocytes was severalfold greater than the phagocytic capacity of cells adherent to plastic and similar to that of macrophages cultured for 7 days (54). In addition, more bacteria were killed by the collagen-adherent cells. The enhanced phagocytic bactericidal activity of the collagen-adherent cells is probably mediated by the activation of complement receptors 1 and 3 for phagocytosis and also by the enhancement of immunoglobulin Fc receptor-mediated phagocytosis without apparent amplification of the respiratory burst. Similarly, adherence of human monocytes to fibronectin- or laminin-coated surfaces stimulated C3b-mediated phagocytosis (9, 83), whereas binding to vitronectin enhanced Fc receptor- but not complement receptor-mediated phagocytosis (27).

Neutrophils

Neutrophils represent over 90% of circulating granulocytes and play a major role in host defense against bacterial infections. The functions of neutrophils include phagocytosis, degranulation, and secretion, all of which are important for the clearance of foreign material or debris at sites of inflammation. Neutrophils have two main types of granules: the primary (azurophilic) granules contain acid hydrolases, myeloperoxidase, and muraminidase (lysozyme), while the secondary or specific granules contain lactoferrin in addition to lysozyme. When activated, neutrophils generate reactive oxygen intermediates, i.e., the respiratory burst, that are essential for the destruction of microorganisms.

In vitro, neutrophils adhere to and spread on surfaces coated with several of the ECM glycoproteins, including fibronectin, laminin, vitronectin, and fibrinogen (44, 46, 53). This binding appears to be mediated by the β_2 subfamily of integrins (CD11/CD18). In vivo, neutrophils must migrate from the circulation to sites of tissue injury. In vitro, surfaces coated with fibronectin or vitronectin provide a substratum on which, in the presence of Ca^{2+}, neutrophils migrate in response to chemoattractants (46). Thus, these ECM glycoproteins may also promote neutrophil migration in vivo. The stimulation by TNF of neutrophils adherent to ECM glycoproteins results in a massive and prolonged respiratory burst (52, 53). Adherence by itself does not trigger the release of oxidative metabolites, and in nonadherent cells, TNF has a weak or no effect on the respiratory burst. This TNF-induced respiratory burst in adherent cells is similar in magnitude to that induced by microbes that are phagocytized. Furthermore, this response is far longer and larger than the respiratory burst in nonadherent neutrophils and requires concentrations of TNF that are orders of magnitude lower. In neutrophils adherent to ECM, there is enhanced TNF-induced degranulation of the specific granules; for example, secretion of lactoferrin in response to TNF is four times greater in adherent cells than in cells in suspension (45).

The intracellular events induced by the adherence of neutrophils to the ECM are not fully understood. Binding to the ECM has intracellular effects, including changes in the level of cytosolic free calcium $[Ca^{2+}]_i$, the concentration of cyclic AMP (cAMP), and the phosphorylation of proteins on tyrosine. Attachment of neutrophils to the ECM induced spontaneous oscillation of $[Ca^{2+}]_i$ (37, 38, 58, 59). These $[Ca^{2+}]_i$ oscillations are important for cell activation. By reducing the extracellular Ca^{2+} level or loading the cells with Ca^{2+} chelators, a close correlation was found between a reduced oscillatory activity and a reduced ability of TNF to induce degranulation (59). The preincubation of neutrophils with antibodies to CD11b or CD18 inhibited adherence-induced oscillation of $[Ca^{2+}]_i$ and TNF-induced secretion of lactoferrin (58). Thus, this experiment demonstrated a direct role of integrins in inducing these oscillations in $[Ca^{2+}]_i$ and TNF-induced secretion. The importance of Ca^{2+} for the effects of the ECM can also be deduced from studies in which the presence of Ca^{2+} in the medium was essential for neutrophil migration on ECM proteins (46). In contrast, migration of neutrophils on surfaces coated with purified albumin was independent of the presence or absence of external Ca^{2+}. Nathan and Sanchez showed that exposure to TNF induced a sustained decline in the concentration of cAMP in adherent neutrophils (51). In contrast, TNF had no effect on cAMP levels in nonadherent cells. Thus, TNF and integrins act synergistically to lower cAMP levels. A decrease in cAMP preceding the addition of TNF was essential for induction of the respiratory burst because drugs that elevated cAMP blocked the respiratory burst if they were added any time before but not after onset of the burst. Recently, it has been observed that TNF-induced tyrosine phosphorylation of several proteins is dependent on the adherence of neutrophils to the ECM proteins (25). Tyrosine kinase inhibitors suppressed TNF-induced tyrosine phosphorylation and blocked the respiratory burst. This suggests that phosphorylation of tyrosine in proteins may be an intracellular signal that is linked to the respiratory burst; however, the specificity of the inhibitors employed must be questioned.

Eosinophils

Mature eosinophils are located predominantly in the extravascular spaces. Increased levels of circulating eosinophils and local accumulation of eosinophils at sites of inflammation have long been associated with parasitic infestations, allergic reactions, and other acute and chronic inflammatory diseases. Although eosinophils are capable of phagocytosis, they are primarily secretory cells. They have granules that contain powerful toxic proteins such as major basic protein but also contain enzymes such as histaminase and aryl sulfatase. Eosinophils adhere to fibronectin via the integrin receptor $\alpha_4\beta_1$ (4, 8). Such adherence significantly prolongs the survival of the cells compared with that of cells cultured on plastic alone (4). At 96 h, almost all of the eosinophils cultured on uncoated surfaces had died, whereas 60% of the eosinophils cultured on fibronectin-coated surfaces excluded trypan blue. Fibronectin-induced survival of eosinophils was significantly inhibited by MAb to $\alpha_4\beta_1$ integrins. In these studies, fibronectin-adherent cells released both IL-3 and GM-CSF. Furthermore, antibodies against both IL-3 and GM-CSF inhibited the survival of the cells cultured on fibronectin. Thus, this effect of fibronectin could be due to the capacity of fibronectin to induce the release of these cytokines.

Mast Cells and Basophils

Mast cells and basophils have many similarities, although mast cells are found in connective tissues and at mucosal surfaces whereas basophils are in the circulation (67). Both cells are derived from bone marrow precursors and are secretory cells that store within their cytoplasmic granules potent inflammatory mediators such as histamine, proteases, and chemotactic factors. In addition to releasing their granule contents, when these cells are activated, they release several cytokines and arachidonic acid metabolites. Both types of cells express on their surfaces receptors with a high affinity for immunoglobulin E (FceRI) (47). Cross-linking surface-bound immunoglobulin E with a multivalent antigen (allergen) activates the cells for secretion. Mast cells and basophils can also be triggered for the release of inflammatory mediators by nonimmunologic stimuli such as neuropeptides and anaphylatoxins. Both types of cells have integrin receptors that can mediate adherence to the ECM (6, 8, 29, 32, 68, 77). Adherence of several different mast cell lines to the ECM induces changes in cell shape, reorganization of the cytoskeleton, and redistribution of the secretory granules (15, 33, 74, 79). Adherence of rat basophilic leukemia (RBL-2H3) cells, a cell line widely used as an in vitro model for mast cells and basophils, to surfaces coated with fibronectin markedly enhanced immunologically and nonimmunologically induced secretion of histamine from these cells (33). The enhancement lasted for at least 6 h, suggesting that this is not a transient intracellular phenomenon. In related studies, adherence of RBL-2H3 cells to fibronectin and activation of the cells through FceRI synergistically regulated tyrosine phosphorylation of several proteins (31, 32). Among these proteins was pp125^{FAK}, a newly described protein tyrosine kinase. Studies suggest that the phosphorylation of these proteins is essential for optimal secretion from these cells. Thus, adherence may regulate mast cell/basophil secretory functions by modulating intracellular protein tyrosine phosphorylation.

Adherence of mast cells also modulates their proliferation. The IL-3 induced proliferation of bone marrow-derived cultured mast cells was augmented by integrin-mediated adherence of the cells to vitronectin (6). Cells adherent to vitronectin had an increased rate of DNA synthesis and an increased rate of proliferation. However, plating the cells on vitronectin in the absence of IL-3 did not support cell growth. Therefore, adherence by itself was not a mitogenic signal but only enhanced the IL-3-induced cell proliferation.

CONCLUDING REMARKS

There is an increase in ECM glycoprotein concentrations at sites of inflammation. In vitro, the ECM influences several aspects of inflammatory cell functions, including adherence, migration, phagocytosis, secretion, and survival in culture (Table 2). Thus, the ECM is important for recruitment, localization, and function of inflammatory cells at areas of tissue injury. The enhanced secretion from ECM-adherent inflammatory cells can have contrasting effects. Increased secretion can recruit more inflammatory cells to the sites of injury, thereby amplifying the inflammatory response. This can lead to a more effective elimination of the causative agent and can accelerate wound healing. In contrast, the increased recruitment of cells may lead to up-regulated and uncontrolled inflammatory responses that can have a devastating effect on host tissues and subsequently lead to chronic inflammation as it exists in chronic joint diseases or periodontitis. Therefore, manipulating the ECM-inflammatory cell interaction could be an important approach for controlling inflammation and subsequent tissue destruction.

Several methods may be employed to modulate ECM-cell interactions. The

Table 2. Cellular processes and functions influenced by the ECM

Cells	Cellular processes or functions	Reference(s)
Monocytes/macrophages	Adhesion	3, 44, 57, 86
	Migration	19, 55
	Differentiation	40, 50, 65
	Gene transcription	21, 24, 34, 65, 69, 86
	Secretion	16, 21, 34, 86
	Phagocytosis	9, 27, 54, 83
Neutrophils	Adhesion	44, 46, 53
	Migration	46
	Oxygen burst	52, 53
	Degranulation	45, 58
	Intracellular Ca^{2+} level	37, 38, 58, 59
	Intracellular cAMP level	51
	Tyrosine phosphorylation	25
Eosinophils	Adhesion	4, 8
	Growth	4
	Cytokine release	4
Mast cells and basophils	Adhesion	6, 8, 15, 29, 32, 33, 68, 74, 77, 79
	Migration	75
	Cytoskeleton organization	15, 33, 74, 79
	Secretion	33
	Tyrosine phosphorylation	31, 32
	Growth	6

adherence of inflammatory cells and their activation through binding are potential points for inhibition. The binding of cells could be inhibited by synthetic peptides containing the RGD or the CS1 sequence. In vitro, these peptides abolish the binding of different cells to the ECM glycoproteins (6, 15, 32, 33, 42, 57, 60). However, there are major problems with such peptides, including lack of specificity, short half-life in vivo, and problems of administration (62). Short synthetic peptides containing the RGD sequence can be designed to exhibit various integrin specificities by restricting the conformation of the peptide. Another approach to modulate cell-ECM binding is the use of MAb specific to integrins. Using an in vivo test system, Weg et al. showed that treating guinea pig eosinophils with MAb to $\alpha_4\beta_1$ integrin and then injecting the cells intravenously inhibited the accumulation of these cells at sites of inflammation (81). Eosinophil accumulation was inhibited to the same extent when the MAb was administered intravenously. The MAb could be inhibiting the binding of the eosinophils to fibronectin or to vascular cell adhesion molecule-1 on endothelial cells. Another approach to regulating cell-ECM interaction will be the design of drugs that can specifically suppress or down-regulate integrins on the surfaces of inflammatory cells or inhibit the activation sequence that is induced by adherence of cells to the ECM glycoproteins.

REFERENCES

1. **Akiyama, S., and K. M. Yamada.** 1983. Fibronectin in disease, p. 55–96. *In* B. M. Wagner, R. Fleischmajer, and N. Kaufman (ed.) *Connective Tissue Diseases.* The Williams & Wilkins Co., Baltimore.

2. **Alitalo, K., T. Hovi, and A. Vaheri.** 1980. Fibronectin is produced by human macrophages. *J. Exp. Med.* **151:**602–613.

3. **Altieri, D. C., F. R. Agbanyo, J. Plescia, M. H. Ginsberg, T. S. Edgington, and E. F. Plow.** 1990. A unique recognition site mediates the interaction of fibrinogen with the leukocyte integrin Mac-1 (CD11b/CD18). *J. Biol. Chem.* **265:**12119–12122.

4. **Anwar, A. R., R. Moqbel, G. M. Walsh, A. B. Kay, and A. J. Wardlaw.** 1993. Adhesion to fibronectin prolongs eosinophil survival. *J. Exp. Med.* **177:**839–843.

5. **Balzac, F., A. M. Belkin, V. E. Koteliansky, Y. V. Balabanov, F. Altruda, L. Silengo, and G. Tarone.** 1993. Expression and functional analysis of a cytoplasmic domain variant of the beta 1 integrin subunit. *J. Cell Biol.* **121:**171–178.

6. **Bianchine, P. J., P. R. Burd, and D. D. Metcalfe.** 1992. IL-3-dependent mast cells attach to plate-bound vitronectin. Demonstration of augmented proliferation in response to signals transduced via cell surface vitronectin receptors. *J. Immunol.* **149:**3665–3671.

7. **Bienenstock, J., A. D. Befus, and J. A. Denburg.** 1986. Mast cell heterogeneity: basic questions and clinical implications, p. 379–402. *In* A. D. Befus, J. Bienenstock, and J. A. Denburg, (ed.), *Mast Cell Differentiation and Heterogeneity.* Raven Press, Inc., New York.

8. **Bochner, B. S., F. W. Luscinskas, M. A. Gimbrone, Jr., W. Newman, S. A. Sterbinsky, C. P. Derse-Anthony, D. Klunk, and R. P. Schleimer.** 1991. Adhesion of human basophils, eosinophils, and neutrophils to interleukin 1-activated human vascular endothelial cells: contributions of endothelial cell adhesion molecules. *J. Exp. Med.* **173:**1553–1557.

9. **Bohnsack, J. F., H. K. Kleinman, T. Takahashi, J. J. O'Shea, and E. J. Brown.** 1985. Connective tissue proteins and phagocytic cell function. Laminin enhances complement and Fc-mediated phagocytosis by cultured human macrophages. *J. Exp. Med.* **161:**912–923.

10. **Brandes, M. E., U. E. Mai, K. Ohura, and S. M. Wahl.** 1991. Type I transforming growth factor-beta receptors on neutrophils mediate chemotaxis to transforming growth factor-beta. *J. Immunol.* **147:**1600–1606.

11. **Brown, D. L., D. R. Phillips, C. H. Damsky, and I. F. Charo.** 1989. Synthesis and expression of the fibroblast fibronectin receptor in human monocytes. *J. Clin. Invest.* **84:**366–370.

12. **Burn, P., A. Kupfer, and S. J. Singer.** 1988. Dynamic membrane-cytoskeletal interactions: specific association of integrin and talin arises *in vivo* after phorbol ester treatment of peripheral blood lymphocytes. *Proc. Natl. Acad. Sci. USA* **85:**497–501.

13. **Burridge, K., K. Fath, T. Kelly, G. Nuckolls, and C. Turner.** 1988. Focal adhesions: transmembrane junctions between the extracellular matrix and the cytoskeleton. *Annu. Rev. Cell Biol.* **4:**487–525.

14. **Clark, R. A., C. R. Horsburgh, A. A. Hoffman, H. F. Dvorak, M. W. Mosesson, and R. B. Colvin.** 1984. Fibronectin deposition in delayed-type hypersensitivity. Reactions of normals and a patient with afibrinogenemia. *J. Clin. Invest.* **74:**1011–1016.

15. **Dastych, J., J. J. Costa, H. L. Thompson, and D. D. Metcalfe.** 1991. Mast cell adhesion to fibronectin. *Immunology* **73:**478–484.

16. **Dayer, J. M., S. Ricard-Blum, M. T. Kaufmann, and D. Herbage.** 1986. Type IX collagen is a potent inducer of PGE2 and interleukin 1 production by human monocyte macrophages. *FEBS Lett.* **198:**208–212.

17. **Deuel, T. F., R. M. Senior, D. Chang, G. L. Griffin, R. L. Heinrikson, and E. T. Kaiser.** 1981. Platelet factor 4 is chemotactic for neutrophils and monocytes. *Proc. Natl. Acad. Sci. USA* **78:**4584–4587.

18. **Doherty, D. E., C. Haslett, M. G. Tonnesen, and P. M. Henson.** 1987. Human monocyte adherence: a primary effect of chemotactic factors on the monocyte to stimulate adherence to human endothelium. *J. Immunol.* **138:**1762–1771.

19. **Doherty, D. E., P. M. Henson, and R. A. Clark.** 1990. Fibronectin fragments containing the RGDS cell-binding domain mediate monocyte migration into the rabbit lung. A potential mechanism for C5 fragment-induced monocyte lung accumulation. *J. Clin. Invest.* **86:**1065–1075.

20. **Ducharme, L. A., and J. H. Weis.** 1992. Modulation of integrin expression during mast cell differentiation. *Eur. J. Immunol.* **22:**2603–2607.

21. **Eierman, D. F., C. E. Johnson, and J. S. Haskill.** 1989. Human monocyte inflammatory mediator gene expression is selectively regulated by adherence substrates. *J. Immunol.* **142:**1970–1976.

22. **Faccioli, L. H., S. Nourshargh, R. Moqbel, F. M. Williams, R. Sehmi, A. B. Kay, and T. J.**

Williams. 1991. The accumulation of 111In-eosinophils induced by inflammatory mediators, *in vivo. Immunology* **73**:222–227.

23. **Fan, S. T., and T. S. Edgington.** 1993. Integrin regulation of leukocyte inflammatory functions, CD11b/CD18 enhancement of the tumor necrosis factor-alpha responses of monocytes. *J. Immunol.* **150**:2972–2980.

24. **Fuhlbrigge, R. C., D. D. Chaplin, J. M. Kiely, and E. R. Unanue.** 1987. Regulation of interleukin 1 gene expression by adherence and lipopolysaccharide. *J. Immunol.* **138**:3799–3802.

25. **Fuortes, M., W. W. Jin, and C. Nathan.** 1993. Adhesion-dependent protein tyrosine phosphorylation in neutrophils treated with tumor necrosis factor. *J. Cell Biol.* **120**:777–784.

26. **Gleich, G. J.** 1982. The late phase of the immunoglobulin E-mediated reaction: a link between anaphylaxis and common allergic disease? *J. Allergy Clin. Immunol.* **70**:160–169.

27. **Gresham, H. D., L. T. Clement, J. E. Lehmeyer, F. M. Griffin, Jr., and J. E. Volanakis.** 1986. Stimulation of human neutrophil Fc receptor-mediated phagocytosis by a low molecular weight cytokine. *J. Immunol.* **137**:868–875.

28. **Griffin, J. D., O. Spertini, T. J. Ernst, M. P. Belvin, H. B. Levine, Y. Kanakura, and T. F. Tedder.** 1990. Granulocyte-macrophage colony-stimulating factor and other cytokines regulate surface expression of the leukocyte adhesion molecule-1 on human neutrophils, monocytes, and their precursors. *J. Immunol.* **145**:576–584.

29. **Guo, C. B., A. Kagey-Sobotka, L. M. Lichtenstein, and B. S. Bochner.** 1992. Immunophenotyping and functional analysis of purified human uterine mast cells. *Blood* **79**:708–712.

30. **Gurish, M. F., A. F. Bell, T. J. Smith, L. A. Ducharme, R. K. Wang, and J. H. Weis.** 1992. Expression of murine beta 7, alpha 4, and beta 1 integrin genes by rodent mast cells. *J. Immunol.* **149**:1964–1972.

31. **Hamawy, M. M., S. Mergenhagen, and R. P. Siraganian.** 1993. Tyrosine phosphorylation of pp125^{FAK} by the aggregation of high affinity immunoglobulin E receptors requires cell adherence. *J. Biol. Chem.* **268**:6851–6854.

32. **Hamawy, M. M., S. E. Mergenhagen, and R. P. Siraganian.** 1993. Cell adherence to fibronectin and the aggregation of the high affinity IgE receptor synergistically regulate tyrosine phosphorylation of 105-115 kDa proteins. *J. Biol. Chem.* **268**:5227–5233.

33. **Hamawy, M. M., C. Oliver, S. E. Mergenhagen, and R. P. Siraganian.** 1992. Adherence of rat basophilic leukemia (RBL-2H3) cells to fibronectin-coated surfaces enhances secretion. *J. Immunol.* **149**:615–621.

34. **Haskill, S., C. Johnson, D. Eierman, S. Becker, and K. Warren.** 1988. Adherence induces selective mRNA expression of monocyte mediators and proto-oncogenes. *J. Immunol.* **140**:1690–1694.

35. **Hay, E. D.** 1992. *Cell Biology of Extracellular Matrix.* Plenum Press, New York.

36. **Hynes, R. O.** 1992. Integrins: versatility, modulation, and signaling in cell adhesion. *Cell* **69**:11–25.

37. **Jaconi, M. E., R. W. Rivest, W. Schlegel, C. B. Wollheim, D. Pittet, and P. D. Lew.** 1988. Spontaneous and chemoattractant-induced oscillations of cytosolic free calcium in single adherent human neutrophils. *J. Biol. Chem.* **263**:10557–10560.

38. **Jaconi, M. E., J. M. Theler, W. Schlegel, R. D. Appel, S. D. Wright, and P. D. Lew.** 1991. Multiple elevations of cytosolic-free Ca^{2+} in human neutrophils: initiation by adherence receptors of the integrin family. *J. Cell Biol.* **112**:1249–1257.

39. **Juliano, R. L., and S. Haskill.** 1993. Signal transduction from the extracellular matrix. *J. Cell. Biol.* **120**:577–585.

40. **Kaplan, G., and G. Gaudernack.** 1982. *In vitro* differentiation of human monocytes. Differences in monocyte phenotypes induced by cultivation on glass or on collagen. *J. Exp. Med.* **156**:1101–1114.

41. **Kay, A. B.** 1970. Studies on eosinophil leucocyte migration. I. Eosinophil and neutrophil accumulation following antigen-antibody reactions in guinea-pig skin. *Clin. Exp. Immunol.* **6**:75–86.

42. **Komoriya, A., L. J. Green, M. Mervic, S. S. Yamada, K. M. Yamada, and M. J. Humphries.** 1991. The minimal essential sequence for a major cell type-specific adhesion site (CS1) within the alternatively spliced type III connecting segment domain of fibronectin is leucine-aspartic acid-valine. *J. Biol. Chem.* **266**:15075–15079.

43. **Krissansen, G. W., M. J. Elliott, C. M. Lucas, F. C. Stomski, M. C. Berndt, D. A. Cheresh, A.**

F. Lopez, and G. F. Burns. 1990. Identification of a novel integrin beta subunit expressed on cultured monocytes (macrophages). Evidence that one alpha subunit can associate with multiple beta subunits. *J. Biol. Chem.* **265:**823–830.

44. Loike, J. D., B. Sodeik, L. Cao, S. Leucona, J. I. Weitz, P. A. Detmers, S. D. Wright, and S. C. Silverstein. 1991. CD11c/CD18 on neutrophils recognizes a domain at the N terminus of the A alpha chain of fibrinogen. *Proc. Natl. Acad. Sci. USA* **88:**1044–1048.

45. Luedke, E. S., and J. L. Humes. 1989. Effect of tumor necrosis factor on granule release and LTB4 production in adherent human polymorphonuclear leukocytes. *Agents Actions* **27:**451–454.

46. Marks, P. W., B. Hendey, and F. R. Maxfield. 1991. Attachment to fibronectin or vitronectin makes human neutrophil migration sensitive to alterations in cytosolic free calcium concentration. *J. Cell. Biol.* **112:**149–158.

47. Metzger, H., G. Alcaraz, R. Hohman, J. P. Kinet, V. Pribluda, and R. Quarto. 1986. The receptor with high affinity for immunoglobulin E. *Annu. Rev. Immunol.* **4:**419–470.

48. Ming, W. J., L. Bersani, and A. Mantovani. 1987. Tumor necrosis factor is chemotactic for monocytes and polymorphonuclear leukocytes. *J. Immunol.* **138:**1469–1474.

49. Mosher, D. F. 1989. *Fibronectin.* Academic Press, Inc., San Diego, Calif.

50. Musson, R. A. 1983. Human serum induces maturation of human monocytes *in vitro*. Changes in cytolytic activity, intracellular lysosomal enzymes, and nonspecific esterase activity. *Am. J. Pathol.* **111:**331–340.

51. Nathan, C., and E. Sanchez. 1990. Tumor necrosis factor and CD11/CD18 (beta 2) integrins act synergistically to lower cAMP in human neutrophils. *J. Cell Biol.* **111:**2171–2181.

52. Nathan, C., S. Srimal, C. Farber, E. Sanchez, L. Kabbash, A. Asch, J. Gailit, and S. D. Wright. 1989. Cytokine-induced respiratory burst of human neutrophils: dependence on extracellular matrix proteins and CD11/CD18 integrins. *J. Cell Biol.* **109:**1341–1349.

53. Nathan, C. F. 1987. Neutrophil activation on biological surfaces. Massive secretion of hydrogen peroxide in response to products of macrophages and lymphocytes. *J. Clin. Invest.* **80:**1550–1560.

54. Newman, S. L., and M. A. Tucci. 1990. Regulation of human monocyte/macrophage function by extracellular matrix. Adherence of monocytes to collagen matrices enhances phagocytosis of opsonized bacteria by activation of complement receptors and enhancement of Fc receptor function. *J. Clin. Invest.* **86:**703–714.

55. Norris, D. A., R. A. Clark, L. M. Swigart, J. C. Huff, W. L. Weston, and S. E. Howell. 1982. Fibronectin fragment(s) are chemotactic for human peripheral blood monocytes. *J. Immunol.* **129:**1612–1618.

56. Otey, C. A., F. M. Pavalko, and K. Burridge. 1990. An interaction between alpha-actinin and the beta 1 integrin subunit *in vitro*. *J. Cell Biol.* **111:**721–729.

57. Owen, C. A., E. J. Campbell, S. L. Hill, and R. A. Stockley. 1992. Increased adherence of monocytes to fibronectin in bronchiectasis. Regulatory effects of bacterial lipopolysaccharide and role of CD11/CD18 integrins. *Am. Rev. Respir. Dis.* **145:**626–631.

58. Richter, J., J. Ng-Sikorski, I. Olsson, and T. Andersson. 1990. Tumor necrosis factor-induced degranulation in adherent human neutrophils is dependent on CD11b/CD18-integrin-triggered oscillations of cytosolic free Ca^{2+}. *Proc. Natl. Acad. Sci. USA* **87:**9472–9476.

59. Richter, J., I. Olsson, and T. Andersson. 1990. Correlation between spontaneous oscillations of cytosolic free Ca^{2+} and tumor necrosis factor-induced degranulation in adherent human neutrophils. *J. Biol. Chem.* **265:**14358–14363.

60. Roth, P., and R. A. Polin. 1990. Lipopolysaccharide enhances monocyte adherence to matrix-bound fibronectin. *Clin. Immunol. Immunopathol.* **57:**363–373.

61. Ruoslahti, E. 1991. Integrins. *J. Clin. Invest.* **87:**1–5.

62. Ruoslahti, E., and M. D. Pierschbacher. 1987. New perspectives in cell adhesion: RGD and integrins. *Science* **238:**491–497.

63. Shaw, L. M., and A. M. Mercurio. 1989. Interferon gamma and lipopolysaccharide promote macrophage adherence to basement membrane glycoproteins. *J. Exp. Med.* **169:**303–308.

64. Shaw, L. M., J. M. Messier, and A. M. Mercurio. 1990. The activation dependent adhesion of macrophages to laminin involves cytoskeletal anchoring and phosphorylation of the alpha 6 beta 1 integrin. *J. Cell Biol.* **110:**2167–2174.

65. Shaw, R. J., D. E. Doherty, A. G. Ritter, S. H. Benedict, and R. A. Clark. 1990. Adherence-

dependent increase in human monocyte PDGF(B) mRNA is associated with increases in c-fos, c-jun, and EGR2 mRNA. *J. Cell. Biol.* **111**:2139–2148.

66. **Shimizu, Y., and S. Shaw.** 1991. Lymphocyte interactions with extracellular matrix. *FASEB J.* **5**:2292–2299.

67. **Siraganian, R. P.** 1988. Mast cells and basophils, p. 513–542. *In* J. I. Gallin, I. M. Goldstein, and R. Snyderman R (ed.), *Inflammation: Basic Principles and Clinical Correlates.* Raven Press, Inc., New York.

68. **Sperr, W. R., H. Agis, K. Czerwenka, W. Klepetko, E. Kubista, G. Boltz-Nitulescu, K. Lechner, and P. Valent.** 1992. Differential expression of cell surface integrins on human mast cells and human basophils. *Ann. Hematol.* **65**:10–16.

69. **Sporn, S. A., D. F. Eierman, C. E. Johnson, J. Morris, G. Martin, M. Ladner, and S. Haskill.** 1990. Monocyte adherence results in selective induction of novel genes sharing homology with mediators of inflammation and tissue repair. *J. Immunol.* **144**:4434–4441.

70. **Tapley, P., A. Horwitz, C. Buck, K. Duggan, and L. Rohrschneider.** 1989. Integrins isolated from Rous sarcoma virus-transformed chicken embryo fibroblasts. *Oncogene* **4**:325–333.

71. **Terranova, V. P., R. DiFlorio, E. S. Hujanen, R. M. Lyall, L. A. Liotta, U. Thorgeirsson, G. P. Siegal, and E. Schiffmann.** 1986. Laminin promotes rabbit neutrophil motility and attachment. *J. Clin. Invest.* **77**:1180–1186.

72. **Thompson, H. L., P. D. Burbelo, G. Gabriel, Y. Yamada, and D. D. Metcalfe.** 1991. Murine mast cells synthesize basement membrane components. A potential role in early fibrosis. *J. Clin. Invest.* **87**:619–623.

73. **Thompson, H. L., P. D. Burbelo, and D. D. Metcalfe.** 1990. Regulation of adhesion of mouse bone marrow-derived mast cells to laminin. *J. Immunol.* **145**:3425–3431.

74. **Thompson, H. L., P. D. Burbelo, B. Segui-Real, Y. Yamada, and D. D. Metcalfe.** 1989. Laminin promotes mast cell attachment. *J. Immunol.* **143**:2323–2327.

75. **Thompson, H. L., P. D. Burbelo, Y. Yamada, H. K. Kleinman, and D. D. Metcalfe.** 1989. Mast cells chemotax to laminin with enhancement after IgE-mediated activation. *J. Immunol.* **143**:4188–4192.

76. **Thorens, B., J. J. Mermod, and P. Vassalli.** 1987. Phagocytosis and inflammatory stimuli induce GM-CSF mRNA in macrophages through posttranscriptional regulation. *Cell* **48**:671–679.

77. **Valent, P., and P. Bettelheim.** 1992. Cell surface structures on human basophils and mast cells: biochemical and functional characterization. *Adv. Immunol.* **52**:333–423.

78. **Wahl, S. M., D. A. Hunt, L. M. Wakefield, N. McCartney-Francis, L. M. Wahl, A. B. Roberts, and M. B. Sporn.** 1987. Transforming growth factor type beta induces monocyte chemotaxis and growth factor production. *Proc. Natl. Acad. Sci. USA* **84**:5788–5792.

79. **Walsh, L. J., M. S. Kaminer, G. S. Lazarus, R. M. Lavker, and G. F. Murphy.** 1991. Role of laminin in localization of human dermal mast cells. *Lab. Invest.* **65**:433–440.

80. **Wardlaw, A. J., R. Moqbel, O. Cromwell, and A. B. Kay.** 1986. Platelet-activating factor. A potent chemotactic and chemokinetic factor for human eosinophils. *J. Clin. Invest.* **78**:1701–1706.

81. **Weg, V. B., T. J. Williams, R. R. Lobb, and S. Nourshargh.** 1993. A monoclonal antibody recognizing very late activation antigen-4 inhibits eosinophil accumulation *in vivo. J. Exp. Med.* **177**:561–566.

82. **Worthen, G. S., N. Avdi, S. Vukajlovich, and P. S. Tobias.** 1992. Neutrophil adherence induced by lipopolysaccharide *in vitro.* Role of plasma component interaction with lipopolysaccharide. *J. Clin. Invest.* **90**:2526–2535.

83. **Wright, S. D., L. S. Craigmyle, and S. C. Silverstein.** 1983. Fibronectin and serum amyloid P component stimulate C3b- and C3bi-mediated phagocytosis in cultured human monocytes. *J. Exp. Med.* **158**:1338–1343.

84. **Yamada, K. M.** 1991. Adhesive recognition sequences. *J. Biol. Chem.* **266**:12809–12812.

85. **Yoon, P. S., L. A. Boxer, L. A. Mayo, A. Y. Yang, and M. S. Wicha.** 1987. Human neutrophil laminin receptors: activation-dependent receptor expression. *J. Immunol.* **138**:259–265.

86. **Yurochko, A. D., D. Y. Liu, D. Eierman, and S. Haskill.** 1992. Integrins as a primary signal transduction molecule regulating monocyte immediate-early gene induction. *Proc. Natl. Acad. Sci. USA* **89**:9034–9038.

Molecular Pathogenesis of Periodontal Disease
Edited by Robert Genco et al.
© 1994 American Society for Microbiology, Washington, DC 20005

Chapter 21

The Platelet as an Inflammatory Cell in Periodontal Diseases: Interactions with *Porphyromonas gingivalis*

Mark C. Herzberg, Gordon D. MacFarlane,
Peixin Liu, and Pamela R. Erickson

Platelets aggregate in response to hemostatic agents such as collagen and thrombotic agents such as thrombin (20). During this process of thromboregulation, platelets share proinflammatory functions with both monocytes and polymorphonuclear leukocytes (for a review, see Weksler [27]). Platelets are abundant (2×10^5 to 4×10^5 mm^3) compared to leukocytes (4.5×10^3 to 11×10^3) and if activated at an inflammatory site may make a substantial contribution to the host response. The inflammatory functions of platelets include release of adhesive proteins, complement activation and regulation, binding to microorganisms, alteration of vascular permeability, and production of chemotactic and growth factors. As a consequence, leukocyte function can be modulated. When activated, platelets release lipids (including platelet-activating factor and arachidonate metabolites); proinflammatory proteins such as platelet factor 4, β-thromboglobulin, and platelet-derived growth factor; biogenic amines such as serotonin; nucleotides such as ATP and ADP; and lysosomal enzymes. Platelet reactivity in hemostasis and thrombosis may result in a multicellular response. For example, chemotactic activity induced by released platelet 12-hydroxyeicosatetraenoic acid (12-HETE) may recruit neutrophils to the injured site (21). The interactions between neutrophils and platelets may be stabilized by the expression of P-selectin on the surface of the platelet (16).

While agents such as collagen, which is exposed to blood in injured tissue, will activate platelets, certain microorganisms in dental plaque will do so also. *Streptococcus sanguis,* the predominant organism in dental plaque, expresses a collagen-like platelet aggregation-associated protein (PAAP) (5). This collagen-like antigen contains a consensus platelet-interactive domain, XPGP/QGPX, found in most common types of collagen (6, 7). Strains of *S. sanguis* that express this protein epitope activate platelets directly in a complement-independent manner (9). These

Mark C. Herzberg, Gordon D. MacFarlane, Peixin Liu, and Pamela R. Erickson • School of Dentistry, University of Minnesota, Minneapolis, Minnesota 55455.

Agg$^+$ (*paap$^+$*) strains may activate and aggregate platelets in bleeding periodontal tissues, responses that may be potentiated by epinephrine (11).

Unlike *S. sanguis*, some putative periodontal pathogens can invade the gingival tissues. *Porphyromonas gingivalis*, for example, invades oral epithelial cells in vitro (4). Among putative periodontal pathogens, invading microorganisms may contribute to the inflammatory response at injured sites within the periodontal tissues by directly inducing platelets to activate and aggregate. To begin to test this hypothesis, activation and aggregation of human platelets in response to strains of *P. gingivalis*, *Actinobacillus actinomycetemcomitans*, *Prevotella intermedia*, *Campylobacter rectus*, *Fusobacterium nucleatum*, and *Eikenella corrodens* were tested.

S. sanguis I133-79, a positive control, was isolated from a patient with a confirmed case of bacterial endocarditis and has been extensively described in earlier studies (5, 9). These bacteria were grown overnight in Todd-Hewitt broth at 37°C in 5% CO_2. Representative putative periodontal pathogens, including *P. gingivalis* strains (33277 from the American Type Culture Collection [ATCC]; 18/10 from G. Bowden, University of Manitoba; W50 from R. Schifferle, State University of New York at Buffalo; and WT40 and WT46, fresh clinical isolates from G. Germaine, University of Minnesota), *A. actinomycetemcomitans* strains (29522 and 29523 from ATCC; Y4 originally from J. Slots, State University of New York at Buffalo), *P. intermedia* 25611 (from the ATCC), and *F. nucleatum* 10953 (from the ATCC) were all grown anaerobically in Todd-Hewitt broth supplemented with 5 μg of hemin and 500 ng of menadione per ml at 37°C. *C. rectus* 33238 (from the ATCC) was grown anaerobically at 37°C in broth described by Tanner et al. (26). *E. corrodens* 23834 (from the ATCC) was cultured at 37°C on blood agar plates and then grown at 37°C in broth described by Tanner et al. (26). All bacteria were harvested and washed twice in cold 0.01 M sodium phosphate buffer (pH 7.4) with 0.9% sodium chloride (PBS). The washed bacteria were suspended to an optical density of 1.5 at 620 nm in PBS. This density was used in all platelet aggregometry experiments.

PLATELET AGGREGATION

To prepare platelet-rich plasma (PRP) for aggregometry, fresh platelets were obtained by venipuncture from a single, healthy, medication-free donor after informed consent under a protocol approved by the University of Minnesota Committee on the Use of Human Subjects. The blood was immediately mixed with 0.1 M citrate-citric acid glucose (10:1 vol/vol) and centrifuged at 100 × g for 20 min at room temperature. The PRP layer was aspirated and pipetted into another tube. The remaining blood was recentrifuged at 500 × g for 20 min to obtain platelet-poor plasma.

Platelet lumiaggregometry in response to each bacterial strain was performed as described previously (9, 11). Stirred suspensions (0.45 ml) of PRP (approximately 2.5 × 10^8 platelets per ml) were challenged with suspensions of bacteria (0.05 ml) at 37°C in a recording lumiaggregometer, and aggregation was indicated by an increase in light transmission. The lag time in minutes from addition of bacteria to onset of aggregation was measured. Secretion of dense granule ATP was determined by the addition of 0.03 ml of luciferin/luciferase (40-mg/ml stock solution; Sigma Chemical Co., St. Louis, Mo.). Stable baselines were established in both visible light and luminescence channels, and then bacteria were added.

Of the strains and species listed in Table 1, only *P. gingivalis* induced aggregation of human platelets in plasma within 25 min (Fig. 1). All strains of *P. gingivalis* tested induced the secretion of platelet-dense granules and the aggregation of platelets (Fig. 1, paired upper and lower tracings for each strain, respectively). For example, within 30 s of incubation with strain 33277, platelets aggregated, as indicated by the rapid rate of increase in light transmission. When light transmission reached its maximum, the widely fluctuating signal suggested that a stable in vitro thrombus had formed. Detectable amounts of ATP were secreted only after the onset of platelet aggregation, as indicated by the downward deflection of the pen in the luminescence channel. The strains of *P. gingivalis* differed in their lag times to onset of secretion and aggregation from less than 1 min to several minutes. Strain W50 expressed strong procoagulant activity resulting in fibrin clot formation that preceded secretion and aggregation. Strain 18/10, which has been passaged repeatedly in the laboratory over a 3-year period, rapidly induced secretion and aggregation, suggesting that the expression of this feature is stable over time.

As cells of *S. sanguis* were diluted, the lag time to onset of platelet aggregation increased and the extent of aggregation (increasing percent transmission of light) changed little (Fig. 2A). Similarly, the lag time to onset of aggregation was also related directly to the concentration of *P. gingivalis* (Fig. 2B). Serial twofold dilutions of *P. gingivalis* resulted in increased lag times that had little effect on the extent of aggregation.

CROSS-REACTIVITY OF *P. GINGIVALIS* WITH PAAP

To determine whether *S. sanguis* and *P. gingivalis* express surface proteins homologous for PAAP, DNA from *P. gingivalis* 33277 was probed in Southern

Table 1. Bacterial strains

Gram negative
  P. gingivalis
    ATCC 33277
    WT40
    18/10
    W50
  A. actinomycetemcomitans
    ATCC 29522
    ATCC 29523
    Y4
  P. intermedia ATCC 25611
  C. rectus ATCC 33238
  F. nucleatum ATCC 10953
  E. corrodens ATCC 23834
Gram positive
  Streptococcus sanguis I133-79

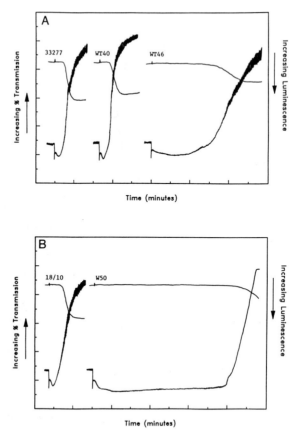

Figure 1. Strains of *P. gingivalis* induce aggregation of human platelets in vitro. As described in the text, cells of *P. gingivalis* were added to warm (37°C) stirred suspensions of PRP at a final ratio of about 0.75:1. The time course of aggregation was monitored by the change in percent light transmission as time progressed from left to right. Simultaneously in the same cuvette, ATP release from platelet-dense granules was assayed by luminescence. The lag time is measured from the addition of bacteria (deflection early in each tracing) to the change in slope reflecting onset of a rapid increase in light transmission or luminescence. Each pair of tracings in panels A and B is labeled with the strain tested.

blots with a 3.6-kb fragment of *paap* from *S. sanguis*. Whole chromosomal DNA from *P. gingivalis* was isolated by the method of Das (as described in reference 24), digested with *Hind*III, and electrophoresed on 0.8% agarose gels. The DNA was blotted and probed at high stringency with a 3.6-kb *Hind*III-*Dra*I DNA fragment from *S. sanguis* I133-79. This probe contains the gene *(paap)* for the collagen-like PAAP (18, 19). At high stringency, no hybridization was detectable.

S. *sanguis* and *P. gingivalis* may still express immunologically cross-reactive PAAP. To learn whether the two species express a common PAAP epitope(s), surface macromolecules were isolated from *P. gingivalis* 33277 by sonication or by lysozyme or trypsin digestion and compared by Western blot (immunoblot) analysis (5–7). Cells were washed three times in cold 0.01 M PBS (pH 7.0) and suspended

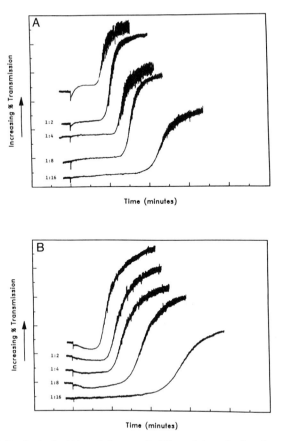

Figure 2. Doses of *P. gingivalis* (A) and *S. sanguis* (B) modulate the lag time to onset of platelet aggregation. Twofold dilutions of *P. gingivalis* 33277 and *S. sanguis* I133-79 were compared for changes in the course of platelet aggregation as described in the text.

to 2×10^9 cells per ml. For sonication, washed cells were chilled to 4°C and sonicated on ice intermittently for 5 min at 50-W output. For trypsin digestion, washed cells were treated at 37°C with 50 μg of L-(tosylamido-2-phenyl) ethyl chloromethyl ketone (TPCK)-trypsin per ml of suspension. Digestion was stopped after 30 min by adding 25 μg of lima bean trypsin inhibitor per ml and immersing the mixture in ice. For lysozyme digestion, washed cells were incubated for 60 min at 37°C with 0.5 mg of lysozyme per ml. After each digestion, cells of *P. gingivalis* were pelleted by centrifugation, and the supernatants were adjusted to equivalent protein concentrations based on the A_{280}. The surface macromolecule preparations were electrophoresed on 8% nonreducing sodium dodecyl sulfate (SDS)-polyacrylamide gels. Unstained SDS-polyacrylamide gels were then electroeluted to nitrocellulose and stained for total protein with colloidal gold (Aurodye Forte; Amersham International). Alternatively, nitrocellulose blots were quenched with hemoglobin and incubated with rabbit antibody to the 23-kDa platelet-interactive fragment of the

S. sanguis PAAP (5–7). The binding of rabbit immunoglobulin G antibodies to *P. gingivalis* antigens was detected by incubation with goat anti-rabbit immunoglobulin G conjugated to alkaline phosphatase and reaction with 5-bromo-4-chloro-3-indoyl phosphate in the presence of Nitro Blue Tetrazolium (6).

Rabbit anti-p23 PAAP from *S. sanguis* reacted with a surface antigen from *P. gingivalis* 33277 of approximately 100 kDa (Fig. 3, right panel, lanes B, C, and D). The *P. gingivalis* antigen was recovered in preparations obtained by sonication and by lysozyme and tryptic digestions. The p100 cross-reactive antigen was the only form noted in the *P. gingivalis* preparations. Like the p115 form of *S. sanguis* PAAP recovered from *S. sanguis* protoplast culture supernatants (10) and other characterized fragments isolated directly from cells (5), the 100-kDa *P. gingivalis* antigen was a relatively minor constituent of the surface macromolecule preparations, staining poorly in the Aurodye transfers (Fig. 3, left panel, lanes B, C, and D).

These data suggest strongly that *P. gingivalis* expresses a 100-kDa surface protein that contains a collagen-like platelet-interactive domain. This protein is immunologically cross-reactive with the PAAP of Agg$^+$ (*paap$^+$*) strains (7) of *S. sanguis*. Like those of *S. sanguis,* interactions between *P. gingivalis* and platelets induce the secretion of platelet granular contents. As a result of platelet activation and aggregation, biologically active proteins and lipid components are released. In vivo, these released products may contribute to the local inflammatory host response in periodontitis.

THE PLATELET AS AN INFLAMMATORY CELL
IN PERIODONTAL DISEASES

Activated platelets may play an important role in modulating the host response in periodontitis. Interactions with platelets may modify endothelial expression of

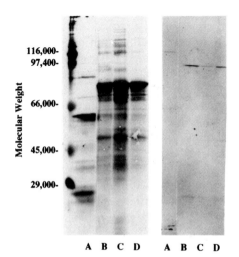

Figure 3. *P. gingivalis* 33277 and *S. sanguis* I133-79 express cross-reactive PAAP antigens. The samples in the left and right panels contained 5 or 20 μg of protein, respectively, each. The left panel was stained for total protein with Aurodye colloidal gold. The right panel was stained after reaction with rabbit immunoglobulin G antibodies to the 23-kDa platelet-interactive fragment from *S. sanguis* PAAP. All samples were solubilized in 1% (wt/vol) SDS sample buffers, subjected to electrophoresis on 8% gels, and immunoblotted as described in the text. Lane A in each panel contains *S. sanguis* protoplast culture supernatants. Lanes B, C, and D contain *P. gingivalis* sonicate and lysozyme and trypsin digest preparations, respectively.

growth factors, such as transforming growth factor alpha/beta and platelet-derived growth factor, and receptors responsible for the homing of lymphocytes and the transmigration of neutrophils. Platelets can modulate the activation of complement and increase the tissue content of biogenic amines such as serotonin and arachidonate metabolites such as thromboxane E_2 and 12-HETE. Each is considered important in our current view of host responses in periodontal diseases (8). Since *P. gingivalis* is an invading microorganism, it has the potential to activate platelets at the interface of the vasculature. Experimental models suggest that activation of platelets in close proximity to the vasculature may contribute to inflammatory disease. For example, neutrophil-dependent immune complex nephritis is mediated by platelets in at least one model (14), contributing to the proliferation of the mesangium (14, 15). In this mechanism, the activation of complement plays a key role. It is important to note that complement is activated on the surface of the platelet coincident with its function in binding proteins from the coagulation cascade (3, 27). When complement is activated on the surface of the platelet, additional platelet activation and aggregation occur (3). While platelet activation and aggregation in response to *P. gingivalis* were complement independent since they occurred in the presence of citrated plasma, which effectively complexes divalent cations needed for activation of the complement cascade, it is likely that complement will be activated locally in the tissues. Activation of complement as a consequence of platelet activation is believed to play an important role in the localization of antibody-dependent cell-mediated cytotoxicity (25). It is clear that activation of platelets by invading periodontal pathogens such as *P. gingivalis* may contribute in numerous ways to the host inflammatory response.

The identity of the 100-kDa *P. gingivalis* PAAP is unclear. *P. gingivalis* expresses hemagglutinins of various sizes from 24 to 115 kDa (1, 12, 23, 24). A 40-kDa form has been cloned, and epitopes of a family of hemagglutinins of *P. gingivalis* have been mapped by DesLauriers and Mouton (2). While these hemagglutinins appear to aggregate into macromolecular complexes and may be exported into culture medium (12), these extrafimbrial surface proteins do not appear to be similar to the *P. gingivalis* PAAP. *P. gingivalis* also appears to express a 43-kDa protein that mediates adhesion to *Streptococcus gordonii* and a 75-kDa protein, fimbrillin, that may mediate binding to saliva-coated hydroxyapatite (17). A 24-kDa polypeptide isolated from outer membrane vesicles of *P. gingivalis* activates fibroblasts (22). While each of these may be surface-expressed proteins, the relationship to *P. gingivalis* PAAP is uncertain. The data suggest, therefore, that *P. gingivalis* among these putative pathogens may contribute to the inflammatory host response in periodontitis by inducing platelets to activate and aggregate. In this model, *P. gingivalis* would invade the connective tissue, where it would encounter and interact with platelets that had been released from injured or perturbed capillaries and venules of the gingival vasculature. Expression of the p100 PAAP by *P. gingivalis* appears to promote the inflammatory potential of platelets in periodontal diseases.

Acknowledgments. We acknowledge the support of NIH grants and aid DE05501, DE08489, and DE09737. We thank Urve Daigle for expert secretarial support.

REFERENCES

1. **Chandad, F., and C. Mouton.** 1990. Molecular size variation of the hemagglutinating adhesin HA-Ag2, a common antigen of *Bacteroides gingivalis. Can. J. Microbiol.* **36:**690–696.
2. **DesLauriers, M., and C. Mouton.** 1992. Epitope mapping of hemagglutinating adhesion HA-Ag2 of *Bacteroides (Porphyromonas) gingivalis. Infect. Immun.* **60:**2791–2799.
3. **Devine, D. V.** 1992. The effects of complement activation on platelets. *Curr. Top. Microbiol. Immunol.* **178:**101–113.
4. **Duncan, M. J., S. Nakao, Z. Skobe, and H. Xie.** 1993. Interactions of *Porphyromonas gingivalis* with epithelial cells. *Infect. Immun.* **61:**2260–2265.
5. **Erickson, P. R., and M. C. Herzberg.** 1987. A collagen-like immunodeterminant on the surface of *Streptococcus sanguis* induces platelet aggregation. *J. Immunol.* **138:**3360–3366.
6. **Erickson, P. R., and M. C. Herzberg.** 1992. Crossreactive immunodeterminants on *Streptococcus sanguis* and collagen: predicting a structural motif of platelet-interactive domains. *J. Biol. Chem.* **267:**10018–10023.
7. **Erickson, P. R., and M. C. Herzberg.** 1993. The *Streptococcus sanguis* platelet aggregation-associated protein: identification and characterization of the minimal platelet interactive domain. *J. Biol. Chem.* **268:**1646–1649.
8. **Genco, R. J.** 1992. Host responses in periodontal diseases: current concepts. *J. Periodontol.* **63:**338–355.
9. **Herzberg, M. C., K. L. Brintzenhofe, and C. C. Clawson.** 1983. Cell-free released components of *Streptococcus sanguis* inhibit human platelet aggregation. *Infect. Immun.* **42:**394–401.
10. **Herzberg, M. C., P. R. Erickson, P. K. Kane, D. J. Clawson, C. C. Clawson, and F. A. Hoff.** 1990. Platelet interactive products of *Streptococcus sanguis* protoplasts. *Infect. Immun.* **58:**4117–4125.
11. **Herzberg, M. C., L. K. Krishnan, and G. D. MacFarlane.** 1993. Involvement of α_2-adrenoreceptors and G-proteins in the modulation of platelet secretion in response to *Streptococcus sanguis. Crit. Rev. Oral Biol. Med.* **4:**435–442.
12. **Inoshita, E., A. Amano, T. Hanioka, H. Tamagawa, S. Shizukuishi, and A. Tsunemitsu.** 1986. Isolation and some properties of exohemagglutinin from the culture medium of *Bacteroides gingivalis* 381. *Infect. Immun.* **52:**421–427.
13. **Johnson, R. J.** 1991. Platelets in inflammatory glomerular injury. *Semin. Nephrol.* **11:**276–284.
14. **Johnson, R. J., C. E. Alpers, P. Pritzl, M. Schulze, P. Baker, C. Pruchno, and W. G. Couser.** 1988. Platelets mediate neutrophil-dependent immune complex nephritis in the rat. *J. Clin. Invest.* **82:**1225–1235.
15. **Johnson, R. J., P. Pritzl, H. Iida, and C. E. Alpers.** 1991. Platelet-complement interactions in mesangial proliferative nephritis in the rat. *Am. J. Pathol.* **138:**313–321.
16. **Larsen, E., A. Celi, G. E. Gilbert, B. C. Furie, J. K. Erban, R. Bonfanti, D. D. Wagner, and B. Furie.** 1989. PADGEM protein: a receptor that mediates the interaction of activated platelets with neutrophils and monocytes. *Cell* **59:**305–312.
17. **Lee, J.-Y., H. T. Sojar, G. S. Bedi, and R. J. Genco.** 1992. Synthetic peptides analogous to the fimbrillin sequence inhibit adherence of *Porphyromonas gingivalis. Infect. Immun.* **60:**1662–1670.
18. **Liu, P., P. P. Cleary, and M. C. Herzberg.** Unpublished data.
19. **Liu, P., and M. C. Herzberg.** 1993. Polymorphism of *paap* gene among strains of *Streptococcus sanguis,* abstr. 1762. *J. Dent. Res.* **72:**(Special Issue):323.
20. **Marcus, A. J., and L. B. Safier.** 1993. Thromboregulation: multicellular modulation of platelet reactivity in hemostasis and thrombosis. *FASEB J.* **7:**516–522.
21. **Marcus, A. J., L. B. Safier, H. L. Ullman, N. Islam, M. J. Broekman, J. R. Falck, S. Fischer, and C. von Schacky.** 1988. Platelet-neutrophil interactions. (12S)-hydroxyeicosatetraen-1,20-dioic acid: a new eicosanoid synthesized by unstimulated neutrophils from 12S-20-dihydroxyeicosatetraenoic acid. *J. Biol. Chem.* **263:**2223–2229.
22. **Mihara, J., and S. C. Holt.** 1993. Purification and characterization of fibroblast-activating factor isolated from *Porphyromonas gingivalis* W50. *Infect. Immun.* **61:**588–595.
23. **Nishikata, M., and F. Yoshimura.** 1991. Characterization of *Porphyromonas (Bacteroides) gingivalis* hemagglutinin as a protease. *Biochem. Biophys. Res. Commun.* **178:**336–342.

24. **Progulske-Fox, A., S. Tumwasorn, and S. C. Holt.** 1989. The expression and function of a *Bacteroides gingivalis* hemagglutinin gene in *Escherichia coli*. *Oral Microbiol. Immunol.* **4:**121–131.
25. **Slezak, S., S. E. Symer, and H. S. Shing.** 1987. Platelet-mediated cytotoxicity. *J. Exp. Med.* **166:**489–505.
26. **Tanner, A. C. R., J. L. Dzink, J. L. Ebersole, and S. S. Socransky.** 1987. *Wolinella recta, Campylobacter concisus, Bacterioides gracilis,* and *Eikenella corrodens* from periodontal lesions. *J. Periodontal Res.* **22:**327–330.
27. **Weksler, B. B.** 1992. Platelets, p. 727–746. *In* J. I. Gallin, I. M. Goldstein, and R. Snyderman (ed.), *Inflammation: Basic Principles and Clinical Correlates,* 2nd ed. Raven Press, Inc., New York.

Molecular Pathogenesis of Periodontal Disease
Edited by Robert Genco et al.
© 1994 American Society for Microbiology, Washington, DC 20005

Chapter 22

Role of Antibodies in Periodontopathic Bacterial Infections

Katsuji Okuda, Atsushi Saito, Kaname Hirai,
Koichi Harano, and Tetsuo Kato

Microbiological and immunological studies have clearly demonstrated that *Porphyromonas gingivalis* is the predominant organism in human adult periodontitis lesions and that patients possess elevated levels of antibody to this microorganism (8, 9, 11, 18, 22). Accordingly, numerous studies of the immune response to *P. gingivalis* have focused on helping elucidate the infections in periodontal lesions (8, 9, 11, 12, 24). However, the potentially protective features of humoral immune responses are still uncharacterized.

The relationship between systemic immunoglobulin G (IgG) antibodies to *Actinobacillus actinomycetemcomitans* and infection by these microorganisms has been studied in order to determine the functional activities of such immune responses (11, 19, 20). These studies have indicated that IgG responses to *A. actinomycetemcomitans* play protective roles in eliminating or inhibiting colonization by the microorganisms in periodontal regions. Better understanding will contribute to design of a vaccine to inhibit colonization by *A. actinomycetemcomitans*.

IMMUNE RESPONSES TO *P. GINGIVALIS* AND INFECTION

We examined serum IgG and IgM responses to *P. gingivalis* and the correlations of these responses with age and infection in the periodontal region (11, 12). A total of 122 individuals were studied: 40 gingivitis patients, 49 gingivitis-free subjects, and 33 adult periodontitis subjects. Subjects in the gingivitis group and some of the gingivitis-free individuals were divided into four groups based on their physiological ages: early childhood, school age (prepuberty), puberty, and adulthood. In the gingivitis group, there was a positive correlation between increase in age and increase of levels of IgG antibody against *P. gingivalis* in serum until puberty.

Katsuji Okuda, Kaname Hirai, Koichi Harano, and Tetsuo Kato • Department of Microbiology, Tokyo Dental College, 1-2-2 Masago, Chiba 261, Japan.   *Atsushi Saito* • Department of Periodontics, Tokyo Dental College, 1-2-2 Masago, Chiba 261, Japan.

As shown in Table 1, there were positive correlations between elevated IgG levels and infection by the microorganisms in the group with puberty gingivitis and the adult groups with gingivitis and periodontitis (11). Thirty of 33 patients with adult periodontitis possessed elevated levels of IgG antibody to *P. gingivalis* and were infected by the microorganisms (11, 12). This positive correlation between the cultivable cell numbers of *P. gingivalis* in periodontal samples and IgG responses to the microorganism indicates that synthesis of the antibody against *P. gingivalis* is strongly enhanced by colonization of periodontal lesions by the microorganisms and that elevation of systemic IgG against *P. gingivalis* cannot eliminate specific *P. gingivalis* strains in periodontal regions.

ENCAPSULATED STRAINS OF *P. GINGIVALIS*

Encapsulated strain *P. gingivalis* 16-1 produces experimental abscesses and remains viable for more than 4 weeks in the abscess (14). Encapsulated strain 16-1 is more resistant to complement-mediated bactericidal activities than *P. gingivalis* ATCC 33277. Also, invasive strains of *P. gingivalis* are less hydrophobic than noninvasive strains (10). This low hydrophobicity of the capsular structure may help these bacteria evade phagocytic cells. Van Steenbergen et al. (23) also pointed out the protective role of encapsulated *P. gingivalis* against host defense mechanisms.

Cutler et al. (4) demonstrated that virulent *P. gingivalis* strains in the mouse chamber model (6) were not opsonized by healthy serum but required strain-specific IgG antibody. Most serum samples from adult periodontitis patients lacked opsonin capability against virulent strains. They indicated that the specific antibody against an invasive *P. gingivalis* strain was required for opsonization of the microorganisms by the antibody-dependent alternate pathway (5). They also demonstrated that serum killing and extracellular killing of *P. gingivalis* were each less effective alone than intracellular killing alone.

Serum antibodies to *P. gingivalis* were capable of complement antibody-mediated cytosis of certain strains of the microorganisms in vitro (15) and of eliminating the bacteria from periodontal regions in experimental animals (16). We also ex-

Table 1. Correlation between elevated IgG responses to *P. gingivalis* and infection in periodontal lesions of subjects with gingivitis or adult periodontitis

Disease and life stage	Spearman rank correlation coefficient
Gingivitis	
Early childhood	No infection
School age	NS[a]
Puberty	0.699[b]
Adulthood	0.611[b]
Periodontitis, adult	0.870[b]

[a] NS, not significant.
[b] $P < 0.01$.

amined the intracellular killing of *P. gingivalis* by perioteneal leukocytes of guinea pig in vitro and compared invasive *P. gingivalis* 16-1 with the noninvasive strain ATCC 33277. Ingested cells were counted electron microscopically, and killing was determined by viable cell counting. Although cells of both invasive and noninvasive strains of *P. gingivalis* were ingested by macrophages in the absence of specific antibody against the microorganisms, rabbit antiserum against whole cells of invasive *P. gingivalis* 16-1 enhanced the phagocytosis. As shown in Fig. 1, rabbit antiserum enhanced the intracellular killing of noninvasive *P. gingivalis* ATCC 33277 but did not significantly alter that of invasive 16-1. As shown in Fig. 2, the fibrous outer surface capsular structure of *P. gingivalis* 16-1 cells is distinct, even in phagosomes in leukocytes. Our results suggest that the capsular structure of this invasive *P. gingivalis* strain may inhibit complement activation and resist intracellular killing by macrophages. However, this whole scheme of the role of the capsular structure of invasive *P. gingivalis* strains in intracellular killing is still debatable.

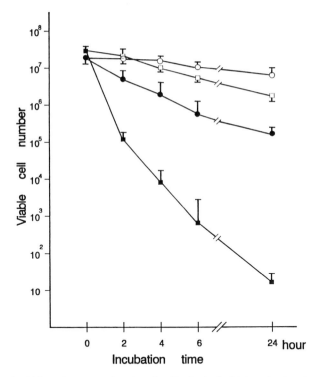

Figure 1. Effect of rabbit antiserum against invasive *P. gingivalis* 16-1 on in vitro bactericidal activity by peritoneal leukocytes from guinea pig. Values are the means of duplicate tubes and the standard errors of the means. Each tested tube contained 5.1×10^6 intraperitoneal leukocytes in 400 μl of RPMI 1640 medium, 100 μl of bacterial suspension, 200 μl of fresh guinea pig serum, and 200 μl of inactivated calf serum, all of which were absorbed with cells of *P. gingivalis* ATCC 33277 and 16-1. A 100-μl aliquot of rabbit antiserum against whole cells of *P. gingivalis* 16-1 was added to suspensions of strains 16-1 (●) and ATCC 33277 (■). A 100-μl aliquot of normal rabbit serum absorbed with both kinds of cells was added to *P. gingivalis* 16-1 (○) and ATCC 33277 (□).

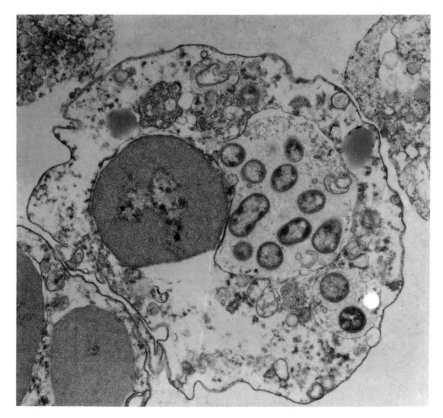

Figure 2. Guinea pig intraperitoneal leukocyte incubated with live *P. gingivalis* 16-1 cells, antiserum against whole cells of *P. gingivalis*, and fresh serum of guinea pig for 24 h in an anaerobic chamber. Note the fibrillar outer surface structure of *P. gingivalis* 16-1 in the leukocyte.

EFFECT OF IMMUNIZATION ON *P. GINGIVALIS* INFECTION

We attempted to determine whether immunization with invasive *P. gingivalis* 16-1 can protect mice from abscess formation and from spreading infection after challenges with the same strain (16). A group of BALB/c mice was immunized subcutaneously four times with formalin-killed cells to a total of 2 mg (wet weight) of cells emulsified in complete Freund's adjuvant. Another group of mice was immunized four times by peritoneal injection of 1.4×10^7 to 4.3×10^7 *P. gingivalis* 16-1 live cells. A control group was injected intraperitoneally with 0.2 ml of phosphate-buffered saline (PBS).

All these mice were challenged after the final immunization with 7.1×10^8 *P. gingivalis* 16-1 cells subcutaneously injected into the abdomen. All the sham-immunized mice developed spreading infections on the abdomen. Seven of the eight mice in this group died by 18 days postchallenge. Mice immunized with sterilized cells inhibited the spreading infection after injection of *P. gingivalis* 16-1. Localized abscesses in these mice ruptured in 5 to 7 days and were cured within 4 to 7 weeks.

Immunization with live *P. gingivalis* 16-1 cells prevented the spreading infections; however, the abscesses in these mice gradually increased in size from 2 to 10 weeks after the challenge. *P. gingivalis* 16-1 cells were isolated from the pus in the abscesses and found within leukocytes. The abscesses increased in size and finally killed these mice. Our results were consistent with findings by Chen et al. (3). They demonstrated that immunization of mice with live cells of invasive *P. gingivalis* ATCC 53977 inhibited secondary lesions and septicemia due to subcutaneous injection of this same strain, but the size of the abscesses increased gradually and killed the mice. Furthermore, Chen et al. showed that immunization with an invasive *P. gingivalis* isolate resulted in localization of a challenge infection with the homologous strain, while immunization with a noninvasive isolate did not (2, 3). These results suggest that cell surface structures such as capsular antigens in invasive strains of *P. gingivalis* are resistant to intracellular killing in immunized animals. It is possible that cells of invasive *P. gingivalis* strains can survive in leukocytes. Yoshimura et al. (25) reported that a 75-kDa protein of a *P. gingivalis* strain was recognized by serum samples from periodontitis patients and immunized rabbits. Chen et al. (2) demonstrated that 57- and 40-kDa proteins are the major proteins among surface antigens of invasive *P. gingivalis* ATCC 53977. We demonstrated that invasive *P. gingivalis* strains possess a 57-kDa surface antigen (unpublished data). Schifferle et al. (21) isolated and characterized a polysaccharide antigen from invasive *P. gingivalis* ATCC 53977. Future efforts will be directed at demonstrating which specific antigenic component will induce protective antibodies against pathogenesis by invasive *P. gingivalis* strains.

IMMUNE RESPONSES TO *A. ACTINOMYCETEMCOMITANS* AND INFECTION

We studied serum IgG responses to three serotypes of *A. actinomycetemcomitans* and the correlations with age and serotype of infection (11). There was a positive correlation between elevated IgG levels against the antigen of serotype c *A. actinomycetemcomitans* and against infection by the serotype found in the school age and childhood gingivitis groups and the adult periodontitis group (Table 2). The highest IgG response against serotype b *A. actinomycetemcomitans* occurred in the puberty gingivitis group (Fig. 3). None of the isolates of *A. actinomycetemcomitans* from adult periodontitis patients were serotype b (19). It is possible that high IgG responses to the surface antigen of serotype b *A. actinomycetemcomitans* specifically eliminates this serotype of the microorganism in the periodontal regions but does not eliminate serotype c.

ROLE OF ANTIBODY TO FIMBRIAL ANTIGEN OF *A. ACTINOMYCETEMCOMITANS*

Recently, fresh isolates of *A. Actinomycetemcomitans* were found to possess fimbriae (7, 17). Antifimbria antibodies are believed to inhibit the colonization of many bacterial species. To determine the role of IgG responses to *A. actinomy-*

Table 2. Correlation between elevated IgG responses to three serotypes of
A. actinomycetemcomitans and their infections in periodontal regions

| Serotype | Spearman rank correlation coefficient[a] | | | | |
| | Gingivitis | | | | Adult periodontitis |
	Early childhood	School age	Puberty	Adulthood	
a	—	—	NS	NS	NS
b	—	NS	NS	NS	NS
c	NS	0.710[b]	0.627[b]	NS	0.568[b]

[a] —, Not detected in plaque sample; NS, not significant.
[b] $P < 0.01$.

cetemcomitans, we measured the relative avidity of serum IgG antibody elicited in response to the fimbria antigen of *A. actinomycetemcomitans* (20). Fimbria antigen was prepared from *A. actinomycetemcomitans* 310-a (donated by H. Ohta, School of Dentistry, Okayama University, Okayama, Japan) by the method of Inoue et al. (7). The extracted fimbriae formed a single main band at approximately 54 kDa in sodium dodecyl sulfate-polyacrylamide gel electrophoresis analysis. The relative avidity of IgG antibody for this fimbria antigen was determined by the method described by Chen et al. (1). The values for IgG titers and avidities obtained by enzyme-linked immunosorbent assay with purified antigen are summarized in Table 3. Differences between the groups of titers or avidity data were determined by using Wilcoxon's rank sum test. The sera from patients with adult periodontitis

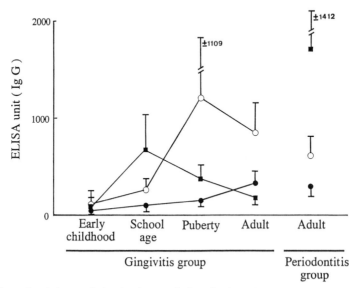

Figure 3. Age-related changes in levels of serum IgG antibody against sonicated antigens of serotype a, b, and c *A. actinomycetemcomitans.* Values are the means of 10 samples from gingivitis groups and 10 from adult patients with periodontitis. Bars indicate standard errors of the means. Symbols: ●, serotype a; ○, serotype b; ■, serotype c. ELISA, enzyme-linked immunosorbent assay.

Table 3. Serum IgG titers and avidity for fimbria antigen of
A. actinomycetemcomitans

Subject	Titer	Avidity index
Controls (n = 10)	116.5 ± 81.2	72.1 ± 18.0
A. actinomycetemcomitans negative (n = 18)	677.4 ± 16.7[a]	59.6 ± 8.2[a]
A. actinomycetemcomitans positive (n = 24)	454.1 ± 111.3[a]	48.6 ± 7.1[a]

[a] P < 0.05 compared with controls.

had statistically significantly higher IgG titers against *A. actinomycetemcomitans* 310-a fimbria antigen than did sera from periodontally healthy individuals, regardless of the presence of infection by the microorganisms in periodontal pockets. No statistically significant differences in titers against fimbria antigen were observed between patients with cultivable *A. actinomycetemcomitans* and patients without *A. actinomycetemcomitans*. The patient sera showed significantly lower avidities for the fimbria antigen than did sera from periodontally healthy individuals. Sera from *A. actinomycetemcomitans* culture-negative patients with elevated titers of IgG against fimbria antigen had significantly (P < 0.01) higher avidity for the fimbria antigen than did sera from *A. actinomycetemcomitans* culture-positive patients with elevated titers. These results led us to the hypothesis that the IgG responses to fimbria antigen elicited by the initial contact with *A. actinomycetemcomitans* played a role in clearing the organisms from the periodontal pockets of these patients. This line of thought suggests that fimbria antigen is a potent vaccine for inducing antibody responses that prevent *A. actinomycetemcomitans* colonization.

MACROMOLECULAR SYNTHESIZED PEPTIDE ANTIGEN OF *A. ACTINOMYCETEMCOMITANS* FIMBRIAE

We synthesized peptides according to the amino acid sequence of fimbria antigen from *A. actinomycetemcomitans* 310-a. To increase the immunogenicity of the synthesized peptide, the peptide antigen was conjugated by the branched lysine polymer resin beads method described by Nardelli et al. (13). This method has the advantage of increasing the molecular size and inducing high levels of antibody without the use of carrier proteins. The high-molecular-weight synthetic peptide antigen was suspended in PBS, emulsified with incomplete Freund's adjuvant, and injected into rabbits. The resultant rabbit antiserum reacted with the 54-kDa protein of crude fimbrial antigens from 310-a and other freshly isolated strains of *A. actinomycetemcomitans*. We found that the rabbit antiserum inhibited the attachment of fimbriated *A. actinomycetemcomitans* strains to human buccal epithelial cells and to saliva-coated hydroxyapatite beads in vitro (6a).

SUMMARY

We have studied the relationship between serum immune responses to *P. gingivalis* and *A. actinomycetemcomitans* and infections by these microorganisms in periodontal lesions (11, 12, 19, 20). With *P. gingivalis,* a positive correlation between elevated IgG level and infection by the microorganisms was found. The high level of IgG antibody to *P. gingivalis* in sera from periodontitis patients and experimental animals played protective roles in colonization inhibition (16), complement antibody-mediated cytosis (15), and phagocytic killing (14) of certain strains of the microorganisms. However, encapsulated strains of *P. gingivalis* resisted these protective roles of serum antibody (14, 15). The elevated level of IgG antibody to serotype b of *A. actinomycetemcomitans* may eliminate this serotype of the microorganisms, but it does not eliminate serotype c (11). The high IgG antibody avidities for the fimbrial antigen of *A. actinomycetemcomitans* in sera from adult periodontitis patients not harboring the microorganisms may afford protection against continued infection (20). These findings clearly indicate that a specific antigen can function as a vaccine that induces protective antibodies against these periodontopathic bacteria.

REFERENCES

1. **Chen, H. A., B. D. Johnson, T. J. Sims, T. P. Darveau, B. J. Moncla, C. W. Whitney, D. Engel, and R. C. Page.** 1991. Humoral immune responses to *Porphyromonas gingivalis* before and following therapy in rapidly progressive periodontitis patients. *J. Periodontol.* **62:**781–791.
2. **Chen, P. B., L. B. Davern, R. Schifferle, and J. J. Zambon.** 1990. Protective immunization against experimental *Bacteroides (Porphyromonas) gingivalis* infection. *Infect. Immun.* **58:**3394–3400.
3. **Chen, P. B., M. E. Neiders, S. J. Millar, H. S. Reynolds, and J. J. Zambon.** 1987. Effect of immunization on experimental *Bacteroides gingivalis* infection in a murine model. *Infect. Immun.* **55:**2534–2537.
4. **Cutler, C. W., J. R. Kalmar, and R. R. Arnold.** 1991. Phagocytosis of virulent *Porphyromonas gingivalis* by human polymorphonuclear leukocytes requires specific immunoglobulin G. *Infect. Immun.* **59:**2097–2104.
5. **Cutler, C. W., J. R. Kalmar, and R. R. Arnold.** 1991. Antibody-dependent alternate pathway of complement activation in opsophagocytosis of *Porphyromonas gingivalis. Infect. Immun.* **59:**2105–2109.
6. **Genco, C. A., C. W. Cutler, D. Kapczynski, K. Maloney, and R. R. Arnold.** 1991. A novel mouse model to study the virulence of and host response to *Porphyromonas (Bacteroides) gingivalis. Infect. Immun.* **59:**1255–1263.
6a. **Harano, K., A. Yamanaka, T. Kato, and K. Okuda.** 1993. Molecular biological study of *A. actinomycetemcomitans* fimbriae, abstr. no. 32. *Abstr. Jpn. Assoc. Dent. Res.*
7. **Inoue, T., H. Ohta, S. Kokeguchi, K. Fukui, and K. Kato.** 1990. Colonial variation and fimbriation of *Actinobacillus actinomycetemcomitans. FEMS Microbiol. Lett.* **69:**13–18.
8. **Mouton, C., P. G. Hammond, J. Slots, and R. J. Genco.** 1981. Serum antibodies to *Bacteroides asaccharolyticus (Bacteroides gingivalis):* relationship to age and periodontal disease. *Infect. Immun.* **31:**182–192.
9. **Naito, Y., K. Okuda, and I. Takazoe.** Immunoglobulin G response to subgingival gram-negative bacteria in human subjects. *Infect. Immun.* **45:**47–51.
10. **Naito, Y., H. Tohda, K. Okuda, and I. Takazoe.** 1993. Adherence and hydrophobicity of invasive and noninvasive strains of *Porphyromonas gingivalis. Oral Microbiol. Immunol.* **8:**195–202.
11. **Nakagawa, S., Y. Machida, T. Nakagawa, H. Fujii, S. Yamada, I. Takazoe, and K. Okuda.** 1994. Infection by *Porphyromonas gingivalis* and *Actinobacillus actinomycetemcomitans,* and antibody responses at different ages in humans. *J. Periodontal Res.* **29:**9–16.
12. **Nakagawa, T., S. Yamada, M. Tsunoda, T. Sato, Y. Naito, I. Takazoe, and K. Okuda.** 1990. Clinical, microbiological, and immunological studies following initial preparation in adult periodontitis. *Bull. Tokyo Dent. Coll.* **31:**321–331.

13. **Nardelli, B., Y.-A. Lu, D. R. Shiu, C. Delpierre-Defoort, A. T. Profy, and J. P. Tam.** 1992. A chemically defined synthetic vaccine model for HIV-1. *J. Immunol.* **148:**912–920.

14. **Okuda, K., and T. Kato.** 1991. Immunity to infection by periodontopathic bacteria in experimental animals, p. 237–250. *In* S. Hamada, S. C. Holt, and J. R. McGhee (ed.), *Periodontal Disease: Pathogens and Host Immune Responses.* Quintessence Publishing Co., Ltd. Tokyo.

15. **Okuda, K., T. Kato, Y. Naito, M. Ono, Y. Kikuchi, and I. Takazoe.** 1986. Susceptibility of *Bacteroides gingivalis* to bactericidal activity of human serum. *J. Dent. Res.* **65:**1024–1027.

16. **Okuda, K., T. Kato, Y. Naito, I. Takazoe, Y. Kikuchi, T. Nakamura, T. Kiyoshige, and S. Sasaki.** 1988. Protective efficacy of active and passive immunizations against experimental infection with *Bacteroides gingivalis* in ligated hamsters. *J. Dent. Res.* **67:**807–811.

17. **Rosan, B., J. Slots, R. J. Ramont, M. A. Listgarten, and G. M. Nelson.** 1988. *Actinobacillus actinomycetemcomitans* fimbriae. *Oral Microbiol. Immunol.* **3:**58–63.

18. **Rosenburg, J. P., A. J. van Winkelhoff, E. G. Winkel, R. J. Goene, F. Abbas, and J. de Graaff.** 1990. Occurrence of *Bacteroides gingivalis, Bacteroides intermedius* and *Actinobacillus actinomycetemcomitans* in severe periodontitis in relation to age and treatment history. *J. Clin. Periodontol.* **17:**392–399.

19. **Saito, A., Y. Hosaka, T. Nakagawa, K. Seida, S. Yamada, I. Takazoe, and K. Okuda.** 1993. Significance of serum antibody against surface antigens of *Actinobacillus actinomycetemcomitans* in patients with adult periodontitis. *Oral Microbiol. Immun.* **8:**146–153.

20. **Saito, A., Y. Hosaka, T. Nakagawa, S. Yamada, and K. Okuda.** 1993. Relative avidity of serum immunoglobulin G antibody for the fimbria antigen of *Actinobacillus actinomycetemcomitans* in patients with adult periodontitis. *Infect. Immun.* **61:**332–334.

21. **Schifferle, R. E., M. S. Reddy, J. J. Zambon, R. J. Genco, and M. J. Levine.** 1989. Characterization of a polysaccharide antigen from *Bacteroides gingivalis. J. Immunol.* **143:**3035–3042.

22. **Slots, J., L. Bragd, M. Wikstrom, and G. Dahlen.** 1986. The occurrence of *Actinobacillus actinomycetemcomitans, Bacteroides gingivalis* and *Bacteroides intermedius* in destructive periodontal disease in adults, *J. Clin. Periodontol.* **13:**570–577.

23. **van Steenbergen, T. J. M., P. Kastelein, J. J. A. Touw, and J. de Graaff.** 1982. Virulence of black-pigmented Bacteroides strains from periodontal pockets and other sites in experimentally induced skin lesions in mice. *J. Periodontal Res.* **17:**41–49.

24. **Whitney, C., J. Ant, B. Moncla, B. Johnson, R. C. Page, and D. Engel.** 1992. Serum immunoglobulin G antibody to *Porphyromonas gingivalis* in rapidly progressive periodontitis: titer, avidity, and subclass distribution. *Infect. Immun.* **60:**2194–2200.

25. **Yoshimura, F., K. Watanabe, T. Takasawa, M. Kawanami, and H. Kato.** 1989. Purification and properties of a 75-kilodalton major protein, an immunodominant surface antigen, from oral anaerobe *Bacteroides gingivalis. Infect. Immun.* **57:**3646–3652.

Molecular Pathogenesis of Periodontal Disease
Edited by Robert Genco et al.
© 1994 American Society for Microbiology, Washington, DC 20005

Chapter 23

Immunization with *Porphyromonas gingivalis* Fimbriae and a Synthetic Fimbrial Peptide

R. T. Evans, B. Klausen, H. T. Sojar, N. S. Ramamurthy, M. J. Evans, D. W. Dyer, and R. J. Genco

Vaccination against infectious agents has greatly reduced the prevalence of diseases throughout the world. This is especially true for viral diseases. After nearly a century of vaccine development, there are, however, still only seven bacterial diseases for which effective vaccines are available (17). Development of successful vaccines against bacterial diseases has been difficult, in part because of the multiplicity and complexity of the antigens found on bacterial cells. New approaches resulting in improved methods of identifying and purifying critical protective antigens and the ability to sequence and synthesize relevant antigenic determinants will allow investigators to set new standards for safe and effective vaccines.

At the outset, immunization against periodontal diseases would appear difficult, since these diseases may be caused by possibly six or more diverse microorganisms (6, 14, 15). However, reports have shown that *Porphyromonas gingivalis* has a dominant role in the development of adult periodontitis. It is not unreasonable, therefore, to assume that a vaccine conferring protection against this organism would markedly reduce the incidence of periodontitis. It is also possible that several periodontal pathogens may need to be included in a polyvalent vaccine to immunize against other recognized forms of periodontal diseases. The focus of this chapter will be on immunization against experimentally induced periodontal destruction in gnotobiotic rats infected with *P. gingivalis* by using either *P. gingivalis* fimbriae or a synthetic fimbrial peptide.

THE GNOTOBIOTIC RAT MODEL

At present, immunization studies with *P. gingivalis* cannot be done in humans, and therefore the development of useful animal models is crucial. Several animal

R. T. Evans, H. T. Sojar, D. W. Dyer, and R. J. Genco • State University of New York at Buffalo, Buffalo, New York 14214-3092. *N. S. Ramamurthy* • State University of New York at Stony Brook, Stony Brook, New York 11790. *B. Klausen* • University of Copenhagen, Copenhagen, Denmark. *M. J. Evans* • Roswell Park Cancer Institute, Buffalo, New York 14263.

models are in current use. Nonhuman primates are similar to humans with respect to periodontal anatomy and microflora (13), but it is difficult to establish *P. gingivalis* in their oral cavities (11). Furthermore, the natural progression of periodontal disease is very slow in monkeys, thus limiting the usefulness of the model. Limitations on infection and progression of the disease can be overcome by tying ligatures around the teeth, but this procedure detracts from the fidelity of the model. Mechanisms responsible for colonization and tissue destruction in naturally occurring periodontitis may be different from disease mechanisms that occur in the presence of ligatures (12). We therefore sought a model in which ligatures would not be needed.

The use of the gnotobiotic rat suggested itself. Several studies have shown that the infection of rats with *P. gingivalis* causes alveolar bone loss. This infection may be of only short duration, but it is sufficient to induce measurable alveolar bone loss without ligation (2–4, 8). Unlike the subcutaneous abscess model in mice (16) and the subcutaneous chamber model in mice and rabbits (1, 5), which allow study of bacterial invasion mechanisms of tissue, the gnotobiotic rat model enables us to study oral colonization, gingival tissue changes, and destruction of alveolar bone. For these reasons, we selected the gnotobiotic rat model monoinfected with human strains of *P. gingivalis* to study the effects of immunization.

METHODS

Animals

The animals were housed in flexible plastic film isolators, with each unit containing eight 3-week-old male Sprague-Dawley rats (Taconic Farms, Germantown, N.Y.). Drinking water and sterilized food pellets (diet L-485; Teklad, Madison, Wis.) were supplied ad libitum. Bedding consisted of quarter-inch granules (approximately 0.64-cm) granules (Bed o'Cobs; Anderson, Inc., Maumee, Ohio), which were selected to reduce impaction and trauma (8).

Infection

Animals used to study infection were given *P. gingivalis* 381 suspended (0.5 ml; 10^{12} cells per ml) in 5% carboxymethyl cellulose (Sigma Chemicals, St. Louis, Mo.). Delivery was by gastric intubation (gavage) three times at 48-h intervals. The animals were terminated by exsanguination under anesthesia at predetermined intervals. Microbiological and tissue samples were taken to evaluate the infection status. Alveolar bone destruction was measured on defleshed jaws (see below).

Immunization

Antigens (10^{10} heat-killed cells; 20 µg of protein or 20 µg of peptide conjugate per animal) in Freund's incomplete adjuvant were given subcutaneously and periglandularly in the neck region (4, 8). Details regarding the antigens and their preparation are given by Evans et al. (2, 4), Klausen et al. (8), and Lee et al. (9). Briefly, intact *P. gingivalis* 381 cells were heat killed for 30 min at 60°C to provide the whole-cell antigen. The partially purified cell surface components, i.e., crude fimbriae, the 43-kDa structural subunit of the fimbriae, and the 75-kDa component, were obtained from *P. gingivalis* 2561 by mild sonication, 40% ammonium sulfate precipitation, and either repetitive differential precipitation with 1% sodium dodecyl sulfate or isoelectric precipitation. Peptides were synthesized on the basis of previously deduced fimbrial structure and were coupled to thyroglobulin to provide hapten carrier activity (2, 9). Challenge infections with *P. gingivalis* were begun 1 week after immunization and consisted of three challenges given every other day (see above). The animals were sacrificed at 42 days postinfection. For each animal, measurements were made of both bone level (horizontal bone loss) and bone support (vertical bone loss). In addition, gingivae were surgically removed in order to measure tissue proteolytic-enzyme activity (3, 4, 8). Serum and saliva samples were obtained for antibody determination. Antibody was measured by a particle concentration fluorescence immunoassay (PCFIA) and expressed as relative fluorescence units (RFU). Details of the enzyme and antibody measurement methods are given by Evans et al. (4) and Klausen et al. (8).

Statistics

One-way analysis of variance was used to analyze the data. Multiple comparisons between groups were made by using the Student-Newman-Keuls method. For rejection of hypotheses, $P < 0.05$ was chosen (18).

RESULTS

Infection with *P. gingivalis* 381

Active bone loss requires infection with viable cells. Figure 1 compares sham-infected animals with animals receiving either viable *P. gingivalis* 381 cells or heat-killed cells of the same strain. Animals receiving living cells had significantly greater bone loss ($P < 0.001$) than animals receiving either of the other two treatments. When bone loss was measured as a function of time, as in Fig. 2, the highest rate occurred between day 10 and day 21, with a further gradual loss to day 42. When sham-infected animals were compared with *P. gingivalis*-infected animals, there were significant differences in bone loss from day 14 ($P < 0.01$) and continuing through day 42 ($P < 0.001$).

When serum and saliva antibody levels (Fig. 3) were measured, no increases due to infection with *P. gingivalis* were noted until day 14, when both titers rose sharply. Serum antibody titers continued to gradually rise to day 42. Salivary antibody titers, on the other hand, declined on day 21 and rose again on day 28. This cycle continued to the end of the experiment on day 42. It is of interest that *P. gingivalis* could be cultured (indicated in Fig. 3 by the small plus signs) from the gingival region only from initial infection until the first appearance of antibody in

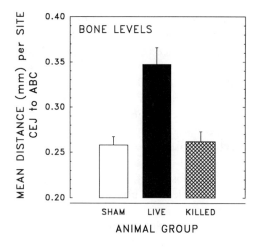

Figure 1. Animal groups were sham infected (□), or infected with viable *P. gingivalis* (■) or heat-killed *P. gingivalis* (▨). Horizontal bone loss was evaluated 42 days after the last infection.

the serum. The presence of *P. gingivalis* at the site of infection was confirmed by using a PCR specific for the fimbriae of strain 381 (1a). These findings suggest that immunity may be responsible for the disappearance of the infecting bacteria.

Infection with Nonfimbriated *P. gingivalis* Strains

The role of *P. gingivalis* fimbriae in infection was investigated by preparing mutants of strain 381 lacking this cell surface structure (10). When these mutant strains were used to infect the gnotobiotic rat, a markedly diminished capacity to induce bone loss was apparent (Fig. 4). Comparisons of horizontal bone loss between the wild-type 381 strain and sham-infected animals were significant at the $P < 0.001$ level. Comparisons between wild-type 381 and the mutant strains DPG-3 and DPG-5 were also significant at the $P < 0.001$ level. There was a significant difference in bone loss when DPG-3 was compared to DPG-5 ($P < 0.05$), which suggests a difference between the two mutants in their abilities to cause bone loss.

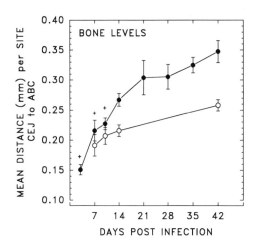

Figure 2. Animal groups were infected with *P. gingivalis* (●) or were sham infected (○). Horizontal bone loss was evaluated on the days indicated. Animal groups positive by culture and PCR for *P. gingivalis* are indicated by the small plus signs.

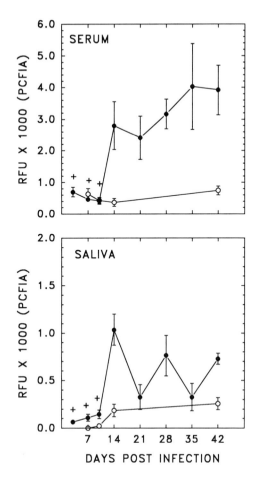

Figure 3. Serum and saliva antibody titers of the animal groups of Fig. 2 were determined by PCFIA. Titers are expressed as RFU.

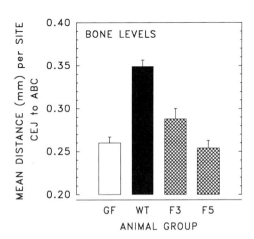

Figure 4. Animal groups were sham infected (GF) or infected with the mutant's parent strain (*P. gingivalis* 381 [WT]) or the fimbrial mutants DPG-3 (F3) and DPG-5 (F5). Horizontal bone loss was evaluated 42 days after the last infection.

Immunization Studies with Antigens from *P. gingivalis* 381

A variety of antigens were used to immunize rats against the effects of *P. gingivalis*. Figure 5 summarizes the results obtained with whole *P. gingivalis* cells, fimbrillin, a 75-kDa cell surface antigen, and a crude cell surface extract that contains both the fimbriae and the 75-kDa fraction. When bone levels (horizontal bone loss) were measured and compared to those in infected-only animals, immunizations with whole cells, cell surface extract, and the 43-kDa fraction were protective against bone destruction ($P < 0.001$). The 75-kDa component did not give evidence of protection. Similar results were seen when bone support (vertical

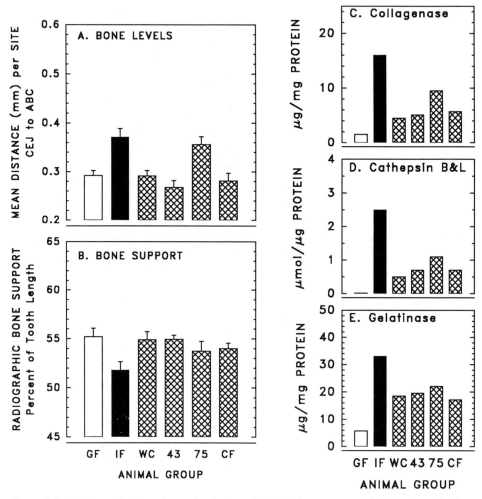

Figure 5. Protection against bone loss and reduction of gingival tissue enzyme activity by immunization. Animal groups were sham immunized and sham infected (GF), sham immunized and *P. gingivalis* infected (IF), whole-cell immunized and infected (WC), fimbrillin immunized and infected (43), 75-kDa-protein immunized and infected (75), or cell extract immunized and infected (CF).

bone loss) was measured, although in this instance there may be some marginal protection by the 75-kDa fraction (Fig. 5B). Gingival tissue enzymes were markedly affected by immunization with these antigens, with the greatest reductions noted for collagenase and the cysteine proteases cathepsin B&L (Fig. 5C, D, and E). When viewed together, these results suggest protection of soft tissue and alveolar bone.

When serum antibody reactions to fimbrillin (43 kDa) were measured (Fig. 6A), good responses were seen in animals immunized with whole cells and fimbrillin. Animals immunized with the 75-kDa component or cell surface extract had lesser reactions. When saliva was examined (Fig. 6C), animals receiving the 43-kDa protein gave the greatest response. When serum antibodies to the 75-kDa cell surface component were measured (Fig. 6B), animals immunized with whole cells or 75-kDa protein gave good responses, while animals receiving the 43-kDa protein and the cell surface extract gave lesser reactions. Interestingly, animals receiving the 75-kDa protein did not respond with high salivary antibody titers (Fig. 6D)

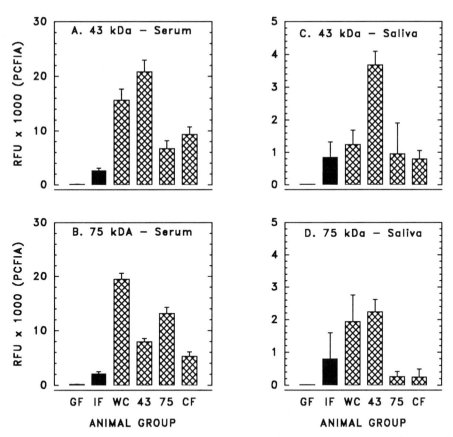

Figure 6. Serum and saliva titers against fimbriae (43 kDa) and the 75-kDa cell surface protein. Abbreviations for the animal groups are given in the legend to Fig. 5. Titers were determined by PCFIA and are expressed as RFU (see Fig. 3).

against this component. Because fimbriae are considered key virulence factors and because of the protective effect of fimbrillin (43-kDa protein) against bone loss and gingival tissue proteolytic enzymes, fimbrillin was chosen for further examination.

Protection against bone loss is antigen concentration dependent. Animals given graded doses of fimbrillin (Fig. 7) were fully protected at 20 and 6 μg per animal ($P < 0.001$). Animals receiving 0.2 μg of the antigen were also protected ($P < 0.05$) but to a lesser degree than animals receiving the higher antigen doses.

The protein structure of fimbrillin has been deduced from information obtained from genetic analysis of the fimbrillin gene of *P. gingivalis* 2561, which for infection is similar to strain 381 used in this study (9). Synthetic peptides based on this structure have been examined for their abilities to inhibit the in vitro attachment of *P. gingivalis* to saliva-coated hydroxyapatite crystals (9). One, a 20-mer peptide, was selected for use as an antigen. Immunizing gnotobiotic rats with the fimbrial peptide alone or with peptide plus a carrier protected the animals against horizontal bone loss (Fig. 8A). However, in contrast to results with the intact fimbrillin molecule, only marginal protection was noted against vertical bone loss (Fig. 8B). When used alone as an immunogen, the carrier did not affect either horizontal or vertical bone loss measurements. Gingival tissue enzyme levels were markedly reduced compared to those in the infected-only controls (Fig. 8C, D, and E). The greatest effect was on collagenase and cathepsin B&L.

Serum titers against fimbrillin, the 75-kDa component, and thyroglobulin were measured in animals immunized with the peptide preparation (Fig. 9). An anomalous antibody pattern was revealed. Contrary to expectations, the peptide-carrier-immunized animals did not have elevated titers. In each instance, the titer was lower than with peptide alone. The animals were fully competent to produce antibody, as seen with animals that received the thyroglobulin carrier only (Fig. 9C). As noted above, the animals respond to cellular antigens because of the infection alone, and the antibody responses seen because of immunization with the peptide may be additive to those responses.

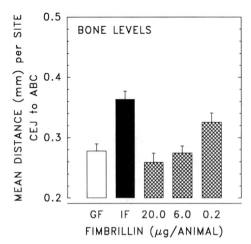

Figure 7. Horizontal bone loss in animals given graded doses (0.2, 6.0, or 20.0 μg per animal) of fimbrillin antigen.

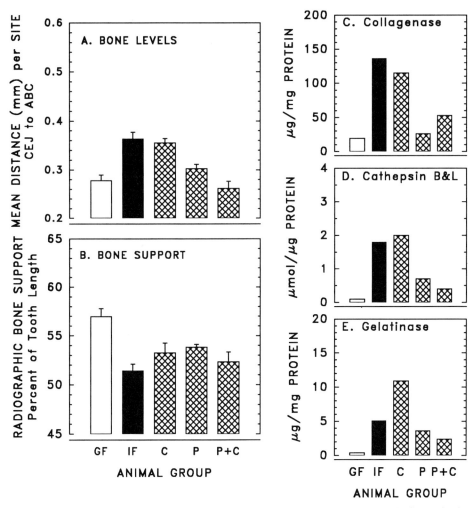

Figure 8. Protection against bone loss and reduction of gingival tissue enzyme activity by immunization with carrier (C), peptide (P), or peptide conjugated to the thryoglobulin carrier (P + C). GF and IF are germfree and infected controls (see Fig. 5).

DISCUSSION

Infection studies show clearly that living fimbriated *P. gingivalis* is required to induce periodontal destruction in the gnotobiotic rat. If the fimbriae are not present, the bacteria may not establish themselves, thus aborting the infection. Although difficult to culture, the wild-type bacterial cells are present at least 10 days after the final infection step, as shown by both direct recovery and PCR. However, when immunity to the infection is established, as evidenced by the appearance of antibody in serum and saliva, recovery of viable bacteria becomes very difficult. The cyclic nature of the salivary antibody response suggests the possibility of reestablishment

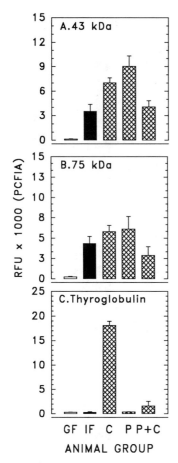

Figure 9. Serum antibody titers of animal groups of Fig. 8. Titers are expressed as RFU obtained by PCFIA.

of the bacteria in the oral cavity, with a concurrent restimulation of the antibody response.

When antigen preparations rich in fimbrial protein or a synthetic peptide based on the fimbrillin structure are used to immunize the rats, a high degree of protection against bone loss is obtained. This protection against bone loss is not seen when the 75-kDa surface component is used as antigen. The protective effect of immunization is also seen when the proteolytic enzymes in gingival tissues are examined. Previous studies showed that increases in collagenase and gelatinase activities in gingival tissues are of mammalian and not bacterial origin (3, 4, 8). When the animals are successfully immunized with the fimbrial proteins or the peptide, there is a marked reduction in this activity.

SUMMARY

The development of a safe and effective vaccine against *P. gingivalis* oral infection will require the identification of critical antigens that induce immunity specifically targeted to the infecting bacterium. The antigen must be free of unwanted side effects to gain acceptance as a safe public health measure. This goal can be achieved by using a completely synthetic molecule with only those epitopes required to induce protection. Progress in this direction is shown in the preliminary studies reported here using a peptide derived from *P. gingivalis* fimbriae.

Acknowledgments. This work was supported in part by grants DE 04898, DE 07034, DE 08240, and DE 03987 from the National Institutes of Health and by The Procter and Gamble Co. and Dandyfonden.

We thank Cornelia Sfintescu and Alice Wendt for their expert technical contributions.

REFERENCES

1. **Dahlen, G., and J. Slots.** 1989. Experimental infections by *Bacteroides gingivalis* in nonimmunized and immunized rabbits. *Oral Microbiol. Immunol.* **4:**6–11.

1a. **Evans, M. J., et al.** Unpublished results.

2. **Evans, R. T., B. Klausen, and R. J. Genco.** 1992. Immunization with fimbrial protein and peptide protects against *Porphyromonas gingivalis*-induced periodontal tissue destruction, p. 255–262. *In* J. Ciardi, J. Keith, and J. McGhee (ed.), *Genetically Engineered Vaccines: Prospects for Oral Disease Prevention.* Plenum Press, New York.

3. **Evans, R. T., B. Klausen, N. S. Ramamurthy, L. M. Golub, C. Sfintescu, and R. J. Genco.** 1992. Periodontopathic potential of two strains of *Porphyromonas gingivalis* in gnotobiotic rats. *Arch. Oral Biol.* **37:**813–819.

4. **Evans, R. T., B. Klausen, H. T. Sojar, G. S. Bedi, C. Sfintescu, N. S. Ramamurthy, L. M. Golub, and R. J. Genco.** 1992. Immunization with *Porphyromonas* (*Bacteroides*) *gingivalis* fimbriae protects against periodontal destruction. *Infect. Immun.* **60:**2926–2935.

5. **Genco, C. A., C. W. Cutler, D. Kapczynski, K. Maloney, and R. R. Arnold.** 1991. A novel mouse model to study the virulence of the host response to *Porphyromonas* (*Bacteroides*) *gingivalis. Infect. Immun.* **59:**1255–1263.

6. **Genco, R. J., J. J. Zambon, and L. A. Christersson.** 1988. The origin of periodontal infections. *Adv. Dent. Res.* **2:**245–259.

7. **Klausen, B.** 1991. Microbiological and immunological aspects of experimental periodontal disease in rats. *J. Periodontol.* **62:**59–73.

8. **Klausen, B., R. T. Evans, N. S. Ramamurthy, L. M. Golub, C. Sfintescu, J.-Y. Lee, G. Bedi, J. J. Zambon, and R. J. Genco.** 1991. Periodontal bone level and gingival proteinase activity in gnotobiotic rats immunized with *Bacteroides gingivalis. Oral Microbiol. Immunol.* **6:**193–201.

9. **Lee, J.-Y., H. T. Sojar, G. S. Bedi, and R. J. Genco.** 1992. Synthetic peptides analogous to the fimbrillin sequence inhibit adherence of *Porphyromonas gingivalis. Infect. Immun.* **60:**1662–1670.

10. **Malek, R., J. G. Fisher, A. Caleca, M. Stinson, C. J. van Oss, J.-Y. Lee, M.-I. Cho, R. J. Genco, R. T. Evans, and D. W. Dyer.** 1994. Inactivation of the *Porphyromonas gingivalis fimA* gene blocks periodontal damage in gnotobiotic rats. *J. Bacteriol.* **176:**1052–1059.

11. **McArthur, W. P., I. Magnusson, R. G. Marks, and W. B. Clark.** 1989. Modulation of colonization by black-pigmented *Bacteroides* species in squirrel monkeys by immunization with *Bacteriodes gingivalis. Infect. Immun.* **57:**2313–2317.

12. **Okuda, K., T. Kato, Y. Naito, I. Takazoe, Y. Kikuchi, T. Nakamura, T. Kiyoshige, and S. Sasaki.** 1988. Protective efficacy of active and passive immunization against experimental infection with *Bacteroides gingivalis* in ligated hamsters. *J. Dent. Res.* **67:**807–811.

13. **Page, R. C., and H. E. Schroeder.** 1982. *Periodontitis in Man and Other Animals: a Comparative Review*, p. 187–212. Karger, Basel.

14. **Slots, J., and G. Dahlen.** 1985. Subgingival microorganisms and bacterial virulence factors in periodontitis. *Scand. J. Dent. Res.* **93:**119–127.

15. **Slots, J., and R. J Genco.** 1984. Black pigmented *Bacteroides* species, *Capnocytophaga* species,

and *Actinobacillus actinomycetemcomitans* in human periodontal disease: virulence factors in colonization, survival, and tissue destruction. *J. Dent. Res.* **63:**412–421.

16. **van Steenbergen, T. J. M., P. Kastelein, J. J. A. Touw, and J. DeGraaf.** 1982. Virulence of black-pigmented *Bacteroides* strains from periodontal pockets and other sites in experimentally induced skin lesions in mice. *J. Periodontal Res.* **17:**41–49.

17. **Warren, K. S.** 1986. New scientific opportunities and old obstacles in vaccine development. *Proc. Natl. Acad. Sci. USA* **83:**9275–9277.

18. **Zar, J. H.** 1984. *Biostatistical Analysis,* 2nd ed., p. 186–190. Prentice Hall, Inc., Englewood Cliffs, N.J.

Molecular Pathogenesis of Periodontal Disease
Edited by Robert Genco et al.
© 1994 American Society for Microbiology, Washington, DC 20005

Chapter 24

T-Cell and B-Cell Epitope Mapping and Construction of Peptide Vaccines

T. Lehner, J. K.-C. Ma, G. Munro, P. Walker, A. Childerstone, S. Todryk, H. Kendal, and C. G. Kelly

Rapid developments in molecular biotechnology, in the cloning and sequencing of proteins, have led to important advances in immunology. The precise localization of T- and B-cell epitopes on antigens has been the subject of many investigations (54). The T-cell epitope is the smallest peptide recognized by T cells, and the B-cell epitope is the smallest peptide recognized by antibodies. However, in oral microbial infection, an additional requirement is for the microorganism to adhere to a salivary glycoprotein on the tooth surface or to the epithelial cells (19). Adherence between a microorganism and host tissue is a ligand-receptor interaction mediated by adhesion epitopes on the microorganism. Mapping the T-cell, B-cell, and adhesion epitopes (EPITAB mapping) is essential in the development of vaccines.

This chapter has four objectives: (i) to discuss the structure and function of synthetic peptides as applied to T- and B-cell immune responses; (ii) to apply these principles to oral bacteria and bacterial adhesion; (iii) to discuss studies of the antigenicity and immunogenicity of synthetic peptide constructs; and (iv) to discuss EPITAB mapping of a cell surface streptococcal antigen.

GENERAL PROPERTIES OF SYNTHETIC PEPTIDES

Role of Size and Shape of Peptides

Although antibodies can recognize very small peptides (two or three residues), functional epitopes are usually larger than 6 amino acids. Most synthetic peptides are linear, but antibodies recognize conformational structures, so that discontinuous epitopes with a folded peptide are likely to be more important than linear peptides.

T. Lehner, J. K.-C. Ma, G. Munro, P. Walker, A. Childerstone, S. Todryk, H. Kendal, and C. G. Kelly • Department of Immunology, United Medical and Dental Schools, Guy's and St. Thomas' Hospitals, Guy's Tower, Floor 28, London Bridge, London SE1 9RT, United Kingdom.

For this reason, some synthetic peptides are cyclized in order to attempt to impart a conformational structure to the peptides and improve their immunogenicity (1). While recognition of a peptide by antibodies is dependent on the conformation of the peptide, T-cell receptors are less stringent and recognize denatured proteins. Some residues within a peptide are essential and cannot be replaced with another amino acid without loss of immunogenicity (17). Other residues are not essential; they can be replaced with other amino acids, and they may be involved in the scaffolding of the peptide.

Antigenicity and Immunogenicity of Peptides

Recognition of a peptide by antibodies or T cells is referred to as antigenicity of the peptide. However, it should be appreciated that antigenic peptides do not necessarily induce an immune response; i.e., antigenicity does not imply immunogenicity of the peptide. This is particularly important in vaccine design, as T- and B-cell epitopes are first mapped by being recognized by T cells or antibodies. Whether these epitopes are immunogenic must then be established by immunization experiments in animals. Small peptides (less than 10 to 15 residues) are generally poor immunogens and may require a carrier molecule, such as albumin.

T- and B-cell epitopes may be distinct, as they are in hepatitis B surface antigen, in which the T-cell (residues 120 to 132) and B-cell (residues 133 to 145) epitopes are contiguous (42). Other epitopes can be overlapping, as with glycoprotein D of herpes simplex virus, in which a B-cell epitope (residues 1 to 23) overlaps two T-cell epitopes (residues 1 to 16 and 8 to 23) (21). Depending on the antigen, there may be several T- and B-cell epitopes, some of them immunodominant and others minor (or silent), leading to a hierarchy of epitopes (15, 49). When synthetic peptide T- and B-cell epitopes are constructed, the sequence of T and B or the reverse B and T determinants may be important in eliciting an immune response, as demonstrated with the hepatitis B surface antigen (42). The immunogenicity of some peptides can be enhanced by polymerization of the B- or T-cell epitope, as was found with the tetramer (NANP) of the circumsporozoite protein of *Plasmodium falciparum* (43), the polymer of the T-cell epitope of foot-and-mouth disease virus (14), and the dimer of streptococcal peptide 1-17 (34).

T-Cell Epitope Restriction in MHC Class I and Promiscuity in MHC Class II Responses

T cells do not recognize a peptide alone but only in association with the major histocompatibility complex (MHC) class I or class II antigen. Indeed, the MHC restriction epitope (referred to as an aggretope) of a peptide binding to an MHC class I antigen is a prerequisite in class I-restricted $CD8_+$ cytotoxic cell reactions. The T-cell epitope consists of about 9 residues and is anchored to defined residues in the groove of the class I molecule, which is closed at both ends (2, 12, 40). Although a similar scheme is envisaged in MHC class II antigen-peptide interaction, there are at least three significant differences. The peptide is longer (about 12 to 15 residues) and shows heterogeneity at the carboxy terminus, suggesting that the class II peptide-binding groove may be open at one end (50). Furthermore, unlike class I antigens, class II antigens tend to be promiscuous in that several MHC class II antigens will bind a peptide, as shown with a 21-residue peptide from the circumsporozoite antigen of *P. falciparum* (56), tetanus toxin (TT) (48), and streptococcal antigen (SA) (4).

Functional Epitopes

In addition to the T- and B-cell epitopes recognized by T cells and antibodies, adhesion epitopes (adhesins) have been defined on microbial surfaces (22). Several adhesion molecules have been described for *Escherichia coli* (22) and *Bordetella pertussis* (53), and in some instances, the domains that recognize the receptors have been determined (35, 36). Adhesins commonly recognize carbohydrate receptors such as galactose, mannose, or sialic acid on eucaryotic glycoconjugates (24). Other epitopes are defined operationally by their functions, e.g., neutralizing epitope on toxins recognized by antibodies and cytotoxic epitopes recognized by T cells.

Damaging Epitopes

Recognition of tissue cross-reactive epitopes is important in designing peptide vaccines. The best examples are M proteins of group A streptococci, which may induce rheumatic carditis and glomerulonephritis by tissue cross-reactive epitopes (3). These common heart or glomerular epitopes have been identified and can be differentiated from protective epitopes so that a vaccine can be designed to elicit only protective immunity.

Designer Vaccine

Advances in recognition of a variety of functional epitopes, matched with the ease of peptide synthesis, make synthetic peptides an attractive field of investigation. For any one organism, the desired epitopes need to be identified; thus, T-helper, cytotoxic, and B-neutralizing epitopes for human immunodeficiency virus might have the right components to prevent infection by this virus. The problem with peptide vaccines is HLA class I restriction, which may require a number of peptides to cater for different HLA class I gene products. However, this may prove to be less of a problem if class I-restricted cytotoxic cells are not essential and if the less stringent HLA class II helper/inducer function for antibody synthesis is the desired response in protection against the microbial agent. Thus, peptide 141-160 from VP1 of foot-and-mouth disease viruses induces protection in cattle (13).

Any tissue cross-reactive epitope can be determined and excluded from the peptide vaccine so as to avoid an autoimmune reaction. The T- and B-cell epitopes can be coupled to adhesion epitopes with the objective of preventing bacterial adhesion to mucosal surfaces. Problems of immunogenicity of peptide vaccines can be overcome by judicious selection of immunodominant epitopes, determination of the correct N- and C-terminal orientation of the elective epitopes, and selection of an appropriate adjuvant. Mucosal immunization can now be pursued by utilizing cholera toxin B subunit (23), liposomes (16), or live carriers such as salmonellae (9). A powerful means of enhancing the immunogenicity of a peptide vaccine is to polymerize some of its component epitopes, as has been achieved with the malarial peptide (43), foot-and-mouth disease peptide (13), and streptococcal peptide (33). An objective that must be strictly achieved is that the T- and B-cell responses are directed not only to the peptides but also to the native antigen and the intact microorganism.

SYNTHETIC PEPTIDES APPLIED TO ORAL MICROORGANISMS

We wish to propose the principle of EPITAB mapping, which might prove significant in the development of peptide or recombinant vaccines against dental caries and periodontal disease. EPITAB mapping identifies the desired immunodominant T- and B-cell epitopes as well as adhesion epitopes against which both immunoglobulin G (IgG) and secretory IgA antibodies may be directed to prevent adherence of oral bacteria to salivary glycoproteins or mucosal receptors. The T-cell epitope is essential, as both IgG and IgA antibodies are T cell dependent. For an immune response to be promptly recalled when the bacterial agent is encountered, an effective T- and B-cell memory is essential. However, the critical factor in designing vaccines against bacterial adhesion has so far not been addressed. It is not known whether antibodies generated by *Porphyromonas gingivalis* or *Streptococcus mutans* are primarily directed against adhesion epitopes or some other B-cell epitopes or indeed whether adhesion epitopes are specialized components of the B- or T-cell epitope repertoire. We have been investigating the relationship between T-cell, B-cell, and adhesion epitopes and will illustrate some of the findings with reference to cell surface antigens of *S. mutans*. The objectives are to construct an immunogenic polypeptide or recombinant protein expressing the three immunodominant epitopes that will elicit an effective T- and B-cell response to the organism and demonstrate a prompt memory.

SYNTHETIC PEPTIDE CONSTRUCT OF A 3.8-kDa SA

A 3.8-kDa SA that copurifies with the 185-kDa SAI/II has been isolated from *S. mutans* (18). A series of experiments established that the 3.8-kDa SA elicits IgG antibodies and T-cell helper function and, like the 185-kDa antigen, prevents colonization of *S. mutans* and the development of dental caries (30). The amino-terminal portion of the 3.8-kDa antigen was sequenced, and peptides with 15, 17, and 21 residues were synthesized (20). Naturally sensitized human subjects showed T-cell proliferative responses and serum IgG antibodies to the synthetic peptides (4–6).

Determination of B- and T-Cell Epitopes

A series of small peptides with progressive deletions at their amino and carboxy termini (residues 1 to 15) were prepared and used in a radioassay with human sera or in a proliferative assay with human mononuclear cells (4). The smallest peptide (residues 8 to 13) that bound significant antibodies was considered to constitute a B-cell epitope and was then used in an inhibition assay (5). Residues 8 to 13 consistently inhibited antibodies to peptide 1-15 ($P < 0.001$). The shortest synthetic peptide able to induce a significant T-cell proliferative response was peptide 6-15, which was designated a T-cell epitope (Fig. 1).

Immunogenicity of T- and B-Cell Epitopes

Immunization with the linear or cyclized peptide (residues 1 to 15) failed to elicit serum antibodies or a significant T-cell proliferative response (34). The synthetic peptide was then covalently linked to TT, which induced significant serum IgG antibodies not only to the peptide but also to the native SA (and TT) and not to a random peptide. The failure to elicit antibodies and T-cell responses by the free peptide might be because the peptide was nonimmunogenic or because it induced tolerance. In order to render the peptide immunogenic, it was dimerized by peptide or disulfide linkage and the dimers were used to immunize nonhuman primates without a carrier (34). The dimers elicited significant serum antibodies, and [³H]thymidine uptake of lymphocytes was stimulated in vitro by the dimers or the monomers but not by the random synthetic peptide. There was no difference in the response to immunization between the peptide-linked and disulfide-linked dimers. Truncated peptides were used in order to determine the shortest peptide that would bind the specific antipeptide antibodies. The shortest peptide that reacted with the antibodies was residues 8 to 13 but not residues 1 to 7 (Fig. 1). Inhibition studies with a number of synthetic peptides and the native SA confirmed that only residues 8 to 13 were capable of inhibiting antibodies to the peptide and the native antigen. This suggests that peptide 8-13 is recognized not only by antibodies to the synthetic peptide but also by antibodies to the native SA.

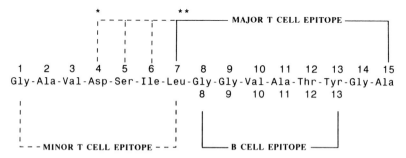

Figure 1. Mapping of T- and B-cell epitopes of a peptide (residues 1 to 15) derived from an SA (3,800 M_r) and comparison of natural immunization in humans with artificial immunization in nonhuman primates.*, Human DR2 and DR3 require Ile (residue 6); DR1 and DR7 require Ser (residue 5); and DR4, DR5, and DR6 require Asp (residue 4).**, nonhuman primates.

Peripheral blood mononuclear cells were then stimulated with peptide 1-15 and the truncated peptides. The shortest peptide that elicited a significant uptake of [^{3}H]thymidine was residues 7 to 15, so this peptide was designated a T-cell epitope (Fig. 1). A comparative investigation of natural immunity in human subjects and artificial immunity in nonhuman primates revealed significant concordance in the T- and B-cell epitopes. The T-cell epitope following artificial immunization in nonhuman primates consisted of residues 7 to 15, which lacks only isoleucine when compared with the residues 6 to 15 T-cell epitope found in naturally immunized humans (Fig. 1). The additional isoleucine at the N terminus appeared to be required for antigenicity of the T-cell epitope after natural immunization in humans. This is probably the first instance in which the requirement of a T-cell epitope has been compared in natural and artificial immunization in primates.

Restriction by HLA class II antigens was studied in human subjects (4). T cells from subjects expressing each of the alleles from DR1 to DR7 responded to synthetic peptides of 15 amino acid residues. Furthermore, inhibition studies with monoclonal antibodies (MAb) to HLA class I and class II antigen showed that the DR antigen was a restriction molecule for the proliferative responses. Mouse L cells transfected with DR1, DR2, DR4, DR5, and DR7 were used to confirm the permissive nature of the responses. An analysis of the fine specificity of the responses showed that the smallest peptides capable of stimulating T cells from subjects with different DR types varied by 1 to 3 residues. For DR2 and DR3, the shortest peptide was residues 6 to 15; an additional serine (residue 5) was required for DR1 and DR7; and an aspartic acid (residue 4) was required for DR4, DR5, and DR6 (Fig. 1). HLA restriction was also studied in macaques immunized with the dimerized synthetic peptide, and restriction within the macaque MHC type A, B, and DR antigens was not detected (34). Successful oral mucosal colonization of humans by a largely commensal streptococcus might be associated with the permissive nature of HLA-DR restriction of the response to a major streptococcal cell surface peptide. The peptide recognized in association with the HLA-DR molecule may induce an immune response that prevents central entry of the organism from the peripheral mucosal site.

Mapping of Major and Minor T-Cell Epitopes In Vitro and Their Immunogenic or Tolerogenic Effect In Vivo

In vitro and in vivo evidence suggests that peptide 1-15 contains T-cell (residues 7 to 15) and B-cell (residues 8 to 13) epitopes but is immunogenic only if dimerized [(1-15$_2$)] or linked to the carrier TT [(1-15TT)] (34). Monomers and dimers of T- and B-cell epitopes were prepared and used to immunize macaques. Immunogenicity was assayed in lymphocytes by the proliferative response and for serum antibodies by a solid-phase radioimmunoassay. Macaques immunized with the dimerized (1-15)$_2$ or carrier-linked (1-15)TT peptide exhibited in vitro T-cell proliferative responses to peptides 1-15 and 7-15 (Tables 1 and 2). T cells from animals immunized with peptide 1-15 or 7-15 or the dimer (7-15)$_2$ failed to elicit an in vitro proliferative response (61). In order to establish whether these nonimmunogenic peptides might induce tolerance, the same macaques were challenged with the

Table 1. Immunogenicity of T- and B-cell epitopes

Epitope	Peptide	Immunogenicity	
		T cells[a]	B cells[b]
T + B	1-15	1.3	0.7
T + B	7-15	1.1	0.5
B	8-13	1.1	0.5
Cryptic T	1-7	0.7	0.4
$(T + B)_2$	$(1\text{-}15)_2$	7.7	10.4
$(T + B)_2$	$(7\text{-}15)_2$	1.3	0.4
$(B)_2$	$(8\text{-}13)_2$	0.6	0.3
Cryptic T	$(1\text{-}7)_2$	1.1	0.6

[a] Proliferative response (stimulation index) to peptide 1-15.
[b] IgG antibody response to peptide 1-15, as evidenced by percent ^{125}I binding.

immunogenic peptide (1-15)TT. The results suggest that T-cell responses to peptide 1-15 were reestablished but that instead of responding to peptide 7-15, they were stimulated by a hitherto silent epitope (residues 1 to 7). Tolerance to the major T-cell epitope (residues 7 to 15) and the expression of a minor (cryptic) T-cell epitope (residues 1 to 7) were associated with B-cell tolerance, suggesting that T-cell help for antibodies resides in the major T-cell epitope (residues 7 to 15). However, short-term T-cell lines revealed T-cell responses to peptides 1-7 and 7-15 in both tolerized and immunized macaques, but the relative frequency of the minor epitope (residues 1 to 7) reactive lines was significantly higher in tolerized animals, while that for the major epitope (residues 7 to 15) was higher in immunized animals (61). These findings suggest that the silent epitope (residues 1 to 7) is really cryptic in that it can be detected if the cell lines are first expanded in vitro with the whole of peptide 1-15 and then stimulated with the truncated peptide 1-7 or 7-15. The results are consistent with the concept of a hierarchy of major and minor T-cell epitopes (55), now demonstrated in nonhuman primates in which tolerance to the major T-cell epitope is associated with tolerance to antibody formation and the emergence of a minor T-cell epitope.

Gingival Immunization with Synthetic Peptide Construct of T- and B-Cell Epitopes

Gingival immunization in macaques was first attempted with the native SA (30). The 3.8-kDa SA but not the 185-kDa SA elicited both salivary IgA and

Table 2. Tolerization of major T-cell epitope

Epitope	Peptide	Stimulation index		Frequency of STCL[a]	
		Peptide 7-15	Peptide 1-7	Peptide 7-15	Peptide 1-7
$(T + B)_2$	$(1\text{-}15)_2$	9.4	1.5	10 / 68 (14.7)	6 / 68 (8.8)
(T + B)	(1-15)	0.9	0.4	—	—
(T + B)TT	(1-15)TT	3.4	1.0	—	—
(T + B)	(1-15)				
(T + B)TT	(1-15)TT	0.8	3.0	4 / 58 (6.9)	18 / 87 (20.7)

[a] STCL, short-term cell reactive lines/total no. of lines (%). —, not done.

gingival IgG antibodies to the SA. This was associated with a significant decrease in colonization of S. mutans and a decreased incidence of dental caries. The peptide construct of T- and B-cell epitopes was then studied by topical gingival application (six times) of three synthetic peptides consisting of 17 or 21 residues and a dimer of 35 residues (34). Significant antibodies to these peptides as well as to the native SA were elicited in the gingival fluid and saliva. The functional significance of this immune response was examined by studying its effect on oral colonization of S. mutans following feeding of animals with a carbohydrate-rich diet. Whereas control animals sham immunized with a random peptide of 11 residues showed increased colonization of teeth by S. mutans, there was no colonization or a significant reduction in colonization of animals immunized with the cyclized 17-mer, linear 21-mer, or dimerized 35-mer synthetic peptides. These experiments suggest that local immunization with peptides derived from the sequences of a cell surface SA induces a dual local immune response of gingival IgG and salivary IgA antibodies against the peptide and native SA. These antibodies may be involved in preventing colonization of S. mutans, which is the principal agent in the development of dental caries.

EPITAB MAPPING OF 185-kDa SA (SA I/II)

The 185-kDa SA I/II has been identified (51), cloned, and sequenced (25, 28, 46, 47); it mediates adherence of S. mutans in vitro (29, 41) and functions as a virulence factor (8). The SA also prevents colonization of S. mutans and the development of caries after active immunization (33), as do epitope-specific MAb to the SA after passive immunization (31, 32, 37, 39). This SA has been extensively investigated and can be utilized as a prototype in designing a vaccine against oral bacteria. The gene encoding SA I/II from S. mutans serotype c, termed spaP, has been cloned and sequenced from two strains of S. mutans (25, 47), and only minor differences between the strains were demonstrated. The deduced amino acid sequence of SA I/II from S. mutans comprises 1,561 residues (Fig. 2). The homologous genes from other mutans streptococci have been cloned and sequenced (27, 45, 60), as has the gene from Streptococcus gordonii (10, 11). The deduced amino acid sequences of SA I/II from these streptococcal species show considerable conservation (11, 27), suggesting a family of cell surface proteins. Indeed, hybridization studies using DNA probes generated by PCR showed that regions of the gene encoding SA I/II were highly conserved, not only in mutans streptococci but also in nonmutans α-hemolytic streptococci (38). In particular, a probe from the region of the SA I/II gene that encodes residues 857 to 1207 hydridized with DNA from representative strains of 19 streptococcal species.

We suggested a model for the structure of SA I/II (Fig. 2) in which the N-terminal region, consisting of four tandem repeats of an alanine-rich sequence, adopts an α-helical coiled coil conformation to form a stalk supporting one or more membrane-distal domains formed by the C-terminal region (27). The adhesion binding site for salivary receptors is suggested to be located in the membrane-distal region. This region may be further subdivided on the basis of sequence comparison

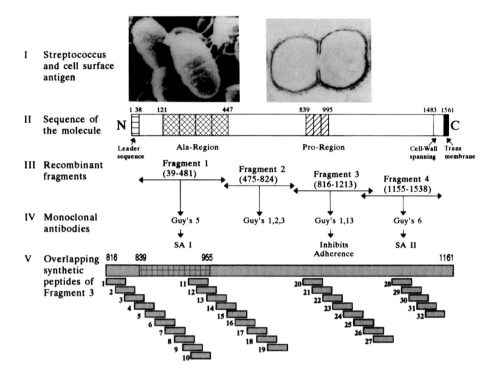

Figure 2. Strategy of EPITAB mapping in bacterial antigens.

(27), with the least conserved portion occurring between residues 448 and 815 and the most conserved region between residues 816 and 1213 (Fig. 2).

In order to identify B-cell and adhesion epitopes, four fragments of the SA I/II gene from *S. mutans* serotype c (strain NG5), which encode overlapping regions of the sequence, were cloned and expressed (44). The four recombinant polypeptides were then examined by a panel of MAb that were raised against SA I/II to determine the pattern of recognition (58, 59) and characterized in vitro and in vivo (31, 32, 37, 39). Fragment 1 was recognized by the MAb Guy's 5, which recognizes the N-terminal antigen I component of SA I/II (57). Fragment 2 was recognized by MAb Guy's 1, 2, and 3 (Fig. 2). Fragment 3 was recognized by Guy's 1 and 13. Fragment 4 was recognized by Guy's 6, which was raised against the C-terminal antigen II component of SA I/II (57). It is noteworthy that Guy's 11, which does not inhibit adherence of *S. mutans* in vivo (37), failed to react with fragment 3 or indeed with any of the other three fragments, although it reacted with SA I/II. These investigations established that Guy's 1 and 13, which prevent adhesion of *S. mutans* and the development of caries in vivo (30, 32, 37), recognize conserved fragment 3 (residues 816 to 1213). In contrast, Guy's 11 does not prevent *S. mutans* adhesion in vivo and does not recognize fragment 3.

The four recombinant polypeptides were then assayed for adhesion to salivary receptors (44). A competitive inhibition assay was used in which the recombinant polypeptides or control proteins were incubated with saliva-coated hydroxyapatite prior to addition of bacteria. SA I/II inhibited binding of *S. mutans* in a dose-dependent manner, with maximal inhibition of 80% at a concentration of approximately 6 μM. Streptococcal antigen III (51), also termed antigen A (52), a surface component of M_r 39,000 derived from *S. mutans* (serotype c), failed to inhibit adhesion. Similarly, incubation of saliva-coated hydroxyapatite with ovalbumin, which has an M_r similar to those of the recombinant fragments, failed to inhibit *S. mutans* binding. Of the recombinant fragments, only fragment 3 (residues 816 to 1213) inhibited *S. mutans* binding in a dose-dependent manner, with maximal inhibition of 65% at a concentration of approximately 25 μM (Fig. 3). The results of this adhesion inhibition assay are consistent with the findings when the panel of MAb was used, that the adhesion epitope(s) resides within the recombinant fragment 3 polypeptide. Two independent approaches have thus identified the region comprising residues 816 to 1213 as important for colonization.

Fragments of relatively high M_r were chosen for expression, since these may be more likely to adopt a conformation similar to that in the intact native protein. The involvement of sites other than fragment 3 in adhesion cannot be ruled out on the basis of the data presented in this study, since the fragments may not adopt the same conformation as in the intact protein. Furthermore, binding domains may be destroyed as a result of proteolytic digestion during fragment purification. Fragment 3 is formed in part by a series of three tandem repeats of a 39-residue proline-rich sequence. Although the conformation adopted by this sequence is not known, it is likely to be a well-defined structure and so may contribute to the folding of fragment 3 to form a functional adhesion binding site. That recombinant fragments of SA I/II may fold to form the adhesion binding site was also demonstrated by

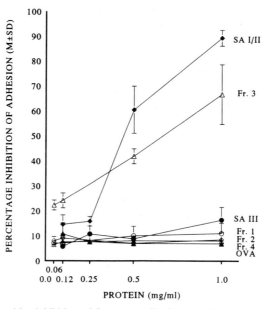

Figure 3. Competitive inhibition of *S. mutans* adhesion to saliva-coated hydroxyapatite.

recent studies in which the fusion protein of residues 18 to 469 linked to the maltose binding protein of *E. coli* (7) inhibited binding of *S. mutans* to saliva-coated hydroxyapatite.

Localization of adhesion to fragment 3 led to the preparation of 32 overlapping synthetic peptides of 20 residues, of which each overlaps by 10 residues (Fig. 2). These peptides are now being used to map the precise localization of T-cell, B-cell, and adhesion epitopes by means of proliferative responses, antibody binding, and adhesion inhibition, respectively (26). Preliminary results have confirmed that several T-cell, B-cell, and adhesion epitopes are found within the polypeptide residues of fragment 3 (26). As soon as mapping of these epitopes is completed, the T-A-B polypeptide or its recombinant counterpart will be constructed and its immunogenicity will be examined. The three-component designer polypeptide will then be ready to be tested for prevention of colonization of *S. mutans* and the development of caries.

The strategy of identifying T-cell, B-cell, and adhesion epitopes by a combination of well-characterized MAb and overlapping synthetic peptides derived from the sequence of the protein might be applied to other microbial pathogens. However, a successful outcome of this approach will depend on careful screening of peptides and MAb preventing adhesion of the pathogen in vitro, followed by prevention of colonization of the pathogen and development of the disease in vivo.

Acknowledgments. This work was carried out with the support of a number of Medical Research Council project grants.

We thank I. Moro for the electron micrographs used in Fig. 2.

REFERENCES

1. **Arnon, R. E., E. Maron, M. Sela, and C. B. Anfinsen.** 1971. Antibodies reactive with native lysozyme elicited by a completely synthetic antigen. *Proc. Natl. Acad. Sci. USA* **68:**1450–1455.

2. **Bjorkman, P. J., M. A. Saper, B. Samraoui, W. S. Bennett, J. L. Strominger, and D. C. Wiley.** 1987. Structure of the human class I histocompatibility antigen HLA-A2. *Nature* (London) **329:**506–512.

3. **Bronze, M. S., E. H. Beachy, and J. B. Dale.** 1988. Protective and heart-cross reactive epitopes located within the NH_2 terminus of type 19 streptococcal M protein. Structural basis of internal image expression. *J. Exp. Med.* **167:**1849–1859.

4. **Childerstone, A., D. Altmann, J. A. Haron, D. Wilkinson, J. Trowsdale, and T. Lehner.** 1991. An analysis of synthetic peptide restriction by HLA-DR alleles in T cells from human subjects, naturally sensitized by *Streptococcus mutans*. *J. Immunol.* **146:**1463–1469.

5. **Childerstone, A., J. Haron, and T. Lehner.** 1990. The reactivity of naturally sensitized human CD4 cells and IgG antibodies to synthetic peptides derived from the amino terminal sequences of a 3800 MW *Streptococcus mutans* antigen. *Immunology* **69:**177–183.

6. **Childerstone, A., J. A. Haron, and T. Lehner.** 1989. T cell interactions generated by synthetic peptides covalently linked to a carrier. *Eur. J. Immunol.* **19:**169–176.

7. **Crowley, P. J., L. J. Brady, D. A. Piacentini, and A. S. Bleiweis.** 1993. Identification of a salivary agglutinin-binding domain within cell surface adhesion P1 of *Streptococcus mutans*. *Infect. Immun.* **61:**1547–1552.

8. **Curtiss, R., III.** 1985. Genetic analysis of *Streptococcus mutans* virulence. *Curr. Top. Microbiol. Immunol.* **118:**253–277.

9. **Curtiss, R., III, R. M. Goldschmidt, N. B. Fletchall, and S. M. Kelly.** 1988. A virulent *Salmonella*

typhimurium δcya δcrp oral vaccine strains expressing streptoccoccal colonization and virulence antigen. *Vaccine* **6:**155–160.

10. **Demuth, D. R., C. A. Davis, A. M. Corner, R. J. Lamont, P. S. Leboy, and D. Malamud.** 1988. Cloning and expression of a *Streptococcus sanguis* surface antigen that interacts with a human salivary agglutinin. *Infect. Immun.* **56:**2484–2490.

11. **Demuth, D. R., E. E. Golub, and D. Malamud.** 1990. Streptococcal-host interactions. Structural and functional analysis of a *Streptococcus sanguis* receptor for a human salivary glycoprotein. *J. Biol. Chem.* **265:**7120–7126.

12. **Falk, K., O. Rotzschke, S. Stevanovic, G. Jong, and H. G. Rammensee.** 1991. Allele specific motifs revealed by sequencing of self-peptides eluted from MHC molecules. *Nature* (London) **351:**290–296.

13. **Francis, M. J., C. Fry, D. J. Rowlands, F. Brown, J. L. Bittle, R. A. Houghton, and R. A. Lerner.** 1985. Immunological priming with synthetic peptides of foot and mouth disease virus. *J. Gen. Virol.* **66:**2347.

14. **Francis, M. J., C. Fry, D. J. Rowlands, F. Brown, J. L. Bittle, R. A. Houghton, and R. A. Lerner.** 1987. Immune responses to uncoupled peptides of foot and mouth disease. *Immunology* **61:**1.

15. **Gammon, G., and E. Sercarz.** 1989. How some T cells escape tolerance induction. *Nature* (London) **342:**183–189.

16. **Garcon, N. M., and H. R. Six.** 1991. Universal vaccine carrier: liposomes that provide T-dependent help to weak antigens. *J. Immunol.* **146:**3697–3702.

17. **Getzoff, E. D., J. A. Tainer, R. A. Lerner, and H. M. Geysen.** 1988. The chemistry and mechanism of antibody binding to protein antigens. *Adv. Immunol.* **43:**1–98.

18. **Giasuddin, A. S. N., T. Lehner, and R. W. Evans.** 1983. Identification, purification and characterisation of a streptococcal protein with a molecular weight of 3800. *Immunology* **50:**651.

19. **Gibbons, R. J., and J. van Houte.** 1980. *In* E. H. Beachy (ed.), *Bacterial Adherence*, p. 61–104. Chapman and Hall, London.

20. **Haron, J., C. Bohart, L. Staffileno, P. Van Hook, and T. Lehner.** 1988. Humoral response to synthetic peptides representing the amino terminus of the 3800-molecular weight antigen from *Streptococcus mutans*, p. 105–110. *In* R. Chanock et al. (ed.), *Vaccines 88*. Cold Spring Harbor Laboratory, Cold Spring Harbor, N.Y.

21. **Heber-Katz, E., S. Valentine, B. Dietzschold, and C. Burns-Purzycki.** 1988. Overlapping T cell antigenic sites on a synthetic peptide fragment from herpes simplex virus glycoprotein D, the degenerate MHC restriction elicited, and functional evidence for antigen-1a interaction. *J. Exp. Med.* **167:**275.

22. **Hoepelman, A. I. M., and E. I. Tuomanen.** 1992. Consequences of microbial attachment: directing host cell functions with adhesins. *Infect. Immun.* **60:**1729–1733.

23. **Holmgren, J.** 1981. Actions of cholera toxin and the prevention and treatment of cholera. *Nature* (London) **292:**413–417.

24. **Karlsson, K. A.** 1989. Animal glycosphingolipids as membrane attachment sites for bacteria. *Annu. Rev. Biochem.* **58:**309–350.

25. **Kelly, C., P. Evans, L. Bergmeier, S. F. Lee, A. Progulske-Fox, A. C. Harris, A. Aitken, A. S. Bleiweis, and T. Lehner.** 1989. Sequence analysis of the cloned streptococcal cell surface antigen I/II. *FEBS Lett.* **258:**127–132.

26. **Kelly, C., G. Munro, S. Todryk, H. Kendal, and T. Lehner.** Unpublished data.

27. **LaPolla, R. J., J. A. Haron, C. G. Kelly, W. R. Taylor, C. Bohart, M. Hendricks, J. Pyati, R. T. Graff, J. K-C. Ma, and T. Lehner.** 1991. Sequence and structural analysis of surface protein antigen I/II (SpaA) of *Streptococcus sobrinus*. *Infect. Immun.* **59:**2677–2685.

28. **Lee, S. F., A. Progulske-Fox, and A. S. Bleiweis.** 1988. Molecular cloning and expression of a *Streptococcus mutans* major surface protein antigen P1 (I/II) in *Escherichia coli*. *Infect. Immun.* **56:**2114–2119.

29. **Lee, S. F., A. Progulske-Fox, G. W. Erdos, D. A. Piacentini, G. Y. Ayakawa, P. J. Crowley, and A. S. Bleiweis.** 1989. Construction and characterization of isogenic mutants of *Streptococcus mutans* deficient in major surface protein antigen P1 (I/II). *Infect. Immun.* **57:**3306.

30. **Lehner, T., J. Caldwell, and A. S. M. Giasuddin.** 1985. Comparative immunogenicity and protective effect against dental caries of a low (3800) and a high (185,000) molecular weight protein in rhesus monkeys (*Macaca mulatta*). *Arch. Oral Biol.* **30:**207.

31. **Lehner, T., J. Caldwell, and R. Smith.** 1985. Local passive immunization by monoclonal antibodies against streptococcal antigen I/II in the prevention of dental caries. *Infect. Immun.* **50:**796.

32. **Lehner, T., J. Ma. K. Grant, and R. Smith.** 1986. Passive dental immunization by monoclonal antibodies against *Streptococcus mutans* antigen in the prevention of dental caries, p. 347. *In* T. Lehner and G. Cimisoni (ed.), *Borderland between Caries and Periodontal Disease,* Editions Medicine et Hygiene, Geneva.

33. **Lehner, T., M. W. Russell, J. Caldwell, and R. Smith.** 1981. Immunization with purified protein antigens from *Streptococcus mutans* against dental caries in rhesus monkeys. *Infect. Immun.* **34:**407–415.

34. **Lehner, T., P. Walker, L. A. Bergmeier, and J. A. Haron.** 1989. Immunogenicity of synthetic peptides derived from the sequences of a *Streptococcus mutans* cell surface antigen in nonhuman primates. *J. Immunol.* **143:**2699–2705.

35. **Leininger, E., C. A. Ewanowich, A. Bhargava, M. S. Peppler, J. G. Kenimer, and M. J. Brennan.** 1992. Comparative roles of the Arg-Gly-Asp sequence present in the *Bordetella pertussis* adhesins pertactin and filamentous haemagglutinin. *Infect. Immun.* **60:**2380–2385.

36. **Leininger, E., M. Roberts, J. G. Kenimer, I. G. Charles, N. Fairweather, P. Novotny, and M. J. Brennan.** 1991. Pertactin, an Arg-Gly-Asp-containing *Bordetella pertussis* surface protein that promotes adherence of mammalian cells. *Proc. Natl. Acad. Sci. USA* **88:**345–349.

37. **Ma, J. K.-C., M. Hunjan, R. Smith, and T. Lehner.** 1989. Specificity of monoclonal antibodies in local passive immunization against *Streptococcus mutans. Clin. Exp. Immunol.* **77:**331–337.

38. **Ma, J. K.-C., C. G. Kelly, G. T. Munro, R. A. Whiley, and T. Lehner.** 1991. Conservation of the gene encoding streptococcal antigen I/II in oral streptococci. *Infect. Immun.* **59:**2686–2694.

39. **Ma, J. K.-C., R. Smith, and T. Lehner.** 1987. Use of monoclonal antibodies in local passive immunization to prevent colonization of human teeth by *Streptococcus mutans. Infect. Immun.* **55:**1274.

40. **Maddon, D. R., J. C. Gorga, J. L. Strominger, and D. C. Wiley.** 1991. The structure of HLA-B27 reveals nonamer self-peptides bound in an extended conformation. *Nature* (London) **353:**321–325.

41. **McBride, B. C., M. Song, B. Krasse, and J. Olssen.** 1984. Biochemical and immunological differences between hydrophobic and hydrophilic strains of *Streptococcus mutans. Infect. Immun.* **44:**68.

42. **Millich, D. R., A. McLachian, F. V. Chisari, and G. B. Thornton.** 1986. Nonoverlapping T and B cell determinants on a hepatitis B surface antigen preS[2] region synthetic peptide. *J. Exp. Med.* **164:**532.

43. **Munesinghe, D. Y., P. Clavijo, M. C. Calle, R. S. Nussenzweig, and E. Nardin.** 1991. Immunogenicity of multiple antigen peptides (MAP) containing T- and B-cell epitopes of the repeat region of the P. falciparum circumsporozoite protein. *Eur. J. Immunol.* **21:**3015–3020.

44. **Munro, G. H., P. Evans, S. Todryk, P. Buckett, C. G. Kelly, and T. Lehner.** 1993. A protein fragment of the streptococcal cell surface antigen I/II which prevents adhesion of *Streptococcus mutans. Infect. Immun.* **61:**4590–4598.

45. **Ogier, J. A., M. Scholler, Y. Lepoivre, A. Pini, P. Sommer, and J. P. Klein.** 1990. Complete nucleotide sequence of the sr gene from *Streptococcus mutans* OMZ 175. *FEMS Microbiol. Lett.* **68:**223–228

46. **Okahashi, N., C. Sasakawa, M. Yoshikawa, S. Hamada, and T. Koga.** 1989. Cloning of a surface protein antigen gene from serotype c *Streptococcus mutans. Mol. Microbiol.* **3:**221–228.

47. **Okahashi, N., C. Sasakawa, M. Yoshikawa, S. Hamada, and T. Koga.** 1989. Molecular characterization of a surface protein antigen gene from serotype c *Streptococcus mutans,* implicated in dental caries. *Mol. Microbiol.* **3:**673–678.

48. **Pamina-Bordington, P., A. Tan, A. Termijtelen, S. Demotz, G. Corradin, and A. Lanzavecchia.** 1989. Universally immunogenic T cell epitopes: promiscuous binding to MHC class II and promiscuous recognition by T cells. *Eur. J. Immunol.* **19:**2237.

49. **Ria, F., B. M. C. Chan, M. T. Scherer, J. A. Smith, and M. L. Gefter.** 1990. Immunological activity of covalently linked T-cell epitopes. *Nature* (London) **343:**381–383.

50. **Rudensky, A. Y., P. Preston-Hurlbert, S. C. Hong, A. Barlow, and C. A. Janeway.** 1991. Sequence analysis of peptides bound to MHC class II molecules. *Nature* (London) **353:**622–627.

292   Lehner et al.

51. **Russell, M. W., and T. Lehner.** 1978. Characterisation of antigens extracted from cells and culture fluids of *Streptococcus mutans* serotype c. *Arch. Oral Biol.* **23:**7–15.

52. **Russell, R. R. B.** 1979. Wall associated antigens of *Streptococcus mutans*. *J. Gen. Microbiol.* **114:**109–115.

53. **Saukkonen, K., W. N Burnette, V. Mar, H. R. Masure, and E. Tuomanen.** 1992. Pertussis toxin has eukaryotic-like carbohydrate recognition domains. *Proc. Natl. Acad. Sci. USA* **89:**118–122.

54. **Schwartz, R. H., B. S. Fox, E. Fraga, C. Chen, and B. Singh.** 1985. The T lymphocyte response to cytochrome c. V. determination of the minimal peptide size required for stimulation of T cell clones and assessment of the contribution of each residue beyond this size for antigenic potency. *J. Immunol.* **135:**2598.

55. **Sercarz, E. E., P. V. Lehmann, A. Ametani, G. Benichou, A. Miller, and K. Moudgil.** 1993. Dominance and crypticity of T cell antigenic determinants. *Annu. Rev. Immunol.* **11:**729–766.

56. **Sinigaglia, F., M. Guttinger, J. Kilgus, D. M. Doran, H. Matile, H. Etlinger, A. Trzeciak, D. Gillessen, and T. J. L. Pink.** 1988. A malaria T cell epitope recognised in association with most mouse and human MHC class II molecules. *Nature* (London) **336:**778.

57. **Smith, R., and T. Lehner.** 1986. Preparation and characterization of monoclonal antibodies to antigenic determinants of *Streptococcus mutans* and *sobrinus,* p. 357–365. *In* T. Lehner and G. Cimasoni (ed.), *Borderland between Caries and Periodontal Disease III.* Editions Medicine et Hygiene, Geneva.

58. **Smith, R., and T. Lehner.** 1989. Characterisation of monoclonal antibodies to common protein epitopes on the cell surface of *Streptococcus mutans* and *Streptococcus sobrinus*. *Oral Microbiol. Immunol.* **4:**153–158.

59. **Smith, R., T. Lehner, and P. C. L. Beverley.** 1984. Characterization of monoclonal antibodies to *Streptococcus mutans* antigenic determinants I/II, and III and their serotype specificities. *Infect. Immun.* **46:**168.

60. **Tokuda, M., N. Okahashi, I. Takahashi, M. Nakai, S. Nagaoka, M. Kawagoe, and T. Koga.** 1991. Complete nucleotide sequence of the gene for a surface protein antigen of *Streptococcus sobrinus*. *Infect. Immun.* **59:**3309–3312.

61. **Walker, P., R. Smerdon, J. Haron, and T. Lehner.** 1993. Mapping major and minor T cell epitopes *in vitro* and their immunogenicity or tolerogenicity *in vivo* in non-human primates. *Immunology* **80:**209–216.

Molecular Pathogenesis of Periodontal Disease
Edited by Robert Genco et al.
© 1994 American Society for Microbiology, Washington, DC 20005

Chapter 25

Regulation of Immunoglobulin A Responses for Oral Mucosal Immunity

Takachika Hiroi, Hiroshi Kiyono, Kohtaro Fujihashi, Junichi Mega, Ichiro Takahashi, Seiji Morishima, Taku Fujiwara, Shigeyuki Hamada, and Jerry R. McGhee

For optimal induction of mucosal immunity including secretory IgA (S-IgA) responses, it is necessary to take into consideration the common mucosal immune system (CMIS). In order to explain the concept of the CMIS, it is helpful to indicate that the mucosal immune system can be divided into two broad compartments, i.e., IgA inductive sites and mucosal effector tissues, according to their functional role in immune responses to mucosally encountered antigens (21). The concept of a CMIS was originally suggested by Heremans and Bazin (15), and experimental evidence was first provided by Craig and Cebra (6). In the latter study, it was shown that the transfer of lymphoid cells isolated from rabbit Peyer's patches (PP) and appendix (example of mucosal inductive tissues) to irradiated allogenic recipients resulted in the repopulation of donor IgA plasma cells in the small intestine lamina propria (LP) of recipient rabbits (IgA effector tissues). On the other hand, lymphocytes from systemic tissues did not migrate to these mucosa-associated effector tissues. During the intervening 22 years, numerous studies have provided strong evidence for the existence of the CMIS in both experimental animal models and humans (19, 21, 23, 24). Mucosal delivery of antigens (e.g., oral or intranasal immunization) induces antigen-specific IgA responses in remote mucosal effector sites via the CMIS (21).

It is generally accepted that antigen-stimulated surface IgA-positive B cells in PP migrate to IgA effector tissues, including the salivary, lachrymal, and mammary glands, and to other mucosal sites (e.g., the LP region of the small intestine) via the CMIS. In this regard, oral immunization with whole *Streptococcus mutans* or

Takachika Hiroi, Hiroshi Kiyono, Kohtaro Fujihashi, Junichi Mega, and Ichiro Takahashi • Mucosal Immunization Research Group, Immunobiology Vaccine Center, Research Center in Oral Biology, and Department of Oral Biology, University of Alabama at Birmingham, Birmingham, Alabama 35294. *Seiji Morishima, Taku Fujiwara, and Shigeyuki Hamada* • Department of Oral Microbiology, Osaka University, Osaka, Japan. *Jerry R. McGhee* • Mucosal Immunization Research Group, Immunobiology Vaccine Center, Research Center in Oral Biology, and Department of Microbiology, University of Alabama at Birmingham, Birmingham, Alabama 35294.

with surface-associated proteins or carbohydrates induced antigen-specific S-IgA responses in saliva (25, 37). Oral immunization with *S. mutans* protein antigen I/II conjugated to cholera toxin B subunit resulted in the induction of antigen-specific IgA-producing cells in salivary glands (7, 33). In addition, oral immunization of mice with *Porphyromonas gingivalis* fimbriae together with an acyl derivative of muramyl dipeptide (GM-53) as adjuvant resulted in the induction of fimbria-specific IgA antibodies in saliva (28, 29).

In the oral cavity, salivary glands are an essential IgA effector tissue for the continuous production of antigen-specific IgA antibodies in saliva. Thus, it is important to characterize T and B cells that reside in this tissue in order to understand the precise nature of the mucosal immune system in the oral region. When the isotype distribution of Ig-producing cells was examined among murine salivary gland-associated tissues (SGAT), the submandibular glands (SMG) contained large numbers of IgA-secreting cells but few IgM- and IgG-producing cells (22). On the other hand, the periglandular lymph nodes (PGLN) and cervical lymph nodes (CLN) had few or no IgA-producing cells, respectively. Thus, our previous findings clearly supported the concept that the SMG are major mucosal effector tissues in the oral region. Regarding regulatory T cells in SMG, the CD3$^+$ CD4$^+$ CD8$^-$ T cells in the SMG contained T cells spontaneously producing gamma interferon (IFN-γ) and interleukin-5 (IL-5) (22). However, very limited information is currently available in terms of the precise regulatory T-cell network that supports and regulates these predominant IgA responses in the SMG.

Murine CD3$^+$ CD4$^+$ T cells are subgrouped as type 1 or type 2 T helper (Th) cells according to the profile of cytokines they produce (26, 34). Type 1 Th cells selectively secrete IFN-γ, IL-2 and tumor necrosis factor beta upon stimulation via the $\alpha\beta$ T-cell receptor (TCR)-CD3 complex, whereas type 2 Th cells produce IL-4, IL-5, IL-6, and IL-10, a group of cytokines important for B-cell responses. The type 1 subset of Th cells is designated Th1, and the type 2 cells are designated Th2 (26, 34). In comparison to mucosal inductive sites, the immunological responses of the effector tissues are much less well characterized. In this regard, single-cell analyses of immunocompetent T and B cells that reside in the salivary glands and the small intestine for their abilities to produce cytokines and IgA, respectively, in response to mucosally administered antigen have not yet been done extensively. This is unfortunate, since the most readily obtained external secretions are saliva and gut washes (or fecal samples) for analysis of mucosal S-IgA immune responses, and numerous past studies of humans and experimental systems have relied on these secretions as a source for analysis of antigen-specific mucosal S-IgA antibody responses.

To this end, we have recently begun to analyze lymphocytes, including T and B cells isolated from salivary glands and intestine, at the single-cell level in order to understand their potential roles for formation of S-IgA immune responses in these IgA effector sites. We summarize here our past studies as well as ongoing work designed to help optimize induction of optimal CD4$^+$ Th cells for induction and regulation of S-IgA responses in the oral cavity. The principles discussed should also apply to other mucosa-associated tissues for generation of appropriate immunity to pathogenic viruses and bacteria that infect distinct mucosal effector sites.

METHODS

Isolation of Lymphocytes from Murine Salivary Glands

C3H/HeN and C57BL/6 mice obtained from the Frederick Cancer Research Facility (National Cancer Institute, Frederick, Md.) were maintained in horizontal-laminar-flow cabinets and provided autoclaved food and water ad libitum. Experiments were performed by using young adult mice between 7 and 9 weeks of age.

In order to understand the characteristics of T and B lymphocytes involved in the regulation and secretion of IgA in the salivary glands, lymphocytes were enzymatically isolated from SGAT. The murine SGAT are located just under the mandible and consist of the SMG, the PGLN, and the CLN (Fig. 1). Mononuclear cells (MC) from SMG were obtained by a modification of the enzymatic dissociation procedure (22) that was originally described by others (8, 30, 31). The SMG were carefully dissected, cut into small fragments (approximately 1.0 by 2.0 mm), and then dissociated into single cells by use of type IV collagenase (Fig. 2). This enzymatic dissociation process was performed at least three or four times. To obtain purified MC populations, Percoll discontinuous gradients were employed (Fig. 2). The isolated cells were suspended in 75% Percoll and carefully transferred to a centrifuge tube. The 55 and 40% Percoll solutions were then carefully layered on 75% Percoll. After centrifugation, the interface between the 75 and 55% layers was removed carefully, and the MC-enriched fraction was washed. This procedure resulted in greater than 95% viable lymphocytes with a cell yield for SMG of 1×10^6 to 2×10^6 cells per mouse.

Functional Studies of SGAT Lymphocytes

The proliferative responses of MC isolated from SGAT to T- and B-cell mitogens were determined by [^{3}H]thymidine uptake as previously described (20, 22). The MC (10^5 cells per well) were incubated with various mitogens (concanavalin A [Con A; 2 μg/ml], phytohemaglutinin [PHA; 5 μg/ml], and pokeweed mitogen [PWM; 1/1000]) for 48 h at 37°C in a humidified atmosphere of 5% CO_2 in air. During the last 6 h of incubation, 0.5 μCi of [^{3}H]thymidine was added, and [^{3}H]thymidine incorporation was assessed in order to determine levels of DNA replication by scintillation counting (20).

Analysis of IgM-, IgG-, and IgA-Producing B Cells in SGAT

Since the SGAT and especially the SMG are considered the major mucosal effector tissues in the oral cavity, it was important to examine and compare the isotypes and frequency of occurrence of antibody-secreting cells with those at other IgA effector sites (e.g., the LP of the small intestine). Antibody-producing cells isolated from various lymphoid tissues were quantitated by use of the Ig isotype-specific enzyme-linked immunospot (ELISPOT) assay (22). Nitrocellulose microtiter plates were coated with 5 μg of affinity-purified goat anti-mouse Ig (Southern Biotechnology Associates [SBA], Birmingham, Ala.) per ml in phosphate-buffered

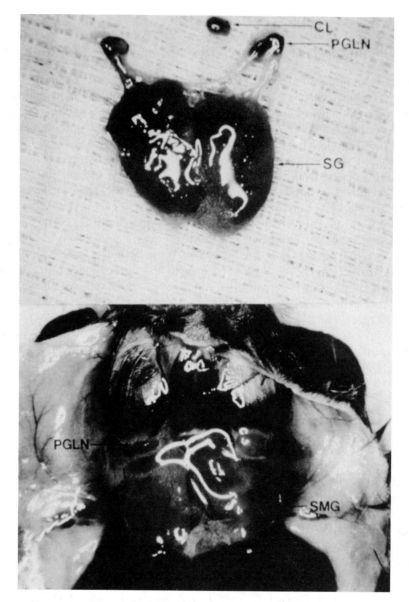

Figure 1. Picture of SGAT. CL, cervical lymph node; PGLN, periglandular lymph nodes; SG or SMG, submandibular gland.

saline (PBS; 100 µl per well) overnight at 4°C. The wells were washed with PBS and blocked with PBS containing 1% bovine serum albumin. Different concentrations of MC were then added to individual wells. The plates were incubated for 4 h at 37°C in 5% CO_2 in air (20). The plates were then washed three times sequentially with PBS followed by PBS containing 0.05% Tween. A 100-µl aliquot of a 1:1,000

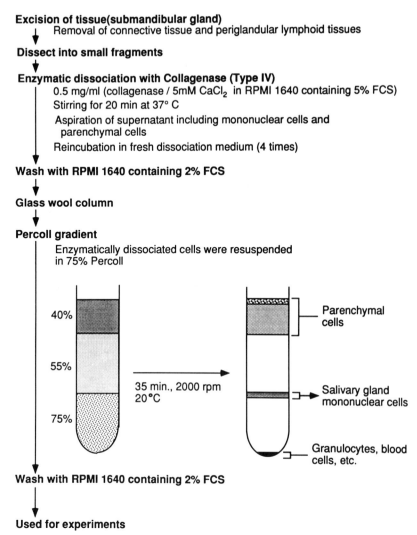

Excision of tissue(submandibular gland)
  Removal of connective tissue and periglandular lymphoid tissues

Dissect into small fragments

Enzymatic dissociation with Collagenase (Type IV)
  0.5 mg/ml (collagenase / 5mM CaCl$_2$ in RPMI 1640 containing 5% FCS)
  Stirring for 20 min at 37° C
  Aspiration of supernatant including mononuclear cells and
    parenchymal cells
  Reincubation in fresh dissociation medium (4 times)

Wash with RPMI 1640 containing 2% FCS

Glass wool column

Percoll gradient
  Enzymatically dissociated cells were resuspended
  in 75% Percoll

40%

55%

75%

35 min., 2000 rpm
20°C

Parenchymal
cells

Salivary gland
mononuclear cells

Granulocytes, blood
cells, etc.

Wash with RPMI 1640 containing 2% FCS

Used for experiments

Figure 2. Method for the isolation of MC from murine salivary glands. FCS, fetal calf serum.

dilution of horseradish peroxidase-conjugated goat anti-mouse μ, γ, or α heavy-chain-specific antibody (SBA in PBS–0.05% Tween containing 1% fetal calf serum was added to each well. Following incubation at 4°C overnight, the plates were washed four times with PBS and developed with 1.6 mM 3-amino-9 ethyl carbazole in 0.1 M sodium acetate buffer (pH 5.0) containing 0.05% H$_2$O$_2$. The spot-forming cells (SFC) were enumerated with the aid of a stereomicroscope.

Characterization of T-Lymphocyte Subsets

For the analysis of CD4, CD8, αβ TCR, and γδ TCR expression on T cells isolated from SGAT, multicolor flow cytometry analysis was performed by using

FACStar[Plus] (Becton Dickinson & Co., Sunnyvale, Calif.) (11, 22). For the staining of cells, a panel of immunofluorescence-conjugated monoclonal antibodies including anti-CD3 (145-2C11), anti-CD4 (GK1.5), anti-CD8 (53.6-72), anti-α/β TCR (H57-597), and anti-$\gamma\delta$ TCR (GL3) was employed as described previously (11, 22). In these multicolor flow cytometry analyses, at least 1×10^5 to 5×10^5 viable cells were analyzed in individual samples.

RESULTS AND DISCUSSION

DNA Replication by SGAT Lymphocytes

When the DNA replication abilities of these isolated lymphocytes were examined by the cell proliferation assay using T- or B-cell mitogens, significant proliferative responses were induced in SMG cells by ConA and PHA (T-cell mitogens) and by PWM (T-cell-dependent B-cell mitogen) (Table 1). The MC from PGLN and CLN gave higher proliferative responses to ConA and PHA than to PWM (Table 1). A similar pattern of DNA replication was seen in MC isolated from peripheral lymph nodes and mesenteric lymph nodes. These results show that MC isolated from SGAT are responsive to T- and B-cell mitogens. A comparison

Table 1. T- and B-cell mitogens induce DNA replication in MC isolated from SGAT[a]

Source of MC	Mitogen	Stimulation index (E/C)[b]
SGAT		
SMG	ConA	23–31
	PHA	14–20
	PWM	19–24
PGLN	ConA	23–32
	PHA	22–27
	PWM	13–17
CLN	ConA	27–36
	PHA	29–39
	PWM	14–17
LN		
PLN	ConA	35–46
	PHA	32–37
	PWM	15–19
MLN	ConA	34–43
	PHA	31–36
	PWM	17–21
PP	ConA	29–43
	PHA	30–41
	PWM	20–25

[a] MC enzymatically isolated from SGAT (10^5 per well) were incubated with ConA (2 μg/ml), PHA (2 μg/ml), or PWM (1/1,000) for 48 h at 37°C in the presence of 5% CO_2.
[b] Data are expressed as the stimulation index, where E/C is the ratio of cpm of [^{3}H]thymidine incorporated with mitogen/cpm of [^{3}H]thymidine incorporated by cells only.

of the proliferative responses among SGAT shows that the MC that reside in the PGLN and CLN resembled other organized systemic lymph nodes, whereas MC isolated from SMG gave responses similar to those of T- and B-cell mitogens (Table 1).

Ig-Producing Cells in SGAT

When the isotype of Ig-producing cells isolated from SGAT was analyzed, the major isotype of Ig-producing cells in SMG was IgA, followed by small numbers of IgM and IgG SFC (Table 2) (22). Two different isotype distribution patterns of Ig-producing cells were noted in the PGLN and CLN of SGAT. For example, the former tissue contained high numbers of IgA and IgM but not IgG SFC. However, essentially no Ig-specific spots were developed by MC obtained from CLN. Thus, this was the first study to show that organized PGLN associated with the oral region contain IgA- and IgM-producing cells (22). When the frequency of IgM-, IgG-, and IgA-producing cells in SMG and PGLN was compared with that in the intestinal LP, the isotype pattern was IgA $>>$ IgM $\geq$ IgG in SMG and in LP. However, it should be noted that the numbers of IgA-producing cells in the two SGAT tissues were lower than the numbers in LP of the gastrointestinal tract.

In summary, among SGAT, the SMG should be considered a typical mucosal effector tissue, while CLN possess characteristics of an organized systemic lymph node. Interestingly, PGLN could be a unique, organized lymph node in the oral region, since this tissue possessed immunological characteristics of both mucosal effector tissue and organized systemic lymph nodes. It might be comparable to the mesenteric lymph node in this regard.

TCR Expression by Salivary Gland T Cells

Mammalian T cells recognize nominal antigen via TCR, which occur as heterodimer chains of either $\alpha\beta$ or $\gamma\delta$. In general, mature T cells that reside in the organized secondary lymphoid tissues harbor the $\alpha\beta$ TCR, which recognizes peptide together with major histocompatibility class I or II (3, 18). The $\alpha\beta$ TCR$^+$ T cells are separated into two subsets based on the coexpression of either CD4 or CD8.

Table 2. IgM-, IgG-, and IgA-secreting cells in murine SGAT

Mucosa-associated tissue	Isotype and no. of SFC/10⁶ cells[a]		
	IgM	IgG	IgA
SGAT			
SMG		<50	~10,000
PGLN	<100	<100	1,000–3,000
CLN	1,500–2,000	<10	<10
Intestine-associated tissues			
LP	3,000–4,000	<100	30,000–40,000
PP	2,000–3,000	1,500–2,000	~6,000

[a] Numbers are from three different experiments.

Thus, T helper cells generally express the αβ TCR[+] and CD4[+], which are stimulated by foreign peptide and major histocompatibility complex class II expressed on antigen-presenting cells, including macrophages, dendritic cells, and B cells. The αβ TCR-bearing CD8[+] T cells behave as cytotoxic lymphocytes and recognize target cells through major histocompatibility complex class I restriction. In contrast to the αβ TCR[+] T cells, the γδ TCR[+] T cells arise earlier in ontogeny, and small numbers of T cells (1–5%) express their receptors in peripheral lymphoid tissues (32). In most cases, γδ TCR-bearing T cells are categorized as double-negative (DN) cells, since they do not express either CD4 or CD8. The immunological function of γδ TCR[+] T cells remains unknown. However, our separate studies have provided evidence that at least two subsets of γδ TCR-bearing T cells in mucosa-associated tissues are capable of producing type 1 and type 2 cytokines, which can serve regulatory functions for IgA B-cell responses (10, 12, 35).

When different subsets of SGAT T cells were assessed according to their expression of CD4 and CD8 molecules, three distinct subsets of T cells including CD4[+] CD8[-] T cells, CD4[-] CD8[+] T cells, and DN T cells were observed in SMG, PGLN, and CLN (22). No CD4[+] CD8[+] (double positive) T cells were found in any of the SGAT. Concerning CD4- and CD8-bearing T cells, approximately equal frequencies of CD4[+] CD8[-] and CD4[-] CD8[+] T cells were seen in the SMG. This finding was very similar to that of studies of rat salivary glands in which the MC were shown to contain approximately 60% W3/13[+] T cells with a CD4/CD8 ratio of 1.0 to 1.3 (8, 31). This equal distribution of CD4[+] CD8[-] and CD4[-] CD8[+] T cells in SMG was unique, since other mucosa-associated tissues, including those of IgA inductive (e.g., PP) and effector (e.g., intestinal LP) tissues and systemic lymphoid tissue (e.g., spleen [SP]) as well as organized lymph nodes in SGAT (e.g., CLN and PGLN), all contained higher frequencies of CD4[+] CD8[-] T cells than of CD4[-] CD8[+] T cells. Further, a relatively higher number of DN T cells was found in SMG than in other tissues. In this regard, up to 15% of CD3[+] T cells were DN T cells in SMG, while other tissues contained lower numbers of CD3[+] DN T cells.

The expression of αβ TCR or γδ TCR by three different subsets of T cells in SGAT was also examined by flow cytometry (22). All CD4[+] CD8[-] T cells harbored the αβ TCR. On the other hand, it was interesting that approximately 25% of CD4[-] CD8[+] T cells isolated from SMG expressed γδ TCR, while the other CD4[-] CD8[+] T cells bore αβ TCR. Prior to this finding, γδ TCR[+] CD8[+] T cells had been found associated only with intraepithelial lymphocytes (IELs) in the small intestine (4, 13). Studies of the function of these SMG γδ TCR[+] CD8[+] T cells in comparison with that of the CD8[+] γδ TCR[+] IEL T cells in the gastrointestinal tract are currently in progress. In contrast to SMG, the CD4[-] CD8[+] T cells that reside in the PGLN and CLN of SGAT expressed the αβ TCR. As one might expect, γδ TCR[+] cells were also found in the DN-T-cell fraction of SMG. In addition, although PGLN and CLN possessed small numbers of DN T cells, all of these cells used the γδ form of TCR. These observations were consistent with those of previous studies, in which DN T cells always associated with γδ TCR (10, 12, 17, 35). Taken together, our results have provided the first evidence that CD4[-] CD8[+] T cells that reside in glandular tissue associated with the oral region contain both γδ TCR[+] and αβ TCR[+] cells.

Although the biological function of $CD8^+$ $\gamma\delta$ TCR^+ T cells in mucosa-associated tissues is not yet known, these T lymphocytes in IELs possess a cytolytic function (10, 13, 16). Further, our studies recently demonstrated a regulatory function of $CD8^+$ $\gamma\delta$ TCR^+ T cells in intestinal IELs, where these T cells from mice orally immunized with T-cell-dependent antigen possessed the ability to convert systemic unresponsiveness (or oral tolerance) to antigen-specific immune responses, including those of the IgA isotype, upon adoptive transfer to orally tolerized mice (10, 12). Thus, $CD8^+$, $\gamma\delta$ TCR^+ T cells could be important regulatory T cells that protect (or enhance) $CD4^+$ Th cells for maximum IgA responses in mucosal effector tissues in an active state of oral tolerance. This type of cell may contribute to the maintenance of mucosal IgA responses in the presence of systemic unresponsiveness. The presence of $CD8^+$ $\gamma\delta$ TCR^+ T cells in salivary glands might be an essential feature for the resultant IgA antibody production associated with this tissue. Our current experiments are being directed to determine whether $\gamma\delta$ TCR^+ $CD8^+$ T cells in the SMG regulate IgA responses.

Role of Th1 and Th2 Cells in Regulation of IgA Responses and Their Frequency of Occurrence in SGAT

$CD3^+$ $CD4^+$ Th cells and their derived cytokines play central roles in the terminal differentiation of surface IgA^+ B cells into IgA plasma cells (21). These $CD3^+$ $CD4^+$ Th cells are subgrouped into Th1 and Th2 cell subsets according to the profiles of cytokines that are produced by the respective T cells (26, 34). $CD3^+$ $CD4^+$ Th cells that synthesize IFN-γ, IL-2, and tumor necrosis factor beta upon stimulation via the TCR-CD3 complex are classified as Th1 cells (26, 34). On the other hand, Th2 cells are capable of producing IL-4, IL-5, IL-6, and IL-10, an array of cytokines essential for B-cell responses. In this regard, cytokines secreted by Th2-type cells, notably IL-5 and IL-6, directly act on surface IgA^+ B cells from gut-associated lymphoid tissue without any costimulation in vitro and induce them to differentiate into IgA-producing cells (1, 2). Further, it should be noted that IL-2 produced by Th1-type cells can also be involved in differentiation of IgA^+ B cells to become plasma cells (5, 27). Therefore, it was beneficial to examine the exact profiles of Th1 and Th2 cells in various mucosa-associated tissues.

When the frequency of Th1- and Th2-type cells was assessed in mucosal inductive and effector compartments by using an IFN-γ (Th1)- and IL-5 (Th2)-specific ELISPOT assay, freshly isolated $CD3^+$ $CD4^+$ T cells from PP contained neither IFN-γ- nor IL-5-producing cells unless these T cells were stimulated with T-cell mitogens in vitro (36). Following T-cell mitogen stimulation, an approximately equal frequency of Th1 (IFN-γ)- and Th2 (IL-5)-type cells was seen in this IgA inductive tissue. In contrast to gut-associated lymphoid tissue, freshly isolated $CD3^+$ $CD4^+$ Th cells from murine intestinal LP contained higher numbers of IL-5 SFC than of IFN-γ-secreting cells (36). These results suggest that Th2-type cells are predominant in mucosal effector regions for IgA responses.

Inasmuch as a distinct isotype distribution of Ig-producing cells occurred among tissues associated with SGAT (Table 2), it was interesting to examine the array of cytokines produced by T cells that reside in the respective SGAT in order to understand the contribution of Th1- and Th2-type cells for IgA B cell responses in the salivary glands. When lymphocytes were isolated from SGAT (e.g., the SMG) and examined for Th1 (IL-2 and IFN-γ)- and Th2 (IL-4, IL-5, and IL-6)-type cytokine production by using the respective cytokine-specific ELISPOT assays, both tissues contained T cells that spontaneously produced IFN-γ, IL-5, and IL-6 (15a). However, neither IL-2- nor IL-4-producing cells were detected in these freshly isolated SMG T cells (Table 3). Among these cytokine-secreting T cells, Th2-type cells were always more numerous than Th1-type cells in SMG. This finding was in complete agreement with those of our previous studies, in which increased numbers of IL-5-secreting Th2-type cells were consistently found in IgA effector tissues such as LP of intestine (36). Stimulation of these T cells from SMG with a T-cell mitogen (e.g., ConA) resulted in the induction of an array of type 1 (IFN-γ and IL-2) and type 2 (IL-4, IL-5, and IL-6) cytokine-producing T cells (Table 3). Thus, these findings suggested that although Th1 and Th2 cells in SMG were capable of producing all of the respective cytokines for these two subsets, Th1 and Th2 cells residing in the SMG might be programmed to secrete IFN-γ as type 1 cytokine and IL-5 and IL-6 as type 2 cytokines, respectively (15a). Further, these results demonstrated that high numbers of IL-5 and IL-6 producing Th2-type cells were always associated with increased numbers of IgA-producing cells in the salivary gland (e.g., SMG).

Cytokine-Specific Reverse Transcriptase PCR Analysis

The findings of the cytokine-specific ELISPOT assay were also confirmed at the mRNA level by reverse transcriptase PCR with cytokine-specific primers. Significant levels of IFN-γ-, IL-5-, and IL-6-specific messages were noted in RNA isolated from freshly isolated SMG T lymphocytes (15a). On the other hand, ConA stimulation induced IL-2 and IL-4 messages in addition to IFN-γ-, IL-5-, and IL-6-specific mRNA in RNA prepared from SMG T cells. These findings further support the notion that elevated numbers of IgA-producing cells in mucosal effector tissues are always associated with a higher frequency of Th2 (IL-5 and IL-6)-type cells than of Th1 (IFN-γ and IL-2)-type CD4$^+$ Th cells.

Table 3. Frequency of Th1- and Th2-type cells in MC isolated from SMG[a]

Th-cell subset	Cytokine tested	Frequency of Th1- and Th2-type cells	
		Freshly isolated	ConA-stimulated
Type 1	IFN-γ	2+	4+
	IL-2	0	2+
Type 2	IL-4	0	2+
	IL-5	4+	5+
	IL-6	2+	4+

[a] T lymphocytes from murine SMG were subjected to cytokine-specific ELISPOT assay. 1+, 20 to 25 SFC per 10^4 cells; 2+, 25 to 50 SFC; 3+, 50 to 75 SFC; 4+, 75 to 125 SFC; and 5+, >125 SFC.

Antigen-Specific Th1 and Th2 Cells and B-Cell Responses in SGAT

We recently established an effective oral immunization regimen that induces large numbers of antigen-specific IgA and Th-2-type cell responses in IgA effector tissues. In these studies, oral administration of protein vaccine (e.g., tetanus toxoid) together with cholera toxin induced a high frequency of antigen-specific Th2-type cells and IgA-producing cells in mucosa-associated tissues (38). Therefore, we have recently adapted this mucosal immunization regimen to derive antigen-specific CD4$^+$ T cells in SMG by immunizing mice with periodontal-disease-associated antigen (*P. gingivalis* fimbriae) in the presence of cholera toxin (10 μg) given via the oral route. After three doses of oral immunization, T cells were isolated from PP and SP of orally immunized mice and then cultured with feeder cells, IL-2, and fimbria-conjugated microspheres in order to expand antigen-specific CD4$^+$Th cells. At the same time, we examined the profiles of antigen-specific B-cell responses in SMG and SP. When the Ig isotype and numbers of fimbria-specific antibody-producing cells in SMG and SP of mice orally immunized with fimbriae plus cholera toxin were examined, the SMG contained large numbers of anti-fimbrial SFC, which were of the IgA isotype (Table 4). On the other hand, the SP contained fimbria-specific IgM and IgG antibody-producing cells. Of interest was the observation that no fimbria-specific IgA-producing cells were seen in SP.

In a recent and separate study, we determined the nature of antigen-specific Th1- and Th2-cell responses induced by oral immunization of mice with fimbriae plus cholera toxin (15b). Fimbria-stimulated CD4$^+$ T cells from PP and SP of orally immunized mice were examined for the Th1 (IFN-γ and IL-2) and Th2 (IL-4, IL-5, and IL-6) types by using cytokine-specific ELISPOT assays. In vitro stimulation with antigen induced increased numbers of fimbria-specific IL-4-, IL-5-, and IL-6-producing Th2-type T cells in both PP and SP cultures, while IL-2 and IFN-γ-secreting Th1-type cells were the same as background. The predominance of Th2-type responses was maintained throughout the culture period. The numbers of IL-4, IL-5, and IL-6 SFC were generally higher in PP than in SP. These findings should provide new information regarding induction of antigen-specific Th2-type cells over Th1-type cells in SMG of mice orally immunized with *P. gingivalis* fimbriae.

Table 4. Antigen-specific B-cell responses in SMG of mice orally immunized with fimbriae

Source of MC	Level of fimbria-specific antibody-producing cells[a]		
	IgM	IgG	IgA
SMG	0	0	4+
SP	2+	4+	±

[a] Freshly isolated MC from SMG and SP of mice orally immunized with fimbriae and from nonimmunized controls were subjected to fimbria-specific ELISPOT assay. SMG and SP of nonimmunized mice did not contain any antigen-specific antibody-producing cells. ±, 0 to 80 SFC per 10^6 cells; 1+, 80 to 100 SFC; 2+, 160 to 200 SFC; and 4+, 320 to 400 SFC.

SUMMARY

Saliva has been used extensively as a convenient secretion for monitoring the induction of antigen-specific IgA immune responses via the CMIS, because this fluid contains S-IgA as a major antibody isotype. Thus, the SMG are considered important compartments for the mucosal IgA immune system in the oral region. We have characterized different subsets of T cells in the SGAT according to their expression of CD4, CD8, $\alpha\beta$ TCR, and $\gamma\delta$ TCR and their abilities to produce cytokines such as IFN-γ, IL-2, IL-4, IL-5, and IL-6. Further, the isotype distribution of Ig-producing cells was also examined. Freshly isolated CD3$^+$ T cells from the SMG harbored T cells that spontaneously produced Th1 (IFN-γ)- and Th2 (IL-5 and IL-6)-type cytokines. Interestingly, a high frequency of IL-5- and IL-6-producing Th2-type cells was always associated with increased numbers of IgA plasma cells in the SMG. Further, the salivary gland CD3$^+$ T cells could be divided into three distinct subsets, including CD4$^+$ CD8$^-$ T cells, CD4$^-$ CD8$^+$ T cells, and DN T cells. In terms of TCR expression, CD4$^+$ CD8$^-$ and DN T cells exclusively expressed $\alpha\beta$ TCR and $\gamma\delta$ TCR, respectively. On the other hand, CD4$^-$ CD8$^+$ T cells contained both $\alpha\beta$ TCR- and $\gamma\delta$ TCR-bearing cells. Further, our separate study showed that the numbers of antigen-specific IL-5- and IL-6-producing Th2-type cells were induced in the SMG together with increased numbers of fimbria-specific IgA-producing cells by oral immunization with *P. gingivalis* fimbriae. Taken together, our results have now provided the first compelling evidence that high levels of IgA antibodies in SMG directly correlate with the presence of antigen-specific Th2-type cells.

Acknowledgments. This study was supported in part by United States Public Health Service grants DE 09837, DE 04217, DE 08182, AI 30366, AI 19674, and DK 44240 and contract AI 15128. Hiroshi Kiyono is a recipient of NIH RCDA DE 00237.
We thank Cindi Eastwood for preparation of the manuscript.

REFERENCES

1. **Beagley, K. W., J. H. Eldridge, H. Kiyono, M. P. Everson, W. J. Koopman, T. Honjo, and J. R. McGhee.** 1988. Recombinant murine IL-5 induces high rate IgA synthesis in cycling IgA-positive Peyer's patch B cells. *J. Immunol.* **141:**2035–2042.
2. **Beagley, K. W., J. H. Eldridge, F. Lee, H. Kiyono, M. P. Everson, W. J. Koopman, T. Hirano, T. Kishimoto, and J. R. McGhee.** 1989. Interleukins and IgA synthesis. Human and murine interleukin 6 induce high rate IgA secretion in IgA-committed B cells. *J. Exp. Med.* **169:**2133–2148.
3. **Bierer, B. E., B. P. Sleckman, S. E. Ratnofsky, and S. J. Burakoff.** 1989. The biologic roles of CD2, CD4 and CD8 in T-cell activation. *Annu. Rev. Immunol.* **7:**579–599.
4. **Bonnevile, M., C. A. Janeway, Jr., K. Ito, W. Haser, I. Ishida, N. Nakanishi, and S. Tonegawa.** 1988. Intestinal intraepithelial lymphocytes are a distinct set of γ/δ T cells. *Nature* (London) **336:**479–481.
5. **Coffman, R. L., B. Shrader, J. Carty, T. R. Mosmann, and M. W. Bond.** 1987. A mouse T cell product that preferentially enhances IgA production. I. Biologic characterization. *J. Immunol.* **139:**3685–3690.
6. **Craig, S. W., and J. J. Cebra.** 1971. Peyer's patches: an enriched source of precursors for IgA-producing immunocytes in the rabbit. *J. Exp. Med.* **134:**188–200.
7. **Czerkinsky, C., M. W. Russell, N. Lycke, M. Lindblad, and J. Holmgren.** 1989. Oral administration

of streptococcal antigen coupled to cholera toxin B subunit evokes strong antibody responses in salivary glands and extramucosal tissues. *Infect. Immun.* **57:**1072–1077.

8. **Ebersole, J. L., M. L. Steffen, and J. Pappo.** 1988. Secretory immune responses in ageing rats. II. Phenotype distribution of lymphocytes in secretory and lymphoid tissues. *Immunology* **64:**289–294.

9. **Elson, C. O.** 1987. The immunology of inflammatory bowel disease, p. 97–164. *In* J. B. Kirsner and R. G. Shorter (ed.), *Inflammatory Bowel Disease,* 3rd ed. Lea & Febiger, Philadelphia.

10. **Ernest, P. B., D. A. Clark, K. L. Rosenthal, A. D. Befus, and J. Bienenstock.** 1986. Detection and characterization of cytotoxic T lymphocyte precursors in the murine intestinal intraepithelial leukocyte population. *J. Immunol.* **136:**2121–2126.

11. **Fujihashi, K., T. Taguchi, W. K. Aicher, J. R. McGhee, J. A. Bluestone, J. H. Eldridge, and H. Kiyono.** 1992. Immunoregulatory functions for murine intraepithelial lymphocytes: γ/δ TCR⁺ T cells abrogate oral tolerance, while α/β TCR⁺ T cells provide B cell help. *J. Exp. Med.* **175:**695–707.

12. **Fujihashi, K., T. Taguchi, J. R. McGhee, J. H. Eldridge, M. G. Bruce, D. R. Green, B. Singh, and H. Kiyono.** 1990. Regulatory function for murine intraepithelial lymphocytes. Two subsets of CD3⁺, T cell receptor-1⁺ intraepithelial lymphocyte T cells abrogate oral tolerance. *J. Immunol.* **145:**2010–2019.

13. **Goodman, T., and L. Lefrancois.** 1988. Expression of the γ/δ Tcell receptor on intestinal CD8⁺ intraepithelial lymphocytes. *Nature* (London) **333:**855–858.

14. **Guy-Grand, D., N. Cerf-Bensussan, B. Malissen, M. Malassis-Seris, C. Briottet, and P. Vassalli.** 1991. Two gut intraepithelial CD8⁺ lymphocyte populations with different T cell receptors: a role for the gut epithelium in T cell differentiation. *J. Exp. Med.* **173:**471–481.

15. **Heremans, J. F., and H. Bazin** 1971. Antibodies induced by local antigenic stimulation of mucosal surfaces. *Ann. N.Y. Acad. Sci.* **190:**268–275.

15a. **Hiroi, T., K. Fujihashi, J. R. McGhee, and H. Kiyono.** Cytokine producing T cells in mucosal effector tissues: mRNA expression and synthesis of Th1 and Th2 cytokines by murine γδ TCR⁺ and αβ TCR⁺ T cells in salivary glands. Submitted for publication.

15b. **Hiroi, T., K. Fujihashi, J. R. McGhee, and H. Kiyono.** Submitted for publication.

16. **Klein, J. R.** 1986. Ontogeny of the Thy-1⁻, Lyt-2⁺ murine intestinal intraepithelial lymphocyte. Characterization of a unique population of thymus-independent cytotoxic effector cells in the intestinal mucosa. *J. Exp. Med.* **164:**309–314.

17. **Lew, A. M., D. W. Pardoll, W. L. Maloy, B. J. Fowlkes, A. Kruisbeek, S. F. Cheng, R. N. Germain, J. A. Bluestone, R. H. Schwartz, and J. E. Coligan.** 1986. Characterization of T cell receptor gamma chain expression in a subset of murine thymocytes. *Science* **234:**1401–1405.

18. **Marrack, P., and J. Kappler.** 1986. The antigen-specific, major histocompatibility complex-restricted receptor on T cells. *Adv. Immunol.* **38:**1–30.

19. **McGhee, J. R.** 1991. Mucosal immunology. *Encycl. Hum. Biol.* **5:**137.

20. **McGhee, J. R., H. Kiyono, S. M. Michalek, J. L. Babb, D. L. Rosenstreich, and S. E. Mergenhagen.** 1980. Lipopolysaccharide (LPS) regulation of the immune response: T lymphocytes from normal mice suppress mitogenic and immunogenic responses to LPS. *J. Immunol.* **124:**1603–1611.

21. **McGhee, J. R., J. Mestecky, C. O. Elson, and H. Kiyono.** 1989. Regulation of IgA synthesis and immune response by T cells and interleukins. *J. Clin. Immunol.* **9:**175–199.

22. **Mega, J., J. R. McGhee, and H. Kiyono.** 1992. Cytokine and Ig producing cells in mucosal effector tissues: analysis of IL-5 and IFN-γ producing T cells, TCR expression, and IgA plasma cells from mouse salivary gland-associated tissues. *J. Immunol.* **148:**2030–2039.

23. **Mestecky, J.** 1987. The common immune system and current strategies for induction of immune responses in external secretions. *J. Clin. Immunol.* **7:**265–276.

24. **Mestecky, J., and J. R. McGhee.** 1987. Immunoglobulin A (IgA): molecular and cellular interactions involved in IgA biosynthesis and immune response. *Adv. Immunol.* **40:**153–245.

25. **Michalek, S. M., and N. K. Childers.** 1990. Development and outlook for a caries vaccine. *Crit. Rev. Oral Biol. Med.* **1:**37–54.

26. **Mosmann, T. R., and R. L. Coffman.** 1989. Th 1 and Th 2 cells: different patterns of lymphokine secretion lead to different functional properties. *Annu. Rev. Immunol.* **7:**145–173.

27. **Murray, P. D., S. L. Swain, and M. F. Kagnoff.** 1985. Regulation of the IgM and IgA anti-dextran B1355S response: synergy between IFN-γ, BCGF-II and IL-2. *J. Immunol.* **135:**4015–4020.

28. **Ogawa, T., H. Kushumoto, H. Shimauchi, H. Kiyono, J. R. McGhee, and S. Hamada.** 1992. Occurrence of antigen-specific B cells following oral or parenteral immunization with *Porphyromonas gingivalis* fimbriae. *Int. Immunol.* **4:**1003–1010.

29. **Ogawa, T., H. Shimauchi, and S. Hamada.** 1989. Mucosal and systemic immune responses in BALB/c mice to *Bacteroides gingivalis* fimbriae and administered orally. *Infect. Immun.* **57:**3466–3471.

30. **Oudghiri, M., J. Seguin, and N. Deslauries.** 1986. The cellular basis of salivary immunity in the mouse: incidence and distribution of B cells, T cells and macrophages in single cell suspensions of the major salivary glands. *Eur. J. Immunol.* **16:**281–285.

31. **Pappo, J., J. L. Ebersole, and M. A. Taubman.** 1988. Phenotype of mononuclear leukocytes resident in rat major salivary and lacrimal glands. *Immunology* **64:**295–300.

32. **Raulet, D. H.** 1989. The structure, function and molecular genetics of the γ/δ T cell receptor. *Annu. Rev. Immunol.* **7:**175–207.

33. **Russell, M. W., and H.-Y. Wu.** 1991. Distribution, persistence and recall of serum and salivary antibody responses to peroral immunization with protein antigen I/II of *Streptococcus mutans* coupled to the cholera toxin B subunit. *Infect. Immun.* **59:**4061–4070.

34. **Street, N. E., and T. R. Mosmann.** 1991. Functional diversity of T lymphocytes due to secretion of different cytokine patterns. *FASEB J.* **5:**171–177.

35. **Taguchi, T., W. K. Aicher, K. Fujihashi, M. Yamamoto, J. R. McGhee, J. A. Bluestone, and H. Kiyono.** 1991. Novel function for intestinal intraepithelial lymphocytes: murine CD3⁺, γ/δ TCR⁺ T cells produce interferon gamma and interleukin 5. *J. Immunol.* **147:**3736–3744.

36. **Taguchi, T., J. R. McGhee, R. L. Coffman, K. W. Beagley, J. H. Eldridge, K. Takatsu, and H. Kiyono.** 1990. Analysis of Th 1 and Th 2 cells in murine gut-associated tissues. Frequencies of CD4⁺ and CD8⁺ cells that secrete IFN-γ and IL-5. *J. Immunol.* **145:**68–77.

37. **Taubman, M. A., and D. J. Smith.** 1989. Oral immunization for the prevention of dental disease. *Curr. Top. Microbiol. Immunol.* **146:**187–195.

38. **Xu-Amano, J., H. Kiyono, R. J. Jackson, H. Staats, K. Fujihashi, P. D. Burrows, C. O. Elson, S. Pillai, and J. R. McGhee.** 1993. Helper T cell subsets for immunglobulin A responses. Oral immunization with tetanus toxoid and cholera toxin as adjuvant selectively induces Th2 cells in mucosa-associated tissues. *J. Exp. Med.* **178:**1309.

Molecular Pathogenesis of Periodontal Disease
Edited by Robert Genco et al.
© 1994 American Society for Microbiology, Washington, DC 20005

Summary of Chapters 12 to 17

Patricia A. Murray and Roy C. Page

Unusually rapid progress has been made recently in our understanding of the role of cytokines and other effector molecules in mediating, perpetuating, and eventually suppressing the events making up acute and chronic inflammation. Studies in this area have been greatly facilitated by the availability of recombinant human and rodent cytokines, which are essential for developing assays as well as for in vitro studies of specific biological effects. Furthermore, the finding of under- or over-expression of a particular cytokine provides insight into pathogenic mechanisms of disease at the molecular level. Recent information in these areas is particularly relevant to understanding the pathogenesis of periodontitis.

Chapters 12 through 17 elucidate and support the general principle that components of periodontopathic bacteria activate resident and infiltrating inflammatory cells of the periodontium directly and indirectly to produce and release cytokines, growth factors, and other effector molecules, such as the metalloproteinases, that initiate, mediate, and modulate the progressive destruction of alveolar bone and the connective tissue matrix of the periodontium in periodontitis.

A dominant feature of advanced periodontitis is the predominance of plasma cells in the inflammatory infiltrate. In chapter 12, Kiyono and colleagues investigate mechanisms that may account for elevated B-cell responses at the local disease site. Using PCR, they demonstrated that mononuclear cells harvested from the gingivae of patients with severe adult periodontitis produced mRNA for several cytokines, including gamma interferon, interleukin-5 (IL-5), and IL-6 but not IL-2 or IL-4, as well as synthesizing these polypeptides. The producing cells were $CD3^+$ T^- lymphocytes. These results suggest that CD3 T cells are responsible for the production of select cytokines that in turn are essential for the B-cell responses in tissues from patients with periodontitis.

The role of T cells in regulating B-cell activities in periodontitis is further elucidated in chapter 13 by Taubman and colleagues, who adoptively transferred cloned Th1 and Th2 lymphocytes specific for antigens of *Actinobacillus actino-*

Patricia A. Murray • Department of Periodontology, New Jersey Dental School, University of Medicine and Dentistry, Newark, New Jersey 07103. *Roy C. Page* • Center for Research in Oral Biology, University of Washington, Seattle, Washington 98195.

mycetemcomitans into rodents that had been orally infected or not infected with this organism. The transferred cells stimulated production of *A. actinomycetem-comitans*-specific antibody and suppressed periodontitis. The Th2 cells but not the Th1 cells migrated to the gingival tissues of *A. actinomycetemcomitans*-infected but not uninfected animals. This finding demonstrates antigen-directed lymphocytic migration and retention at the local site. Thus, trafficking of T-lymphocyte subsets appears to be influenced by antigenic specificity and the clonal type of the cells. The same investigators demonstrated that mononuclear cells harvested from gingivae of patients with periodontitis were enriched in CD8$^+$ cells and that CD4$^+$ cells expressing CD29 and CD45RA, markers characteristic of memory cells, were also prominent. The data demonstrated that cytokines mediate communication between various subsets of lymphocytes in the periodontium. Additional studies to elucidate the details and exact roles of each subset are now needed.

Wahl et al. describe in chapter 16 the interaction of cytokines, integrins, and nitric oxide in the evolution and regulation of chronic inflammatory responses such as periodontitis. Cytokines and bacterial components such as lipopolysaccharide can initiate inflammatory responses by activating expression of specific cell surface receptors on circulating leukocytes and vascular endothelial cells, resulting in adhesion and directed migration of the leukocytes into the tissues. Cytokines also up-regulate the α_5 and β_1 integrins that interact with cell surface and matrix fibronectin in determining cell-cell and cell-matrix interactions, thereby directing leukocytes to and mediating their retention at sites of inflammation. Activated leukocytes and the molecules they release participate in destruction of the invading pathogens as well as in destruction of the tissue. In long-standing inflammation, the continuous recruitment of leukocytes that occurs can be interrupted by fibronectin-derived synthetic peptides. Various mediators trigger the release of nitric oxide, which can exert cytotoxic and cytostatic effects on the infiltrating inflammatory and resident connective-tissue cells. An inhibitor of nitric oxide, N^G-monoethyl-L-arginine can prevent accumulation of inflammatory leukocytes and contribute to resolution of the inflammatory lesions.

In chapter 14, Shenker and colleagues describe a model system in which some of the unresolved issues raised by the experiments described above may be resolved. They reconstituted mice with severe combined immunodeficiency by using circulating lymphocytes harvested from patients with localized juvenile periodontitis (LJP) who were infected with *A. actinomycetemcomitans* and manifested serum antibodies specific for antigens of this organism and by using cells from periodontally healthy control subjects. Animals in both groups began producing human immunoglobulins within 2 weeks and continued to do so for about 8 weeks. Sera from mice reconstituted with LJP patient cells produced antibodies to the same repertoire of *A. actinomycetemcomitans* antigens recognized by the sera of human patient donors, including antibody to *A. actinomycetemcomitans* leukotoxin, which was functional in neutralizing leukotoxin activity. Animals reconstituted with LJP patient cells were better able to tolerate infection by live *A. actinomycetemcomitans* than were animals reconstituted with cells from healthy control subjects. Thus, the antibodies produced appeared to be functional.

Cytokines play a determinative role in bone resorption and regeneration in periodontitis. Both IL-1 and, to a lesser extent, tumor necrosis factor (TNF) are capable of mediating bone resorption. Cytokine levels are elevated in periodontal pocket fluid and diseased gingival tissue. These mediators are modulated by and act synergistically with prostaglandin E_2. In chapter 15, Stashenko et al. report on the mechanism by which IL-1 and TNF down-regulate bone matrix formation by osteoblasts and inhibit bone formation. Osteocalcin is the major noncollagenous component of bone matrix. These investigators demonstrated that TNF-α inhibited production of mRNA for osteocalcin at the transcriptional level, while IL-1 inhibited production at the translational level. TNF inhibition is controlled by a response element in the osteocalcin gene promotor. Thus, multiple mechanisms are probably involved in cytokine-mediated alterations in bone.

A large body of evidence indicates that a family of nine or more Zn^{2+}-dependent enzymes known as the matrix metalloproteinases (MMP) plays a major role in destruction of collagenous tissues at the sites of acute and chronic inflammation. Birkedal-Hansen describes this phenomenon and his own work in this area in chapter 17. Transcription of the genes for this family of enzymes is activated by the proinflammatory cytokines and by growth factors, including IL-1, TNF-α, epidermal growth factor, and transforming growth factor α (TGF-α), and is repressed by IL-4, gamma interferon, and anti-inflammatory steroids. TGF-β is especially interesting inasmuch as it activates transcription in some cells but suppresses it in others. The newly synthesized and secreted enzymes are maintained in a latent state by reversible coordinate bonding of a propeptide Cys (-SH) moiety to the active-site Zn^{2+}, which, when catalytically altered, results in conversion to active enzyme. Two of the MMPs are true collagenases and are considered to be responsible for initial cleavage of the collagen triple helix and subsequent collapse of the stromal matrix. Immunolocalization studies and analysis of gingival crevicular fluids have shown that the MMPs are abundantly synthesized in gingival polymorphonuclear leukocytes, fibroblasts, and macrophages. Immunoreactivity was demonstrated most frequently in fibroblasts and macrophages and occasionally in epithelial and vascular endothelial cells.

Molecular Pathogenesis of Periodontal Disease
Edited by Robert Genco et al.
© 1994 American Society for Microbiology, Washington, DC 20005

Summary of Chapters 18 to 25

J. L. Ebersole and R. P. Ellen

Periodontitis represents a destructive disease of the soft and hard tissue supporting structures of the tooth that demonstrates aspects of both acute and chronic inflammation. On the basis primarily of clinical findings, studies over the past 15 years have suggested that a subset of subjects with progressing periodontitis show a disease pattern consistent with a burst of activity and exacerbation of the disease process followed by periods of remission or quiescence. This pattern suggests that active disease episodes might be defined by intervals of acute inflammation and accompanying tissue destruction superimposed on a background of more chronic inflammation and remodeling in response to the accumulation of subgingival plaque. While inflammation may be a contributor to the deleterious outcomes noted in periodontitis, it is clear that the host inflammatory response is a natural consequence of innate protective mechanisms defending the host against pathogenic microorganisms. Furthermore, the inflammatory response is a required prelude to the induction of a specific immune response.

In chapter 18, Offenbacher et al. report on prostaglandin E_2 (PGE_2), one of the cyclooxygenase pathway mediators that has classically been implicated in both gingival inflammatory responses and mineralized connective tissue resorption. The novel approaches in their presentation are their use of subjects grouped into different disease classifications according to demonstrated susceptibility or resistance to tissue destruction and their documentation of subjects' inherent differences in PGE_2 responses to treatment. Their hypothesis is that the PGE_2 response is a crucial determinant in the host reaction to bacteria that culminates in tissue destruction and that PGE_2 levels respond to treatment of periodontitis. Furthering their previous findings that PGE_2 levels in crevicular fluid are associated with periods of gingival attachment loss, they used prospective, longitudinal models of experimental gingivitis and monitoring of gingival attachment loss in adults with adult or early-onset periodontitis. They demonstrated elevated PGE_2 levels in pockets of patients with progressive periodontitis, especially at sites undergoing attachment loss. In-

J. L. Ebersole • Department of Periodontics, University of Texas Health Science Center at San Antonio, San Antonio, Texas 78284. *R. P. Ellen* • Department of Periodontics, Faculty of Dentistry, University of Toronto, Toronto, Ontario M5G 1G6, Canada.

dividuals with early-onset types of periodontitis had greatly elevated PGE_2 levels. While there was a positive association of PGE_2 level and pocket depth in adult periodontitis subjects, the concentration of PGE_2 was equally high in pockets of different depths in the early-onset subjects. After infections in subjects with different clinical case types had been treated, all subjects improved clinically, but the mean PGE_2 level was reduced significantly only in the adult periodontitis cases.

Offenbacher's data seem to imply that elevated PGE_2 crevicular fluid levels might have greater significance as a measure of connective tissue destruction and healing responses than as a mere reflection of the degree of marginal gingival inflammation. Yet the closer match between PGE_2 fluctuations posttreatment and pocket depth improvement in adult cases in contrast to early-onset cases might imply that PGE_2 levels represent underlying tissue conditions and responses much more complex than those reflected in attachment level or pocket depth changes alone. Presumably, the improved clinical condition of early-onset cases should have also led to decreased mean PGE_2 levels, at least in the pockets with depths equivalent to those in the adult periodontitis cases. Thus, there must be inherent differences between the pathways and mechanisms by which PGE_2 is stimulated, generated, and/or metabolized in lesions of persons with early-onset periodontitis and of those with adult periodontitis. The findings also raise the possibility that early-onset cases are relatively intractable owing to lingering elevated levels of mediators such as PGE_2 despite treatment. Alternatively, strategies for treating these two case types might need to differ, as therapy aimed at removing the insult of infection evidently does not uniformly affect the mean level of a mediator, PGE_2, considered significant in pathways leading to (and presumably maintaining) connective tissue resorption in persons with different periodontal diagnoses.

Murayama et al., in chapter 19, scrutinize the underlying signaling mechanisms that guide the rapid emigration and accumulation of polymorphonuclear leukocytes (PMNs) at inflammatory sites, including the periodontium. Previous studies have aimed at defining the soluble molecules responsible for recruitment and describing the inflammatory cell receptors for these chemoattractants, the changes in phenotypes of the inflammatory cells during directed migration, and the molecules necessary for cell-cell interactions presaging migration. Other studies have indicated that alterations in the chemotactic ability of the PMNs increase susceptibility to various infections, including periodontitis. PMN defects represent alterations in the expression of chemotactic receptors (i.e., for N-formyl-l-methionyl-l-leucyl-l-phenylalanine, C5a, and interleukin-8 [IL-8]) on the surface of PMNs. Ongoing research is directed toward delineating a genetic and/or environmental basis for these functional alterations.

More recently, it has been demonstrated that expression of specific cell surface receptors on both endothelial cells and PMNs is required for the process of margination, whereby PMNs line the endothelial vascular surface prior to diapedesis into the tissues in inflammatory sites. Descriptions of patients with inherent defects in which subjects lack the ability to express specific receptors have provided information about additional mechanisms resulting in functional defects of PMNs. Such patients evidently exhibit leukocyte adhesion deficiencies. Investigating these deficiencies in depth is the objective of Murayama et al.'s investigations. Their chapter specifically addresses alterations in the CD11/CD18 complex of receptors, which is required for PMN adherence to endothelial cells. Defects in this receptor complex result in severe complications in bacterial infections. The chapter addresses the potential for changes in this complex to contribute to our understanding of susceptibility to periodontitis. Specifically, expression of the CD18 antigen was documented for generalized prepubertal periodontitis and localized juvenile periodontitis patients. Various mutations are known to exist in CD18. Murayama's laboratory compared three patients having defects in the expression of CD18 with controls. The patients with CD18 defects had more severe early-onset inflammation and bone loss. Some deficiencies in PMN adhesion functions of CD18-deficient generalized juvenile periodontitis and localized juvenile periodontitis subjects were shown, but the small sample limits the interpretation of these findings. Although a few of Murayama's localized juvenile periodontitis subjects evidently had normal transcription and translation of the CD18 gene, the possibility of repressed transcription of CD18 in the generalized juvenile periodontitis subjects was raised. These subjects evidently had mutations in the same region, which was reflected in the transcribed mRNA.

Murayama's work concentrates on "experiments of nature," documenting deficiencies in inflammatory cells collected from patients with perhaps rather rare conditions. There is a more general growing interest in cell recognition molecules significant for PMN-endothelial adhesion and diapedesis. Moreover, there have been recent reports in the literature on infectious diseases of bacterial mimicry of host cell adhesion molecules that might be significant for microbial invasion through endothelial barriers, such as in the progression of respiratory *Bordetella pertussis* infections and abrogation of the blood-brain barrier in meningitis. It is possible that work such as Murayama's on severe early-onset diseases will lead to parallel situations in which fundamental PMN recognition molecules not only are found to be significant for guiding chemotactic functions of PMNs but also might serve as templates for the evolution of complex microbe-inflammatory cell interactions at endothelial surfaces.

In chapter 20, Mergenhagen et al. continue the emphasis on inflammatory cell recognition and communication functions and their effects on cell responses leading to release of inflammatory mediators. They concentrate on the effects of receptor-mediated cellular adhesion to extracellular matrix glycoproteins, which has been implicated in the localization of inflammatory cells at sites of tissue injury. Cell-matrix interactions also influence the functions of mast cells, PMNs, eosinophils, and monocytes. In particular, binding of macrophages to the extracellular matrix up-regulates transcription of a variety of cytokines and molecules that respond to cytokines.

Mergenhagen et al. describe in vitro models for exploring mast cell adhesion to fibronectin (Fn) and other matrix proteins and the effect of adhesion on cell functions. They demonstrate that R-G-D peptides inhibit the binding of mast cells to Fn, implicating integrins as putative mast cell receptors for Fn. They show that cell shape and the distribution of immunolabeled microfilaments differ between mast cells in suspension and those adherent to Fn, suggesting that adhesion stimulates cytoskeletal reorganization. Moreover, the amount of histamine released in response to cross-linking of surface immunoglobulin E (IgE) receptors virtually doubled in mast cells plated on Fn. Likewise, there appeared to be qualitative differences in arachidonic acid metabolite profiles of these cells. Mast cell changes similar to those seen following adhesion to Fn were also found in mast cells stimulated by calcium ionophores, suggesting a similarity in responses to transmembrane signals or in the signaling networks themselves. Furthermore, cells adherent to Fn had some different proteins phosphorylated, as detected in Western blots (immunoblots) of membranes probed with antiphosphotyrosine antibodies. Stimulating cells immunologically and by adhesion to Fn seemingly led to synergistic phorphorylation.

Mergenhagen et al.'s work on mast cells is a pointed demonstration of the growing interest in cell-extracellular matrix communication. As an experimental model exemplifying concepts reviewed by Nathan and Sporn (J. Cell Biol, 1991), the work of Mergenhagen et al. demonstrates multiple levels through which cellular messages pass, tying together intracellular, plasma membrane, cytokine, and extracellular matrix signaling networks.

While chapters 18 through 20 center on extravascular cell responses, Herzberg et al. describe in chapter 21 the role of platelets at sites of inflammation, with particular attention to the periodontium. Platelets aggregate at sites of inflammation and wound healing and are central in initiating vascular responses to immune complexes. Platelets share a variety of proinflammatory functions with monocytes and PMNs. These include the release of arachidonate metabolites, platelet-derived growth factor, lysosomal enzymes, and platelet-activating factor. These factors are produced and released as a consequence of platelet aggregation and activation by thrombin, other intravascular stimulants, and specific bacteria like *Streptococcus sanguis* that Herzberg has studied for several years. Signal transduction pathways in platelets are evidently similar to those in other cells. However, Herzberg stresses that although G proteins are well known to be important for platelet responses to extracellular agonists, G proteins are evidently not involved in signal transduction secondary to *S. sanguis* stimulation. Moreover, since platelets have no nucleus, signal transduction is independent of regulation at the gene level. Platelets have preformed messengers. Since they do respond to some bacteria, platelets might represent an opportunity for investigators to study fundamental pathways of signal transduction induced by periodontal pathogens in a less complex cell model.

Herzberg's laboratory has previously shown that strains of *S. sanguis* express three different systems for activating platelets, including a surface antigen that mimics a consensus platelet-interactive domain of collagen and can conceivably cause platelet aggregation in the gingival sulcus. They have more recently conducted experiments using a library of transposon mutants in the *S. sanguis paap* gene and have thus identified an amino acid sequence significant for signaling platelets. Revertants of the mutants were able to stimulate platelets just as the parent *S. sanguis* strain did.

In their chapter, Herzberg et al. turn their attention to species considered to be more significant in periodontitis. The hypothesis is that putative pathogens share the capacity and perhaps the same mechanism for aggregating and activating platelets. Neither *Actinobacillus actinomycetemcomitans, Campylobacter rectus, Pre-*

votella intermedia, Eikenella corrodens, nor *Fusobacterium nucleatum* was able to aggregate platelets. However, all four strains of *Porphyromonas gingivalis* tested yielded time- and dose-dependent aggregation of platelets. The activity in *P. gingivalis* ATCC 33277 appeared to be associated with a 90-kDa surface antigen that cross-reacted immunologically with the platelet-interactive domain of *S. sanguis*, although *P. gingivalis* DNA failed to hybridize at high stringency with a 3.6-kb *paap* gene probe. These findings suggest that fragments of *P. gingivalis* or other invasive bacteria might contribute to inflammation in periodontitis partly by the local direct activation of platelets. Herzberg's findings should encourage more investigators to study pathways by which periodontal pathogens might stimulate vascular and other responses via direct or immune activation of platelets.

Chapters 18 through 21 in this section emphasize aspects of the inflammatory responses that contribute to the inflammatory signaling network processes in periodontitis. If the inflammatory response is not sufficient to eliminate the noxious material (i.e., bacterial infection), a more specific immune response is activated. This response can take the form of cellular and/or humoral responses. However, as the pathogens associated with periodontitis have been primarily identified as extracellular pathogens, the emphasis in host defenses resides in the production of specific antibody molecules to interfere with bacterial virulence factors. Chapters 22 through 25 thus direct their attention to the characteristics and functions of antibody produced in both human and animal models of periodontitis, molecular concepts of antigen structure for the induction of protective immunity to bacterial antigens, and characteristics of the cell types and cytokine profiles that contribute to the induction of IgA immune responses at mucosal surfaces.

Previous studies by many laboratories have identified a complex of antigens expressed by *A. actinomycetemcomitans* or *P. gingivalis* to which humans with periodontitis demonstrate specific and elevated serum antibody responses. In chapter 22 by Okuda et al., results that attempt to better define the function and specificity of antibodies to these pathogens are reported. It is shown that *P. gingivalis* ATCC 33277 is susceptible to antibody-mediated killing with either immune rabbit or human periodontitis sera. However, a virulent strain of *P. gingivalis*, 16-1, is resistant to this activity. The authors suggest the existence of a 60-kDa antigen that may be related to the lipopolysaccharide or capsular material present on invasive strains like 16-1 and that may confer a phenotype of resistance to these strains.

Okuda et al. further describe the preparation of a monoclonal antibody that reacted with a 150-kDa antigen from *C. rectus* and has been described as an S layer produced by this microorganism. The S or A layers of bacteria have been described as extracellular proteinaceous structures that alter the bacterial interactions with host cells and contribute to resistance to phagocytosis. Thus, Okuda et al. propose that antibody to the *P. gingivalis* antigen described and the S layer from *C. rectus* may be involved in eliminating the bacteria.

Okuda et al. then describe their findings related to levels of antibody to the fimbriae of *A. actinomycetemcomitans*. In these studies, levels of antibody to fimbriae were higher in periodontitis patients than in healthy subjects; however, these levels were independent of infection with the microorganism. The outcomes indicate that the fimbrial antigen is shared among *A. actinomycetemcomitans* strains, and it was proposed that antibody to this structure may inhibit colonization by *A. actinomycetemcomitans*. Additionally, they report findings demonstrating a higher avidity of antibody to the fimbriae in non-*A. actinomycetemcomitans*-infected patients than in infected patients. This was also noted with rabbit antibodies reactive with synthetic peptides prepared from the fimbriae. Recently, similar studies of antibody avidity have been undertaken by others attempting to more clearly elucidate the potential function of systemic and local antibodies elicited by periodontopathogens. Reports describing avidity of human antibody to *P. gingivalis* and *A. actinomycetemcomitans* have generally suggested that the host response to this microorganism does not appear to mature during the infection and that the low avidity of the antibody may contribute to continuing susceptibility to disease. Furthermore, treatment of periodontitis patients with low-avidity antibody to *P. gingivalis* increases the antibody avidity and presumably the functioning of the immune system. Therefore, a potential strategy for future research might be to address augmenting the maturation of active immune responses to antigens critical for survival of the pathogens in the subgingival microbial environment.

Chapter 23 by Evans et al. deals with a gnotobiotic animal model for studying the role of virulence factors and immunization against periodontitis. They emphasize use of the gnotobiotic rat in studies using bone loss as an outcome measure. These studies with rats monoinfected with *P. gingivalis* indicate that *P. gingivalis* lacking fimbriae is incapable of causing bone loss in this model. *P. gingivalis* was detectable by cultivation for 10 days postinoculation, with maximum bone loss occurring shortly after loss of detection of the organism. Reduction in levels of organism was also related to a rise in serum and salivary antibody levels. When fimbria-positive *P. gingivalis* colonized the rats, there was a local increase in crevicular fluid levels of collagenase, cathepsins B and L, and gelatinase at sites of infection, potentially indicative of a destructive process. Immunization led to a marked diminution of the local enzyme concentrations. The chapter describes a design to test the hypothesis that induction of immune responses to the fimbriae or fimbrial peptides would provide protection against periodontitis caused by *P. gingivalis*. The results indicate a significant difference in bone loss between immunized and control rats with fimbrillin peptides as immunogens.

One shared aspect of the approaches used by both Okuda and Evans is a proposal to use synthetic peptides for critically examining bacterial components as antigens. Such a use is also the topic of chapter 24 by Lehner et al. This report provides a description of the strategies used to identify significant epitopes of antigens from *S. mutans* with the potential for induction of protective immunity. The techniques of gene cloning and DNA sequencing and predictions of amino acid sequences have made possible the generation of polypeptide models and allowed an indication of probable domains as immunogens for vaccine development. Lehner emphasizes the studies of a streptococcal antigen of 185 kDa that is expressed on the surface of *S. mutans*. The epitope in this protein that is involved in adhesion has been identified by using monoclonal antibodies. These antibodies also inhibit *S. mutans*-induced dental caries in nonhuman primates and adhesion of this bacterium to human teeth. Lehner et al. report the procedures necessary to map the functional adhesion epitope as well as those epitopes that determine T- and B-cell recognition of the antigen. Using overlapping peptides, this group has mapped three determinants. The goal is to prepare a designer peptide vaccine consisting of the functional epitope and those epitopes that would enhance immune recognition of the immunogen. This vaccine could then be tested using a gingival, oral, or systemic immunization route.

The final contribution to this section, by McGhee et al., addresses the various cellular interactions controlling the development of mucosal IgA immune responses. It has been clear for many years that the IgA system is under important T-cell regulatory influences. These T cells and IgA plasmacytes derive from antigen stimulation of mucosa-associated lymphoid tissue, including the gut-associated lymphoreticular tissue, upper respiratory tract, and exocrine glands. Data that support the concept that mucosal immunity is manifested by specific T-cell responses and IgA antibody formation are available. McGhee et al.'s chapter focuses on the characteristics of the T helper (Th) cells that contribute to these immune responses. They present evidence that the Th cells principally regulating the mucosal IgA responses are Th2 cells characterized by production of IL-4 and IL-5 in the absence of IL-2 and gamma interferon. Thus, the results support the idea that mucosal vaccines should be optimized to activate Th2 cells for an up-regulation of IgA-producing B lymphocytes. McGhee et al. further describe studies in which a murine system was utilized to delineate the characteristics of salivary immune responses to fimbriae from *P. gingivalis*. The administration of the fimbriae in conjunction with cholera toxin as an adjuvant induced significant numbers of IgA-producing cells in the salivary glands. The T cells associated with this response also appear to be of the Th2 type. Thus, it appears that the mucosal immune system is carefully controlled by T cells and specific cytokine profiles of these particular subsets. Successful development of oral vaccines will require this knowledge and the manipulation of the host system to best activate these particular immune cell phenotypes.

Overall, the chapters in this section describe interesting concepts of inflammatory cell-extracellular signaling networks and exciting strategies to enhance the development of actively and passively acquired immunity to interfere with the pathogenic potential of bacteria causing dental caries and periodontitis. The initiation of studies utilizing unique peptides and adjuvants should enhance our capabilities for the future development of a vaccine against dental caries. The identification of potential antigens and structures that may be critical components for the expression of virulence of periodontal pathogens will also stimulate continued scientific progress on mechanisms of disease and methods for interference with this destructive process.

HOST FACTORS: NEUTROPHILS, MAPPING T- AND B-CELL EPITOPES, AND GENETIC FACTORS

Molecular Pathogenesis of Periodontal Disease
Edited by Robert Genco et al.
© 1994 American Society for Microbiology, Washington, DC 20005

Chapter 26

New Ideas about Neutrophil Antimicrobial Mechanisms: Antibiotic Peptides, Postphagocytic Protein Processing, and Cytosolic Defense Factors

Kenneth T. Miyasaki, Amy L. Bodeau, William M. Shafer, Jan Pohl, A. Rekha K. Murthy, and Robert I. Lehrer

Neutrophils are short-lived, terminally differentiated end cells that protect the host against periodontal diseases and systemic infection by bacteria originating from subgingival plaque. Daily, about 1% of our circulating neutrophils leave the blood and emigrate into the junctional epithelium and gingival crevice, where they sweep over the sulcular and pocket epithelium to establish a defensive barrier against offending bacteria.

Effects of neutrophil interactions with bacteria within the gingival crevice that may promote periodontal health include decreasing the colonization, growth, and viability of periodontal pathogens (37) and perhaps favoring the colonization, growth, and viability of commensal bacteria. Additionally, neutrophils within the subjacent connective tissues may serve to maintain periodontal tissues directly by altering the activities of chronic inflammatory cells, including monocytes and lymphocytes (31), as well as other host cells. Our research focuses primarily on the interaction of neutrophils with bacteria.

Neutrophils maintain a complex cellular compartmentalization that includes a prominent segmented nucleus, at least two types of membrane-enclosed storage organelles (referred to as "granules"), cytosol, and cell membrane. Neutrophils accomplish their antimicrobial effects by (i) phagocytosis and the release of antimicrobial substances into a membrane-delimited structure, the phagolysosome; (ii)

Kenneth T. Miyasaki and Amy L. Bodeau • Section of Oral Biology and Dental Research Institute, School of Dentistry, University of California, Los Angeles, California 90024. *William M. Shafer* • Department of Microbiology and Immunology, School of Medicine, Emory University, Atlanta, Georgia 30322, and Laboratories of Microbial Pathogenesis and Research Sciences, Veterans Affairs Medical Center (Atlanta), Decatur, Georgia 30033. *Jan Pohl* • Microchemical Facility, Winship Cancer Center, School of Medicine, Emory University, Atlanta, Georgia 30322. *A. Rekha K. Murthy and Robert I. Lehrer* • Department of Medicine, School of Medicine, University of California, Los Angeles, California 90024.

the release of antimicrobial substances into an external milieu (such as the gingival crevice) by secretion, lysis, or apoptosis; or (iii) the exposure of microbes to antimicrobial substances within their cytosols or nuclei.

ANTIMICROBIAL SUBSTANCES OF HUMAN NEUTROPHILS

Neutrophils possess oxidative and nonoxidative mechanisms of controlling microorganisms (Fig. 1). The nonoxidative antimicrobial (microbicidal and microbiostatic) mechanisms are likely to be more important in the anaerobic gingival crevice. Phagocytic microbicidal activity has been related to the secretion of granule components into an endocytic vacuole, referred to as a phagosome, with the resultant formation of a phagolysosome. Many of the most potent antimicrobial substances reside within the azurophil granules that sequester defensins, cathepsin G, elastase, proteinase 3, azurocidin, lysozyme, and bactericidal/permeability-increasing protein. Specific granules contain several antimicrobial substances that may function extracellularly, including lactoferrin, lysozyme, and B_{12}-binding protein (cobalophilin).

Although the granule compartments remain important repositories of antimicrobial substances, other cellular compartments may also play a prominent role, since neutrophils are terminally differentiated. The cytosolic compartment of neutrophils has received recent attention as a source of an antimicrobial factor called calprotectin (43, 60–62). The segmented nucleus of the neutrophil remains enigmatic. Why should circulating neutrophils retain their nuclei? Some insight is pro-

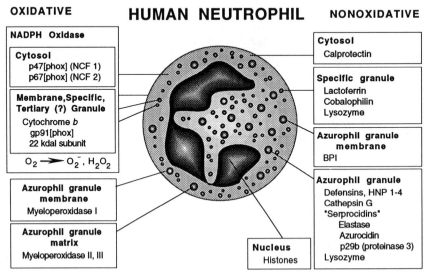

Figure 1. Schematic diagram of the human neutrophil, which contains antimicrobial weapons in virtually every cellular compartment. The azurophil granule contains a variety of enzymes, including myeloperoxidase and the neutral serine proteases, that are known to be capable of converting proprotein storage molecules into functional proteins. BPI, bactericidal/permeability-increasing protein.

vided by the recent isolation of three microbicidal proteins from mouse macrophage granules (24). These proteins were designated murine microbicidal proteins 1, 2, and 3, or Mump-1–3. Mump-1 and Mump-2 appear to be members of the highly variable H1 linker histone family. Mump-3 is identical to the mouse H2b core histone. The antimicrobial activities of histones may reflect an ancient biologic role. Perhaps neutrophils retain their nuclei because, unlike macrophages, they are terminally differentiated and do not need to package antimicrobial histones within the granule compartment.

Three emerging concepts regarding neutrophils and host defense will be developed here. First, neutrophils and other innate defense cells use small, metabolically inexpensive antibiotic peptides to deliver lethal blows to microorganisms. Second, these antibiotic peptides may be sequestered by small propieces or within larger molecular structures to prevent autotoxicity. Third, exposure of cytosolic (or nuclear) components to microbes does not necessarily represent the end of the neutrophil's contribution to host defense.

ANTIBIOTIC PEPTIDES

Antibiotic peptides are widely distributed throughout the animal kingdom and are produced not only in leukocytes but also in many other cells, including glandular, mucosal, and skin epithelium (5, 6, 10, 33, 54, 71). Several general comments may be made regarding the antibiotic peptides from the animal kingdom: (i) they are extremely diverse; (ii) they are composed of common L-amino acids, unlike microbial peptide antibiotics, which often possess unusual amino acids and D-isomers; (iii) they often possess unusual sequences; (iv) no consistent motif or domain is shared by all of them; and (v) many are quite ancient and span tissues, species, genera, and phyla (Table 1). Some antibiotic peptide sequences are highly conserved, as in the striking case of cecropins 1 and 1A, in which there is 100% amino acid sequence homology but only 73% homology at the nucleotide level (6). Cecropins, originally characterized in insects, have unexpectedly been isolated from the intestines of pigs (6). Bee venom mellitin exhibits modest homology with the frog magainins, and insects and vertebrates possess cysteine-rich antibiotic peptides with three internal disulfide bonds (22, 33, 34, 55, 67).

DEFENSINS

Defensins are small (approximately 3,200 to 4,000 Da), cationic, amphipathic, arginine- and cysteine-rich peptides composed of 29 to 38 amino acids. They make up 5 to 7% of the total protein and 30 to 50% of the azurophil granule content of human neutrophils (33). Defensins are membrane-permeabilizing molecules that can kill gram-negative bacteria, gram-positive bacteria, fungi, and eucaryotic cells (33). Defensins were originally found in leukocytes but have since been found in the intestinal Paneth cells of laboratory mice (cryptdin) and humans (human defensin 5) (16, 27). The human leukocyte defensins HNP-1, HNP-2, HNP-3, and

Table 1. Some peptide antibiotics from the animal kingdom

Source and molecular family	Source	Size (amino acids)	Representative sequence[a]	Reference
Invertebrates				
Cecropins	Hemolymph; *Cecropia, Sarcophaga, Drosophila* spp.; other insects, pigs	31–39	SWLSKTAKKLENSAKKRISEGIAIAIQGGPR (pig intestine)	6
Andropin (cecropin-like)	*Drosophila* males	34	VFIDILDKVENAIHNAAQVGIG-FAKPFEKLINPK	52
Melitin	*Apis mellifera* venom	26	GIGAVLKVLTTGLPALISWIKRKRQQ-amide	67
Apidaecins	*Apis mellifera* lymph fluid	18	GNNRPVYIPQPRPPHPRI (apidaecin Ia)	9
Tachyplesins	*Tachypleus tridentatus* hemocytes	17	KWCFRVCYRGICYRRCR-amide[b] (tachyplesin I)	49
Polyphemusin	*Limulus polyphemus* hemocytes	18	RRWCFRVCYRGFCYRKCR-amide[b] (polyphemusin I)	46
Sarcotoxins	*Sarcophaga perigrina* larva hemolymph	39	GWLKKIGKKIERVGQHTRDATIQGLGIAQQAAN-VAATAR-amide (sarcotoxin 1A)	34
Royalisin	*Apis mellifera* royal jelly	51	VTCDLLSFKGQVNDSACAANCLSLGAGGHCE-KGVCICRKTSFKDLWDKYF[c]	22
"Insect defensins"	*Phormia terranovae* (phormicins)	40	ATCDLLSGTGINHSACAAHCLLRGNRGGYCN-GKGVCVCRN[d] (phormicin A)	32
	Sarcophaga perigrina (sapecin)	40	ATCDLLSGTGINHSACAAHCLLRGNRGGYCN-GKAVCVCRN[d]	35
Vertebrates				
Magainin	*Xenopus laevis* skin		GIGKFLHSAGKFGKAFVGEIMKS (magainin-1)	5
Bombinin	*Bombina variegata* skin	27	GIGGALLSAAKVGLKGLAKGLAEHFN (unnamed member)	59

Name	Source		Sequence[a]	
Indolicidin	Bovine neutrophils	13	ILPWKWPWWPWPR-amide	54
Bactenecin 7 (PRP motif)	Bovine neutrophils	59	RRIRPRPPRLPRPRPRPLPFPRGPRPIPRPLPFP-RPGPRPRP	20
Bactenecin 5 (RPP motif)	Bovine neutrophils	42	RFRPPIRRPPIRPPFYPPFRPPIRPPIFPPIR-PPFRPPLRFP	20
Porcine PR-39 (RPP, PPR)	Pig intestines	39	RRRPRPPYLPRPRPPPFFPPRLPPRIPPG-FPPRFPPRFP-amide	1
Bactenecin (dodecapeptide)	Bovine neutrophils	12	RLCAIVVIRVCR[c]	51
eNAP-1 and -2	Equine neutrophils	50–70	DVQCGEGHFCHDxQTCCRASQGGxACC-PYSQGVCCADQRHCCPVGF . . . (eNAP-1)	10
β-Defensins	Bovine neutrophils	38–42	GPLSCGRNGGVCIPIRCPVPMRQIGTCFGRPVK-CCRSW[f] (BNBD-12)	55, 63
Tracheal antimicrobial peptide (defensin-like)	Bovine tracheal epithelium	38	NPVSCVRNKGICVPIRCPGSMKQIGTCVGRAVK-CCRKK[g] (TAP)	12
Defensins	Neutrophils, some macrophages (human, guinea pig, rabbit, rat)	29–38	ACYCRIPACIAGERRYGTCIYQGRLWAFCC[h] (HNP-1)	33
Cryptdin, HD5 (defensin-like)	Mouse and human intestinal Paneth cells	35	LDRLVCYCRSRGCKGRERMNGTCRKGHLLYTL-CCR[i] (cryptdin)	16, 27
Histatins	Human saliva	38	DSHEKRHHGYRRKFHEKHHSHREFPFYGDYG-SNYLYDN (HRP-2)	30

[a] Single-letter amino acid codes are as follows: A, alanine; C, cysteine; D, aspartic acid; E, glutamic acid; F, phenylalanine; G, glycine; H, histidine; I, isoleucine; K, lysine; L, leucine; M, methionine; N, asparagine; P, proline; Q, glutamine; R, arginine; S, serine; T, threonine; V, valine; W, tryptophan; Y, tyrosine.
[b] Disulfide bonds between cysteines 1 and 4 and cysteines 2 and 3.
[c] Disulfide bonds between cysteines 1 and 4, 2 and 5, and 3 and 6.
[d] Disulfide bonds unknown but assumed to be royalisin-like.
[e] One disulfide bond.
[f] Disulfide bonds between cysteines 1 and 5, 2 and 4, and 3 and 6.
[g] Disulfide bonds unknown but assumed to be β-defensin-like.
[h] Disulfide bonds between cysteines 1 and 6, 2 and 4, and 3 and 5.
[i] Disulfide bonds unknown but assumed to be defensin-like.

HNP-4 possess net charges at neutral pH of $+3$, $+3$, $+2$, and $+4$, respectively (33, 69). It is remarkable that laboratory mice that lack leukocyte defensins nevertheless possess the intestinal defensin cryptdin (16, 17).

Defensins may not kill all bacteria. We have observed that *Actinobacillus actinomycetemcomitans* is highly sensitive to the bactericidal effects of rabbit neutrophil defensin NP-1 but relatively resistant to the bactericidal effects of the human defensins HNP-1, HNP-2, and HNP-3 in vitro (40, 42). In contrast, the *Capnocytophaga* spp. are more sensitive to the human defensins and are killed under both aerobic and anaerobic conditions by these peptides.

In general, antibiotic peptides are synthesized as larger, nontoxic precursors that are referred to as prepropeptides (23, 65, 67). Human neutrophil defensins HNP-1, HNP-2, HNP-3, and HNP-4 originate as 93- to 95-amino-acid preprodefensins (Fig. 2) that undergo a series of cotranslational and posttranslational

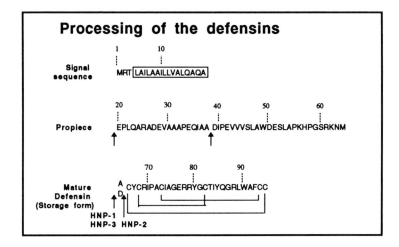

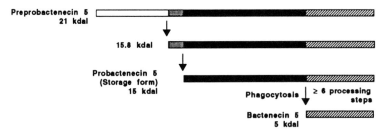

Figure 2. Defensin and bactenecin processing. Both molecules are translated as prepropeptides. The defensins are processed by proteases within the endoplasmic reticulum (suspected to include the signal peptidase, proline-directed proteases, and dipeptidyl aminopeptidases). The major difference between the defensins and the bactenecins in respect to processing is that the defensins are stored as mature antibiotic peptides, whereas the bactenecins are stored as propeptides virtually lacking antimicrobial activity.

modifications involving the removal of a 19-amino-acid signal sequence and a multiple-step removal of the 45-amino-acid propiece (65). In the rabbit system, the prodefensins are charge neutralized (36). It has been proposed that one of the initial events in the interaction of defensins with the target membrane is electrostatic. Thus, by charge neutralization, the anionic propiece may prevent defensin-mediated membrane disruption and leukocyte autocytotoxicity (36). Alternatively, the propiece may interact with the defensin by charge interaction and sequester sites involved in later membrane disruptive events. Either scenario would enable the propiece to maintain immature defensins in a biologically inactive form (36, 65). The processing of the prodefensins is essentially complete in mature neutrophils, and less than 0.25% of the total defensin content of neutrophils can be isolated as prodefensins (23). The toxic properties of defensins are presumably neutralized by storage in association with granule matrix proteoglycans.

Unlike defensins, certain bovine neutrophil antibiotics, the proline- and arginine-rich bactenecins Bac5 and Bac7, are packaged in granules in an immature, propeptide form (Fig. 2; 71) that is inert with respect to antimicrobial activities (53). Bactenecins are processed "postphagocytically" by neutral serine proteases. As evidence, diisopropyl fluorophosphate, a serine protease inhibitor that can cross biologic membranes, blocks the maturation of bactenecins in intact, phagocytosing bovine neutrophils (72), and conversely, purified leukocyte elastase (but not cathepsin G or proteinase 3) converts probactenecins to the active form (53).

The strategies used to process these simple peptides provide insight in several ways. Some multifunctional molecules require considerable additional processing after they have been released in order to exert multiple biologic effects (e.g., complement factor C3). Neutral serine proteases, often involved in the extracellular processing of plasma proteins, may be viewed as autoreactive enzymes that maintain proteolytic activity primarily for such late extracellular or intraphagolysosomal processing events.

CATHEPSIN G

Cathepsin G is a neutral serine protease found in the granules of neutrophils. In common with T-cell granzyme B and mast cell chymase, its structure features three disulfide bonds (most neutral serine proteases feature four highly conserved disulfide bonds). Clearly, it is a member of the granzyme B family (25, 58) and should not be grouped with leukocyte elastase, proteinase 3, and azurocidin in the coordinately regulated serprocidin gene family (73). Although cathepsin G is not remarkably active against laboratory substrates, the specificity of its esterolytic and proteolytic activities resembles that of chymotrypsin.

Our interest in this molecule is fueled by several observations. First, cathepsin G is the most potent nonoxidative antimicrobial agent against periodontal pathogens that we have isolated from the human neutrophil. Second, cathepsin G has two modes of killing bacteria: one that is enzyme independent and another that is enzyme dependent (38, 39, 41). The enzyme-independent mechanism, known since the 1970s (14), is active against *A. actinomycetemcomitans* but not *Capnocytophaga* spp. For many years, the functional significance of the chymotryptic activity of cathepsin G remained enigmatic, and the demonstration of its enzyme-dependent killing mechanism was novel. Our finding that cathepsin G required an active enzyme site to kill *Capnocytophaga* spp. was reinforced by others, who observed enzyme-dependent killing of *Pseudomonas* spp. (69). Moreover, azurocidin, an enzymatically inactive member of the neutral serine protease family, was found to exert no bactericidal activity against oral and nonoral bacteria unless it was supplemented with enzymatically intact leukocyte elastase or cathepsin G (40).

Although cathepsin G is a mature enzyme within the azurophil granule, it may be a propeptide with respect to certain antibiotic peptide domains. We believe that some antibiotic sites are sequestered within the cathepsin G molecule (some may be sequestered within azurocidin) and are exposed postphagocytically in an autoproteolytic manner when the enzyme approaches a target surface.

Using synthetic peptides (58), we have begun to map the primary sequence substructures (sites) of cathepsin G for antimicrobial activities against the oral bacteria (Fig. 3, top). Two sites are important for the sake of comparison. The N-terminal 20-mer oligopeptide CG01-20 and another sequence, CG61-80, are believed to be buried and exposed, respectively, in the intact molecule. These two sequences contain IIGGR and a portion of HPQYNQR, small peptides whose antimicrobial activities (3, 57) are presently under investigation as potential dental therapeutic agents (7, 44). Consistent with our hypothesis, the buried sequence kills both *A. actinomycetemcomitans* and *Capnocytophaga sputigena,* but the exposed sequence kills only *A. actinomycetemcomitans* (Fig. 3, bottom). A third sequence, CG77-96, which contains the entire HPQYNQR heptamer, has little antimicrobial activity against any bacteria. The precise significance of this observation is unclear and may represent a pitfall in the synthetic peptide approach, or it may be trying to tell us something very important in terms of how we protect ourselves from our own weapons. Other fragments of cathepsin G are intensely antimicrobial. For example, CG117-136 and CG198-223 also exhibit antimicrobial effects (58; Fig. 3, bottom). It is uncertain whether the synthetic peptides derived from cathepsin G represent functional sites on the molecule or actual peptides released by autoproteolysis.

The bactericidal activities of the defensins, the neutral serine proteases, and the synthetic peptides derived from cathepsin G are optimal under stringent, hypotonic conditions. Both the defensins and the serine proteases are neutralized by molecules such as α-1-antitrypsin, α-1-antichymotrypsin, and α-2-macroglobulin (38, 56). This may indicate that these bactericidal mechanisms are designed to function within a membrane-delimited space (i.e., a phagolysosome or a sealed cleft).

CALPROTECTIN

Calprotectin (a molecular complex that deserves the sobriquet "polynomin" for its many alternative names: L1 antigen, calgranulins A and B, migration inhibition factor-related proteins 14 and 8 or MRP14 and MRP8, and p8,14) makes

DESIGNATION	SEQUENCE	FORMULA WEIGHT
CG 01-20	IIGGRESRPHSRPYMAYLQI	2344
CG 21-40	QSPAGQSRCGGFLVREDFVL	2166
CG 41-60	TAAHCWGSNINVTLGAHNIQ	2107
CG 61-80	RRENTQQHITARRAIRHPQY	2531
CG 77-96	HPQYNQRTIQNDIMLLQLSR	2468
CG 97-116	RVRRNRNVNPVALPRAQEGL	2315
CG 117-136	RPGTLCTVAGWGRVSMRRGT	2161
CG 137-156	DTLREVQLRVQRDRQCLRIF	2544
CG 157-176	GSYDPRRQICVGDRRERKAA	2333
CG177-197	FKGDSGGPLLCNNVAHGIVSY	2148
CG 198-223	GKSSGVPPEVFTRVSSFLPWIRTTMR	2935

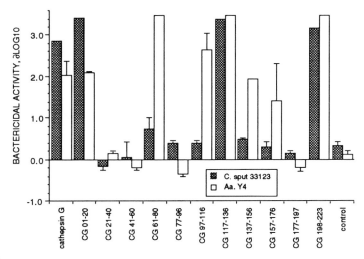

Figure 3. Synthetic 20-mer peptides derived from human neutrophil cathepsin G: formula weights, primary structures, and antimicrobial activities against two periodontal bacteria. *A. actinomycetemcomitans* FDC-Y4, and *C. sputigena* ATCC 33123. The decrease in viability after 2 h at 37°C in 1% Hank's balanced salt solution (pH 7.0) is shown as a $\delta\log_{10}$ value. The $\delta\log_{10}$ is the $\log_{10}$ decrease in viability relative to the viability of the control reaction (i.e., no antimicrobial peptide added) at $t = 0$ h (45). Bars and lines represent the means and standard deviations, respectively, of a quadruplicate assay.

up about 45% of the cytosolic protein of the neutrophils, which translates to an intracellular concentration of about 5 to 15 mg/ml (11, 15)! We and others have observed that calprotectin exerts microbiostatic activity against fungi, such as *Candida albicans,* and bacteria, including *C. sputigena* (43, 60–62). Calprotectin exerts profound microbiostatic activity in isotonic, nutritionally rich environments such as the extracellular milieu or within the cytosol of the host cell. Were a neutrophil to lyse, its cytosolic calprotectin could exert a microbiostatic effect within a sphere with a radius equal to about five leukocyte diameters.

Calprotectin has a specific and limited distribution, being found normally in neutrophils and certain monocytic phagocytes. Additionally, the complex has been detected immunochemically in oral/gingival keratinocytes (8, 18). It has not been found in Langerhans cells, interdigitating cells, lymphocytes, or any nonmyeloid cells of the pancreas, intestines, spleen, kidney, brain, bladder, placenta, lung, or healthy skin (2, 19, 66, 70). This remarkable distribution pattern suggests that calprotectin is a molecular complex of unique importance to oral health and mucosal defense.

Calprotectin, which is expressed during terminal cellular differentiation stages (2, 48), is composed of two subunits. The MRP14 subunit has an M_r of 14 kDa, and the MRP8 subunit (once referred to as the cystic fibrosis antigen) has an M_r of 10 kDa (once thought to be 8 kDa). In cells able to differentiate locally, the protein appears to be induced by inflammation (25). MRP14 (but not MRP8) is expressed in perivascular macrophages in lesions of acute inflammation, including gingivitis, but not in health (50). Both MRP8 and MRP14 are expressed in perivascular macrophages in chronically inflamed tissue (50). Calprotectin is also expressed in inflamed (psoriatic) epidermal keratinocytes (8).

MRP14 contains 114 amino acids, 25 of which are shared by most other members of the S-100 family. MRP8 contains 93 amino acids, with 30 in common with most other members of the S100 family. Neither the MRP8 nor MRP14 subunits contain signal or membrane anchor sequences, and neither contains consensus sequences for N-linked glycosylation (50). The S-100 proteins possess four known functional domains: two calcium-binding domains (two "EF hands," one basic, one acidic) and two hydrophobic N-terminal and C-terminal domains (Fig. 4). Hydropathy plots drawn using the SOAP7 algorithm of Kyte and Doolittle (29) reveal the striking similarities between MRP14 and MRP8 for the first 90 amino acids. Both molecules exhibit extensive hydrophilic regions between their two hydrophobic domains, a characteristic of the S-100 proteins that might be expected from small cytosolic molecules, which should possess a relatively large surface-to-volume ratio.

The carboxy termini and hinge regions of S-100 proteins exhibit the greatest divergence (28). The hinge region may bind these molecules to other specific effector molecules, perhaps protein kinases, in the cell cytoplasm. MRP14 possesses a unique, 26-amino-acid C-terminal domain (residues 88 to 114) that exhibits a sequence not shared by other members of the S-100 family but instead is identical to the N-terminal sequences of neutrophil-immobilizing factors 1 and 2 (NIF-1 and NIF-2 are 41- and 39-amino-acid peptides, respectively) (21). In general, the basic EF hand of S-100 proteins has a low calcium-binding affinity (K_d = 200 to 500 μM), and the acidic EF hand has a higher calcium-binding affinity (K_d = 20 to 50 μM), suggesting that S-100 proteins can exhibit several distinct behaviors depending on the concentration of Ca^{2+} (28). The basic and acidic EF hands of MRP14 and MRP8 (residues 24 to 37 and 65 to 77 in MRP14 and residues 20 to 33 and 57 to

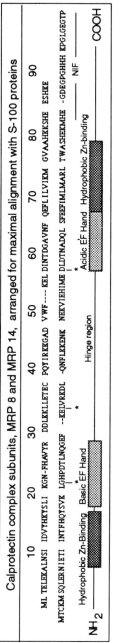

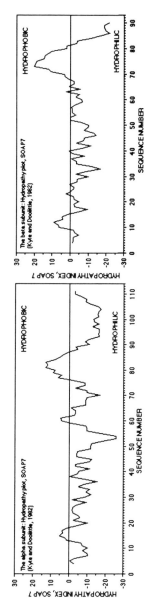

Figure 4. Schematic diagram of calprotectin subunits MRP14 and MRP8: primary sequence, hydropathy plots, and SOAP7 algorithm (29). S-100 proteins are characterized by two hydrophobic domains and one acidic and one basic EF hand.

74 in MRP8, respectively) are well conserved, suggesting that they should be fully functional (13).

Calprotectin located from neutrophils is a highly heterogeneous mixture composed of MRP8, MRP14, and homodimeric, heterodimeric, and multimeric species (2, 4, 11, 15, 43, 61, 64). Both MRP8 and MRP14 possess a single cysteinyl residue, which limits the number of disulfide-based covalent isomeric forms of calprotectin to two homodimers and a heterodimer. There are at least two forms of MRP14, which have pIs of 5.5 and 6.0 and M_rs of 13.3 and 12.5 kDa (4). MRP8 may also possess two forms. We have observed that MRP8 exhibits a doublet banding pattern by sodium dodecyl sulfate-polyacrylamide gel electrophoresis that is apparent on Coomassie-stained polyvinylidene difluoride membranes (43). It has been suggested that the calprotectin isoforms result from posttranslational processing, but this was not proven (4).

We have observed that calprotectin exerts profound microbiostatic activity against the periodontal bacterium *C. sputigena* (43). The mechanism of action of calprotectin is unknown. Calprotectin appears to bind weakly to fungi (62) and is capable of blocking protein kinase activity in eucaryotic cells (47). The antimicrobial activity of calprotectin can be blocked by MAC387 antibody at a ratio of 1:5, MAC387:calprotectin (61). However, because MAC387 reacts against both MRP8 and MRP14 (there are conflicting reports as to whether this is true), it remains to be resolved as to whether one or both subunits are required for antimicrobial effects to occur.

SUMMARY

Neutrophils are economical cells. They sequester proteins with antimicrobial activities within every major subcellular compartment. Among the cheapest (smallest) units neutrophils use to control microbes are defensins, peptides with broad-spectrum antibiotic properties. The defensins are among the diverse antibiotic peptides in the animal kingdom. Bacteria that are relatively resistant to the human defensins may be controlled by more elaborate and expensive "multiple-warhead" weapons such as cathepsin G. Similar to the bactenecins, some antibiotic sites of cathepsin G may require postphagocytic processing by neutral serine proteases for activation prior to delivery to the pathogen. Calprotectin is a cytosolic molecule that can exert antimicrobial effects outside of the phagolysosome. It is able to exert antimicrobial effects in a nutritionally rich environment, a characteristic that should enable it to play an important, nonphagocytic, defensive role against mucocutaneous, intracellular, and gingival crevicular infections. Future research of the antimicrobial systems of neutrophils will enable us to identify individuals at risk for specific periodontal infections and to understand the biology of a truly remarkable cell.

Acknowledgments. This work was supported in part by U.S. Public Health Service grant DE00282 from NIDR (K.T.M), CFAR seed grant NIAID 28697 (K.T.M.), NIAID AI-21105 (W.M.S.), and NIAID AI 22839 (R.I.L.). It was also funded by the Veteran's Affairs Research Service (W.M.S.).

REFERENCES

1. **Agerberth, B., J.-Y. Lee, T. Bergman, M. Carlquist, H. G. Boman, V. Mutt, and H. Jörnvall.** 1991. Amino acid sequence of PR-39. Isolation from pig intestine of a new member of the family of proline-arginine-rich antibacterial peptides. *Eur. J. Biochem.* **202:**849–854.

2. **Andersson, K. B., K. Sletten, H. B. Berntzen, I. Dale, P. Brandtzaeg, E. Jellum, and M. K. Fagerhol.** 1988. The leukocyte L1 protein: identity with the cystic fibrosis antigen and the calcium-binding MRP-8 and MRP-14 macrophage components. *Scand. J. Immunol.* **28:**241–245.

3. **Bangalore, N., J. Travis, V. C. Onunka, J. Pohl, and W. M. Shafer.** 19909. Identification of the primary antimicrobial domains in human neutrophil cathepsin G. *J. Biol. Chem.* **265:**13584–13588.

4. **Berntzen, H. B., and M. K. Fagerhol.** 1988. L1, a major granulocyte protein: antigenic properties of its subunits. *Scand. J. Clin. Lab. Invest.* **48:**647–652.

5. **Bevins, C. L., and M. Zasloff.** 1990. Peptides from frog skin. *Annu. Rev. Biochem.* **59:**395–414.

6. **Boman, H. G.** 1991. Antibacterial peptides: key components needed in immunity. *Cell* **65:**205–207.

7. **Boomer, S., W. A. Coulter, and D. J. S. Guthrie.** 1991. Effects on oral bacteria of a peptide derived from human cathepsin G. *Biochem. Soc. Trans.* **20:**61S.

8. **Brandtzaeg, P., I. Dale, and M. K. Fagerhol.** 1987. Distribution of a formalin-resistant myelo-monocytic antigen (L1) in human tissues. II. Normal and aberrant occurrence in various epithelia. *Am. J. Clin. Pathol.* **87:**700–707.

9. **Casteels, P., C. Ampe, F. Jacob, M. Vaeck, and P. Tempst.** 1989. Apidaecins: antibacterial peptides from honeybees. *EMBO J.***8:**2387–2391.

10. **Couto, M. A., S. S. L. Harwig, J. S. Cullor, J. P. Hughes, and R. I. Lehrer.** 1992. Identification of eNAP-1, an antimicrobial peptide from equine neutrophils. *Infect. Immun.* **60:**3065–3071.

11. **Dale, I., M. K. Fagerhol, and I. Naesgaard.** 1983. Purification and partial characterization of a highly immunogenic human leukocyte protein, the L1 antigen. *Eur. J. Biochem.* **134:**1–6.

12. **Diamond, G., M. Zasloff, H. Eck, M. Brasseur, W. L. Maloy, and C. L. Bevins.** 1991. Tracheal antimicrobial peptide, a cysteine-rich peptide from mammalian tracheal mucosa: peptide isolation and cloning of cDNA. *Proc. Natl. Acad. Sci. USA* **88:**3952–3956.

13. **Dorin, J. R., M. Novak, R. E. Hill, D. J. H. Brock, D. S. Secher, and V. Van Heyningen.** 1987. A clue to the basic defect in cystic fibrosis from cloning the CF antigen gene. *Nature* (London) **326:**614–617.

14. **Drazin, R. E., and R. I. Lehrer.** 1977. Fungicidal properties of a chymotrypsin-like cationic protein from human neutrophils: adsorption to *Candida parapsilosis. Infect. Immun.* **17:**382–388.

15. **Edgeworth, J., M. Gorman, R. Bennett, P. Freemont, and N. Hogg.** 1991. Identification of p8,14 as a highly abundant heterodimeric calcium binding protein complex of myeloid cells. *J. Biol. Chem.* **266:**7706–7713.

16. **Eisenhauer, P. B., S. S. L. Harwig, and R. I. Lehrer.** 1992. Cryptdins: antimicrobial defensins of the murine small intestine. *Infect. Immun.* **60:**3556–3565.

17. **Eisenhauer, P. B., and R. I. Lehrer.** 1992. Mouse neutrophils lack defensins. *Infect. Immun.* **60:**3446–3447.

18. **Eversole, L. R., K. T. Miyasaki, and R. E. Christensen.** 1992. The distribution of the antimicrobial protein, calprotectin, in normal oral keratinocytes. *Arch. Oral Biol.* **37:**963–968.

19. **Flavell, D. J., D. B. Jones, and D. H. Wright.** 1987. Identification of tissue histiocytes on paraffin sections by a new monoclonal antibody. *J. Histochem. Cytochem.* **35:**1217–1226.

20. **Frank, R. W., R. Gennaro, K. Schneider, M. Przybylski, and D. Romeo.** 1990. Amino acid sequences of two proline-rich bactenecins. Antimicrobial peptides from bovine neutrophils. *J. Biol. Chem.* **265:**18871–18874.

21. **Freemont, P., N. Hogg, and J. Edgeworth.** 1989. Sequence identity. *Nature* (London) **339:**516.

22. **Fujiwara, S., J. Imai, M. Fujiwara, T. Yaeshima, T. Kawashima, and K. Kobayashi.** 1990. A potent antibacterial protein in royal jelly. Purification and determination of the primary structure of royalisin. *J. Biol. Chem.* **265:**11333–11337.

23. **Harwig, S. S. L., A. S. K. Park, and R. I. Lehrer.** 1992. Characterization of defensin precursors in mature human neutrophils. *Blood* **79:**1532–1537.

24. **Heimstra, P. S., P. B. Eisenhauer, S. S. L. Harwig, M. T. van den Barselaar, R. van Furth, and R. I. Lehrer.** 1993. Antimicrobial proteins of murine macrophages. *Infect. Immun.* **61:**3038–3046.

25. **Hogg, N., C. Allen, and J. Edgeworth.** 1989. Monoclonal antibody 5.5 reacts with p8,14, a myeloid molecule associated with vascular endothelium. *Eur. J. Biochem.* **19:**1053–1061.

26. **Jenne, D. E., and J. Tschopp.** 1988. Granzymes, a family of serine proteases released from the granules of cytolytic T lymphocytes upon T cell receptor stimulation. *Immunol. Rev.* **103:**53–71.

27. **Jones, D. E., and C. L. Bevins.** 1992. Paneth cells of the human small intestine express an anti-microbial peptide gene. *J. Biol. Chem.* **267:**23216–23225.

28. **Kligman, D., and D. C. Hilt.** 1988. The S100 protein family. *Trends Biochem. Sci.* **13:**437–443.

29. **Kyte, J., and R. F. Doolittle.** 1982. A simple method for displaying the hydropathic character of a protein. *J. Mol. Biol.* **157:**105–132.

30. **Lal, K., L. Xu, J. Colburn, A. L. Hong, and J. J. Pollock.** 1992. The use of capillary electrophoresis to identify cationic proteins in human parotid saliva. *Arch. Oral Biol.* **37:**7–13.

31. **Lala, A., R. A. Lindemann, and K. T. Miyasaki.** 1992. Effect of polymorphonuclear leukocyte secretions on lymphokine-activated killer cell activity. *Oral Microbiol. Immunol.* **7:**89–95.

32. **Lambert, J., E. Keppi, J.-L. Dimarcq, C. Wicker, J.-M. Reichhart, B. Dunbar, P. Lepage, A. Van Dorsselaer, J. Hoffman, J. Fothergill, and D. Hoffmann.** 1989. Insect immunity: isolation from immune blood of dipteran *Phomia terranovae* of two insect antibacterial peptides with sequence homology to rabbit macrophage bactericidal peptides. *Proc. Natl. Acad. Sci. USA* **86:**262–266.

33. **Lehrer, R. I., A. K. Lichtenstein, and T. Ganz.** 1993. Defensins: antimicrobial and cytotoxic peptides of mammalian cells. *Annu. Rev. Immunol.* **11:**105–128.

34. **Matsuyama, K., and S. Natori.** 1988. Purification of three antibacterial proteins from the culture medium of NIH-Sape-4, and embryonic cell line of *Sarcophaga peregrina. J. Biol. Chem.* **263:**17112–17116.

35. **Matsuyama, K., and S. Natori.** 1988. Molecular cloning of cDNA for sapecin and unique expression of the sapecin gene during the development of *Sarcophaga peregrina. J. Biol. Chem.* **263:**17117–17121.

36. **Michaelson, D., J. Rayner, M. Couto, and T. Ganz.** 1992. Cationic defensins arise from charge-neutralized propeptides: a mechanism for avoiding leukocyte autotoxicity? *J. Leukocyte Biol.* **51:**634–639.

37. **Miyasaki, K. T.** 1991. The neutrophil: mechanisms of controlling periodontal bacteria. *J. Periodontol.* **62:**761–774.

38. **Miyasaki, K. T., and A. L. Bodeau.** 1991. In vitro killing of *Capnocytophaga* by granule fractions of human neutrophils is associated with cathepsin G activity. *J. Clin. Invest.* **87:**1585–1593.

39. **Miyasaki, K. T., and A. L. Bodeau.** 1991. In vitro killing of *Actinobacillus actinomycetemcomitans* and *Capnocytophaga* spp. by human neutrophil cathepsin G and elastase. *Infect. Immun.* **59:**3015–3020.

40. **Miyasaki, K. T., and A. L. Bodeau.** 1992. Human neutrophil azurocidin synergizes the killing of *Capnocytophaga sputigena* by leukocyte elastase. *Infect. Immun.* **60:**4973–4975.

41. **Miyasaki, K. T., A. L. Bodeau, and T. F. Flemming.** 1991. Differential killing of *Actinobacillus actinomycetemcomitans* and *Capnocytophaga* spp. by human neutrophil granule components. *Infect. Immun.* **59:**3760–3767.

42. **Miyasaki, K. T., A. L. Bodeau, T. Ganz, M. E. Selsted, and R. I. Lehrer.** 1990. Sensitivity of oral, gram-negative, facultative bacteria to the bactericidal activity of human neutrophil defensins. *Infect. Immun.* **58:**3934–3940.

43. **Miyasaki, K. T., A. L. Bodeau, A. R. K. Murthy, and R. I. Lehrer.** 1993. *In vitro* antimicrobial activity of the human neutrophil cytosolic complex, calprotectin, against *Capnocytophaga sputigena. J. Dent. Res.* **72:**517–523.

44. **Miyasaki, K. T., A. L. Bodeau, J. Pohl, and W. M. Shafer.** 1993. Bactericidal activities of synthetic human leukocyte cathepsin G-derived antibiotic peptides and congeners against *Actinobacillus actinomycetemcomitans* and *Capnocytophaga sputigena. Antimicrob. Agents Chemother.* **37:**2710–2715.

45. **Miyasaki, K. T., A. L. Bodeau, M. E. Selsted, T. Ganz, and R. I. Lehrer.** 1990. Killing of oral, gram-negative, facultative bacteria by the rabbit defensin, NP-1. *Oral Microbiol. Immunol.* **5:**315–319.

46. **Miyata, T., F. Tokunaga, T. Yoneya, K. Yoshikawa, S. Iwanaga, M. Niwa, T. Takao, and Y. Shi-monishi.** 1989. Antimicrobial peptides, isolated from horseshoe crab hemocytes, tachyplesin II, and polyphemusins I and II: chemical structures and biologic activity. *J. Biochem.* **106:**663–668.

47. **Murao, S., F. R. Collart, and E. Huberman.** 1989. A protein containing the cystic fibrosis antigen is an inhibitor of protein kinases. *J. Biol. Chem.* **264:**8356–8360.

48. **Murao, S., F. R. Collart, and E. Huberman.** 1990. A protein complex expressed during terminal differentiation of monomyelocytic cells is an inhibitor of cell growth. *Cell Growth Differ.* **1:**447–457.

49. **Nakamura, T., H. Furanaka, T. Miyata, F. Tokunaga, T. Muta, and S. Iwanage.** 1988. Tachyplesin, a class of antimicrobial peptide from hemocytes of the horseshoe crab (*Tachypleus tridentatus*). Isolation and chemical structure. *J. Biol. Chem.* **263:**16709–16713.

50. **Odink, K., N. Cerletti, J. Brüggen, R. G. Clerc, L. Tarcsay, G. Zwadlo, G. Gerhards, R. Schlegel, and C. Sorg.** 1987. Two calcium-binding proteins in infiltrate macrophages of rheumatoid arthritis. *Nature* (London) **330:**80–82.

51. **Romeo, D., B. Skerlavaj, M. Bolognesi, and R. Gennaro.** 1988. Structure and bactericidal activity of an antibiotic dodecapeptide purified from bovine neutrophils. *J. Biol. Chem.* **263:**9573–9575.

52. **Samakovlis, C., P. Kylsten, D. A. Kimbrell, Å. Engström, and D. Hultmark.** 1991. The *andropin* gene and its product, a male-specific antibacterial peptide in *Drosophila melanogaster. EMBO J.* **10:**163–169.

53. **Scocchi, M., B. Skerlavaj, D. Romeo, and R. Gennaro.** 1992. Proteolytic cleavage by neutrophil elastase converts inactive storage proforms to antibacterial bactenecins. *Eur. J. Biochem.* **209:**589–595.

54. **Selsted, M. E., M. J. Novotny, W. L. Morris, Y.-Q. Tang, W. Smith, and J. S. Cullor.** 1992. Indolicidin, a novel bactericidal tridecapeptide amide from neutrophils. *J. Biol. Chem.* **267:**4292–4295.

55. **Selsted, M. E., Y.-Q. Tang, W. L. Morris, P. A. McGuire, M. J. Novotny, W. Smith, A. H. Henschen, and J. S. Cullor.** 1993. Purification, primary structures, and antibacterial activities of β-defensins, a new family of antimicrobial peptides from bovine neutrophils. *J. Biol. Chem.* **268:**6641–6648.

56. **Shafer, W. M., and V. C. Onunka.** 1989. Mechanism of staphylococcal resistance to nonoxidative antimicrobial action of neutrophils: importance of pH and ionic strength in determining the bactericidal action of cathepsin G. *J. Gen. Microbiol.* **135:**825–830.

57. **Shafer, W. M., J. Pohl, V. C. Onunka, N. Bangalore, and J. Travis.** 1991. Human lysosomal cathepsin G and granzyme B share a functionally conserved broad spectrum antibacterial peptide. *J. Biol. Chem.* **266:**112–116.

58. **Shafer, W. M., M. E. Shepherd, B. Boltin, L. Wells, and J. Pohl.** 1993. Synthetic peptides of human lysosomal cathepsin G with potent antipseudomonal activity. *Infect. Immun.* **61:**1900–1908.

59. **Simmaco, M., D. Barra, F. Chiarini, L. Noviello, P. Melchiorri, G. Kreil, and K. Richter.** 1991. A family of bombinin-related peptides from the skin of *Bombina variegata. Eur. J. Biochem.* **199:**217–222.

60. **Sohnle, P. G., C. Collins-Lech, and J. H. Weissner.** 1990. Antimicrobial activity of an abundant calcium-binding protein in the cytoplasm of human neutrophils. *J. Infect Dis.* **163:**187–192.

61. **Sohnle, P. G., C. Collins-Lech, and J. H. Weissner.** 1991. The zinc-reversible antimicrobial activity of neutrophil lysates and abscess fluid supernatants. *J. Infect. Dis.* **164:**136–142.

62. **Steinbakk, M., C.-F. Naess-Andersen, E. Lingaas, I. Dale, P. Brandtzaeg, and M. K. Fagerhol.** 1990. Antimicrobial actions of calcium binding leucocyte L1 protein, calprotectin. *Lancet* **336:**763–765.

63. **Tang, Y.-Q., and M. E. Selsted.** 1993. Characterization of the disulfide motif in BNBD-12, an antimicrobial β-defensin peptide from bovine neutrophils. *J. Biol. Chem.* **268:**6649–6653.

64. **Teigelkamp, S., R. S. Bhardwaj, J. Roth, G. Meinardus-Hager, M. Karas, and C. Sorg.** 1991. Calcium-dependent complex assembly of the myeloic differentiation proteins MRP-8 and MRP-14. *J. Biol. Chem.* **266:**13462–13467.

65. **Valore, E. V., and T. Ganz.** 1992. Posttranslational processing of defensins in immature human myeloid cells. *Blood* **79:**1538–1544.

66. **Van Heyningen, V., and J. Dorin.** 1990. Possible role for two-calcium-binding proteins of the S-

100 family, co-expressed in granulocytes and certain epithelia. *Adv. Exp. Med. Biol.* **269**:139–143.

67. **Wade, D., A. Boman, B. Wåhlin, C. M. Drain, D. Andreu, H. G. Boman, and R. B. Merrifield.** 1990. All-D amino acid-containing channel forming antibiotic peptides. *Proc. Natl. Acad. Sci. USA* **87**:4761–4765.

68. **Wasiluk, K. R., K. M. Skubitz, and B. H. Gray.** 1991. Comparison of granule proteins from human polymorphonuclear leukocytes which are bactericidal toward *Pseudomonas aeruginosa. Infect. Immun.* **59**:4193–4200.

69. **Wilde, C. G., J. E. Griffith, M. N. Marra, J. L. Snable, and R. W. Scott.** 1989. Purification and characterization of human neutrophil peptide 4, a novel member of the defensin family. *J. Biol. Chem.* **264**:11200–11203.

70. **Wilkinson, M. M., A. Bussuttil, C. Hayward, D. J. H. Brock, J. R. Dorin, and V. Van Heyningen.** 1988. Expression pattern of two related cystic fibrosis associated calcium-binding proteins in normal and abnormal tissues. *J. Cell. Sci.* **91**:221–230.

71. **Zanetti, M., L. Litteri, R. Gennaro, H. Hostmann, and D. Romeo.** 1990. Bactenecins, defense polypeptides of bovine neutrophils, are generated from precursor molecules stored in the large granules. *J. Cell Biol.* **111**:1363–1371.

72. **Zanetti, M., L. Litteri, G. Griffiths, R. Gennaro, and D. Romeo.** 1991. Stimulus-induced maturation of probactenecins, precursors of neutrophil antimicrobial polypeptides. *J. Immunol.* **146**:4295–4300.

73. **Zimmer, M., R. L. Medcalf, T. M. Fink, C. Mattmann, P. Lichter, and D. E. Jenne.** 1992. Three human elastase-like genes coordinately expressed in the myelomonocytic lineage are organized as a single genetic locus on 19pter. *Proc. Natl. Acad. Sci. USA* **89**:8215–8219.

Molecular Pathogenesis of Periodontal Disease
Edited by Robert Genco et al.
© 1994 American Society for Microbiology, Washington, DC 20005

Chapter 27

Antimicrobial Dysfunction in Localized Juvenile Periodontitis Neutrophils

John R. Kalmar

The polymorphonuclear neutrophil (PMN) plays a pivotal role in host defense against infectious periodontal disease. Clinical studies of primary and secondary immunodeficiency disorders such as cyclic neutropenia (17), drug-induced agranulocytosis (11), Chediak-Higashi syndrome (25), and leukocyte adhesion deficiency (LAD) disease (67, 81) have clearly demonstrated that conditions affecting granulocytic cell number, maturation, or function are often associated with a markedly increased susceptibility to rapid and often severe periodontal destruction. By contrast, individuals with primary lymphocyte disorders exhibit less gingival inflammation than healthy individuals and periodontal disease of no greater severity (62, 63). Immunosuppressed transplant patients and their age-, sex-, and plaque-matched controls likewise exhibit no difference in periodontal disease incidence or severity (67, 72). Together, these findings suggest that although adaptive immune cells and mechanisms may help modulate disease progression, the PMN is crucial to normal host resistance against periodontopathogenic microorganisms.

LJP

Localized juvenile periodontitis (LJP) represents a subset of early-onset periodontitis occurring in otherwise healthy individuals under the age of 30 that is characterized by circumpubertal onset and vertical alveolar defects localized to the permanent first molar and/or incisor teeth. Males and females are affected equally (27), and most family studies report an autosomal mode of inheritance (13, 27, 47). The latest epidemiologic data suggest that the prevalence of LJP is racially dependent, affecting 2.05% of African-Americans aged 14 to 17 years compared with 0.17% of Caucasian Americans (46). Similar estimates of prevalence have been reported elsewhere, with Caucasian individuals generally less affected than those of African descent (26, 29, 38, 64).

John R. Kalmar • Division of Oral and Maxillofacial Pathology, Eastman Dental Center, 625 Elmwood Avenue, Rochester, New York 14620.

A number of bacteria including *Actinobacillus actinomycetemcomitans, Capnocytophaga ochracea, Prevotella intermedia,* and *Eikenella corrodens* are linked as predominant components of the subgingival microflora associated with LJP (82). To date, however, the most compelling evidence for disease association has been made for *A. actinomycetemcomitans,* a capnophilic, gram-negative coccobacillus related to *Haemophilus* spp. While occurring infrequently in periodontally healthy sites or even in sites affected with other forms of periodontal disease, *A. actinomycetemcomitans* has been recovered from subgingival plaque in 90 to 100% of LJP patients (86). Histopathologic evidence of connective tissue invasion by *A. actinomycetemcomitans* has been demonstrated in LJP disease sites (85). Furthermore, treatment studies have shown that elimination of this organism results in arrest of clinical disease, while continued infection or recolonization with it is associated with disease persistence or recurrence (84). Significant exposure to this bacterium is also supported by reports of markedly elevated antibody titers, often specific to the infecting serotype, in serum, saliva, and gingival crevicular fluid of 70 to 90% of LJP patients (20, 60). By contrast, periodontally healthy individuals and patients with other forms of periodontal disease have a low prevalence of serum antibodies to *A. actinomycetemcomitans* (84). Finally, the organism produces a number of potential virulence factors that could contribute to the pathogenesis of LJP. These include leukotoxin, epitheliotoxin, collagenase, immunoglobulin A protease, fibroblast inhibitory factor, bone resorption-inducing toxin, polyclonal B-cell activator, and a potent lipopolysaccharide (endotoxin) (24, 82).

LJP PMN FUNCTION

Chemotaxis

The evidence linking certain cases of rapidly progressive periodontal disease with abnormal neutrophil numbers or function led several investigators to test the functional capacity of PMN from LJP patients. Defective chemotactic responsiveness of LJP PMN was first reported in abstract form in 1976 by Lavine et al. (44) and then in a full report in early 1977 by Cianciola et al. (14). Since that time, chemotaxis has been the major focus of PMN function studies in LJP patient populations. Depressed chemotactic activity has been confirmed by the majority of authors (1, 9, 12, 14, 16, 18, 23, 39, 43, 56–58, 75–79), although this finding is not universal (21, 36, 37, 42, 59, 61). Approximately 70% of affected patients exhibit decreased responsiveness to various chemoattractants, including formyl-methionyl-Leu-Phe (FMLP), C5a, and leukotriene B_4. Defective PMN chemotaxis does not normalize with disease resolution and is unaffected by serum source (22), although recent evidence supporting immunomodulatory levels of cytokines in certain LJP serum samples has been presented (2). In addition, healthy siblings of LJP patients are occasionally defective, suggesting that this hypofunction is an inherent trait and not merely a result of periodontal disease (22). Chemotactic dysfunction appears to segregate with the disease in families and can precede disease development (12, 78). Among African-American patient populations with LJP, defective chemotaxis has been attributed to a reduction in the number and modulation of receptors for these molecules on the PMN surface (22). Recently, racial differences in chemotaxis have been reported, with black subjects exhibiting significantly lower responses than white subjects (65). This work has important implications, both in the interpretation of past work and in the design of future studies concerning chemotaxis and possibly other cellular functions. DeNardin et al. reported binding differences between control and LJP PMN when a panel of monoclonal antibodies to the FMLP receptor was used, a finding supporting either reduced expression or qualitative differences in the receptor itself (19). Evidence of abnormal intracellular CA^{2+} mobilization (1, 18), elevated diacylglycerol levels coupled with decreased diglyceride kinase activity (73), and reduced activity of calcium-dependent protein kinase C point to defective G-protein-linked membrane transduction pathways in chemotactically defective LJP PMN (39).

Antimicrobial Functions

Since severely affected LJP patients can exhibit normal or even supranormal chemotaxis, the relationship of this response to disease is still uncertain and may reflect heterogeneity in terms of etiology, underlying host susceptibility, or both. However, other aspects of LJP patient PMN function, including antimicrobial activities, have not been rigorously tested.

Respiratory burst

Stimulation of PMN with certain stimuli (i.e., chemoattractants, opsonized bacteria) can result in oxygen radical generation through activation of the NADPH oxidase system. Together with myeloperoxidase, a major component of azurophilic granules, superoxide anion (O_2^-) and H_2O_2 are produced and converted into hydroxyl radical (OH^-) and hypochlorus acid (HOCl), which have potent antimicrobial activities and can also injure host tissues (49). Using a luminol-enhanced chemiluminescence assay, Asman et al. demonstrated increased radical generation by LJP PMN compared to healthy control PMN in response to *Staphylococcus aureus* opsonized with either autologous serum or gamma globulin (8). This was associated with increased release of the azurophilic granule marker elastase (5). More recently, similarly elevated chemiluminescence was reported with opsonized *A. actinomycetemcomitans* and *Porphyromonas (Bacteroides) gingivalis* (6). Those authors suggest that elevated extracellular release of reactive oxygen species may damage host tissues and contribute to the pathogenesis of LJP (7). Zafiropoulos et al., however, detected no difference in chemiluminescence between LJP and matched control PMN in response to opsonized zymosan (83), and similar results were found with this stimulus and superoxide production (see below) (21, 78).

Phagocytosis and killing

Phagocytosis and killing are fundamental activities of the PMN that sequester and eliminate invading microorganisms. Phagocytosis is the engulfment of particulate matter or microbial parasites by the external cell membrane of the phagocyte, resulting in an intracellular, membrane-delimited structure termed the phagosome. This process is aided by opsonization of the microorganism with complement and/ or antibody, both of which can subsequently interact with corresponding receptors on the PMN surface. Killing is accomplished both by oxidative mechanisms associated with respiratory burst activity and by nonoxidative mechanisms associated with components of both azurophilic and specific granules. As phagocytosis proceeds, the granules mobilize, fuse with the phagosome, and release their contents into the developing phagolysosome. Few studies of the phagocytic or microbicidal activity of LJP PMN have been reported. As shown in Table 1, a variety of microbial and particulate targets have been tested, with diverse results.

Using *S. aureus* opsonized with 1% human gamma globulin, Cianciola et al. found significantly less phagocytic activity with PMN from six of seven LJP patients than with age-, sex-, and race-matched healthy controls (14). Contact angles produced by drops of saline on PMN monolayers were consistently higher with patient cells, indicating a less hydrophilic phagocyte surface, a feature previously associated with decreased phagocytic function (80). Lavine et al. used a quantitative nitro-blue tetrazolium test as a screen for phagocytic versus bactericidal function (43). Only one of eight LJP patient PMN samples demonstrated phagocytosis that differed significantly (reduced) from that of control PMN. Killing was assessed using an unspecified *S. aureus* strain and was judged normal in five of five patients, including the patient with defective phagocytic activity. Measuring phagocytosis with fluorescein isothiocyanate-labeled zymosan particles opsonized with fetal calf serum, Ellegaard et al. reported no significant difference between 12 LJP patients and their age- and sex-matched controls (21). In response to zymosan ingestion, patient cells were also equivalent to control cells in oxygen consumption, superoxide production, and extracellular release of the granule markers β-glucuronidase and lysozyme.

The ability of healthy PMN to induce germination of certain bacterial spores in culture was used by Suzuki et al. to examine both phagocytosis and intra-phagosomal metabolism (microbicidal mechanisms) (69). Phagocytosis of ^{45}Ca-labeled, unopsonized *Bacillus cereus* spores was considered defective in 18 (62%) of 29 LJP patients. Depressed spore germination was reportedly detected in 13 (65%) of 20 patients, although the data actually indicate defective function in 15

Table 1. LJP PMN phagocytosis and killing

Target[a]	Phagocytosis/killing[b]	Reference
S. aureus	Decreased/—	14
NBT/*S. aureus*	Normal/normal	43
Zymosan	Normal/—	21
B. cereus spores	Decreased/decreased	69
S. aureus	Decreased/—	79
A. actinomycetemcomitans	Normal/decreased	34
Fluorescent beads	Decreased/—	35

[a] NBT, nitroblue tetrazolium.
[b] —, not tested.

(94%) of 16 patient specimens. No consistent pattern of association was noted among the functional variables of chemotaxis, phagocytosis, and spore germination in LJP patient PMN. Van Dyke et al. used a protein A-negative strain of *S. aureus* to quantitate phagocytosis following opsonization with pooled AB human serum (79). Although the percentages of active PMN were equal in the two groups, slightly but significantly fewer bacteria (10 to 20%) were ingested per patient PMN ($n = 14$) than per PMN of age- and sex-matched healthy controls. Yet opsonized-zymosan-induced superoxide production and lactoferrin secretion in response to 10^{-5} M FMLP and cytochalasin B were equivalent. Recently, Kimura et al. reported that the primarily complement-mediated phagocytosis of fluorescent microspheres in some but not all LJP patient PMN was depressed compared to that in healthy control cells (35). The percentage of phagocytically active PMN was depressed in 8 of 15 patients, while mean particle numbers per cell were significantly reduced in 10 of 15 specimens. LJP PMN phagocytosis was defective in the presence of either autologous or allogeneic plasma. Additionally, there was no difference between plasma source and the percentage of phagocytically active PMN with either patient or control cells.

Phagocytosis and killing of *A. actinomycetemcomitans*

Like patients with LJP, individuals with Chediak-Higashi disease and LAD disease exhibit depressed PMN chemotaxis in addition to an increased risk for precocious and severe periodontitis. Yet, while Chediak-Higashi and LAD disease patients demonstrate increased susceptibility for recurrent, severe bacterial infections at other sites, LJP patients apparently do not suffer from other common forms of infectious disease. The clinical evidence of several periodontal tissue loss in young, otherwise healthy patients and the substantial evidence for *A. actinomycetemcomitans* infection in the pathogenesis of LJP strongly suggest a selective, possibly unique susceptibility of host defense to this parasite in certain affected individuals. These features provide a compelling argument to examine and compare the direct interaction of control and patient PMN with *A. actinomycetemcomitans*.

Previous work had demonstrated that phagocytosis and killing of *A. actinomycetemcomitans* by healthy human PMN was opsonization dependent (48, 51). Using a dual-fluorochrome microassay, Kalmar et al. confirmed that phagocytosis of a serotype b (83a), leukotoxin-producing *A. actinomycetemcomitans* strain (CDC A7154; LJP patient isolate) required serum opsonization (34). Compared with age- and sex-matched (race-matched in four of six cases) control cells, African-American LJP patient PMN were equivalent in terms of percent phagocytosis and number of bacteria ingested per cell (Fig. 1). While control serum without antileukotoxic activity provided effective opsonization at 15 min, patient serum that blocked leukotoxin and possessed elevated reactivity to bacterial sonicate by enzyme-linked immunosorbent assay (185 ± 16 to 53 ± 9 ELISA units) promoted increased bacterial ingestion at 1 h. This result could reflect PMN membrane protection by leukotoxin antibody, increased opsonin density on the bacterial surface, or both. While bacterial activity of control PMN against *A. actinomycetemcomitans* increased significantly over time regardless of serum source ($P < 0.001$), no additional

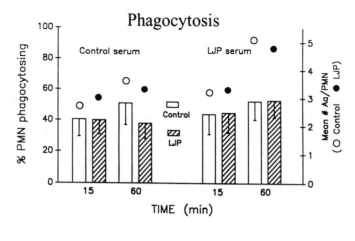

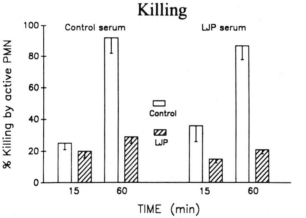

Figure 1. Phagocytosis and killing of *A. actinomycetemcomitans* (Aa) by control and LJP patient PMN (*n* = 6). Fluorescein isothiocyanate-labeled *A. actinomycetemcomitans* was preopsonized with heterologous control or patient serum and mixed with PMN. Propidium iodide was used as a viability factor.

killing was seen with LJP patient cells (Fig. 1). Despite the finding of equivalent sensitivity to leukotoxin cytolysis by patient and control cells in the presence of control serum, significantly greater killing of *A. actinomycetemcomitans* was observed with control PMN when either control or immune patient serum was used (*P* < 0.0001). Together, these findings indicate that although LJP PMN ingested *A. actinomycetemcomitans* normally, their bactericidal mechanisms against this microorganism were defective compared with those of control cells.

A. actinomycetemcomitans versus PMN: Bactericidal Mechanisms and Granule Secretion

Several antimicrobial substances and systems found in normal PMN have been tested against *A. actinomycetemcomitans*. Intact PMN kill these organisms by both oxygen-dependent and oxygen-independent mechanisms, with much of the former activity attributed to myeloperoxidase (49, 51, 52). Yet, as *A. actinomycetemcomitans* lipopolysaccharide has no carbohydrate chain structure analogous to that of the O antigen of enteric bacteria, it has been suggested that nonoxidative mechanisms may play a more important role against this microorganism (49). Specific immunoglobulin may promote PMN killing of *A. actinomycetemcomitans*, but intact complement components appear to be necessary for effective opsonization (10, 28, 32, 48). Most recently, Holm et al. reported that fresh isolates of *A. actinomycetemcomitans* were more resistant to PMN killing than laboratory strains, while both were more resistant than either fresh or laboratory strains of *Haemophilus aphrophilus* (28). In studies with purified PMN components, *A. actinomycetemcomitans* demonstrated relative resistance to lysozyme (30), H_2O_2 (53), xanthine oxidase (54), and elastase (50) but was readily killed by myeloperoxidase (49, 52, 55), lactoferrin (33), and cathepsin G (50). Myeloperoxidase also oxidatively inactivates leukotoxin (15). Using PMN granule fractions, Miyasaki examined oxygen-independent killing of *A. actinomycetemcomitans* and found that a significant proportion of killing was associated with a fraction rich in cathepsin G, especially under anaerobic conditions (49).

Preliminary studies of PMN stimulation and secretion may provide some insight into the failure of LJP PMN to kill *A. actinomycetemcomitans* normally. Previously, McArthur et al. demonstrated that opsonized *A. actinomycetemcomitans* caused extracellular release of lysozyme by normal PMN (48). In our work, patient and control PMN pairs were exposed to various agonists, including opsonized zymosan, opsonized *A. actinomycetemcomitans*, calcium ionophore A23187, and FMLP, in order to examine the extracellular release of selected granule markers (32). As shown in Fig. 2, lactoferrin secretion from control PMN was equivalent for all agonists. By comparison, LJP PMN secreted less lactoferrin in response to *A. actinomycetemcomitans* than to other agonists and a mean of nearly 50% less than control cells. Release of lysozyme, found in both primary and secondary granules, was equivalent in patient and control cells except with A23187 (depressed in two of three LJP patients), and only minimal extracellular release of the azurophilic granule marker myeloperoxidase was seen under these conditions. Total cellular content of all markers was equivalent in control and LJP patient cells.

Decreased lactoferrin secretion by patient PMN would likely correspond to reduced lactoferrin concentrations within the phagolysosome. In addition, as lactoferrin killing of *A. actinomycetemcomitans* requires acidic conditions (<pH 5.8), even normal amounts of lactoferrin would be ineffective if access to the early acidic phase of phagolysosome development was delayed (33). Either condition would favor survival of ingested *A. actinomycetemcomitans*. The abnormal secretion of other granule components with independent or synergistic activity against *A. actinomycetemcomitans* (such as cathepsin G) would contribute to the apparent bactericidal dysfunction of LJP PMN. Supporting the concept that LJP may result from a variety of otherwise subclinical abnormalities, Söderström and Wikström reported that six of eight LJP patient PMN samples were deficient in total lactoferrin (66).

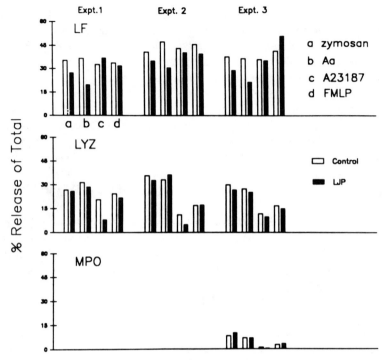

Figure 2. Extracellular release of lactoferrin (LF), lysozyme (LYZ), and myeloperoxidase (MPO) by control and LJP patient PMN with soluble and particulate stimuli. PMN (5×10^6/ml) were exposed to stimulants for 15 min at 37°C and placed on ice. Then supernatants were harvested. Final concentrations of stimuli were 5×10^7/ml for zymosan, 5×10^7/ml for *A. actinomycetemcomitans* (Aa), 10^{-6} M A23187, and 10^{-6} M FMLP plus 5 μg of cytochalasin B per ml.

DISCUSSION

As outlined by Miyasaki, certain periodontal diseases may result from the failure of specific neutrophil antimicrobial substances or systems to control specific periodontopathogens (49). While this concept is attractive in terms of host-parasite immunology, the question of killing defect specificity for the LJP PMN remains unresolved and of obvious significance. Additional study is also needed to determine the role, if any, of the numerous virulence mechanisms produced by *A. actinomycetemcomitans* in disease development. Previously, it has been reported that *A. actinomycetemcomitans* culture supernatant (74) and bacterial extract (3) inhibit PMN chemotaxis, the latter being reversible by proteinase K treatment or phenol extraction. Bacterial extract was also shown to contain a heat-resistant protein component that apparently "primes" PMN activation through Fc, C3b, and FMLP receptors while inhibiting phorbol myristate acetate-mediated activation of protein kinase C (4).

Of particular interest in the context of possible PMN dysfunction are the potential subcytolytic effects of leukotoxin. *A. actinomycetemcomitans* leukotoxin is a member of the RTX (repeat toxin) cytoxin family, with homology to *Escherichia coli* α-hemolysin and *Pasteurella haemolytica* leukotoxin (40). Despite this homology, it exhibits a distinct and limited range of target cell specificity that includes PMN, monocytes, and lymphocytes of humans and neutrophils and monocytes of great apes and Old World monkeys (70, 71). Leukotoxin produces ion-permeable membrane pores in susceptible cells, leading to Ca^{2+}-influx, reduced membrane potential, and cell death (70, 71). However, its effects at subcytolytic concentrations have not been well described and may be of critical importance to the ability of *A. actinomycetemcomitans* to cause disease. For example, leukotoxin binds Ca^{2+} with a binding constant (10^{-7} M) similar to that of parvalbumin, a high-affinity Ca^{2+}-binding muscle protein (41). Since Ca^{2+} fluxes from both intracellular and extracellular sources are critical for chemotaxis (1, 18), granule secretion (45), and phagosome-lysosome fusion (31), leukotoxin could disrupt these membrane events by Ca^{2+} scavenging. By inference, cells with defective Ca^{2+} mobilization such as that described for LJP PMN appear to be especially susceptible to leukotoxin.

SUMMARY

While functional studies of the LJP neutrophil have increased our cellular and biochemical understanding of the disease process over the past 15 years, numerous questions remain. The finding of depressed chemotactic responsiveness has received majority support, yet not all patient cohorts seem affected. For direct and indirect measures of antimicrobial function, including chemiluminescence, superoxide production, phagocytosis, and killing, no consensus has been developed. In order to better understand the relationship of these studies to the pathogenesis of LJP, racial and genetic compositions of study cohorts must be carefully defined. With the known predilection for LJP among descendents of Africans compared with Caucasians, it is logical to suggest that differences in disease prevalence could reflect genotypically distinct forms of disease artifically grouped together by a similar clinical expression or phenotype. Additional emphasis on genetic linkage studies and putative risk markers are indicated to better define disease transmission and heterogeneity within and between different patient populations. Increased awareness of the potential influence of hereditary or environmental factors on measures of cell function and the determination of possible LJP subforms will permit more rapid and coherent correlation of relevant risk elements. Ultimately, indicators that will permit detection of individuals at risk and guide the application of appropriate prophylactic therapy may be identified.

REFERENCES

1. **Agarwal, D., M. A. Reynolds, L. D. Duckett, and J. B. Suzuki.** 1989. Altered free cytosolic calcium changes and neutrophil chemotaxis in patients with juvenile periodontitis. *J. Periodontal Res.* **24:**149–154.
2. **Agarwal, S., and J. B. Suzuki.** 1991. Altered neutrophil function in localized juvenile periodontitis: intrinsic cellular defect or effect of immune mediators? *J. Periodontal Res.* **26:**276–278.
3. **Ashkenazi, M., R. R. White, and D. K. Dennison.** 1992. Neutrophil modulation by *Actinobacillus*

actinomycetemcomitans. I. Chemotaxis, surface receptor expression and F-actin polymerization. *J. Periodontal Res.* **27**:264–273.

4. **Ashkenazi, M., R. R. White, and D. K. Dennison.** 1992. Neutrophil modulation by *Actinobacillus actinomycetemcomitans*. II. Phagocytosis and development of respiratory burst. *J. Periodontal Res.* **27**:457–465.

5. **Asman, B.** 1988. Peripheral PMN cells in juvenile periodontitis. *J. Clin. Periodontol.* **15**:360–364.

6. **Asman, B., and K. Bergström.** 1992. Expression of Fc-γ-RIII and fibronectin in peripheral polymorphonuclear neutrophils with increased response to Fc stimulation in patients with juvenile periodontitis. *Arch. Oral Biol.* **37**:991–995.

7. **Asman, B., K. Bergström, P. Wijkander, and B. Lockowandt.** 1986. Influence of plasma components on luminol-enhanced chemiluminescence from peripheral granulocytes in juvenile periodontitis. *J. Clin. Periodontol.* **13**:850–855.

8. **Asman, B., P.-E. Engström, T. Olsson, and K. Bergström.** 1984. Increased luminol enhanced chemiluminescence from peripheral granulocytes in juvenile periodontitis. *Scand. J. Dent. Res.* **92**:218–223.

9. **Astemborski, J. A., J. A. Boughman, P. O. Myrick, S. B. Goodman, R. K. Wooten, J. W. Vincent, and J. B. Suzuki.** 1989. Clinical and laboratory characterization of early onset periodontitis. *J. Periodontol.* **60**:557–563.

10. **Baker, P., and M. Wilson.** 1989. Opsonic IgG antibody against *Actinobacillus actinomycetemcomitans* in localized juvenile periodontitis. *Oral Microbiol. Immunol.* **4**:98–105.

11. **Bauer, W. H.** 1946. The supporting tissue of the tooth in acute secondary agranulocytosis (Arsphenamin neutropenia). *J. Dent. Res.* **25**:501–508.

12. **Boughman, J. A., J. A. Astemborski, and J. B. Suzuki.** 1992. Phenotypic assessment of early onset periodontitis in sibships. *J. Clin. Periodontol.* **19**:233–239.

13. **Boughman, J. A., S. L. Halloran, D. Roulston, S. Schwartz, J. B. Suzuki, L. R. Weitkamp, R. E. Wenk, R. Wooten, and H. M. Cohen.** 1986. An autosomal-dominant form of juvenile periodontitis: its localization to chromosome 4 and linkage to dentinogenesis imperfecta and Gc. *J. Craniofac. Genet. Dev. Biol.* **6**:341–350.

14. **Cianciola, L. J., R. J. Genco, M. R. Patters, J. McKenna, and C. J. van Oss.** 1977. Defective polymorphonuclear leukocyte function in a human periodontal disease. *Nature* (London) **265**:445–447.

15. **Clark, R. A., K. G. Leidal, and N. S. Taichman.** 1986. Oxidative inactivation of *Actinobacillus actinomycetemcomitans* leukotoxin by the neutrophil myeloperoxidase system. *Infect. Immun.* **53**:252–256.

16. **Clark, R. A., R. C. Page, and G. Wilde.** 1977. Defective neutrophil chemotaxis in juvenile periodontitis. *Infect. Immun.* **18**:694–700.

17. **Cohen, D. W., and A. L. Morris.** 1961. Periodontal manifestation of cyclic neutropenia. *J. Periodontol.* **32**:159–168.

18. **Daniel, M. A., G. McDonald, S. Offenbacher, and T. E. Van Dyke.** 1993. Defective chemotaxis and calcium response in localized juvenile periodontitis neutrophils. *J. Periodontol.* **64**:617–621.

19. **DeNardin, E., C. DeLuca, M. J. Levine, and R. J. Genco.** 1990. Antibodies directed to the chemotactic factor receptor detect differences between chemotactically normal and defective neutrophils from LJP patients. *J. Periodontol.* **61**:609–617.

20. **Ebersole, J. L., M. A. Taubman, D. J. Smith, R. J. Genco, and D. E. Frey.** 1982. Human immune response to oral microorganisms. I. Association of localized juvenile periodontitis (LJP) with serum antibody responses to *Actinobacillus actinomycetemcomitans*. *Clin. Exp. Immunol.* **47**:43–52.

21. **Ellegaard, B., N. Borregaard, and J. Ellegaard.** 1984. Neutrophil chemotaxis and phagocytosis in juvenile periodontitis. *J. Periodontal Res.* **19**:261–268.

22. **Genco, R. J., and J. Slots.** 1984. Host responses in periodontal diseases. *J. Dent. Res.* **63**:441–451.

23. **Genco, R. J., T. E. Van Dyke, B. Park, M. Ciminelli, and H. Horoszewicz.** 1980. Neutrophil chemotaxis impairment in juvenile periodontitis: evaluation of specificity, adherence, deformability, and serum factors. *J. Recticuloendothel. Soc.* **28**:81s–91s.

24. **Gregory, R. L., D. E. Kim, J. C. Kindle, L. C. Hobbs, and D. R. Lloyd.** 1992. Immunoglobulin-degrading enzymes in localized juvenile periodontitis. *J. Periodontal Res.* **27**:176–183.

25. **Hamilton, R. E., and J. S. Giansanti.** 1974. Chediak-Higashi syndrome: report of a case and review of the literature. *Oral Surg.* **37**:754–761.

26. **Harley, A. F., and P. D. Floyd.** 1988. Prevalence of juvenile periodontitis in school children in Lagos, Nigeria. *Community Dent. Oral Epidemiol.* **16**:299–301.

27. **Hart, T. C., M. L. Marazita, H. A. Schenkein, and S. A. Diehl.** 1992. Reinterpretation of the evidence for X-linked dominant inheritance of juvenile periodontitis. *J. Periodontol.* **63**:169–173.

28. **Holm, A., S. Kalfas, and S. E. Holm.** 1993. Killing of *Actinobacillus actinomycetemcomitans* and *Haemophilus aphrophilus* by human polymorphonuclear leukocytes in serum and saliva. *Oral Microbiol. Immunol.* **8**:134–140.

29. **Hoover, J. N., B. Ellegaard, and R. Attstrom.** 1989. Periodontal status of 14- to 16-year-old Danish schoolchildren. *Scand. J. Dent. Res.* **89**:175–179.

30. **Iacono, V. J., P. R. Boldt, B. J. MacKay, M. Cho, and J. J. Pollock.** 1983. Lytic sensitivity of *Actinobacillus actinomycetemcomitans* Y4 to lysozyme. *Infect. Immun.* **40**:773–784.

31. **Jaconi, M. E. E., D. P. Lew, J. L. Carpenter, K. E. Magnusson, M. Sjögren, and O. Stendahl.** 1990. Cytosolic free calcium elevation modulates the phagosome-lysosome fusion during phagocytosis in human neutrophils. *J. Clin. Biochem.* **110**:1555–1564.

32. **Kalmar, J. R.** 1989. Ph.D. thesis. Emory University, Atlanta.

33. **Kalmar, J. R., and R. R. Arnold.** 1988. Killing of *Actinobacillus actinomycetemcomitans* by human lactoferrin. *Infect. Immun.* **56**:2552–2557.

34. **Kalmar, J. R., R. R. Arnold, and T. E. Van Dyke.** 1987. Direct interaction of *Actinobacillus actinomycetemcomitans* with normal and defective (LJP) neutrophils. *J. Periodontal Res.* **22**:179–181.

35. **Kimura, S., T. Yonemura, T. Hiraga, and H. Okada.** 1992. Flow cytometric evaluation of phagocytosis by peripheral blood polymorphonuclear leucocytes in human periodontal diseases. *Arch. Oral Biol.* **37**:492–501.

36. **Kinane, D. F., C. F. Cullen, F. A. Johnston, and C. W. Evans.** 1989. Neutrophil chemotactic behaviour in patients with early-onset forms of periodontitis. I. Leading front analysis in Boyden chambers. *J. Clin. Periodontol.* **16**:242–246.

37. **Kinane, D. F., C. F. Cullen, F. A. Johnston, and C. W. Evans.** 1989. Neutrophil chemotactic behaviour in patients with early-onset forms of periodontitis. II. Assessment using the under agarose technique. *J. Clin. Periodontol.* **16**:247–251.

38. **Kronauer, E., D. Borsa, and N. P. Lang.** 1986. Prevalence of incipient juvenile periodontitis at age 16 years in Switzerland. *J. Clin. Periodontol.* **13**:103–108.

39. **Kurihara, H., Y. Murayama, M. L. Warbington, C. M. E. Champagne, and T. E. Van Dyke.** 1993. Calcium-dependent protein kinase C activity of neutrophils in localized juvenile periodontitis. *Infect. Immun.* **61**:3137–3142.

40. **Lally, E. T., E. E. Golub, I. R. Kieba, N. S. Taichman, J. Rosenbloom, J. C. Rosenbloom, C. W. Gibson, and D. R. Demuth.** 1989. Analysis of the *Actinobacillus actinomycetemcomitans* leukotoxin gene: delineation of unique features and comparison to homologous toxins. *J. Biol. Chem.* **264**:15451.

41. **Lally, E. T., I. R. Kieba, N. S. Taichman, J. Rosenbloom, C. W. Gibson, D. R. Demuth, G. Harrison, and E. E. Golub.** 1991. *Actinobacillus actinomycetemcomitans* leukotoxin is a calcium-bonding protein. *J. Periodontal Res.* **26**:268–271.

42. **Larjava, H., L. Saxén, T. Kosunen, and C. G. Gahmberg.** 1984. Chemotaxis and surface glycoproteins of neutrophil granulocytes from patients with juvenile periodontitis. *Arch. Oral Biol.* **29**:935–939.

43. **Lavine, W. S., E. G. Maderazo, J. Stolman, P. A. Ward, R. B. Cogen, I. Greenblatt, and P. B. Robertson.** 1979. Impaired neutrophil chemotaxis in patients with juvenile and rapidly progressing periodontitis. *J. Periodontal Res.* **14**:10–19.

44. **Lavine, W. S., J. Stolman, E. Maderazo, P. Ward, and R. Cogen.** 1976. Abstr. no. 603. *J. Dent. Res.* **55** (Special Issue):B2112.

45. **Lew, P. O., A. Monod, F. A. Waldvogel, B. Dewald, M. Baggiolini, and T. Pozzan.** 1986. Quantitative analysis of the cytosolic free calcium dependency of exocytosis from three subcellular compartments in intact human neutrophils. *J. Cell Biol.* **102**:2137–2204.

46. **Loe, H., and L. J. Brown.** 1991. Early onset periodontitis in the United States of America. *J. Periodontol.* **62:**608–616.

47. **Marazita, M. L., J. A. Burmeister, J. C. Gunsolley, T. E. Koertge, K. Lake, and H. A. Schenkein.** Evidence of autosomal dominant inheritance and race-specific heterogeneity in early-onset periodontitis. *J. Periodontol.,* in press.

48. **McArthur, W. P., C. Tsai, P. Baehni, B. Shenker, and N. S. Taichman.** 1982. Non-cytolytic effect of *Actinobacillus actinomycetemcomitans* on leukocyte functions, p. 179–192. *In* R. J. Genco and S. E. Mergenhagen (ed.), *Host Parasite Interactions in Human Periodontal Diseases.* American Society for Microbiology, Washington, D.C.

49. **Miyasaki, K. T.** 1991. The neutrophil: mechanisms of controlling periodontal bacteria. *J. Periodontol.* **62:**761–774.

50. **Miyasaki, K. T., and A. L. Bodeau.** 1991. In vitro killing of *Actinobacillus actinomycetemcomitans* and *Capnocytophaga* spp. by human neutrophil cathepsin G and elastase. *Infect. Immun.* **59:**3015–3020.

51. **Miyasaki, K. T., M. E. Wilson, A. J. Brunetti, and R. J. Genco.** 1986. Oxidative and nonoxidative killing of *Actinobacillus actinomycetemcomitans* by human neutrophils. *Infect. Immun.* **53:**154–160.

52. **Miyasaki, K. T., M. E. Wilson, and R. J. Genco.** 1986. Killing of *Actinobacillus actinomycetemcomitans* by the human neutrophil myeloperoxidase-hydrogen peroxide-chloride system. *Infect. Immun.* **53:**161–165.

53. **Miyasaki, K. T., M. E. Wilson, H. S. Reynolds, and R. J. Genco.** 1984. Resistance of *Actinobacillus actinomycetemcomitans* and differential susceptibility of oral *Haemophilus* species to the bactericidal effects of hydrogen peroxide. *Infect. Immun.* **46:**644–648.

54. **Miyasaki, K. T., M. E. Wilson, J. J. Zambon, and R. J. Genco.** 1985. Influence of endogenous catalase activity on the sensitivity of the oral bacterium *Actinobacillus actinomycetemcomitans* and the oral haemophili to the bactericidal properties of hydrogen peroxide. *Arch. Oral Biol.* **30:**843–848.

55. **Miyasaki, K. T., J. J. Zambon, C. A. Jones, and M. E. Wilson.** 1987. Role of high-avidity binding of human neutrophil myeloperoxidase in the killing of *Actinobacillus actinomycetemcomitans*. *Infect. Immun.* **55:**1029–1036.

56. **Offenbacher, S., S. S. Scott, B. M. Odle, C. Wilson-Burrows, and T. E. Van Dyke.** 1987. Depressed leukotriene B$_4$ chemotaxis response of neutrophils from localized juvenile periodontitis patients. *J. Periodontol.* **58:**602–606.

57. **Page, R. C., T. J. Sims, F. Geissler, L. C. Altman, and D. A. Baab.** 1985. Defective neutrophil and monocyte motility in patients with early onset periodontitis. *Infect. Immun.* **47:**169–175.

58. **Page, R. C., G. E. Vandesteen, J. L. Ebersole, B. L. Williams, I. L. Dixon, and L. C. Altman.** 1985. Clinical and laboratory studies of a family with a high prevalence of juvenile periodontitis. *J. Periodontol.* **56:**602–610.

59. **Pedersen, M. M.** 1988. Chemotactic response of neutrophil polymorphonuclear leukocytes in juvenile periodontitis measured by the leading front method. *Scand. J. Dent. Res.* **96:**421–427.

60. **Ranney, R. R., N. R. Yanni, J. A. Burmeister, and J. G. Tew.** 1982. Relationship between attachment loss and precipitating serum antibody to *Actinobacillus actinomycetemcomitans* in adolescents and young adults having severe periodontal destruction. *J. Periodontol.* **53:**1–7.

61. **Repo, H., L. Saxén, M. Jäättelä, M. Ristola, and M. Leirisalo-Repo.** 1990. Phagocyte function in juvenile periodontitis. *Infect. Immun.* **58:**1085–1092.

62. **Robertson, P. B., B. F. Mackler, T. E. Wright, and B. M. Levy.** 1980. Periodontal status of patients with abnormalities of the immune system. II. Observations over a 2-year period. *J. Periodontol.* **51:**70–73.

63. **Robertson, P. B., T. E. Wright, B. F. Mackler, D. M. Lenertz, and B. M. Levy.** 1978. Periodontal status of patients with abnormalities of the immune system. *J. Periodontal Res.* **13:**37–45.

64. **Saxen, L.** 1980. Prevalence of juvenile periodontitis in Finland. *J. Clin. Periodontol.* **7:**177–186.

65. **Schenkein, H. A., A. M. Best, and J. C. Gunsolley.** 1991. Influence of race and periodontal clinical status on neutrophil chemotactic responses. *J. Periodontal Res.* **26:**272–275.

66. **Söderström, T., and M. Wikström.** 1989. Decreased lactoferrin content in granulocytes from subjects with *Actinobacillus actinomycetemcomitans* associated periodontal diseases. *J. Parodontol.* **9:**195–199.

67. **Springer, T. A., W. S. Thompson, L. J. Miller, F. C. Schmalstieg, and D. Anderson.** 1984. Inherited deficiency of the Mac-1, LFA-1, p1so, 9s glycoprotein family and its molecular basis. *J. Exp. Med.* **160:**1901–1918.

68. **Sutton, R. B. O., and F. C. Smales.** 1983. Cross-sectional study of the effects of immunosuppressive drugs on chronic periodontal disease in man. *J. Clin. Periodontol.* **10:**317–326.

69. **Suzuki, J. B., B. C. Collison, W. A. Falker, Jr., and R. K. Nauman.** 1983. Immunologic profile of juvenile periodontitis. II. Neutrophil chemotaxis, phagocytosis and spore germination. *J. Periodontol.* **56:**461–467.

70. **Taichman, N. S., M. Iwase, H. Korchak, P. Berthold, and E. T. Lally.** 1991. Membranolytic activity of *Actinobacillus actinomycetemcomitans* leukotoxin. *J. Periodontal Res.* **26:**258–260.

71. **Taichman, N. S., M. Iwase, E. T. Lally, S. J. Shattil, M. E. Cunningham, and H. M. Korchak.** 1991. Early changes in cytosolic calcium and membrane potential induced by *Actinobacillus actinomycetemcomitans* leukotoxin in susceptible and resistant target cells. *J. Immunol.* **147:**3587–3594.

72. **Tollefsen, T., E. Saltvedt, and H. S. Koppang.** 1978. The effect of immunosuppressive agents on periodontal disease in man. *J. Periodontal Res.* **13:**240–250.

73. **Tyagi, S. R., D. J. Uhlinger, J. D. Lambeth, C. Champagne, and T. E. Van Dyke.** 1992. Altered diacylglycerol level and metabolism in neutrophils from patients with localized juvenile periodontitis. *Infect. Immun.* **60:**2481–2487.

74. **Van Dyke, T. E., E. Bartholomew, R. J. Genco, J. Slots, and M. J. Levine.** 1982. Inhibition of neutrophil chemotaxis by soluble bacterial products. *J. Periodontol.* **53:**502–508.

75. **Van Dyke, T. E., H. U. Horoszewicz, L. J. Cianciola, and R. J. Genco.** 1980. Neutrophil chemotaxis dysfunction in human periodontitis. *Infect. Immun.* **27:**124–132.

76. **Van Dyke, T. E., H. U. Horoszewicz, and R. J. Genco.** 1993. The polymorphonuclear leukocyte (PMNL) locomotor defect in juvenile periodontitis. Study of random migration, chemokinesis and chemotaxis. *J. Periodontol.* **53:**682–687.

77. **Van Dyke, T. E., M. J. Levine, L. A. Tabak, and R. J. Genco.** 1981. Reduced chemotactic binding in juvenile periodontitis: a model for neutrophil function. *Biochem. Biophys. Res. Commun.* **100:**1278–1284.

78. **Van Dyke, T. E., M. Schweinebraten, L. Cianciola, S. Offenbacher, and R. J. Genco.** 1985. Neutrophil chemotaxis in families with localized juvenile periodontitis. *J. Periodontal Res.* **20:**503–514.

79. **Van Dyke, T. E., W. Zinney, K. Winkel, A. Taufiq, S. Offenbacher, and R. R. Arnold.** 1986. Neutrophil function in localized juvenile periodontitis. Phagocytosis, superoxide production and specific granule release. *J. Periodontol.* **57:**703–708.

80. **van Oss, C. J., C. F. Gillman, and A. W. Neumann.** 1975. *Phagocytic Engulfment and Cell Adhesiveness.* Marcel Dekker, Inc., New York.

81. **Waldrop, T. C., D. C. Anderson, W. W. Hallmon, F. C. Schmalstieg, and R. L. Jacobs.** 1987. Periodontal manifestations of the heritable Mac-1, LFA-1, deficiency syndrome. Clinical, histopathologic and molecular characteristics. *J. Periodontol.* **58:**400–416.

82. **Wisner-Lynch, L. A., and W. V. Giannobile.** 1993. Current concepts in juvenile periodontitis. *Curr. Opin. Periodontol.* **1:**28–42.

83. **Zafiropoulos, G.-G. K., L. Flores-de-Jacoby, V.-M. Plate, I. Eckle, and G. Kolb.** 1991. Polymorphonuclear neutrophil chemiluminescence in periodontal disease. *J. Clin. Periodontol.* **18:**634–639.

83a. **Zambon, J. J.** Personal communication.

84. **Zambon, J. J.** 1985. *Actinobacillus actinomycetemcomitans* in human periodontal disease. *J. Clin. Periodontol.* **12:**1–20.

85. **Zambon, J. J., L. A. Christersson, and R. J. Genco.** 1986. Diagnosis and treatment of localized juvenile periodontitis. *J. Am. Dent. Assoc.* **113:**295–299.

86. **Zambon, J. J., L. A. Christersson, and J. Slots.** 1983. *Actinobacillus actinomycetemcomitans* in human periodontal disease: prevalence in patient groups and distribution of biotypes and serotypes within families. *J. Periodontol.* **54:**707–711.

Molecular Pathogenesis of Periodontal Disease
Edited by Robert Genco et al.
© 1994 American Society for Microbiology, Washington, DC 20005

Chapter 28

Neutrophil Receptors: *N*-Formyl-*l*-Methionyl-*l*-Leucyl-*l*-Phenylalanine and Interleukin-8

Ernesto De Nardin

Neutrophils (PMN) constitute approximately 40 to 70% of total circulating leukocytes. As the primary phagocytic cells, PMN play a key role in host defense against extracellular bacteria. They also play a role in the acute phase of inflammatory reactions. The importance of these cells in combating infectious disease is demonstrated by the increased susceptibility to recurrent bacterial infection in patients with defective PMN production and/or function. PMN circulate in the bloodstream and can be recruited in high numbers in infected and inflamed tissues, where invading microorganisms can be detected through chemoattractants either released by the microorganisms themselves or arising from the inflammatory reaction in the infected tissue. Various PMN chemotactic agonists have been identified; these include the complement component fragment C5a (23); the *N*-formylmethionyl peptides such as *N*-formyl-1-methionyl-1-leucyl-1-phenylalanine (FMLP) (51), which are thought to be analogs of bacterial products; platelet-activating factor (PAF); and leukotriene B_4 (LTB_4) (5). In spite of major structural differences, these agonists share similar biologic activities. They stimulate PMN by binding to specific and distinct receptors on the surface of the cell and elicit cellular responses such as cell motility, exocytosis of granule contents, and the respiratory burst.

PMN AND THE FORMYL PEPTIDE RECEPTOR

PMN chemotaxis stimulated by FMLP is initiated by binding the chemotactic peptide to specific PMN plasma membrane receptors (4, 53, 66). Stimulation of PMN secretory responses such as superoxide production or lysozomal enzyme release requires a much higher concentration of formyl peptide than chemotaxis does (41, 56).

Goetzl et al. (28) isolated the FMLP receptor from human PMN and found it

Ernesto De Nardin • Department of Oral Biology, State University of New York at Buffalo, Foster Hall, Buffalo, New York 14214-3092.

to consist of three components, the major one having a molecular weight of 68,000. This was confirmed by Niedel et al. (45), Schmitt et al. (52), and Marasco et al. (40), who found the receptor on human PMN to be a glycoprotein of 55,000 to 70,000 molecular weight. On the other hand, Allen et al. (1), using sedimentation analysis, determined the receptor to be a single glycoprotein of approximately 63 kDa. FMLP receptors on human PMN exist in both high- and low-affinity states (33). Jesaitis et al. (31, 32) suggested that a single receptor can be converted from a low-affinity to a high-affinity form, with the latter becoming inactive and internalized. On the other hand, Yuli et al. (68) proposed that a particular cellular response is associated with a distinct high- or low-affinity receptor and that either receptor can be regulated and, more specifically, favored by exogenous agents. Feltner and Marasco (22) found three different receptor dissociation rates, of which two appear to correspond to the two affinity states detected in equilibrium binding studies. More recently, De Nardin et al. (20) demonstrated that the FMLP receptor complex is composed of more than one component, that each of the receptor moieties binds FMLP with a different affinity, and that the different affinity states described for the FMLP receptor may be associated with distinct ligand-binding sites.

PMN AND IL-8

Interleukin-8 (IL-8; previously described as monocyte-derived chemotactic factor), also known as neutrophil-activating peptide 1 (NAP-1) (6) and its analogs, NAP-2 (65) and melanoma growth-stimulatory activity factor (GRO) (43), are members of a family of cytokines that can attract and activate PMN. They share many properties, such as stimulation of chemotaxis, release of granule enzymes, and triggering of the respiratory burst, with other PMN chemoattractants such as C5a, FMLP, PAF, and LTB$_4$. IL-8, NAP-2, and GRO can thus be classified as classic chemotactic agonists. IL-8 was identified for its ability to stimulate PMN chemotactic activity, release of granule enzymes, and other cellular events characteristic of the response of phagocytes to chemotactic agonists in vitro (6, 37) as well as in vivo (16, 17). Exposure of PMN to IL-8 induces up-regulation of CD18 (LFA-1) expression (54) as well as integrin receptor activation (13, 39). IL-8 exposure also induces rapid shedding of L selectins such as MEL-14 by human PMN (54). This was determined by flow cytometric analysis using anti-L-selectin monoclonal antibodies (LAM 1.2 or anti-DREG 56). The major difference between IL-8 and some of the other chemoattractants is that the former is a "signal from within." Unlike FMLP, which is thought to be of bacterial origin, IL-8 is produced mainly in the tissues upon stimulation with IL-1, tumor necrosis factor (TNF), lipopolysaccharide, and other agents (6). Although monocytes and macrophages are the major source of IL-8, recent studies show that IL-8 can also be produced by fibroblasts (27), keratinocytes (35), endothelial cells (42), and gingival fibroblasts (57) in response to IL-1b and TNF. Interestingly enough, PMN themselves produce IL-8 during phagocytosis (7). Furthermore, IL-8/NAP-1 is more selective for neutrophils than C5a, FMLP, PAF, and LTB$_4$ and has minimal effects on eosinophils, monocytes, and basophils (6, 37). IL-8 also differs from other chemoattractants in its resistance to inactivation and remains active for long periods in culture medium as well as in biologic fluids (16).

PMN AND THE IL-8 RECEPTOR

PMN possess on their plasma membranes unique receptors for IL-8 that are distinct from the receptors for other cytokines and chemotactic agents such as IL-1, TNF, monocyte chemotactic and activating factor, FMLP, C5a, LTB$_4$, and PAF (27). More mature PMN express more IL-8 receptors than promyelocytes, as suggested by the fact that the human promyelocytic cell line HL-60 exhibits increased expression of IL-8 receptors upon differentiating into PMN after treatment with dimethyl sulfoxide (27). On the other hand, phorbal myristic acetate-treated HL-60, U937, and THP-1 cells that differentiate into monocytes do not exhibit specific binding for IL-8 (27). Treatment of cells with lectins such as wheat germ hemagglutinin or with pronase K can diminish IL-8-binding activity (3). These observations suggest that IL-8 receptors are glycoproteins and that sugar moieties may contribute to ligand binding, although such moieties may not be totally responsible for such binding. IL-8 receptors on human PMN have been estimated to be between 59 and 67 kDa in molecular size (27) and to be composed of two distinct, low- and high-affinity binding components (30, 44). After ligand binding, the ligand-receptor complex is internalized within 10 min at 378C and is transported to the lysosomes, where IL-8 is degraded and released from the cells. Once depleted of free ligand, the receptor is then recycled to the surface of the cell (50). This process is not dependent on de novo protein synthesis, suggesting that IL-8 receptor expression is regulated by the ligand and that rapid recycling of the receptor may be necessary for the directed recognition of a chemotactic gradient (50).

CHEMOTACTIC RECEPTORS AND INTERACTION WITH G PROTEINS

FMLP receptors process chemotactic signals via a receptor-G protein-phospholipase C pathway leading to the production of inositolphosphates such as inositol triphosphate and diacylglycerol (55, 62, 63). Inositol triphosphate is generated by phospholipase C-mediated phosphodiester cleavage of a membrane phospholipid, phosphatidylinositol 4,5-bisphosphate (9). IL-8 receptor's signal transduction mechanism is similar to that of receptors for FMLP and C5a, like them, involving the activation of specific G proteins (67). In addition, the FMLP, IL-8, and C5a receptors have hydropathic profiles similar to that of the b2 adrenergic receptor (10), exhibiting the classic seven transmembrane hydrophobic segments seen in other G-protein-coupled receptors.

MOLECULAR BIOLOGY OF FMLP AND IL-8 RECEPTORS

Recently, several investigators reported the isolation and characterization of cDNAs encoding at least one of the components of the FMLP receptor. Boulay et al. (10, 11) isolated clones encoding the FMLP receptor from a CDM8 expression library of differentiated HL-60 cells. These clones could transfer to COS-7 cells the capacity to specifically bind a chemotactic peptide and could induce the expression of the FMLP receptor on the surface of these cells. Northern (RNA) analysis demonstrated that these clones represent transcripts of 1.6, 1.9, 2.3, and 3.1 kb, with reading frames of 1,050 bases almost identical except for two single-base-substitution differences. Following Southern analysis of genomic DNA, Boulet et al. (10) suggested the presence of either a single gene with at least one intron in the coding region or two genes possibly encoding different FMLP receptor isoforms, represented by the single-base substitutions seen in the different clones. De Nardin et al. (21) amplified, cloned, and sequenced an $\approx$1-kb intronless genomic DNA encoding the human formyl peptide receptor component described by Boulay et al. (11). Although the majority of known mammalian genes are known to contain introns, the b adrenergic receptors, which are similar in overall structure to the FMLP receptor, are also encoded by intronless genes (12). However, Haviland et al. (29) reported that the FMLP receptor gene is $\approx$7.5 kb long and is composed of two exons separated by an $\approx$5-kb intron. The first exon encodes 66 bp of the 5' untranslated region, while the second exon contains the coding as well as the 3' untranslated regions.

Recently, several investigators reported the isolation and characterization of cDNAs encoding the human and the rabbit IL-8 receptors (30, 44, 58). Genomic DNA studies with rabbits demonstrated that the IL-8 receptor is encoded by an intronless, $\approx$1-kb genomic DNA fragment (8). This finding is consistent with findings for other G-protein-activating molecules, such as the b_2 adrenergic receptor (12), the FMLP receptor (21, 29), and the C5a receptor (26). It is interesting that the genes for these chemotactic receptors appear to be clustered on chromosomes (26, 29), implying a possible coordinated regulation for these genes.

FUNCTIONAL CHARACTERIZATION OF FMLP AND IL-8 RECEPTORS

The IL-8 receptor has a hydropathic profile similar to those of the FMLP, C5a, and b2 adrenergic receptors (10). A model based on this profile and exhibiting the seven transmembrane hydrophobic segments characteristic of other G-protein-coupled receptors has been proposed (30, 44, 58). The IL-8 receptor also has 29% sequence homology with the human FMLP and C5a receptors (30).

Recently, Radel et al. (48), using synthetic peptides, found that the primary ligand-binding site of the FMLP receptor is localized on a 17-amino-acid stretch on the first extracellular loop of the N-terminal end of the receptor protein, although other portions of the molecule may contribute to ligand binding. More detailed studies indicate that the charged moieties on the N-terminal end of the first extracellular loop of the FMLP receptor protein, specifically Arg-84 and Lys-85, contribute to the active site in formylated peptide binding to the receptor (49). In addition, Lala et al. (34) constructed a recombinant, truncated FMLP receptor in an *Escherichia coli* expression system that lacked the first 86 residues of the full-length molecule. This mutant construct, which did not contain Arg-84 and Lys-85, failed to demonstrate specific ligand binding, unlike the full-length receptor protein, further suggesting that these charged residues could act as important "contacts" in receptor-ligand interaction.

Preliminary data from our laboratory (19) also suggest that the first N-terminal extracellular loop of the IL-8 receptor molecule may contribute to IL-8 binding to its receptor. Gayle et al. (24) proposed that the amino terminus of the IL-8 receptor may contribute to ligand binding. This is somewhat surprising, since, as the authors themselves point out, there is little homology in this region among the different IL-8 receptors. Indeed, one would expect increased homology in the portions of different receptor molecules that bind the same ligand. In lieu of such homology, the authors suggest that other parts of the molecule may play an important role in receptor-ligand interactions. More important, the same group proposes the presence of single "contact point" residues involved in ligand binding. Interestingly, the first N-terminal extracellular loops of the FMLP and IL-8 receptor molecules exhibit over 50% homology, almost exclusively in their C-terminal portions. These observations, our findings on the FMLP receptor, and preliminary data on the IL-8 receptor indicate that a common structural motif of the IL-8 and FMLP receptors may be the ligand-binding activity of the first extracellular loop of the receptor molecule. The observation that the N-terminal portion of this loop has very little homology in different molecules is consistent with the prediction that this segment is the ligand-binding site for both receptors (thus being different for distinct ligands), while the C-terminal end of the loop represents a conserved region between the two receptors. More detailed research is, of course, needed in order to elucidate this point.

PMN AND PERIODONTAL DISEASES

It has become increasingly clear in recent years that host defense mechanisms play a crucial role in protection against a variety of periodontal diseases. Strong supporting evidence for this is the fact that periodontal disease is often seen in patients with dysfunction and/or diminutions in their defense mechanisms. In some cases, control (whether direct or indirect) of the host defense results in diminution of the periodontitis associated with it. It is important to recognize that factors other than host defense play a role in the development of periodontitis. These may include oral hygiene, colonization by particular microorganisms, malnutrition, stress, and various socioeconomic parameters. However, host defense remains the most important innate factor in the prevention of periodontal disease.

Patients with either quantitative (neutropenia) or qualitative (adherence, chemotaxis, microbicidal activity) PMN defects often suffer from oral mucosal ulcerations, gingivitis, and periodontitis. Severe oral disease occurs with both primary and secondary PMN abnormalities. In contrast, patients with mononuclear phagocyte defects do not exhibit an apparent predisposition to severe periodontitis. Individuals with immunodeficiencies in either T or B lymphocytes also do not suffer from more severe periodontal disease than immunocompetent subjects. However, periodontal disease is often observed when the abnormality affects both polymorphonuclear (PMN) and mononuclear phagocyte functions, as in the case of leukocyte adherence deficiency, Chediak-Higashi syndrome, and AIDS. In general, then, patients with systemic diseases associated with PMN abnormalities have more severe periodontal disease than those with normal PMN. This observation has led to the concept that the PMN plays a key role in protection against periodontitis.

ROLE OF IL-8 IN PERIODONTAL INFLAMMATION AND PMN EMIGRATION

Previous studies have demonstrated the important role of IL-8 in control of infection and in inflammation. It is my belief that this cytokine also plays a crucial role in the periodontum. On the basis of numerous published reports on the function of IL-8, its receptor, and the cells and bioactive molecules involved in its functions, I propose the following model for the action of IL-8 in periodontal tissues.

Step 1. Bacteria trigger gingival fibroblasts and other cells to produce IL-8 (35, 57).
Step 2. IL-8 stimulates mast cells to produce histamine and TNF-a.
Step 3. Histamine and TNF-a (and IL-1) stimulate endothelial cells of the venules to produce P selectin (GMP-140 or CD-62) and E selectin (ELAM-1) (46).
Step 4. PMN begin to roll along the endothelial surface after the expression of P and E selectins on the surface. This "rolling" results from binding of sialyl Lewis (L selectins) (46) on the PMN surface to P and E selectins on the endothelial cells.
Step 5. IL-8 stimulates PMN to transport CD11/CD18 (Mac-1/LFA) integrin from the PMN cytoplasm to its cell surface. The integrin binds to ICAM-1 (intercellular adhesion molecule 1), which is expressed on the endothelial cell. This interaction anchors the PMN. At the same time, IL-8 stimulates PMN shape changes and shedding of their L selectin (which contains the ligand for P and E selectins).
Step 6. The PMN then emigrates (haptotaxis) out of the blood vessel by binding to IL-8 (present as a substrate-bound gradient on the endothelial cells) and through the extracellular matrix.
Step 7. Once in the tissue, PMN ingest and kill bacteria. In the process of controlling infection, they release granular enzymes that may be responsible in part for local tissue destruction.

IL-8 is likely an important cytokine in modulating periodontal infections, and understanding the molecular basis of its activity will assist in the design of pharmacotherapeutic intervention and other strategies that may be useful in modulating periodontal diseases.

PMN AND LJP

LJP is a disease that emerges around puberty and leads to a very rapid destruction of supporting tissue and a loss of teeth at localized sites, especially the incisors and the first molars. LJP has been associated with defective PMN function, i.e., altered responses to classical chemoattractants such as FMLP and C5a as well as to LTB$_4$ (14, 15, 25, 36, 59, 60). More recently, my group (18) suggested that the decrease in chemotactic response to FMLP in certain LJP patients may be associated with an alteration of the receptor itself (due to an innate "defect" or to environmental factors) rather than with a quantitative decrease of total receptor number. This concept has been supported by Perez et al. (47). Interestingly, the response of LJP PMN to IL-8 is less than that of PMN from healthy subjects (14, 38). Although these patients do not suffer from more systemic infections than healthy controls, these alterations of chemotactic responses may be a predisposing factor in the onset of the disease.

LJP and the chemotactic defect follow a strong familial pattern of inheritance (61), and various genetic patterns of inheritance have been suggested for this disease. One of the most commonly asked questions about LJP involves how the PMN can have a genetically transmitted altered response to so many different chemoattractants and in addition have altered cellular functions (e.g., Ca$^+$ mobilization) that are usually associated with the unaltered response. Unique as it may seem that a particular cell may have so many genetic "defects" in response to different chemoattractants, there are two basic commonalities among the different chemoattractant receptors: (i) their structural arrangements in the plasma membrane and/or their effects on cell activation (chemotaxis, respiratory burst, etc.) and (ii) their mechanisms of transmembrane signaling. If for some reason a cell cannot generate and/or express membrane receptors that have common structural features, then all receptors sharing those features may be affected. Likewise, if a particular cell has deficiencies in the signal transduction mechanism common to many surface components, then all components utilizing that mechanism may have altered functions. Previous studies of LJP PMN functions have focused on specific cellular defects. To date, no project has addressed the possibility of a common cellular alteration that might result in all the phenomena observed in LJP PMN. My hypothesis is that in LJP, a common underlying cellular defect either at the receptor or at the post-receptor/membrane signal transmission level may contribute to the altered responses in LJP patients.

The inability to express adequate number of integrins is associated with diminished adherence-dependent PMN functions, including chemotaxis and phagocytosis (2), which can be associated with severe periodontitis (64). Considerable evidence indicates that cell adhesion molecules, particularly L selectin and Mac-1/CR3, play a critical role in facilitating leukocyte-endothelial interactions that underlie the migration of circulating phagocytes to the extravascular compartment. Inasmuch as an adequate influx of circulating PMN into the tissues of the periodontium is thought to provide significant protection against invasion by periodontal organisms, changes in cell adhesion molecule expression that derive from altered responses to FMLP and/or IL-8 may predispose individuals to periodontal disease.

The long-term goal is to define a possible "common denominator defect" that results in many of the PMN functional alterations reported for LJP patients. If a genetic defect in cell response to distinct yet functionally similar agents is to be identified and understood, then studies of structural as well as functional common denominators in these agents must be carried out. My laboratory will initiate the search for such a common denominator that may underlie the various PMN dysfunctions reported for LJP. In addition to providing understanding of basic PMN function, the characterization of structural and functional commonalities among similar membrane components such as the IL-8 and FMLP receptors as well as among the receptors for C5a, NAP-2, and GRO (which we are also currently studying) may aid in the elucidation of this common defect. In addition, functional profiles of LJP PMN responses to different chemotactic cytokines will confirm the pattern of observed defects, while the characterization of transmembrane signaling will expand the search for such defects. This novel approach will provide new insight into the molecular and genetic bases of the disease and will contribute to the long-term goal of identifying individuals at high risk for developing periodontal disease.

Acknowledgments. I thank the following for contributing to some of the findings summarized in this report: Steve Radel, A. Lala, Ashu Sharma, A. De Nardin, J. Katancik, H. Sojar, and R. Genco.

Some of the work summarized in this report was supported in part by U.S. Public Health Service grants DE-07926 and DE-04898.

REFERENCES

1. **Allen, R. A., A. J. Jesaitis, L. A. Sklar, C. G. Cochrane, and R. G. Painter.** 1986. Physiochemical properties of the N-formyl peptide receptor on human neutrophils. *J. Biol. Chem.* **261:**1854–1857.
2. **Anderson, D., F. Schmalsteig, M. Finegold, B. Hughes, R. Rothlein, L. Miller, S. Kohl, M. Tosi, R. Jacobs, T. Waldrop, A. Goldman, W. Shearer, and T. Springer.** 1985. The severe and moderate phenotypes of heritable Mac-1, LFA-1 deficiency: their quantitative definition and relation to leukocyte dysfunction and clinical features. *J. Infect. Dis.* **152:**668–689.
3. **Arfors, K. E., C. Lundberg, L. Lindbom, K. Lundberg, P. G. Beatty, and J. M. Harlan.** 1987. A monoclonal antibody to the membrane glycoprotein complex CD 18 inhibits polymorphonuclear leukocyte accumulation and plasma leakage in vivo. *Blood* **69:**338–340.
4. **Aswanikumar, S., B. A. Corcoran, E. Schiffman, R. A. Day, R. J. Freer, H. J. Showell, E. L. Becker, and C. B. Pert.** 1977. Demonstration of a receptor on rabbit neutrophils for chemotactic peptides. *Biochem. Biophys. Res. Commun.* **74:**810–817.
5. **Baggiolini, M., B. Dewald, and M. Thelen.** 1988. Effects of PAP on neutrophils and mononuclear phagocytes. *Prog. Biochem. Pharmacol.* **22:**90–105.
6. **Baggiolini, M., A. Walz, and S. L. Kunkel.** 1989. Neutrophil-activating peptide-1/interleukin 8, a novel cytokine that activates neutrophils. *J. Clin. Invest.* **84:**1045–1049.
7. **Bazzoni, F., M. A. Cassatella, F. Rossi, M. Ceska, B. Dewald, and M. Baggiolini.** 1991. Phagocytosing neutrophils produce and release high amounts of the neutrophil-activating peptide NAP-I/IL-8. *J. Exp. Med.* **173:**771–774.
8. **Beckmann, M. P., W. E. Munger, C. Kozlosky, T. Vanden Bos, V. Price, S. Lyman, N. P. Gerard, C. Gerard, and D. P. Cerretti.** 1991. Molecular characterization of the interleukin-8 receptor. *Biochem. Biophys. Res. Commun.* **179:**784–789.
9. **Berridge, M. J.** 1987. Inositol triphosphate and diacylglycerol: two interacting second messengers. *Annu. Rev. Biochem.* **56:**159–193.
10. **Boulay, F., M. Tardif, L. Brouchton, and P. Vignais.** 1990. The human N-formylpeptide receptor. Characterization of two cDNA isolates and evidence for a new subfamily of G-protein-coupled receptors. *Biochemistry* **29:**11123–11133.
11. **Boulay, F., M. Tardif, L. Brouchon, and P. Vignais.** 1990. Synthesis and use of a novel N-formyl peptide derivative to isolate a human N-formyl peptide receptor cDNA. *Biochem. Biophys. Res. Commun.* **168:**1103–1109.

12. **Buckland, P. R., R. M. Hill, S. F. Tidmarsh, and P. McGuffin.** 1990. Primary structure of the rat beta-2 adrenergic receptor gene. *Nucleic Acid Res.* **18:**682.

13. **Carveth, H. J., J. F. Bohnsack, T. M. McIntyre, M. Baggiolini, S. M. Prescott, and G. A. Zimmerman.** 1989. Neutrophil activating factor (NAP) induces polymorphonuclear leukocyte adherence to endothelial cells and to subendothelial matrix proteins. *Biochem. Biophys. Res. Commun.* **162:**387–393.

14. **Cianciola, L. J., R. J. Genco, M. R. Patters, J. McKenna, and J. C. J. Van Oss.** 1977. Defective polymorphonuclear leukocyte function in a human periodontal disease. *Nature* (London) **265:**445–447.

15. **Clark, R. A., R. C. Page, and G. Wilde.** 1977. Defective neutrophil chemotaxis in juvenile periodontitis. *Infect. Immun.* **18:**694–700.

16. **Colditz, I., R. D. Zwahlen, and M. Baggiolini.** 1989. In vivo inflammatory activity of neutrophil-activating factor, a novel chemotactic peptide derived from human monocytes. *Am. J. Pathol.* **134:**755–760.

17. **Colditz, I. G., R. D. Zwahlen, and M. Baggiolini.** 1990. Neutrophil accumulation and plasma leakage induced in vivo by neutrophil-activating peptide-1. *J. Leukocyte Biol.* **48:**129–137.

18. **De Nardin, E., C. DeLuca, M. J. Levine, and R. J. Genco.** 1990. Antibodies directed to the chemotactic factor receptor detect differences between chemotactically normal and defective neutrophils from LJP patients. *J. Periodontol.* **61:**609–617.

19. **De Nardin, E., S. J. Radel, A. M. De Nardin, and R. J. Genco.** 1993. Characterization of the ligand binding regions of the human IL-8 receptor, abstr. no. 1473. *J. Dent. Res.* **72:**287.

20. **De Nardin, E., S. J. Radel, and R. J. Genco.** 1991. Isolation and partial characterization of the formyl peptide receptor components on human neutrophils. *Biochem. Biophys. Res. Commun.* **174:**84–89.

21. **De Nardin, E., S. J. Radel, and R. J. Genco.** 1992. Identification of a gene encoding for the human formyl peptide receptor. *Biochem. Int.* **26:**381–387.

22. **Feltner, D. E., and W. A. Marasco.** 1989. Regulation of the formyl peptide receptor binding to rabbit neutrophil plasma membranes. Use of monovalent cations, guanine nucleotides, and bacterial toxins to discriminate among different states of the receptor. *J. Immunol.* **142:**3963–3970.

23. **Fernandez, H. N., P. M. Henson, A. Otani, and T. E. Hugli.** 1978. Chemotactic response to human C3a and C5a anaphylatoxins. I. Evaluation of C3a and C5a leukotaxis in vitro and under stimulated in vivo conditions. *J. Immunol.* **120:**109–115.

24. **Gayle, R. B., P. R. Sleath, S. Srinivason, C. W. Birks, K. S. Weerawarna, D. P. Cerretti, C. J. Kozlosky, N. Nelson, T. Vanden Bos, and M. P. Beckmann.** 1993. Importance of the amino terminus of the interleukin-8 receptor in ligand interactions. *J. Biol. Chem.* **268:**7283–7289.

25. **Genco, R. J., T. E. Van Dyke, B. Park, M. Ciminelli, and H. Horoszewicz.** 1980. Neutrophil chemotaxis impairment in juvenile periodontitis: evaluation of specificity, adherence, deformability, and serum factors. *J. Reticuloendothel. Soc.* **28:**81–91.

26. **Gerard, N. P., L. Bao, H. Xiao-Ping, R. L. Addy, T. B. Shows, and C. Gerard.** 1993. Human chemotactic receptor genes cluster at 19.q.13.3-13.4. Characterization of the human C5a receptor gene. *Biochemistry* **32:**1243–1250.

27. **Gimbrone, M. A., Jr., M. S. Obin, A. F. Brock, E. A. Louis, P. E. Hass, C. A. Hébert, Y. K. Kip, D. W. Leung, D. G. Lowe, W. J. Kohr, W. C. Darbonne, K. B. Bechtol, and J. B. Baker.** 1989. Endothelial interleukin-8: a novel inhibitor of leukocyte-endothelial interactions. *Science* **246:**1601–1603.

28. **Goetzl, E. J., D. W. Foster, and D. W. Goldman.** 1981. Isolation and partial characterization of membrane protein constituents of human neutrophil receptors for chemotactic formylmethionyl peptides. *Biochemistry* **20:**5717–5722.

29. **Haviland, D. L., A. C. Borel, D. T. Fleischer, J. C. Haviland, and R. A. Wetsel.** 1993. Structure, 5'-flanking sequence, and chromosome localization of the human n-formyl peptide receptor gene. A single copy gene comprised of two exons on chromosome 19q.13.3 that yields two distinct transcripts by alternative polyadenylation. *Biochemistry* **32:**4168–4174.

30. **Holmes, W. E., J. Lee, W. J. Kuang, G. C. Rice, and W. I. Wood.** 1991. Structure and functional expression of a human interleukin-8 receptor. *Science* **253:**1278–1280.

31. **Jesaitis, A. J., J. R. Naemura, L. A. Sklar, C. G. Cochrane, and R. G. Painter.** 1984. Rapid

modulation of N-formyl chemotactic peptide receptors on the surface of human granulocytes: formation of high affinity ligand-receptor complexes in transient association with cell cytoskeleton. *J. Cell Biol.* **98:**1378–1387.

32. **Jesaitis, A. J., J. O. Tolley, and R. A. Allen.** 1986. Receptor-cytoskeleton interactions with membrane traffic may regulate chemoattractant-induced superoxide production in human granulocytes. *J. Biol. Chem.* **261:**13662–13669.

33. **Koo, C., R. J. Lefkowitz, and R. Snyderman.** 1982. The oligopeptide chemotactic factor receptor on human polymorphonuclear leukocyte membranes exists in two affinity states. *Biochem. Biophys. Res. Commun.* **106:**442–449.

34. **Lala, A., A. Sharma, H. Sojar, S. J. Radel, R. J. Genco, and E. De Nardin.** 1993. Recombinant expression and partial characterization of the human formyl peptide receptor. *Biochem. Biophys. Acta* **1178:**302–306.

35. **Larsen, C. G., A. O. Anderson, J. J. Oppenheim, and K. Matsushima.** 1989. Production of interleukin-8 by human dermal fibroblasts and keratinocytes in response to interleukin-1 or tumor necrosis factor. *Immunology* **68:**31–36.

36. **Lavine, W. S., E. G. Maderazo, J. Stolman, P. A. Ward, R. B. Cogan, I. Greenblatt, and P. B. Robertson.** 1979. Impaired neutrophil chemotaxis in patients with juvenile and rapidly progressing periodontitis. *J. Periodontal Res.* **14:**10–19.

37. **Leonard, E. J., and T. Yoshimura.** 1990. Neutrophil attractant/activation protein-I (NAP-1 (interleukin-8)). *Am. J. Respir. Cell Mol. Biol.* **2:**479–486.

38. **Lester, M., A. Schneider, M. Warbington, and T. E. Van Dyke.** 1992. Defective neutrophil chemotaxis to interleukin-8 in localized juvenile periodontitis patients, abstr. no. 1102. *J. Dent. Res.* **72:**653.

39. **Lo, S. K., P. A. Detmers, S. M. Levin, and S. D. Wright.** 1989. Transient adhesion of neutrophils to endothelium. *J. Exp. Med.* **169:**1770–1793.

40. **Marasco, W. A., K. M. Becker, D. E. Feltner, C. S. Brown, P. A. Ward, and R. Nairn.** 1985. Covalent affinity labeling, detergent solubilization, and fluid-phase characterization of the rabbit neutrophil formyl peptide chemotaxis receptor. *Biochemistry* **24:**2227–2236.

41. **Marasco, W. A., H. J. Showell, R. J. Freer, and E. L. Becker.** 1982. Anti-f-Met-Leu-Phe: similarities in fine specificity with the formyl peptide chemotaxis receptor of the neutrophil. *J. Immunol.* **128:**956–962.

42. **Matsushima, K., and J. J. Oppenheim.** 1989. Interleukin 8 and MCAF: novel inflammatory cytokines inducible by IL 1 and TNF. *Cytokine* **1:**2–13.

43. **Moser, B., I. Clark-Lewis, R. Zwahlen, and M. Baggiolini.** 1990. Neutrophil-activating properties of the melanoma growth-stimulatory activity. *J. Exp. Med.* **171:**1797–1802.

44. **Murphy, P. M., and H. L. Tiffany.** 1991. Cloning of complementary DNA encoding a functional human interleukin-8 receptor. *Science* **253:**1280–1283.

45. **Niedel, J., J. Davis, and P. Cuatrecasas.** 1980. Covalent affinity labeling of the formyl peptide chemotactic receptor. *J. Biol. Chem.* **255:**7063–7066.

46. **Paulson, J.** 1992. Selectin/carbohydrate mediated adhesion of leukocytes, p. 19–42. *In* J. Harlan and D. Liu (ed.), *Adhesion: Its Role in Inflammatory Disease.* W. H. Freeman & Co., New York.

47. **Perez, H. D., E. Kelly, F. Elfman, G. Armitage, and J. Winkler.** 1991. Defective polymorphonuclear leukocyte formyl peptide receptor(s) in juvenile periodontitis. *J. Clin. Invest.* **87:**971–976.

48. **Radel, S., R. J. Genco, and E. De Nardin.** 1991. Localization of the ligand binding regions of the human formyl peptide receptor. *Biochem. Int.* **25:**745–753.

49. **Radel, S., R. J. Genco, and E. De Nardin.** 1994. Structural and functional characterization of the human formyl peptide receptor ligand-binding region. *Infect. Immun.* **62:**1726–1732.

50. **Rosen, H.** 1990. Role of CR3 in induced myelomonocytic recruitment: insights from in vivo monoclonal antibody studies in the mouse. *J. Leukocyte Biol.* **48:**465–469.

51. **Schiffmann, E., B. A. Corcoran, and S. M. Wahl.** 1975. N-Formylmethionyl peptides as chemoattractants for leucocytes. *Proc. Natl. Acad. Sci. USA* **72:**1059–1062.

52. **Schmitt, M., R. G. Painter, A. J. Jesaitis, K. Preissner, L. A. Sklar, and C. G. Cochrane.** 1983. Photoaffinity labeling of the N-formyl peptide receptor binding site of intact human polymorphonuclear leukocytes. *J. Biol. Chem.* **258:**649–654.

53. **Sha'afi, R. I., K. Williams, M. C. Wacholtz, and E. L. Becker.** 1978. Binding of the chemotactic

synthetic peptide [³H]formyl-Nor-Leu-Leu-Phe to plasma membrane of rabbit neutrophils. *FEBS Lett.* **91**:305–309.

54. **Smith, C., T. Kishimoto, O. Abbass, B. Hughes, R. Rothlein, L. McIntire, E. Butcher, and D. Anderson.** 1991. Chemotactic factors regulate lectin adhesion molecule-1 (LECAM-1)-dependent neutrophil adhesion to cytokine-stimulated endothelial cells in vitro. *J. Clin. Invest.* **87**:609–618.

55. **Smith, R., L. Sam, K. Leach, and J. Justen.** 1992. Postreceptor events associated with human neutrophil activation by interleukin-8. *J. Leukocyte Biol.* **52**:17–26.

56. **Snyderman, R.** 1985. Regulatory mechanisms of a chemoattractant receptor on human polymorphonuclear leukocytes. *Rev. Infect. Dis.* **7**:390–394.

57. **Takashiba, S., M. Takigawa, K. Takahashi, F. Myokai, F. Nishimura, T. Chihara, H. Kurihara, Y. Nomura, and Y. Murayama.** 1992. Interleukin-8 is a major neutrophil chemotactic factor derived from cultured human gingival fibroblasts stimulated with interleukin-1b or tumor necrosis factor a. *Infect. Immun.* **60**:5253–5258.

58. **Thomas, K. M., L. Taylor, and J. Navarro.** 1991. The interleukin-8 receptor is encoded by a neutrophil-specific cDNA clone, F3R. *J. Biol. Chem.* **266**:14839–14841.

59. **Van Dyke, T. E., H. U. Horoszewicz, L. J. Cianciola, and R. J. Genco.** 1980. Neutrophil chemotaxis dysfunction in human periodontitis. *Infect. Immun.* **27**:124–132.

60. **Van Dyke, T. E., M. J. Levine, L. A. Tabak, and R. J. Genco.** 1981. Reduced chemotactic peptide binding in juvenile periodontitis: a model for neutrophil function. *Biochem. Biophys. Res. Commun.* **100**:1278–1284.

61. **Van Dyke, T. E., M. Schweinebraten, L. J. Cianciola, S. Offenbacher, and R. J. Genco.** 1985. Neutrophil chemotaxis in families with localized juvenile periodontitis. *J. Periodontal Res.* **20**:503–514.

62. **Verghese, M. W., C. D. Smith, and R. Snyderman.** 1986. Role of guanine nucleotide regulatory protein in polyphosphoinositide degradation and activation of phagocytic leukocytes by chemoattractants. *J. Cell. Biochem.* **32**:59–69.

63. **Verghese, M. W., R. J. Uhing, and R. Snyderman.** 1986. A pertussis toxin/cholera toxin-sensitive N protein may mediate chemoattractant receptor signal transduction. *Biochem. Biophys. Res. Commun.* **138**:887–894.

64. **Waldrop, T., D. Anderson, W. Hallmon, F. Schmalstieg, and R. Jacobs.** 1987. Periodontal manifestations of the heritable Mac-1, LFA-1, deficiency syndrome. Clinical, histopathologic and molecular characteristics. *J. Periodontol.* **58**:400–414.

65. **Walz, A., B. Dewald, V. von Tscharner, and B. Baggiolini.** 1989. Effects of the neutrophil-activating peptide NAP-2, platelet basic protein, connective tissue-activating peptide III and platelet factor 4 on human neutrophils. *J. Exp. Med.* **170**:1745–1750.

66. **Williams, L. T., R. Snyderman, M. C. Pike, and R. J. Lefkowitz.** 1977. Specific receptor site for chemotactic peptides on human polymorphonuclear leukocytes. *Proc. Natl. Acad. Sci. USA* **74**:1204–1208.

67. **Wright, S. D., S. M. Levin, M. T. Jong, Z. Chad, and L. G. Kabbash.** 1989. CR3 (CD11bICD1B) expresses one binding site for Arg-Gly-Asp-containing peptides and a second site for bacterial lipopolysaccharide. *J. Exp. Med.* **169**:175–183.

68. **Yuli, I., A. Tomonaga, and R. Snyderman.** 1982. Chemoattractant receptor functions in human polymorphonuclear leukocytes are divergently altered by membrane fluidizers. *Proc. Natl. Acad. Sci. USA* **79**:5906–5910.

Molecular Pathogenesis of Periodontal Disease
Edited by Robert Genco et al.
© 1994 American Society for Microbiology, Washington, DC 20005

Chapter 29

The OmpA Protein of *Actinobacillus actinomycetemcomitans*: Physicochemical and Immunologic Features

Mark E. Wilson

Actinobacillus actinomycetemcomitans is a capnophilic gram-negative coccobacillus that has been strongly implicated in the etiology and pathogenesis of localized juvenile periodontitis (LPJ) (35) and rapidly progressive periodontitis (22). A number of groups have observed that periodontitis patients (particularly LJP subjects) colonized by *A. actinomycetemcomitans* exhibit a pronounced increase in serum, salivary, and gingival crevicular fluid immunoglobulin G (IgG), IgA, and IgM antibody activity against this organism (10, 14). A number of techniques, such as enzyme-linked immunosorbent assays (ELISA) and immunoblot analysis, have been employed to define the antigenic specificities of *A. actinomycetemcomitans*-reactive antibodies in the sera of periodontitis patients. The collective results of these studies indicate that sera of periodontitis patients colonized by *A. actinomycetemcomitans* contain IgG antibodies that recognize a diversity of antigens derived from this organism, including (i) a leukotoxin related to the RTX family of cytotoxins (10), (ii) a serotype-defining lipopolysaccharide (11, 20, 34), and (iii) a number of integral outer membrane proteins (4, 9, 30).

Because of their surface accessibility, outer membrane proteins of *A. actinomycetemcomitans* are potentially important targets for interaction with host defense systems. The objective of this chapter is to summarize available information concerning one of the principal outer membrane proteins of *A. actinomycetemcomitans*. This protein, which has an apparent mass of 29 kDa on sodium dodecyl sulfate (SDS)-polyacrylamide gels, appears to be a major target for IgG antibodies present in the sera of LJP subjects colonized by *A. actinomycetemcomitans*.

OUTER MEMBRANE PROTEINS OF *A. ACTINOMYCETEMCOMITANS*

A number of investigators have examined the protein profiles of the outer membrane proteins of reference strains and clinical isolates of *A. actinomycetem-*

Mark E. Wilson • Department of Oral Biology, State University of New York at Buffalo, Buffalo, New York 14214.

comitans on SDS-polyacrylamide gels (4, 6, 9, 16, 28). Although different methods were employed in the preparation of outer membrane fractions, most groups have identified 7 to 10 major protein bands (varying in apparent mass from 16 to 70 kDa), with 3 to 5 of these being prominent on Coomassie-stained gels. One such principal outer membrane protein exhibits the heat modifiability observed in many species of gram-negative bacteria. This protein, originally termed Env2 (7), migrates with an apparent mass of approximately 29 kDa following solubilization in sample buffer at reduced temperatures (22 to 50°C) but with a mass of 34 kDa when solubilized at 100°C (4, 30). The heat-modifiable protein was detected in both fresh clinical isolates and reference strains of *A. actinomycetemcomitans* (4).

My laboratory has also examined the characteristics of the outer membrane proteins of *A. actinomycetemcomitans* Y4 on SDS-polyacrylamide gels and obtained similar results (30). As shown in Fig. 1, a total membrane fraction of *A. actinomycetemcomitans* (containing both cytoplasmic and outer membranes) contained numerous bands, whereas the Sarkosyl-insoluble outer membrane fraction contained approximately six bands. An intense protein band migrating with an apparent mass of 29 kDa is evident in the sample solubilized at 50°C (Fig. 1, lane 3). Solubilization of the outer membrane fraction at 100°C resulted in loss of the 29-kDa band and concurrent emergence of a 34-kDa species (lane 4). Consistent with previous observations, the heat-modifiable protein appears to be one of the most abundant proteins present in the outer membrane of *A. actinomycetemcomitans*.

Immunoblot analysis has been employed to determine which of the outer membrane proteins of *A. actinomycetemcomitans* react strongly with antibodies (particularly IgG) in the sera of periodontitis subjects colonized by this organism.

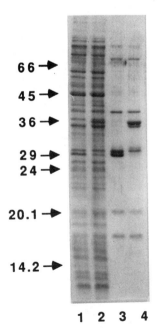

66 →
45 →
36 →
29 →
24 →
20.1 →
14.2 →

1   2   3   4

Figure 1. SDS-polyacrylamide gel electrophoresis protein profiles of a total membrane fraction and a sodium lauroyl sarcosinate (Sarkosyl)-insoluble outer membrane fraction of *A. actinomycetemcomitans* Y4. Lanes 1 and 2, samples of a total-membrane fraction; lanes 3 and 4, samples of an outer membrane fraction. Samples were incubated at 50°C (lanes 1 and 3) or 100°C (lanes 2 and 4) for 10 min in sample buffer prior to gel loading. Following electrophoresis, the gels were stained with Coomassie blue. Reproduced from Wilson (30) with permission.

Watanabe and coworkers (28) observed that sera from patients with LJP and generalized severe periodontitis reacted strongly toward a 35-kDa protein as well as toward a 19-kDa protein. Inasmuch as the outer membrane fractions were solubilized at 100°C prior to being electrophoresed, the 35-kDa protein likely represents the heat-modified form of the 29-kDa protein, although those authors did not evaluate heat-modifiable characteristics of the membrane proteins. Ebersole examined the reactivities of sera from LJP, rapidly progressive periodontitis, and adult periodontitis patients toward the outer membrane proteins of *A. actinomycetemcomitans*, noting that the predominant protein species recognized had apparent masses of 58, 48, 28, and 16 to 18 kDa (9). The conditions for solubilization of the membrane fraction were not indicated. Hence, it is unclear whether the 28-kDa protein represents the unmodified form of the heat-modifiable protein. Bolstad and coworkers (4) performed similar experiments with serum from a patient with Papillon Lefévre syndrome who harbored *A. actinomycetemcomitans*. Strong immunologic reactions toward proteins with apparent masses of 34, 29, and 16.5 kDa were noted. The 34-kDa protein was heat-modifiable and was present in both wild-type and reference strains of *A. actinomycetemcomitans*.

    Immunoblot analysis of the outer membrane of *A. actinomycetemcomitans* Y4, performed with high-titer sera from patients with localized juvenile periodontitis, yielded results (30) similar to those reported by Bolstad and coworkers (4). As depicted in Fig. 2, strong IgG antibody reactivity toward a 16.5- and a 29-kDa protein was observed under conditions in which the outer membrane fraction was solubilized at 50°C. Following solubilization at 100°C, reactivity toward the 16.5-kDa protein was unaltered (results not shown). However, antibody reactivity toward the 29-kDa protein was diminished, with a concomitant increase in antibody reactivity toward a protein with an apparent mass of 34 kDa. The 16.5-kDa protein, which is antigenically related to the peptidoglycan-associated lipoprotein of *Hae-*

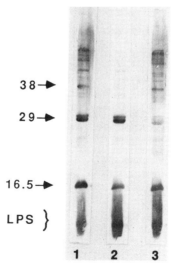

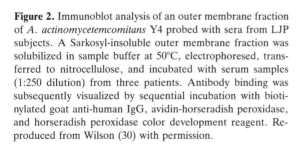

Figure 2. Immunoblot analysis of an outer membrane fraction of *A. actinomycetemcomitans* Y4 probed with sera from LJP subjects. A Sarkosyl-insoluble outer membrane fraction was solubilized in sample buffer at 50°C, electrophoresed, transferred to nitrocellulose, and incubated with serum samples (1:250 dilution) from three patients. Antibody binding was subsequently visualized by sequential incubation with biotinylated goat anti-human IgG, avidin-horseradish peroxidase, and horseradish peroxidase color development reagent. Reproduced from Wilson (30) with permission.

mophilus influenzae (32), stained poorly with Coomassie blue but was highly reactive by immunoblotting (compare Fig. 1 and 2). This characteristic was noted previously (4).

ISOLATION AND CHARACTERIZATION OF THE HEAT-MODIFIABLE OUTER MEMBRANE PROTEIN OF *A. ACTINOMYCETEMCOMITANS*

The 29- and 16.5-kDa proteins appear to represent both quantitatively and immunologically significant constituents of the outer membrane of *A. actinomycetemcomitans*. This has prompted our interest in defining the physicochemical and biological properties of these proteins as well as the functional properties of IgG antibodies directed to these cell envelope antigens. Initial efforts in this regard have been directed to the 29-kDa heat-modifiable protein, inasmuch as this protein appears to be a key antigen recognized by IgG antibodies in sera of *A. actinomycetemcomitans*-infected periodontitis patients (4, 9, 28).

The 29-kDa outer membrane protein of *A. actinomycetemcomitans* Y4 was isolated (30) by using a method employed to purify a similar heat-modifiable protein from *H. influenzae* type b (19). This protein, which was insoluble in octylglucoside-NaCl at pH 8.0, was solubilized from a Sarkosyl-insoluble outer membrane fraction in its heat-unmodified form by incubation at ambient temperature in 20 mM sodium phosphate (pH 7.5) containing 1% SDS. The protein was further purified by hydroxylapatite chromatography. When solubilized in sample buffer at ambient temperature, the isolated protein migrated with an apparent mass of 29 kDa (Fig. 3). Solubilization at higher temperatures (e.g., 50°C) resulted in the emergence of a band with an apparent mass of 34 kDa. Although solubilization at 100°C yielded proportionately greater amounts of 34-kDa species, even prolonged incubation at

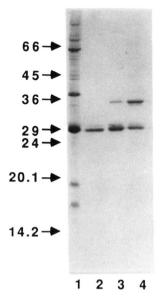

Figure 3. Electrophoretic mobility of the isolated heat-modifiable outer membrane protein of *A. actinomycetemcomitans* Y4 at different temperatures of solubilization. Lane 1, sample of the Sarkosyl-insoluble outer membrane fraction solubilized at 50°C; lane 2, isolated heat-modifiable protein solubilized in sample buffer at ambient temperature; lane 3, isolated protein solubilized in sample buffer at 50°C; lane 4, isolated protein solubilized in sample buffer at 100°C. Reproduced from Wilson (30) with permission.

this temperature did not result in total conversion of the 29-kDa species to its heat-modified form (not shown). The isolated 29-kDa protein contained less than 2 ng of endotoxin mer mg of protein, as determined by *Limulus* lysate gelation assay. Hence, the anomalous electrophoretic mobility of the isolated protein at various temperatures of solubilization did not appear to be attributable to protein interaction with lipopolysaccharide.

Numerous species of enteric and nonenteric gram-negative bacteria contain a heat-modifiable outer membrane protein that is chemically and antigenically similar to the OmpA protein of *Escherichia coli* (3). Amino-terminal sequence and compositional analysis of the 29-kDa protein of *A. actinomycetemcomitans* indicate that this protein is structurally related to the OmpA proteins of *Escherichia coli* and other enteric gram-negative bacteria (6, 31). Moreover, the 29-kDa protein reacted strongly with rabbit antiserum to the *E. coli* OmpA protein (Fig. 4). These studies also provided clear evidence that the 34-kDa protein represents the heat-modified form of the 29-kDa protein. This conclusion was based on (i) the identity of the first 13 amino acid residues of the 29- and 34-kDa proteins, (ii) the virtual identity of the amino acid compositions of these two species, and (iii) the reactivity of both proteins with rabbit antiserum to *E. coli* OmpA.

The amino acid sequences of the OmpA protein of *E. coli* and a number of other species of enteric gram-negative bacteria have been determined (2, 5, 12). These proteins exhibit 67 to 75% homology to the first 12 amino acid residues of the 29-kDa protein of *A. actinomycetemcomitans*. More recently, the N-terminal sequences of OmpA-like membrane proteins of several members of the genus *Haemophilus* have been reported (24, 25). As indicated in Table 1, the first 13 amino acids of the 29-kDa protein of *A. actinomycetemcomitans* exhibit 100% identity with those of the OmpA protein of *Haemophilus somnus* and 85% identity with those of the OmpA proteins of *Haemophilus ducreyi* and *H. influenzae*. These findings are consistent with the close relationship of *A. actinomycetemcomitans* to members of the genus *Haemophilus* (21).

Figure 4. Immunoblot analysis of the heat-modifiable outer membrane protein of *A. actinomycetemcomitans* probed with normal rabbit serum or antiserum to the OmpA protein of *E. coli* K-12. The lanes of each gel were loaded with the following samples: lanes 1, biotinylated molecular mass standards; lanes 2, Sarkosyl-insoluble outer membrane fraction solubilized at 100°C; lanes 3, Sarkosyl-insoluble outer membrane fraction solubilized at ambient temperature; lanes 4, isolated heat-modifiable outer membrane protein solubilized at 100°C; lanes 5, the same protein solubilized at ambient temperature. The electrophoresed proteins were transferred to nitrocellulose and probed with antiserum to *E. coli* K-12 OmpA (left panel) or with normal rabbit serum (right panel) at a 1:500 dilution. Reproduced from Wilson (31) with permission.

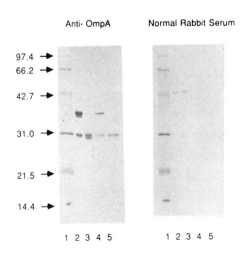

Table 1. Relationship between *A. actinomycetemcomitans* 29-kDa outer membrane protein and other members of the OmpA family: N-terminal sequence data

Bacterial species	N-terminal sequence													Reference
Escherichia coli	A	P	K	D	N	T	W	Y	T	G	A	K	L	2
Salmonella typhimurium	A	P	K	D	N	T	W	Y	A	G	A	K	L	12
Shigella dysenteriae	A	P	K	D	N	T	W	Y	T	G	A	K	L	5
Actinobacillus actinomycetemcomitans	A	P	Q	A	N	T	F	Y	A	G	A	K	A	30
Haemophilus somnus	A	P	Q	A	N	T	F	Y	A	G	A	K	A	25
Haemophilus ducreyi	A	P	Q	A	D		F	Y	V	G	A	K	A	24
Haemophilus influenzae type b	A	P	Q	E	N	T	F	Y	A	G	V	K	A	18a

The 29-kDa protein *A. actinomycetemcomitans* contains a relatively high content of dicarboxylic (25%) and basic (9%) amino acids, similar to that observed with other OmpA proteins (6, 24). The protein is unreactive to staining with periodic acid-Schiff's reagent and is similarly unreactive in an enzyme immunoassay for glycoproteins (33). Only minor amounts of amino sugars (1.5% glucosamine, 1.1% galactosamine) have been detected in the isolated protein. Hence, the OmpA protein of *A. actinomycetemcomitans* is probably not glycosylated, as is also the case for *E. coli* OmpA (6, 13).

The characteristic heat modifiability of OmpA proteins has been attributed to their high contents of antiparallel β structure (approximately 50% in *E. coli* OmpA [8]), resulting in excessive binding of SDS in the absence of heating (6, 15). Incubation of the OmpA protein in the presence of SDS at elevated temperatures extends the β structure to an α helix, resulting in a shift in electrophoretic mobility in SDS-polyacrylamide gels (8). At least in the case of *E. coli* OmpA, the denatured form of the protein migrates with an apparent mass that is consistent with its mass as deduced from amino acid sequence data. The β structure is thought to arise from the N-terminal portion of the molecule, which repeatedly traverses the outer membrane (18). In this manner, a number of surface-accessible domains are generated. The carboxy-terminal region of the OmpA protein, which is structurally conserved, is thought to reside in the periplasmic space. The amino terminus itself does not appear to be exposed on the bacterial surface but rather is "buried" in the outer membrane. The hypothetical structure of the *E. coli* OmpA molecule, as proposed by Morona and coworkers (18), is shown in Fig. 5. This hypothetical structure is based on several lines of evidence, including susceptibility of certain

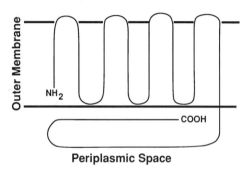

Figure 5. Proposed architecture of the OmpA protein in the outer membrane of *E. coli* K-12. Reproduced in modified form with permission from Morona et al. (18).

domains to pronase cleavage and surface exposure of other domains involved in conjugation and phage receptor activity. However, this structure remains largely hypothetical, as is the extent to which *A. actinomycetemcomitans* OmpA protein structure mimicks that of *E. coli*.

IMMUNOLOGIC PROPERTIES OF THE OmpA PROTEIN OF *A. ACTINOMYCETEMCOMITANS*

As noted above, immunoblot analysis revealed that the 29-kDa OmpA protein of *A. actinomycetemcomitans* is an important target for IgG antibodies in the sera of LJP subjects colonized by this organism. Utilizing an ELISA, my laboratory compared IgG antibody titers to the 29-kDa OmpA protein of *A. actinomycetem-comitans* in the sera of 35 LJP subjects and an equal number of periodontally healthy individuals (30). The isolated 29-kDa protein employed in these studies contained only trace amounts of lipopolysaccharide, as determined by *Limulus* lysate gelation assay (1.7 ng/mg of protein) and by silver staining of SDS-polyacrylamide gels overloaded with the membrane protein (<1%). The geometric mean IgG antibody titer to the 29-kDa protein was roughly 10-fold higher in the LJP group than in the corresponding healthy group. Sixty-three percent (22 of 35 subjects) of the LJP patient sera tested contained IgG titers that were ≥2 standard deviations from the mean of the healthy group.

The pattern of the IgG response of LJP subjects to the 29-kDa OmpA protein was also assessed as a function of age. Such analysis indicated that sera from LJP subjects <18 years of age contained either normal or modestly elevated IgG titers to the 29-kDa protein. On the other hand, 53% (10 of 19) of the LJP subjects ≥18 years of age exhibited markedly elevated titers to the OmpA protein. The geometric mean IgG titer in the latter group was more than threefold greater than the corresponding mean titer for subjects under 18 years of age. Among LJP subjects ≥18 years of age, antibody titers were most elevated in subjects 18 to 21 years of age. Within this subgroup, mean IgG titer was more than fivefold greater than the corresponding mean titer for subjects <18 years of age.

The outer membrane proteins of *A. actinomycetemcomitans* closely resemble those of bacteria in the genus *Haemophilus*, particularly *Haemophilus aphrophilus* (4, 7). This is reflected in the considerable N-terminal sequence homology between the OmpA protein of *A. actinomycetemcomitans* and the corresponding heat-modifiable proteins of *Haemophilus* spp. (Table 1). Moreover, the *A. actinomycetemcomitans* OmpA protein contains an epitope recognized by a monoclonal antibody (2C7) known to react with virtually all members of the family *Pasteurellaceae* (24). Accordingly, experiments were conducted to determine whether LJP patient sera contain IgG antibodies specific for the *A. actinomycetemcomitans* OmpA protein or whether the antibodies recognize determinants common to OmpA proteins found in members of the genus *Haemophilus*. LJP patient serum was subjected to multiple cycles of absorption at 4°C with viable *H. influenzae* (nontypeable), *H. aphrophilus*, or *A. actinomycetemcomitans* Y4, after which residual IgG antibody titer to the

29-kDa OmpA protein of *A. actinomycetemcomitans* was quantified by ELISA (30). Whereas absorption with *A. actinomycetemcomitans* significantly reduced IgG antibody titer to the OmpA protein, similar absorptions with *Haemophilus* spp. had minimal effect on antibody titer. These results indicated that IgG antibodies to the *A. actinomycetemcomitans* OmpA protein recognize a unique determinant(s) in this molecule that is not expressed on the homologous protein of *Haemophilus* spp.

The inability of members of the genus *Haemophilus* to absorb antibodies reactive toward the OmpA protein of *A. actinomycetemcomitans* despite the presence of a strongly homologous N terminus and at least one common epitope (recognized by monoclonal antibody 2C7) is consistent with the proposed organization of the OmpA protein in the outer membrane (Fig. 5). As depicted in this model, the N terminus is inserted into the outer membrane and is therefore not surface accessible (even if this region were sufficiently immunogenic). Monoclonal antibody 2C7 also recognizes a determinant that is not surface accessible (16a). Hence, IgG antibodies in LJP patient sera appear to recognize *A. actinomycetemcomitans* OmpA sequences that are surface accessible (possibly the extracellular domains as depicted in Fig. 5) and are unique to this species.

At present, information regarding the functional properties of IgG antibodies directed to the OmpA protein of *A. actinomycetemcomitans* is limited. LJP sera have been demonstrated to contain opsonic IgG antibodies capable of promoting phagocytosis and killing of *A. actinomycetemcomitans* by human neutrophils (1). The cell envelope determinant(s) recognized by such opsonic antibodies has not been defined. However, opsonic antibodies against *Actinobacillus pleuropneumoniae* appear to recognize outer membrane proteins, possibly including a 29-kDa species (26). For this reason, my laboratory has recently prepared rabbit hyperimmune antiserum to the OmpA protein and is currently evaluating the IgG fraction of this antiserum with respect to putative opsonic and/or bactericidal antibody activity. Efforts are also under way to obtain affinity-purified human IgG antibodies to this protein and to characterize the functional properties of these antibodies.

The biologic functions of the *A. actinomycetemcomitans* OmpA protein are unknown. The OmpA protein of *E. coli* participates in F-mediated conjugation and, together with the lipoprotein, contributes to the structural integrity of the outer membrane (23, 27). Moreover, a recent study by Weiser and Gotschlich (29) indicates that the OmpA protein may be an important factor contributing to the resistance of *E. coli* to complement-mediated killing. Finally, the extracellular domains of *E. coli* OmpA serve a nonphysiologic role as receptors for certain bacteriophages (17, 27). Efforts to determine whether the OmpA protein of *A. actinomycetemcomitans* exhibits similar physiologic and nonphysiologic functions will be greatly facilitated by the acquisition of the complete DNA sequence of this molecule and by the generation of OmpA-defective *A. actinomycetemcomitans* mutants.

Acknowledgments. I express my appreciation to Paul Bronson for technical assistance in the conduct of many of the studies described in this review and to M. Reddy for performing amino sugar analyses.

This work was supported by U.S. Public Health Service grants DE04898 and DE10041 from the National Institute of Dental Research.

REFERENCES

1. **Baker, P. J., and M. E. Wilson.** 1989. Opsonic IgG antibody against *A. actinomycetemcomitans* in localized juvenile periodontitis. *Oral Microbiol. Immunol.* **4:**98–105.

2. **Beck, E., and E. Bremer.** 1980. Nucleotide sequence of the gene *ompA* coding the outer membrane protein II* of *Escherichia coli* K-12. *Nucleic Acids Res.* **8:**3011–3024.

3. **Beher, M. G., C. A. Schnaitman, and A. P. Pugsley.** 1980. Major heat-modifiable outer membrane protein in gram-negative bacteria: comparison with the OmpA protein of *Escherichia coli.* *J. Bacteriol.* **143:**906–913.

4. **Bolstad, A. I., T. Kristoffersen, I. Olsen, H. R. Preus, H. B. Jensen, E. N. Vasstrand, and B. Bakken.** 1990. Outer membrane proteins of *Actinobacillus actinomycetemcomitans* and *Haemophilus aphrophilus* studied by SDS-PAGE and immunoblotting. *Oral Microbiol. Immun.* **5:**155–161.

5. **Braun, G., and S. T. Cole.** 1982. The nucleotide sequence coding for the major outer membrane protein OmpA of *Shigella dysenteriae, Nucleic Acids Res.* **10:**2367–2378.

6. **Chen, R., W. Schmidmayr, C. Krämer, U. Chen-Schmeisser, and U. Henning.** 1980. Primary structure of major outer membrane protein II* (ompA protein) of *Escherichia coli* K-12 *Proc. Natl. Acad. Sci. USA* **77:**4592–4596.

7. **Di Rienzo, J. M., and E. L. Spieler.** 1983. Identification and characterization of the major cell envelope proteins of oral strains of *Actinobacillus actinomycetemcomitans.* *Infect. Immun.* **39:**253–261.

8. **Dornmair, K., H. Kiefer, and F. Jähnig.** 1990. Refolding of an integral membrane protein. OmpA of *Escherichia coli.* *J. Biol. Chem.* **265:**18907–18911.

9. **Ebersole, J. L.** 1990. Systemic humoral immune responses in periodontal disease. *Crit. Rev. Oral Biol. Med.* **1:**283–331.

10. **Ebersole, J. L., M. A. Taubman, D. J. Smith, R. J. Genco, and D. E. Frey.** 1982. Human immune response to oral microorganisms. I. Association of localized juvenile periodontitis (LJP) with serum antibody response to *Actinobacillus actinomycetemcomitans.* *Clin. Exp. Immunol.* **47:**43–52.

11. **Farida, R., M. Wilson, and L. Ivanyi.** 1986. Serum IgG antibodies to lipopolysaccharides in various forms of periodontal disease in man. *Arch. Oral Biol.* **31:**711–715.

12. **Freudl, R., and S. T. Cole.** 1983. Cloning and molecular characterization of the ompA gene from *Salmonella typhimurium.* *Eur. J. Biochem.* **134:**497–502.

13. **Garten, W., I. Hindennach, and U. Henning.** 1975. The major proteins of the *Escherichia coli* outer cell envelope membrane. Characterization of proteins II* and II, comparison of all proteins. *Eur. J. Biochem.* **59:**215–221.

14. **Genco, R. J., J. J. Zambon, and P. A. Murray.** 1985. Serum and gingival fluid antibodies as adjuncts in the diagnosis of *Actinobacillus actinomycetemcomitans*-associated periodontal disease. *J. Periodontol.* **56**(Suppl.)**:**41–50.

15. **Heller, K. B.** 1978. Apparent molecular weights of a heat-modifiable protein from the outer membrane of *Escherichia coli* in gels with different acrylamide concentrations. *J. Bacteriol.* **134:**1181–1183.

16. **Kato, K., S. Kokeguchi, H. Ishihara, Y. Murayama, M. Tsujimoto, H. Takada, T. Ogawa, and S. Kotani.** 1986. Chemical composition and immunobiological activities of sodium dodecyl sulphate extracts from the cell envelopes of *Actinobacillus actinomycetemcomitans, Bacteroides gingivalis* and *Fusobacterium nucleatum.* *J. Gen. Microbiol.* **133:**1033–1043.

16a. **Lesse, A.** Unpublished observations.

17. **Lugtenberg, B., and L. van Alphen.** 1983. Molecular architecture and functioning of the outer membrane of *Escherichia coli* and other gram-negative bacteria. *Biochim. Biophys. Acta* **737:**51–115.

18. **Morona, R., M. Klose, and U. Henning.** 1984. *Escherichia coli* K-12 outer membrane protein (OmpA) as a bacteriophage receptor: analysis of mutant genes expressing altered proteins. *J. Bacteriol.* **159:**570–578.

18a. **Munson, R. S., Jr.** Personal communication.

19. **Munson, R. S., Jr., and D. M. Granoff.** 1985. Purification and partial characterization of outer membrane proteins P5 and P6 from *Haemophilus influenzae* type b. *Infect. Immun.* **49:**544–549.

20. **Page, R. C., T. J. Sims, L. D. Engel, B. J. Moncla, B. Bainbridge, J. Stray, and R. P. Darveau.** 1991. The immunodominant outer membrane antigen of *Actinobacillus actinomycetemcomitans* is located in the serotype-specific high-molecular-mass carbohydrate moiety of lipopolysaccharide. *Infect. Immun.* **59**:3451–3462.

21. **Potts, T. V., J. J. Zambon, and R. J. Genco.** 1985. Reassignment of *Actinobacillus actinomyce-temcomitans* to the genus *Haemophilus* as *Haemophilus actinomycetemcomitans* comb. nov. *Int. J. Syst. Bacteriol.* **35**:337–341.

22. **Slots, J., L. Bragd, M. Wikström, and G. Dahlén.** 1986. The occurrence of *Actinobacillus acti-nomycetemcomitans*, *Bacteroides gingivalis* and *Bacteroides intermedius* in destructive periodontal disease in adults. *J. Clin. Periodontol.* **13**:570–577.

23. **Sonntag, I., H. Schwarz, Y. Hirota, and U. Henning.** 1978. Cell envelope and shape of *Escherichia coli*: multiple mutants missing the outer membrane lipoprotein and other major outer membrane proteins. *J. Bacteriol.* **136**:280–285.

24. **Spinola, S. M., G. E. Griffiths, K. L. Shanks, and M. S. Blake.** 1993. The major outer membrane protein of *Haemophilus ducreyi* is a member of the OmpA family of proteins. *Infect. Immun.* **61**:1346–1351.

25. **Tagawa, Y., M. Haritani, H. Ishikawa, and N. Yuasa.** 1993. Characterization of a heat-modifiable outer membrane protein of *Haemophilus somnus*. *Infect. Immun.* **61**:1750–1755.

26. **Thwaits, R. N., and S. Kadis.** 1991. Immunogenicity of *A. pleuropneumoniae* outer membrane proteins and enhancement of phagocytosis by antibodies to the proteins. *Infect. Immun.* **59**:544–549.

27. **van Alphen, L., L. Havekes, and B. Lugtenberg.** 1977. Major outer membrane protein *d* of *Escherichia coli* K12. Purification and in vitro activity of bacteriophage k3 and f-pilus mediated con-jugation. *FEBS Lett.* **75**:285–290.

28. **Watanabe, H., P. D. Marsh, and L. Ivanyi.** 1989. Antigens of *Actinobacillus actinomycetemcomitans* identified by immunoblotting with sera from patients with localized juvenile periodontitis and generalized severe periodontitis. *Arch. Oral Biol.* **34**:649–656.

29. **Weiser, J. N., and E. C. Gotschlich.** 1991. Outer membrane protein A (OmpA) contributes to serum resistance and pathogenicity of *Escherichia coli* K-1. *Infect. Immun.* **59**:2252–2258.

30. **Wilson, M. E.** 1991. IgG antibody response of localized juvenile periodontitis patients to the 29 kilodalton outer membrane protein of *Actinobacillus actinomycetemcomitans*. *J. Periodontol.* **62**:211–218.

31. **Wilson, M. E.** 1991. The heat-modifiable outer membrane protein of *Actinobacillus actinomyce-temcomitans*: relationship to OmpA proteins. *Infect. Immun.* **59**:2505–2507.

32. **Wilson, M. E.** Unpublished data.

33. **Wilson, M. E., and R. G. Hamilton.** Unpublished data.

34. **Wilson, M. E., and R. E. Schifferle.** 1991. Evidence that the serotype b antigenic determinant of *Actinobacillus actinomycetemcomitans* Y4 resides in the polysaccharide moiety of lipopolysaccha-ride. *Infect. Immun.* **59**:1544–1551.

35. **Zambon, J. J.** 1985. *Actinobacillus actinomycetemcomitans* in human periodontal disease. *J. Clin. Periodontol.* **12**:1–20.

Molecular Pathogenesis of Periodontal Disease
Edited by Robert Genco et al.
© 1994 American Society for Microbiology, Washington, DC 20005

Chapter 30

Genetics of Early-Onset Periodontal Diseases

Harvey A. Schenkein

EARLY-ONSET PERIODONTAL DISEASES

The early-onset periodontal diseases (EOP), including clinical syndromes designated localized juvenile periodontitis (LJP), generalized juvenile periodontitis (GJP), and rapidly progressive periodontitis (RPP), are characterized by their age of onset, which is usually after puberty, and the unusually rapid progression of periodontal attachment loss in affected individuals (16). An additional form of EOP that occurs prior to puberty and usually involves the primary teeth is called prepubertal periodontitis. The localized (LJP) and generalized (GJP, RPP) forms of EOP are distinguished arbitrarily by the pattern of teeth affected by such attachment loss. Classically, LJP has been defined by attachment loss at first molar and incisor teeth, while GJP affects a large number of teeth not limited to first molars and incisors (3). RPP is also a form of generalized EOP that may have some unique characteristics with regard to age of onset and certain laboratory findings (52), but definitive characteristics that distinguish it from GJP have not been established.

An important observation with regard to EOP is that the disease(s) clearly aggregates within families (19, 25). Numerous case reports that document this observation have been published (13, 23, 32, 42, 46, 49–51, 60, 65, 69, 70, 75), and attempts to study families of probands with EOP have verified that this is in fact the case. Interestingly, many reports have documented that localized and generalized forms of EOP may coexist in the same family, implying that the etiology of these relatively rare syndromes may be similar within such families. Furthermore, longitudinal assessment of some subjects has documented the progression of cases of LJP from a localized to a generalized affectation pattern, again implying that the two clinical entities may be related in some individuals.

The aggregation of cases of EOP in families implies heritable risk factors for this disease. Since it is accepted that development of EOP requires bacterial plaque and that the disease is therefore "infectious," such heritable factors may be related

Harvey A. Schenkein • Clinical Research Center for Periodontal Diseases, School of Dentistry, Virginia Commonwealth University, MCV Station Box 980566, Richmond, Virginia 23298-0566.

to inflammatory or immunologic mechanisms that, if rendered ineffective or hyperactive by defective or inactive genes, could enhance the pathogenic potential of plaque bacteria. Studies that examine some of these potential candidates have been carried out and are discussed below.

A second important characteristic of EOP is its demographic distribution. A recent survey of 14- to 17-year-old children in the United States indicated that black Americans are at far higher risk for EOP than are white Americans (40). These prevalence estimates are consistent with those documented in other parts of the world (7, 37, 43, 47, 48, 57, 62, 81). The striking difference in prevalence of EOP between black and white populations raises interesting issues relating to the relative impact of cultural, environmental, and genetic factors that impart risk for EOP. Studies that examine race as a variable in the assessment of pathogenic and etiologic correlates of periodontal disease have been carried out by my group. We detected differences between black and white subjects with regard to bacteriologic factors (67), leukocyte function (66), and immunologic responses (27, 28). It is thus likely that risk factors or protective responses for EOP, whether genetic or environmental, may be enriched in certain racial populations.

ETIOLOGIC AND CLINICAL HETEROGENEITY IN EOP FAMILIES

Studies of the influence of genes on EOP would be greatly facilitated by categorization of etiologic factors within disease categories or within families. A number of such etiologic factors, for example, the presence or amount of bacteria such as *Actinobacillus actinomycetemcomitans*, the presence of leukocyte function defects, and antibody reactivity to periodontal pathogens, are reasonable "markers" and are available for study. Several recent reports of families in which EOP was detected illustrate the difficulty presented to investigators who seek to study the genetics of EOP by focusing on presumptive etiologic correlates of disease. Some examples follow.

Vandesteen and coworkers (75) described a family identified through a white female proband with RPP. All six of her siblings had LJP. Polymorphonuclear neutrophil (PMN) chemotaxis defects were detected in the proband and only two siblings. None of the subjects had *A. actinomycetemcomitans* in their subgingival plaque or detectable antibodies to it in their sera.

Nishimura and coworkers (49) reported on a mother and daughter with EOP. The mother had RPP, and the daughter had LJP. Examination of PMN function revealed depressed chemotaxis and superoxide production in both mother and daughter. The daughter additionally demonstrated defective phagocytosis. Both had elevated antibody titers to *A. actinomycetemcomitans*.

Sbordone et al. (65) described a family in which two siblings had GJP while the dizygous twin of one was periodontally healthy. Both affected siblings were infected with *A. actinomycetemcomitans*, yet only one demonstrated defective PMN chemotaxis.

Lopez (42) reported on a family in which PMN chemotaxis was depressed in two LJP subjects and one healthy sibling and was normal in the prepubertal periodontitis subjects.

Boughman et al. (9) analyzed 39 EOP sibships; in 11 of these sibships there was at least one sib with each of the localized and generalized forms of the disease. When a proband had defective PMN chemotaxis, 71% of affected and 36% of unaffected siblings also had defective chemotaxis. Similar data were reported for antibody responses to *A. actinomycetemcomitans*.

The following points are clear from these reports. (i) The localized and generalized forms of EOP are frequently present within the same family. Given the rarity of these diseases, it is unlikely that the underlying genetic risk factors for EOP in these families are different. (ii) The associated risk factors for EOP may differ between families, indicating that different heritable forms of EOP may exist. Alternatively, the items thought to be risk factors may not be so at all. (iii) There is a remarkable lack of consistency within families and a lack of direct correlation between family members with regard to the presence of demonstrable PMN defects and antibody responses and disease. This may reflect true etiologic heterogeneity, inadequacies in the assay systems used to measure these factors, or a failure to identify bona fide risk factors.

ASSOCIATION OF EOP WITH GENETIC MARKERS

A number of groups have investigated the association of EOP with known genetic markers. Such studies have been quite limited owing to the paucity of such markers that could be rationally associated with risk for EOP and to the small numbers of subjects available for such studies. New genetic methods plus additional insights into the pathogenesis of EOP will provide additional material for such studies.

Most of the association studies reported to date involve the HLA antigens, which play a key role in regulation and mediation of molecular events during immunologic responses. The first such reports by Terasaki, Kaslick, and their co-workers (34, 73) indicated that there was a negative correlation between HLA-A2 and LJP. Subsequently, Reinholdt et al. (58) reported increased frequencies of HLA-A9, HLA-A28, and HLA-Bw15 in LJP subjects and a somewhat decreased frequency of HLA-A2. Both Cullinan et al. (21) and Saxen and Koskimies (63) failed to detect significant association or linkage with HLA antigens. In a study of RPP subjects, Katz et al. (35) reported a higher frequency of HLA-DR4 in affected subjects than in controls. A number of other studies failed to conclusively demonstrate an association with HLA antigens. Data from a number of these studies were recently compiled by Sofaer (68), who concluded that the strongest negative associations with EOP are with HLA-A2 and that subjects with HLA-A9 or HLA-B15 may have increased risk for LJP. However, the small number of subjects studied and the failure to associate different haplotypes with LJP in different families would not substantiate an association of risk for LJP with the HLA complex.

A small number of studies have likewise examined the association of EOP with blood group markers (33, 56). Such studies have failed to demonstrate an association of EOP with such markers.

PMN CHEMOTAXIS: INFLUENCES OF FAMILY AND RACE

In the late 1970s, it was observed that a substantial percentage of individuals with LJP demonstrate a relative defect in the ability of their peripheral blood PMNs to respond to chemotactic agents in a bioassay of chemotaxis (17, 18, 39). This observation has been confirmed and extended by several groups (20, 71), although some investigators have failed to demonstrate the phenomenon (36, 38, 54, 59). The response of LJP patients is heterogeneous in that a significant percentage of subjects, probably around 20 to 30% of subjects with LJP, consistently fail to manifest this defect. Nevertheless, this defective function could be an important etiologic factor in many cases of EOP or at least could serve as an important marker for presumptive inflammatory dysfunction.

Examination of PMNs from patients with the chemotactic defect indicates that such cells have a relative deficiency of surface receptors for chemotactic agents such as formyl peptides and the complement-derived chemotactic factor C5a (77, 78) as well as of a chemotaxis-related surface glycoprotein designated GP 150 (80). It is likely that the biochemical defect is intrinsic to the cell itself, judging from the observation that it cannot be induced in cells of healthy subjects by sera from LJP subjects. Furthermore, many subjects with verifiable histories of previously active LJP retain this dysfunction (76). It has been hypothesized, however, that the defect is in fact acquired by the cells following in vivo exposure to various desensitizing cytokines (1). In view of the likely importance of this cell type in protection against periodontal infections, defective PMN chemotaxis could be an important risk factor in individuals with such a defect.

A number of lines of evidence imply a genetic component to the chemotactic defect seen in LJP. (i) Further examination of this phenomenon, in particular by Van Dyke and coworkers, has shown that the defect apparently persists in older patients with inactive disease, implying that the defect is intrinsic and independent of the disease process (76). (ii) In a study of members of 21 families with LJP probands, the defect was demonstrable only in affected family members in 19 of the 21 families. In addition, inspection of the pedigrees demonstrated consistency with a dominant mode of inheritance of the trait (79). (iii) Perez and coworkers demonstrated a diminution of high-affinity receptors for formyl peptides and decreased amounts of formyl peptide receptor isoforms in a patient with LJP, implying a possible defect in the structural gene for such receptors in this individual (55). (iv) Comparison of the chemotactic responsiveness of cells from black and white subjects indicated that in all categories of EOP as well as in periodontally healthy subjects, the chemotactic response was lower in black subjects than in white subjects. This difference could relate to the relative risk for EOP seen in demographic studies and could be, in part, genetically determined (66). (v) In several disorders, genetically determined defects in phagocyte function are associated with the presence of severe periodontitis (2, 12, 22).

The utilization of information about PMN dysfunction and its relationship to EOP has potential with regard to identification of genetic markers that could be associated with risk for disease in some populations. Despite the inconsistencies observed both between and within populations with regard to this defect and despite alternative hypotheses concerning the biochemical foundation of EOP, there are sufficient data to justify further examination of genes known to be associated with PMN function for their association with risk for EOP.

GENETIC ANALYSES OF EOP: SEGREGATION AND LINKAGE ANALYSES

Segregation analysis of pedigree data from EOP families has been used to determine the likely mode(s) of inheritance of this disease, if any. Typically, EOP probands are identified and family members are then studied to determine the presence or absence of disease. As outlined by Boughman et al. (8, 10), a number of factors that relate to the reliable diagnosis of such family members complicate such analyses. For example, there is clinical heterogeneity with regard to the relative extent and severity of disease, leading to problems in defining the periodontal status of family members who may not demonstrate the classic signs and symptoms of EOP. In fact, as explained above, the definitions of the various forms of EOP are themselves overlapping and are based on clinical history and presentation rather than etiology and pathogenesis. The uncertain age of onset of EOP and the presence of adult forms of periodontitis limit our ability to diagnose individuals older than 30 to 35 years, complicates the diagnosis of healthy family members who are within the putative age range for disease onset, and entirely excludes prepubertal family members from analysis. Furthermore, the expression of EOP may be clinically modified in family members via clinical therapeutic intervention, with extremes being tooth loss or resolution of disease with regeneration of periodontal attachment. It is clear that genetic analyses based on family studies are made more difficult by these factors.

Although only a limited number of family studies of EOP have been reported, evidence for heritability of EOP is substantial. Early studies usually reported small numbers of families in which multiple siblings were affected (6, 13, 19, 25). Because of the small number of families, there was insufficient statistical power to perform formal segregation analyses. However, inspection of pedigrees confirmed the notion that EOP aggregates in families.

A number of genetic hypotheses about risk for EOP have been proposed and tested. Most commonly, autosomal recessive inheritance has been posed as the most likely mode of inheritance because of the lack of affected parents in the pedigrees (32, 61, 64). Segregation analyses that support this model have been performed. Long et al. (41) compared the relative likelihoods of autosomal recessive and X-linked models in a study of 33 EOP families and observed that the autosomal recessive model was most likely. However, an autosomal dominant model was not considered. Beaty et al. (5) examined 28 families and found that the data favored the autosomal recessive model, yet transmission of the trait in some families appeared to follow a dominant model.

Some investigators have favored an X-linked model for transmission of EOP (24, 46, 60). This preference was based on the observation that the prevalence of disease in females was greater than that in males and that in some families, transmission through the mother appeared to be the most likely explanation of inheritance. Recent data from a number of groups indicate, however, that the observed female preponderance of EOP may be due to ascertainment bias. For example, Hart et al. (31) demonstrated that in high-density families, there is a preponderance of female probands but an equal prevalence of disease among relatives of the probands. Arguments against X-linked transmission of EOP have recently been proposed by Hart and coworkers (30).

Recent data indicate that autosomal dominant transmission of EOP in many families is likely. The most convincing demonstration of this was in the study by Boughman and coworkers, who demonstrated genetic linkage of EOP (as well as of dentinogenesis imperfecta) in one large family to a site on chromosome 4 (11). A subsequent study by Hart and coworkers (29) failed to reproduce this finding in a more heterogeneous cohort of families, indicating that there is either heterogeneity in genetic risk for EOP or that the family studied by Boughman is unique with respect to this linkage result.

Marazita and coworkers (45) recently performed segregation analysis of EOP in more than 120 families with an affected (i.e., LJP, GJP, or RPP) proband. The results indicated that autosomal dominant inheritance of EOP was the only genetic model consistent with the pedigree data.

Genetic analyses of EOP have therefore demonstrated that there is a strong genetic component to risk for these diseases. There appears to be at least one dominant form that is linked to a region on chromosome 4, and other regions of the genome will likely be found to be linked to EOP in other families. A number of important considerations arise with regard to the design of future studies that aim to further identify the gene(s) associated with EOP. (i) EOP may very well demonstrate a great deal of heterogeneity; the gene(s) linked to disease in one family may differ from that in other families. (ii) Although clues to possible candidate loci for performance of linkage studies exist, there are obvious inconsistencies in their relationship to disease. Thus, analyses of large numbers of families that are more homogeneous for such traits may be required. Alternatively, these traits may be of little use in providing clues to candidate regions. (iii) There are families in which individuals with only LJP or only RPP are found as well as families in which both phenotypes are observed. Genetic analyses may require consideration of these groups of families separately in order to determine whether these disease forms have unique etiologies or represent phenotypic variation of a trait with a common underlying etiology.

ANTIBODY RESPONSES, RACE, AND GENETIC SUSCEPTIBILITY TO EOP

Interest in both genetic risk for EOP and immunologic responses to periodontal microorganisms has led my group to examine the relationships between antibody responses, the epidemiology of EOP with regard to race, and the influence of genes on both the EOP trait and the immune response. These studies will be reviewed below, and a hypothesis that synthesizes our results will be presented.

The antibody response in EOP, and in particular LJP, has some remarkable attributes with respect to the pathogen *A. actinomycetemcomitans*. Measurement of serum antibody to *A. actinomycetemcomitans* in subjects with LJP has revealed very high levels of immunoglobulin G (IgG) reactive with this organism (72). The function of this antibody has not been definitively determined. However, the results of several studies imply that the response to *A. actinomycetemcomitans* is protective in nature. For example, Gunsolley and coworkers (26) demonstrated an inverse correlation between extent and severity of attachment loss and anti-*A. actinomycetemcomitans* antibody in patients with EOP. Individuals lacking positive antibody responses to both *A. actinomycetemcomitans* and *Porphyromonas gingivalis* had the most severe forms of EOP, implying a protective role. Furthermore, in vitro tests of sera from LJP patients with high anti-*A. actinomycetemcomitans* titers indicated the presence of opsonic antibody in such sera (4), indicating that anti-*A. actinomycetemcomitans* could enhance antibacterial leukocyte function. In addition, recent studies by Underwood et al. demonstrate high chemiluminescence responses in high-titer anti-*A. actinomycetemcomitans* antibody sera, indicating the opsonic potential of such sera (74).

A series of studies in our laboratories indicates that antibody responses to *A. actinomycetemcomitans* may be dependent on race. In view of the racial dependency evident in epidemiologic studies of EOP and the observation that white subjects rarely have LJP, the interaction of race and antibody response to clinical forms of EOP has been of great interest to us.

Our initial assessment of IgG antibody titer in response to *A. actinomycetemcomitans* serotype b was that it was uniformly elevated in members of some families with EOP, while members of other families had little or no such antibody. Further analysis of demographic and periodontal disease variables indicated that elevated anti-*A. actinomycetemcomitans* serotype b titers and their magnitude were dependent on race; black subjects in all disease categories had higher titers than white subjects. Other variables such as plaque scores, gingival inflammation, age, gender, and presence of *A. actinomycetemcomitans* could not account for these differences. When a group of periodontally healthy age- and race-matched controls were examined, positive responses were more prevalent in the black subjects. This relationship held regardless of whether the periodontally healthy subjects were members of EOP families or were independently ascertained (28). Further examination of antibody titers to organisms that predominate in the cultivable flora of the generalized EOP group indicated that the racial dependency of this response was evident almost exclusively with *A. actinomycetemcomitans* serotypes b and c (27).

In view of our interest in identifying the antigen(s) of *A. actinomycetemcomitans* that induced the remarkably high antibody responses in EOP subjects, we identified the immunodominant antigen of *A. actinomycetemcomitans* serotype b as the serotype-specific carbohydrate antigen (14), which was later defined by Wilson and Schifferle (85) and Page et al. (53) as a polysaccharide side chain of lipopolysaccharide (LPS). Interestingly, for serotypes b and c, the serotypes associated with race in antibody measurements, the immunodominant antigen appeared to be a carbohydrate. A comparison of the immunodominant antigen in black and white subjects with elevated anti-*A. actinomycetemcomitans* titers demonstrated that the immunodominant antigen was identical in the two racial groups (15). This indicated that there was a quantitative rather than a qualitative effect of race on the response to *A. actinomycetemcomitans*.

Further studies were subsequently carried out by our group and others to determine the class and subclass specificity of the antibody response to *A. actinomycetemcomitans*. Wilson and Hamilton (84), using subclass-specific chimeric antibodies as standards, determined that the dominant response of LJP patients to LPS from *A. actinomycetemcomitans* was the IgG2 subclass. They reported that IgG2 levels of LJP patients reactive with LPS averaged 136 μg/ml, versus only 7.8 μg/ml for IgG1, 0.6 μg/ml for IgG3, and 0.01 μg/ml for IgG4. Our own studies also demonstrated that titers to this antigen are highest with the IgG2 subclass (44). We found that our high-responder patients had about 150 μg of IgG per ml reactive with the immunodominant antigen, and the vast majority of this was IgG2. This observation is consistent with the observed human antibody response to LPS and other carbohydrate antigens and was intriguing, since antibodies of the IgG2 subclass are known to be poor opsonins and are poor activators of the complement system. This observation casts some doubt on the protective quality of the majority of the antibody reactive with this organism. This point has been further illustrated by Whitney and coworkers (83), who examined serum IgG responses and IgG subclass distributions in response to *P. gingivalis* in subjects with RPP. They observed that the predominant response to *P. gingivalis* was also the IgG2 subclass and that the relative avidity of the response in high-titer individuals was low.

Earlier studies by Waldrop and coworkers (82) examined serum IgG subclass levels in a very small number of periodontitis subjects (i.e., nine subjects with juvenile periodontitis and 9 with adult periodontitis) and found no differences from levels in healthy controls. However, results of studies by Wilson and Hamilton (84), also performed on a small number of subjects, implied that individuals with LJP who have high anti-*A. actinomycetemcomitans* antibody titers also have higher total serum IgG2 concentrations than do age- and race-matched controls. However, only a small percentage of the increase in total IgG2 in these sera can be accounted for by the increase in antibody titer against this single organism. Since the LJP subjects are nearly all black and since we have noted significant differences between black and white subjects in antibody titers, we felt that examination of IgG subclass concentrations as a function of both race and periodontal diagnosis merited closer scrutiny.

The high level of IgG2 reactive with *A. actinomycetemcomitans* serotype b LPS prompted us to check the total levels of IgG2 in LJP patient serum to determine whether they were elevated. In view of our results indicating that black EOP patients had higher titers of anti-*A. actinomycetemcomitans* serotype b antibody than white EOP patients, we compared both black and white LJP patients and black and white periodontally healthy subjects and included black and white generalized EOP patients (comprising both GJP and RPP subjects) and adult periodontitis patients as controls in this comparative study. As shown in Table 1, several important points were noted. (i) Black periodontally healthy subjects had about 30% more (900 µg/ml more) IgG2 than age-matched periodontally healthy white subjects. (ii) This tendency for black subjects to have more IgG2 (about 1 mg/ml more) than white subjects is apparent in all the periodontal groups. (iii) LJP subjects have 30 to 40% more serum IgG2 than their race-matched controls. (iv) In contrast, generalized EOP subjects had IgG2 levels that were virtually identical to those of their age-matched periodontally healthy controls, indicating that IgG2 "hyperresponsiveness" is restricted to LJP.

In short, there is a marked relationship between IgG2, periodontal status, and race: serum IgG2 levels were substantially elevated in LJP patients of both races and were higher in black subjects of all diagnostic groups than in white subjects.

These observations prompted us to explore the possible inheritance of the ability to produce high levels of IgG2. To do so, IgG2 levels in 122 families, including nuclear and extended EOP families and twin pairs, plus 511 unrelated individuals were determined, and genetic analyses were performed. By using variance component modeling, about 25% of the total variance in IgG2 levels was estimated to be due to genetic factors and about 75% was due to environmental factors. Segregation analysis indicated that transmission of IgG2 concentration is consistent with a Mendelian major locus and that the best-fitting model is the autosomal codominant major locus model. This genetic model is significant even after correction for LJP status and thus follows a genetic model independent of the segregation of EOP in these families.

Since the inheritance of EOP and IgG2 are apparently independent and since high IgG2 concentrations appear to coincide with cases of LJP (in contrast to generalized EOP), we hypothesized that the ability to hyperrespond by producing high concentrations of IgG may be protective in individuals with EOP. Figure 1 illustrates this hypothesis. In families with EOP, children would be at genetic risk of inheriting a risk factor for this disease, which is transmitted as an autosomal dominant trait (Fig. 1, subjects 1, 2, and 4). Independently of inheritance of risk for EOP, individual serum levels of IgG2 would likewise be genetically controlled and inherited as an autosomal dominant trait. In individuals at risk for EOP, high

Table 1. Serum levels of anti-*A. actinomycetemcomitans* antibody and IgG2 as a function of race and periodontal status in EOP subjects

Diagnosis[a]	Race	Anti-*A. actinomycetemcomitans* serotype b titer (cpm)[b]	Serum IgG2 concn (µg/ml)[c]
NP	White	2,752	3,059
	Black	8,116	3,992
LJP	White	21,800	4,233
	Black	31,501	5,306
SP	White	8,814	2,931
	Black	28,520	4,087

[a] NP, periodontally healthy; SP, generalized EOP.
[b] Antibody titer as determined by radioimmunoassay.
[c] As determined by single radial immunodiffusion.

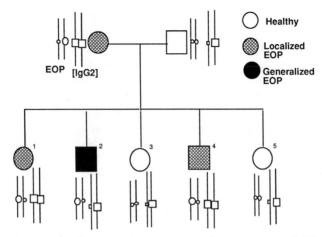

Figure 1. Theoretical pedigree illustrating the hypothesis that the progression of EOP from a localized to a generalized form may be determined by the patient's ability to produce high levels of serum IgG2. The allele predisposing to EOP (○) and the allele inducing high IgG2 levels (□) control the EOP phenotype. Since IgG2 levels are inherited as a codominant trait, two copies of the gene would limit progression of EOP to the generalized form.

concentrations of IgG2 would be protective in the sense that they would prevent extension of localized disease to a generalized form (Fig. 1, subjects 1 and 4). Family members who are at genetic risk for EOP but are unable to develop these protective antibody responses would more likely experience extension of their disease to a more generalized form of EOP (e.g., GJP or RPP; Fig. 1, subject 2).

SUMMARY

Elucidation of genetic risk factors for EOP poses many challenges because the etiologic and pathogenic mechanisms that are responsible for these diseases are ill-defined. Clinical heterogeneity exists in EOP, and heterogeneity likely exists with regard to genetic etiology; the degree of such heterogeneity and the diversity of mechanisms that might lead to disease expression are unknown. On the other hand, defining genes that are associated with risk for EOP may in turn define the etiology of EOP without prior knowledge of such factors. Future studies will examine likely candidates for genetic risk as well as new technologies to scan the genome for regions that are linked to disease expression.

Acknowledgment. This work was supported in part by Public Health Service grants DE-08972 and DE-10703 from the National Institute of Dental Research.

REFERENCES

1. **Agarwal, S., and J. B. Suzuki.** 1991. Altered neutrophil function in localized juvenile periodontitis: intrinsic cellular defect or effect of immune mediators? *J. Periodontal Res.* **26:**276–278.
2. **Anderson, D. C., F. C. Schmalstieg, M. J. Finegold, B. J. Hughes, R. Rothlein, L. J. Miller, S.**

Kohl, M. F. Tosi, R. L. Jacobs, T. C. Waldrop, A. S. Goldman, W. T. Shearer, and T. A. Springer. 1985. The severe and moderate phenotypes of heritable Mac-1, LFA-1 deficiency: their quantitative definitions and relation to leukocyte dysfunction and clinical features. *J. Infect. Dis.* **152:**668–689.

3. **Baer, P. N.** 1971. The case of periodontosis as a clinical entity. *J. Periodontol.* **42:**516–520.
4. **Baker, P. J., and M. E. Wilson.** 1989. Opsonic IgG antibodies against *Actinobacillus actinomycetemcomitans* in localized juvenile periodontitis. *Oral Microbiol. Immunol.* **4:**98–105.
5. **Beaty, T. H., J. A. Boughman, P. Yang, J. A. Astemborski, and J. B. Suzuki.** 1987. Genetic analysis of juvenile periodontitis in families ascertained through an affected proband. *Am. J. Hum. Genet.* **40:**443–452.
6. **Benjamin, S. D., and P. N. Baer.** 1967. Familial patterns of advanced alveolar bone loss in adolescence (periodontosis). *Periodontics* **5:**82–88.
7. **Ben Yehouda, A., A. Shifer, J. Katz, W. Machtei, and M. Shmerling.** 1991. Prevalence of juvenile periodontitis in Israeli military recruits as determined by panoramic radiographs. *Community Dent. Oral Epidemiol.* **19:**359–360.
8. **Boughman, J. A., J. A. Astemborski, and M. G. Blitzer.** 1990. Early onset periodontal disease: a genetics perspective. *Crit. Rev. Oral Biol. Med.* **1:**89–99.
9. **Boughman, J. A., J. A. Astemborski, and J. B. Suzuki.** 1992. Phenotypic assessment of early onset periodontitis in sibships. *J. Clin. Periodontol.* **19:**233–239.
10. **Boughman, J. A., T. H. Beaty, P. Yang, S. B. Goodman, R. K. Wooten, and J. B. Suzuki.** 1988. Problems of genetic model testing in early onset periodontitis. *J. Periodontol.* **59:**332–337.
11. **Boughman, J. A., S. L. Halloran, D. Roulston, S. Schwartz, J. B. Suzuki, L. R. Weltkamp, R. E. Wenk, R. Wooten, and M. M. Cohen.** 1986. An autosomal-dominant form of juvenile periodontitis: its localization to chromosome 4 and linkage to dentinogenesis imperfecta and Gc. *J. Craniofac. Genet. Dev. Biol.* **6:**341–350.
12. **Bruckner, R. J., N. H. Rickles, and D. R. Porter.** 1962. Hypophosphatasia with premature shedding of teeth and aplasia of cementum. *Oral Med. Oral Pathol.* **15:**1351–1369.
13. **Butler, J. H.** 1969. A familial pattern of juvenile periodontitis (periodontosis). *J. Periodontol.* **40:**115–118.
14. **Califano, J., V., H. A. Schenkein, and J. G. Tew.** 1989. Immunodominant antigen of *Actinobacillus actinomycetemcomitans* Y4 (AaY4) in high-responder patients. *Infect. Immun.* **57:**1582–1589.
15. **Califano, J. V., H. A. Schenkein, and J. G. Tew.** 1992. Immunodominant antigens of *Actinobacillus actinomycetemcomitans* serotype b in early onset periodontitis patients. *Oral Microbiol. Immunol.* **7:**65–70.
16. **Caton, J.** 1989. *Periodontal Diagnosis and Diagnostic Aids.* World Workshop in Clinical Periodontics. American Academy of Periodontology, Chicago.
17. **Cianciola, L. J., R. J. Genco, M. R. Patters, J. McKenna, and C. J. Van Oss.** 1977. Defective polymorphonuclear leukocyte function in a human periodontal disease. *Nature* (London) **265:**445–447.
18. **Clark, R., R. C. Page, and G. Wilde.** 1977. Defective neutrophil chemotaxis in juvenile periodontitis. *Infect. Immun.* **18:**694–700.
19. **Cohen, D. W., and H. M. Goldman.** 1960. Clinical observations on the modification of human oral tissue metabolism by local intra-oral factors. *Ann. N.Y. Acad. Sci.* **85:**68.
20. **Coles, R. B., R. R. Ranney, R. J. Freer, and R. A. Carchman.** 1989. Thermal regulation of FMLP receptors on human neutrophils. *J. Leukocyte Biol.* **45:**529–537.
21. **Cullinan, M. P., J. Sachs, E. Wolf, and G. J. Seymour.** 1980. The distribution of HLA-A and -B antigens in patients and their families with periodontosis. *J. Periodontal Res.* **15:**177–184.
22. **Dekker, G., and L. G. Jansen.** 1958. Periodontosis in a child with hyperkeratosis palmo-plantaris. *J. Periodontol.* **29:**266–271.
23. **Fourel, J.** 1972. Periodontosis: a periodontal syndrome. *J. Periodontol.* **43:**240–255.
24. **Fretwell, L. D., T. E. Leinback, and D. O. Wiley.** 1982. Juvenile periodontitis: report of cases. *J. Am. Dent. Assoc.* **105:**1022.
25. **Gorlin, R. J., R. E. Stallard, and B. L. Shapiro.** 1967. Genetics and periodontal disease. *J. Periodontol.* **38:**5.
26. **Gunsolley, J. C., J. A. Burmeister, J. G. Tew, A. M. Best, and R. R. Ranney.** 1987. Relationship

of serum antibody to attachment level patterns in young adults with juvenile periodontitis or generalized severe periodontitis. *J. Periodontol.* **58:**314–320.

27. **Gunsolley, J. C., J. G. Tew, T. Conner, J. A. Burmeister, and H. A. Schenkein.** 1991. Relationship between race and antibody reactive with periodontitis-associated bacteria. *J. Periodontal Res.* **26:**59–63.

28. **Gunsolley, J. C., J. G. Tew, C. M. Gooss, J. A. Burmeister, and H. A. Schenkein.** 1988. Effects of race and periodontal status on antibody reactive with *Actinobacillus actinomycetemcomitans* strain Y4. *J. Periodontal Res.* **23:**303–307.

29. **Hart, T. E., M. L. Marazita, K. M. McCanna, H. A. Schenkein, and S. R. Diehl.** 1993. Reevaluation of the chromosome 4q candidate region for early onset periodontitis. *Hum. Genet.* **91:**416–422.

30. **Hart, T. E., M. L. Marazita, H. A. Schenkein, and S. R. Diehl.** 1992. Re-interpretation of the evidence for X-linked dominant inheritance of juvenile periodontitis. *J. Periodontol.* **63:**169–173.

31. **Hart, T. E., H. A. Schenkein, M. L. Marazita, C. N. Brooks, J. G. Gunsolley, and S. R. Diehl.** 1991. No female predominance in juvenile periodontitis after correction for ascertainment bias. *J. Periodontol.* **62:**745–749.

32. **Jorgenson, R. J., L. S. Levin, S. T. Hutcherson, and C. F. Salinas.** 1975. Periodontitis in sibs. *Oral Surg.* **39:**396–402.

33. **Kaslick, R. S., T. L. West, and A. I. Chasens.** 1980. Association between ABO blood groups, HL-A antigens and periodontal disease in young adults: a follow-up study. *J. Periodontol.* **51:**343–344.

34. **Kaslick, R. S., T. L. West, A. I. Chasens, P. I. Terascki, R. Lazzara, and S. Weinberg.** 1975. Association between HL-A2 antigen and various periodontal diseases in young adults. *J. Dent. Res.* **54:**424.

35. **Katz, J., J. Goultschin, R. Benoliel, and C. Brautbar.** 1987. Human leukocyte antigen (HLA) DR4. Positive association with rapidly progressing periodontitis. *J. Periodontol.* **58:**607–610.

36. **Kinane, D. F., C. F. Cullen, F. A. Johnston, and C. W. Evans.** 1989. Neutrophil chemotactic behaviour in patients with early-onset forms of periodontitis. I. Leading front analysis of Boyden chambers. *J. Clin. Periodontol.* **16:**242–246.

37. **Kronauer, E., G. Borsa, and N. P. Lang.** 1986. Prevalence of incipient juvenile periodontitis at age 16 in Switzerland. *J. Clin. Periodontol.* **13:**103–108.

38. **Larjava, H., L. Saxen, T. Kosunen, and C. G. Gahmberg.** 1984. Chemotaxis and surface glycoproteins of neutrophil granulocytes from patients with juvenile periodontitis. *Arch. Oral Biol.* **29:**935–939.

39. **Lavine, W. S., E. G. Maderazo, J. Stolman, P. A. Ward, R. B. Cogen, I. Greenblatt, and P. B. Robertson.** 1979. Impaired neutrophil chemotaxis in patients with juvenile and rapidly progressing periodontitis. *J. Periodontal Res.* **14:**10–19.

40. **Löe, H., and L. J. Brown.** 1991. Juvenile periodontitis in the United States of America. *J. Periodontol.* **62:**608–616.

41. **Long, J. C., W. E. Nance, P. Waring, J. A. Burmeister, and R. R. Ranney.** 1987. Early onset periodontitis: a comparison and evaluation of two proposed modes of inheritance. *Genet. Epidemiol.* **4:**13–24.

42. **Lopez, N. J.** 1992. Clinical, laboratory, and immunological studies of a family with a high prevalence of prepubertal and juvenile periodontitis. *J. Periodontol.* **63:**457–468.

43. **Lopez, N. J., V. Rios, M. A. Pareja, and O. Fernandez.** 1991. Prevalence of juvenile periodontitis in Chile. *J. Clin. Periodontol.* **18:**529–533.

44. **Lu, H., J. V. Califano, H. A. Schenkein, and J. G. Tew.** 1993. Immunoglobulin class and subclass distribution of antibodies reactive with the immunodominant antigen of *Actinobacillus actinomycetemcomitans* serotype b. *Infect. Immun.* **61:**2400–2407.

45. **Marazita, M. L., K. Lake, and H. A.. Schenkein.** 1992. Segregation analysis of early onset periodontitis. *J. Dent. Res.* **71:**261.

46. **Melnick, M., E. D. Shields, and E. Bixler.** 1976. Periodontitis: a phenotypic and genetic analysis. *Oral Surg.* **42:**32–41.

47. **Melvin, W. L., J. B. Sandifer, and J. L. Gray.** 1991. The prevalence and sex ratio of juvenile periodontitis in a young racially mixed population. *J. Periodontol.* **62:**330–334.

48. **Neely, A. L.** 1992. Prevalence of juvenile periodontitis in a circumpubertal population. *J. Clin. Periodontol.* **19:**367–372.

49. **Nishimura, F., A. Nagai, K. Kurimoto, O. Isoshima, S. Takashiba, M. Kobayashi, I. Akutsu, H. Kurihara, Y. Nomura, Y. Murayama, H. Ohta, and K. Kato.** 1990. A family study of a mother and daughter with increased susceptibility to early-onset periodontitis: microbiological, immunological, host defensive, and genetic analysis. *J. Periodontol.* **61:**755–765.

50. **Ohtonen, S., L. Kontturri-Narki, H. Markkanen, and S. Synjanen.** 1983. Juvenile periodontitis— a clinical and radiological familial study. *J. Periodontics* **8:**28–33.

51. **Okuda, K., Y. Naito, K. Ohta, Y. Fukumoto, Y. Kimura, I. Ishikawa, S. Kinoshita, and I. Takazoe.** 1984. Bacteriological study of periodontal lesions in two sisters with juvenile periodontitis and their mother. *Infect. Immun.* **45:**118–121.

52. **Page, R. C., L. C. Altman, J. L. Ebersole, C. E. Vandersteen, W. H. Dahlberg, B. L. William, and S. K. Osterbag.** 1983. Rapidly progressive periodontitis: a distinct clinical condition. *J. Periodontol.* **54:**197–209.

53. **Page, R. C., T. J. Sims, D. Engel, B. J. Moncla, B. Bainbridge, J. Stray, and R. P. Darveau.** 1991. The immunodominant outer membrane antigen of *Actinobacillus actinomycetemcomitans* is located in the serotype-specific high-molecular-mass carbohydrate moiety of lipopolysaccharide. *Infect. Immun.* **59:**3451–3462.

54. **Pedersen, M. M.** 1989. Chemotactic response of neutrophil polymorphonuclear leukocytes in juvenile periodontitis measured by the leading front method. *Scand. J. Dent. Res.* **96:**421–427.

55. **Perez, H. D., E. Kelley, F. Elfman, G. Armitage, and J. Winkler.** 1991. Defective polymorpho-nuclear leukocyte formyl peptide receptor(s) in juvenile periodontitis. *J. Clin. Invest.* **87:**971–976.

56. **Pradhan, A. C., T. N. Chawla, K. C. Smauel, and S. Pradham.** 1971. The relationship between periodontal disease and blood groups and secretor status. *J. Periodontal Res.* **6:**294–300.

57. **Rao, S. S., and S. V. Tweani.** 1968. Prevalence of periodontosis among Indians. *J. Periodontol.* **39:**27–34.

58. **Reinholdt, J., I. Bay, and A. Svejgaard.** 1977. Association between HLA-antigens and periodontal disease. *J. Dent. Res.* **56:**1261–1263.

59. **Repo, H., L. Saxen, M. Jäätelä, M. Ristola, and M. Leirisalo-Repo.** 1990. Phagocyte function in juvenile periodontitis. *Infect. Immun.* **58:**1085–1092.

60. **Risom, L. L., J. B. Suzuki, and J. A. Boughman.** 1985. Juvenile periodontitis (periodontosis) and inheritance patterns. *Dent. Hyg.* **59:**65.

61. **Saxen, L.** 1980. Heredity of juvenile periodontitis. *J. Clin. Periodontol.* **7:**276–288.

62. **Saxen, L.** 1980. Prevalence of juvenile periodontitis in Finland. *J. Clin. Periodontol.* **7:**165–176.

63. **Saxen, L., and S. Koskimies.** 1984. Juvenile periodontitis—no linkage with HLA-antigens. *J. Periodontal Res.* **19:**441–444.

64. **Saxen, L, and H. R. Nevanlinna.** 1984. Autosomal recessive inheritance of juvenile periodontitis: test of a hypothesis. *Clin. Genet.* **25:**332–335.

65. **Sbordone, L., L. Ramaglia, and E. Bucci.** 1990. Generalized juvenile periodontitis: report of a familial case followed for 5 years. *J. Periodontol.* **61:**590–596.

66. **Schenkein, H. A., A. M. Best, and J. C. Gunsolley.** 1991. The influence of race and periodontal clinical status on neutrophil chemotactic responses. *J. Periodontal Res.* **26:**272–275.

67. **Schenkein, H. A., H. A. Burmeister, T. E. Koertge, C. N. Brooks, A. M. Best, L. V. H. Moore, and W. E. C. Moore.** 1993. The influence of race and gender on periodontal microflora. *J. Periodontol.* **64:**292–296.

68. **Sofaer, J. A.** 1990. Genetic approaches in the study of periodontal diseases. *J. Clin. Periodontol.* **17:**401–408.

69. **Spektor, M. D., G. E. Vandesteen, and R. C. Page.** 1985. Clinical studies of one family manifesting rapidly progressive, juvenile, and pre-pubertal periodontitis. *J. Periodontol.* **56:**93–101.

70. **Sussman, H. I., and P. N. Baer.** 1978. Three generations of periodontosis: case reports. *Ann. Dent.* **37:**8–11.

71. **Suzuki, J. B., B. C. Collison, W. A. Falkler, and R. K. Nauman.** 1984. Immunologic profile of juvenile periodontitis. II. Neutrophil chemotaxis, phagocytosis and spore germination. *J. Periodontol.* **55:**461–467.

72. **Taubman, M. A., J. L. Ebersole, and D. J. Smith.** 1982. Association between systemic and local antibody and periodontal diseases, p. 283–298. *In* R. J. Genco and S. E. Mergenhagen (ed.), *Host-Parasite Interactions in Periodontal Diseases.* American Society for Microbiology, Washington, D.C.

73. **Terasaki, P. I., R. S. Kaslick, T. L. West, and A. I. Chasens.** 1975. Low HL-A-2 frequency and periodontitis. *Tissue Antigens* **5:**286–288.

74. **Underwood, K., K. Sjostrom, R. Darveau, R. Lamont, H. A. Schenkein, J. G. Gunsolley, R. Page, and D. Engel.** 1993. Serum antibody opsonic activity against *Actinobacillus actinomycetemcomitans* in human periodontal diseases. *J. Infect. Dis.* **168:**1436–1443.

75. **Vandesteen, G. E., B. L. Williams, J. L. Ebersole, and L. C. Altman.** 1984. Clinical, microbiological and immunological studies of a family with a high prevalence of early-onset periodontitis. *J. Periodontol.* **55:**159–169.

76. **Van Dyke, T. E., H. V. Horoszewicz, L. J. Cianciola, and R. J. Genco.** 1980. Neutrophil chemotaxis dysfunction in human periodontitis. *Infect. Immun.* **27:**124–132.

77. **Van Dyke, T. E., M. J. Levine, L. A. Tabak, and R. J. Genco.** 1981. Reduced chemotactic peptide binding in juvenile periodontitis: a model for neutrophil function. *Biochem. Biophys. Res. Commun.* **100:**1278–1284.

78. **Van Dyke, T. E., M. J. Levine, L. A. Tabak, and R. J. Genco.** 1983. Juvenile periodontitis as a model for neutrophil function: reduced binding of complement chemotactic fragment, C5a. *J. Dent. Res.* **62:**870–872.

79. **Van Dyke, T. E., M. Schweinebraten, L. J. Cianciola, S. Offenbacher, and R. J. Genco.** 1985. Neutrophil chemotaxis in families with localized juvenile periodontitis. *J. Periodontal Res.* **20:**503–514.

80. **Van Dyke, T. E., C. Wilson-Burrows, S. Offenbacher, and P. Henson.** 1987. Association of an abnormality of neutrophil chemotaxis in human periodontal disease with a cell surface protein. *Infect. Immun.* **55:**2262–2267.

81. **Waerhaug, J.** 1967. Prevalence of periodontal disease in Ceylon. *Acta Odontol. Scand.* **25:**205–231.

82. **Waldrop, T. C., B. F. Mackler, and P. Schur.** 1981. IgG and IgG subclasses in human periodontosis (juvenile periodontitis)—serum concentrations. *J. Periodontol.* **52:**96–98.

83. **Whitney, C., J. Ant, B. Moncla, B. Johnson, R. Page, and D. Engel.** 1992. Serum immunoglobulin G antibody to *Porphyromonas gingivalis* in rapidly progressive periodontitis: titer, avidity, and subclass distribution. *Infect. Immun.* **60:**2194–2200.

84. **Wilson, M. E., and R. G. Hamilton.** 1992. Immunoglobulin G subclass response of localized juvenile periodontitis patients to *Actinobacillus actinomycetemcomitans* Y4 lipopolysaccharide. *Infect. Immun.* **60:**1806–1812.

85. **Wilson, M. E., and R. E. Schifferle.** 1991. Evidence that the serotype b antigenic determinant of *Actinobacillus actinomycetemcomitans* Y4 resides in the polysaccharide moiety of lipopolysaccharide. *Infect. Immun.* **59:**1544–1551.

Molecular Pathogenesis of Periodontal Disease
Edited by Robert Genco et al.
© 1994 American Society for Microbiology, Washington, DC 20005

Summary of Chapters 26 to 30

Sara Grossi and David Engel

Polymorphonuclear leukocytes (PMN) are believed to play a central role in human periodontal diseases. PMN form the primary defense against infection by periodontal bacteria and therefore are of great interest in studies of both resistance and susceptibility to periodontal diseases. The first three chapters in this section provide stimulating new perspectives on PMN antimicrobial mechanisms and on PMN dysfunction and susceptibility in periodontal disease. Opsonization of bacteria by certain subclasses of immunoglobulin G greatly enhances the phagocytosis of such bacteria by PMN. The next two chapters provide new insight into the previously described observation that serum IgG responses to *Actinobacillus actinomycetemcomitans* are highly enhanced in localized juvenile periodontitis (LJP).

PMN produce a variety of antimicrobial molecules, some of which (e.g., lysosomal enzymes) have long been recognized, and some of which have only recently come under study. In chapter 26, Miyasaki et al. describe advances in determining the bactericidal and bacteriostatic mechanisms of cathepsin G (CatG), a lysosomal neutral protease. A synthetic 20-mer peptide corresponding to an exposed domain of CatG (amino acids 61 to 80) was reported to kill bacteria, raising the possibility that a synthetic CatG-peptide antimicrobial drug might be possible in the future. Miyasaki also discusses the growing interest in other PMN peptides and proteins with the potential of becoming therapeutic agents. The defensins, small cationic peptides (M_r = 3,500 to 4,000) that kill bacteria by binding to their surfaces and inducing permeability, can now also be synthesized. In some studies the synthetic molecules have excellent antimicrobial activity in vitro. Several periodontal pathogens are susceptible to the rabbit defensin, NP-1. Calprotectin is a recently described anionic complex of two cytosolic proteins that is found in PMN, some macrophages, and oral keratinocytes. Although calprotectin complex is generally bacteriostatic, not bactericidal, it constitutes a high proportion of the cytosolic protein of the PMN and thus may be of major importance in regulating microbial growth in periodontal lesions. Therapeutics aimed at regulation of expression of

Sara Grossi • Department of Oral Biology, State University of New York at Buffalo, Buffalo, New York 14214. *David Engel* • Department of Periodontics, University of Washington, Seattle, Washington 98195.

calprotectin in gingival keratinocytes may bolster oral defense against microbial pathogens.

In chapter 27, Kalmar describes studies in which he determined that PMN from LJP patients show normal phagocytosis of *A. actinomycetemcomitans* but depress killing of these organisms within the PMN. This difference may be due to altered fusion of lysosomes and the phagocytic vacuole, since lysosomal granule secretion appears to be depressed with respect to release of lactoferrin in LJP PMN. Kalmar suggests that the reduced capacity of LJP PMN to kill *A. actinomycetemcomitans* may enable this periodontal pathogen to at least partially evade healthy host defense mechanisms. There may be a genetic basis for this abnormality, perhaps through host biochemical alterations such as cytokine imbalance. However, other explanations, such as microbial virulence factors for *A. actinomycetemcomitans*, have not been excluded.

In chapter 28, De Nardin expands on PMN abnormality in LJP patients, discussing PMN receptors for formyl peptides and interleukin-8. Both of these receptors are characterized by seven transmembrane regions. An intracellular portion of the molecule is integrally involved in G-protein activation following binding of the specific ligand by the external part of the molecule. De Nardin describes recent studies in which the major FMLP (*N*-formyl-*l*-methionyl-*l*-leucyl-*l*-phenyl-alanine)-binding region at the amino-terminal end of the first extracellular loop of the receptor was characterized by construction of synthetic peptides corresponding to portions of this loop. Amino acid residues in the N terminus of this first extracellular domain may play a critical role in ligand binding. De Nardin's research group has proposed that there may be a common underlying cell-associated defect, either at the receptor level or at the membrane level, that contributes to depressed PMN function in LJP. While this is an intriguing hypothesis, other potential explanations, such as the effects of cytokines on the expression of these receptor molecules, should also be explored.

In chapter 29, Wilson describes the isolation and partial characterization of an outer membrane protein of *A. actinomycetemcomitans* that is structurally related to the OmpA proteins of *Escherichia coli* and other enteric gram-negative bacteria. This protein exhibits heat modifiability, migrating with an apparent M_r of 29 kDa when solubilized at low temperatures (22 or 37°C) but with an M_r of 34 kDa when solubilized at higher temperatures ($\geq$50°C) prior to analysis by sodium dodecyl sulfate-polyacrylamide gel electrophoresis. Serum samples from 35 LJP patients were analyzed by an enzyme-linked immunosorbent assay for the presence of IgG antibodies to the OmpA protein of *A. actinomycetemcomitans*. Twenty-two of the 35 LJP sera showed IgG antibody titers >2 standard deviations from the main titer of a periodontally healthy group. This antibody response was age related and consisted primarily of IgG2 subclass antibodies. Among the LJP sera tested, antibody titers to the OmpA protein were low in subjects $\leq$17 years of age but markedly elevated in 53% of subjects $\geq$18 years of age. Although *A. actinomycetemcomitans* OmpA exhibits significant amino-terminal sequence homology with the OmpA protein of members of the genus *Haemophilus*, absorption studies indicated that IgG antibodies in LJP sera recognize a determinant(s) that is unique to the OmpA of *A. actinomycetemcomitans*. Hence, LJP sera have been shown to contain IgG antibodies directed to a surface-accessible determinant present in the OmpA protein of *A. actinomycetemcomitans*. Studies currently in progress are directed toward analyzing the bactericidal and/or opsonic activities of these anti-OmpA IgG antibodies.

In chapter 30, Schenkein discusses how LJP and other aggressive forms of periodontitis presenting in juveniles and young adults, collectively defined as early-onset periodontal diseases (EOP), have been shown to have a genetic susceptibility. Pedigree analysis of more than 120 EOP families indicates that the EOP trait is most likely inherited as an autosomal dominant. Immunologic studies from Schenkein's group also indicate that race is directly related to the magnitude of the IgG antibody response to *A. actinomycetemcomitans* serotype b, and this response is confined mostly to the IgG2 subclass antibody. Titers of anti-*A. actinomycetemcomitans* serotype b IgG2 antibodies were higher in black subjects than in white subjects irrespective of periodontal status and higher in LJP patients than age- and race-matched controls. Genetic analysis of serum IgG2 concentrations in EOP families revealed that the IgG2 trait is an autosomal dominant inherited independently from that of EOP. Schenkein's research group further hypothesizes that the high IgG2 antibody response to *A. actinomycetemcomitans* serotype b antigens observed in LJP patients confers protection against the organism and limits the periodontal infection to a localized clinical appearance. Future studies will focus on determining the genetic risk for EOP and will eventually identify regions in the genome that are associated with disease expression.

The studies presented in this session collectively support the role of PMN abnormalities in the pathogenesis of LJP. The appealing hypothesis that LJP and all other clinical entities of EOP have a genetic susceptibility gives a new dimension to the basis of the PMN abnormalities observed in LJP patients. More important, it presents new challenges to both treatment and prevention of these diseases.

SECTION IV

PERIODONTAL WOUND HEALING

Molecular Pathogenesis of Periodontal Disease
Edited by Robert Genco et al.
© 1994 American Society for Microbiology, Washington, DC 20005

Chapter 31

Principles of Regenerative Therapy

Jan Gottlow

This chapter is not intended to be a comprehensive review. Instead, it addresses different treatment approaches found in the literature that aim at periodontal regeneration. The ultimate goal of periodontal therapy is the restitution of the attachment apparatus, i.e., cementum, ligament, and bone.

CONVENTIONAL TREATMENT

Histological evaluation of surgical and nonsurgical treatments in an animal study by Caton and Nyman (19) showed re-formation of an epithelial lining (long junctional epithelium) facing the instrumented root surfaces with no new connective tissue attachment. The development of a long junctional epithelium occurred in sites with suprabony pockets as well as in sites with intrabony pockets. New bone formation (bone fill) in intrabony defects was a frequent finding, but a long junctional epithelium was always interposed between the newly formed bone and the root. The formation of a long junctional epithelium was reported by Bowers et al. (11) to occur also in humans following flap curettage of intrabony defects.

PERIODONTAL TISSUE COMPONENTS: WOUND HEALING

In the wound-healing process following surgical treatment, the denuded root surface can be repopulated by cells derived from the dentogingival epithelium, the gingival connective tissue, the alveolar bone, and the periodontal ligament. The dentogingival epithelium seems to be the "sprinter" in the race. A series of animal studies by Karring, Nyman, and coworkers elucidated the effects of individual periodontal tissue components when they were repopulating the root surface during wound healing (37, 39, 43–46, 59, 60). Bone tissue growing into contact with the root resulted in resorption and ankylosis (45). Gingival connective tissue caused root resorption when it was growing into contact with the root or aligned itself

---

Jan Gottlow • Guidor Research Center, Gruvgatan 8, S-421 30 V. Frolunda, Sweden.

parallel to the root without attachment (37, 60). Only granulation tissue originating from the periodontal ligament possessed the ability to form a new connective tissue attachment (39, 43, 44, 59).

Epithelial downgrowth prevents the formation of a new connective tissue attachment by preventing repopulation of the root surface by cells derived from the periodontal ligament (19, 44). However, coverage of the root surface by an epithelial layer also has a beneficial effect, namely, the prevention of root resorption and ankylosis, which otherwise could be induced by gingival connective tissue and bone (45, 46, 60).

GUIDED TISSUE REGENERATION

In guided tissue regeneration (GTR) therapy, a barrier is placed to prevent the soft tissue flap from reaching contact with the root during healing. At the same time, a space is formed between the barrier and the root, allowing the periodontal ligament cells to create a new attachment and the bone cells to form new bone. The first studies that demonstrated the formation of a new connective tissue attachment and bone as a result of GTR treatment were presented by Nyman et al. (59) and Gottlow et al. (39). The results of these animal studies have been confirmed in a number of studies (e.g., see references 4, 5, 15, 17, 20, 21, 36, and 54).

The first report of a human tooth treated according to the principle of GTR was by Nyman et al. (63). At the surgical procedure, a periodontal defect 11 mm long from the cementoenamel junction to the bottom of the defect was diagnosed. The histological analysis demonstrated new connective tissue attachment extending 7 mm coronally as measured from the bottom of the previous defect. Case reports on 12 teeth in 10 patients were presented by Gottlow et al. (40). Five teeth that were histologically evaluated demonstrated new connective tissue attachment. New attachment following GTR treatment in humans has also been shown histologically by Becker et al. (7), Stahl et al. (83), and Stahl and Froum (81).

In the study by Gottlow et al. (40), the result of therapy for seven teeth with furcation class II or angular bony defects was clinically evaluated. The mean gain of clinical attachment amounted to 5.6 mm, and the corresponding bone regrowth was 5.1 mm. The bone regrowth was predominantly associated with the intrabony part of furcation or angular bony defects. Gains in probing attachment level and probing bone level following GTR treatment have been reported in several short-term clinical studies, e.g., those by Becker et al. (6), Schallhorn and McClain (74), and Cortellini et al. (25), and in a long-term clinical study by Gottlow et al. (38).

Controlled clinical studies with intraindividual comparison of GTR therapy and conventional flap surgery in the treatment of class II furcation defects have been presented by Pontoriero et al. (68), Lekovic, et al. (50), and Caffesse et al. (18). All studies showed significantly more gain in probing attachment level at GTR-treated sites.

Results regarding the benefit of combining GTR therapy with root surface conditioning with citric acid and/or bone grafts or bone substitutes have been contradictory (3, 41, 47, 49, 74).

Successful treatment of buccal gingival recession defects has been reported by Tinti et al. (84) and Pini Prato et al. (66).

The principle of GTR has also been used successfully outside the periodontium for guided bone regeneration together with dental implants or ridge augmentation. See, e.g., work by Dahlin et al. (28–30), Nyman et al. (61), Becker et al. (8, 9), and Buser et al. (14).

The most commonly used barrier material has been nonresorbable, which calls for a second surgical procedure to remove the barrier. This is a negative factor with respect both to cost-benefit and to the additional surgical trauma to the patient and the new regenerated tissues. Gingival recession and subsequent exposure of the coronal portion of the barrier during healing have been described as frequent complications (6, 74–76). These are in most cases caused by epithelial migration apically on the connective tissue surface of the flap located on the outer surface of the barrier. Exposure of the barrier and pocket formation caused by epithelial downgrowth allow bacterial contamination (75, 76), which may decrease the regenerative capacity of the periodontal tissues (75).

BONE GRAFTING PROCEDURES

Bone grafting materials or bone substitutes are frequently used in regenerative therapy. It has been hypothesized that if bone regeneration can be induced it, too, will lead to induction of cementum formation (42).

Autogenous intraoral bone grafts have been histologically evaluated in many animal studies. Ellegaard et al. (34) found that the fate of fresh intraoral bone grafts was similar to that of devitalized transplants taken from the same donor site. One week after transplantation, the graft contained only a few osteocytes. Shortly after transplantation, the grafted area was invaded by granulation tissue. Osteoclasts started to resorb bone particles in some areas, while new bone was deposited in other areas (osteoconduction).

Findings from animal studies comparing bone graft sites to nongrafted sites are inconsistent. There is considerable variation in the amounts of bone regeneration and new attachment formation found at test and control sites, respectively, and also when comparing the procedures (26, 32, 33, 58, 64, 73). These findings do not support the hypothesis that bone regeneration leads to induction of cementum formation.

True osteogenesis following transplantation of fresh iliac bone marrow autografts has been demonstrated in many animal studies (1, 19, 32, 33). However, severe root resorption and ankylosis were common complications.

Demineralized freeze-dried bone allograft (DFDBA) should be osteoinductive because of released bone morphogenic protein (85). Bowers et al. (12, 13) have presented human histological evaluation following the use of DFDBA in intrabony defects. The first study (12) compared the healing of intrabony defects with the placement of DFDBA (30 sites) and without the placement of grafts (13 sites) in a submerged environment. Significantly more bone formation was found at grafted sites, but no statistical difference was found regarding the amount of new connective tissue attachment. Both test and control teeth exhibited new attachment formation on previously plaque-exposed root surface areas. In the second study (13) a non-submerged environment was used. Both grafted and nongrafted sites were covered by a free gingival graft to retard epithelial migration. All control sites (no graft) showed a long junctional epithelium and no new attachment formation. The test sites demonstrated both new attachment formation and new bone formation. The apical level of the junctional epithelium always coincided with the coronal level of new attachment formation. In some specimens, the apical level of the junctional epithelium terminated apical to the level of the alveolar crest. No root resorption or ankylosis was found. These studies support the osteoconductive properties of DFDBA. On the other hand, since new attachment formation was evident in both test and control teeth in the submerged environment, the osteoconductive effect was not correlated to the amount of new attachment formation. In the non-submerged environment, the control teeth demonstrated epithelial downgrowth without new attachment formation. Could the new attachment formation at DFDBA-grafted nonsubmerged sites be explained by a "barrier" effect similar to that in GTR therapy? The graft will make it more difficult for the soft tissue flap, i.e., gingival connective tissue and epithelium, to grow or collapse down into the defect.

The actual influence of graft materials on the wound-healing process (e.g., induction, conduction, scaffolding, barrier function) is still not clear and may be different for each graft material.

ROOT SURFACE DEMINERALIZATION

Root surface demineralization with citric acid following removal of the cementum layer has been shown to promote new connective tissue attachment in animals (27, 57, 69, 72). The mechanism by which citric acid conditioning facilitates the formation of a new fibrous attachment has been suggested to involve the exposure of the collagen fibrils of the dentine matrix and a subsequent interdigitation of these fibrils with collagen fibrils of the adjacent soft tissue flap (67, 72, 77). Other animal studies have reported the formation of a long junctional epithelium without new attachment formation (62) or an increased frequency of root resorption (31, 37, 53). Pettersson and Aukhil (65) found that the citric acid applied to the wound damaged the periodontal ligament cells, resulting in delayed cell migration.

Histological evidence of new attachment formation in humans has been reported by Cole et al. (22), Albair et al. (2), and Common and McFall (24). On the other hand, Stahl et al. (82) noted only limited amounts of new attachment formation and only in close proximity with the "old" periodontal ligament.

Studies in humans have failed to detect clinically significant differences between acid-treated and non-acid-treated teeth (23, 56, 70, 71, 79). Root conditioning with citric acid did not enhance the effect of GTR therapy according to results presented by Handelsman et al. (41) on the treatment of furcation degree II defects and by Kersten et al. (47) on the treatment of intrabony defects.

CORONALLY POSITIONED FLAPS

Gingival recession is a limiting factor for new attachment formation; it decreases the root surface area available for regeneration. Gingival recession might be compensated for by placing the flap coronal to its intended final position (48). This might increase the distance to the bottom of the defect for the apically migrating epithelium, thereby facilitating coronal growth of periodontal ligament cells.

Martin et al. (55) used resin-bonded orthodontic brackets on teeth as suture retainers and a special suturing technique for the coronal positioning of the flap in humans. Utilizing this technique, Gantes et al. (35) reported successful closure of furcation class II defects. The closure of the defects was determined by clinical evaluation of bone fill during a reentry procedure. This technique has not been evaluated histologically. Hence, it can only be speculated whether healing had occurred with new attachment formation or with the formation of a long junctional epithelium interposed between the newly formed bone and the root.

GROWTH FACTORS, EXTRACELLULAR MATRIX PROTEINS, AND OTHER BIOCHEMICAL AGENTS

The regenerative wound-healing process is dependent on cellular activities such as migration, proliferation, differentiation, and synthesis of matrix components. This means that growth factors, extracellular matrix proteins, and other biochemical agents may enhance or replace the regenerative procedure discussed above. The majority of these approaches published in the literature are still at the in vitro level.

Much interest has focused on the use of extracellular matrix proteins such as fibronectin and laminin, but in vivo results have been conflicting (16, 78, 80, 86). A combination of platelet-derived and insulin-like growth factors has been reported to accelerate and enhance periodontal regeneration (51, 52) and bone regeneration around dental titanium implants (10).

CAN WE ACCOMPLISH THE GOAL OF REGENERATIVE THERAPY?

Prospective controlled human studies comparing the results of regenerative therapies with those of conventional surgical therapy are still few. Such studies are needed to determine the predictability of the regenerative procedures, both short term and long term, so that we can give recommendations when or for what types of defects regenerative therapy should be used. However, the possibilities of regenerating lost periodontum have definitely become a clinical reality.

Acknowledgments. This chapter is sponsored by Guidor AB, Sweden. Jan Gottlow is Scientific Director at Guidor Research Center, Sweden.

REFERENCES

1. **Adell R.** 1974. Regeneration of the periodontium. An experimental study in dogs. Ph.D. thesis. University of Gothenburg, Göteborg, Sweden.
2. **Albair, W. B., C. M. Cobb, and W. J. Killoy.** 1982. Connective tissue attachment to periodontally diseased roots after citric acid demineralization. *J. Periodontol.* **53:**515–526.
3. **Anderegg, C. R., S. J. Martin, J. L. Gray, J. T. Mellonig, and M. E. Gher.** 1991. Clinical evaluation of the use of decalcified freeze-dried bone allograft with guided tissue regeneration in the treatment of molar furcation invasions. *J. Periodontol.* **62:**264–268.
4. **Aukhil, I., E. Petterson, and C. Suggs.** 1986. Guided tissue regeneration. An experimental procedure in Beagle dogs. *J. Periodontol.* **57:**727–734.
5. **Aukhil, I., D. M. Simson, and T. V. Schaberg.** 1983. An experimental study of new attachment procedure in Beagle dogs. *J. Periodontal Res.* **18:**643–654.
6. **Becker, W., B. Becker, L. Berg, J. Prichard, R. Caffesse, and E. Rosenberg.** 1988. New attachment after treatment with root isolation procedures: report for treated class III and class II furcations and vertical osseous defects. *Int. J. Periodontal Rest. Dent.* **3:**9–23.
7. **Becker, W., B. Becker, J. Prichard, R. Caffesse, E. Rosenberg, and J. Gian-Grasso.** 1987. Root isolation for new attachment procedures: a surgical and suturing method. Three case reports. *J. Periodontol.* **58:**819–826.
8. **Becker, W., and B. E. Becker.** 1990. Guided tissue regeneration for implants placed into extraction sockets and for implant dihiscences: surgical techniques and case reports. *Int. J. Periodontal Rest. Dent.* **10:**377–391.
9. **Becker, W., B. E. Becker, M. Handelsman, R. Celleti, C. Ochsenbein, R. Hardwick, and B. Langer.** 1990. Bone formation at dehisced dental implant sites treated with implant augmentation material: a pilot study in dogs. *Int. J. Periodontal Rest. Dent.* **10:**93–101.
10. **Becker, W., S. Lynch, U. Lekholm, B. Becker, R. Caffesse, K. Donath, and R. Sanchez.** 1992. A comparison of ePTFE membranes alone or in combination with platelet-derived growth factors and insulin-like factor-I or demineralized freeze-dried bone in promoting bone formation around immediate extraction socket implants. *J. Periodontol.* **63:**929–940.
11. **Bowers, G. M., B. Chadroff, R. Carnevale, J. Mellonig, R. Corio, J. Emerson, M. Stevens, and E. Romberg.** 1989. Histologic evaluation of new attachment apparatus formation in humans. Part I. *J. Periodontol.* **60:**664–674.
12. **Bowers, G. M., B. Chadroff, R. Carnevale, J. Mellonig, R. Corio, J. Emerson, M. Stevens, and E. Romberg.** 1989. Histologic evaluation of new attachment apparatus formation in humans. Part II. *J. Periodontol.* **60:**675–682.
13. **Bowers, G. M., B. Chadroff, R. Carnevale, J. Mellonig, R. Corio, J. Emerson, M. Stevens, and E. Romberg.** 1989. Histologic evaluation of new attachment apparatus formation in humans. Part III. *J. Periodontol.* **60:**683–693.
14. **Buser, D., U. Brägger, N. P. Lang, and S. Nyman.** 1990. Regeneration and enlargement of jaw bone using guided tissue regeneration. *Clin. Oral Implant. Res.* **1:**22–32.
15. **Caffesse, R. G., L. E. Dominguez, C. E. Nasjleti, W. A. Castelli, E. C. Morrison, and B. A. Smith.** 1990. Furcation defects in dogs treated by guided tissue regeneration GTR. *J. Periodontol.* **61:**45–50.
16. **Caffesse, R. G., M. J. Holden, S. Kon, and C. E. Nasjleti.** 1985. The effect of citric acid and fibronectin application on healing following surgical treatment of naturally occurring periodontal disease in beagle dogs. *J. Clin. Periodontol.* **12:**578–590.
17. **Caffesse, R. G., B. A. Smith, W. A. Castelli, and C. E. Nasjleti.** 1988. New attachment achieved by guided tissue regeneration in Beagle dogs. *J. Periodontol.* **59:**589–594.
18. **Caffesse, R. G., B. A. Smith, B. Duff, E. C. Morrison, D. Merril, and W. Becker.** 1990. Class II functions treated by guided tissue regeneration in humans: case reports. *J. Periodontol.* **61:**510–514.
19. **Caton, J., and S. Nyman.** 1980. Histometric evaluation of periodontal surgery. I. The modified Widman flap procedure. *J. Clin. Periodontol.* **7:**212–223.
20. **Caton, J. G., C. Wagener, A. Polson, S. Nyman, B. Franz, O. Bouwsma, and T. Blieden.** 1992.

Guided tissue regeneration in interproximal defects using Gore-Tex® periodontal material. *Int. J. Periodontal Rest. Dent.* **12**:267–277.

21. **Claffey, N., S. Motsinger, J. Ambruster, and J. Egelberg.** 1989. Placement of a porous membrane underneath the mucoperiosteal flap and its effect on periodontal wound healing in dogs. *J. Clin. Periodontol.* **16**:12–16.

22. **Cole, R. T., M. Crigger, G. Bogle, J. Egelberg, and K. A. Selvig.** 1980. Connective tissue regeneration to periodontally diseased teeth. *J. Periodontal Res.* **15**:1–9.

23. **Cole, R. T., R. Nilvéus, J. Ainamo, G. Bogle, M. Crigger, and J. Egelberg.** 1981. Pilot clinical studies on the effect of topical citric acid application on healing after replaced periodontal flap surgery. *J. Periodontal Res.* **16**:117–122.

24. **Common, J., and W. T. McFall.** 1983. The effect of citric acid on attachment of laterally positioned flaps. *J. Periodontol.* **54**:9–18.

25. **Cortellini, P., G. Pini Prato, C. Baldi, and C. Clauser.** 1990. Guided tissue regeneration with different materials. *Int. J. Periodontal Rest. Dent.* **10**:137–151.

26. **Coverly, L., P. Toto, and A. Gargiulo.** 1975. Osseous coagulum: a histologic evaluation. *J. Periodontol.* **46**:596.

27. **Crigger, M., G. Bogle, R. Nilvéus, J. Egelberg, and K. A. Selvig.** 1978. The effect of topical citric acid application on the healing of experimental furcations in dogs. *J. Periodontal Res.* **13**:538–549.

28. **Dahlin, C., J. Gottlow, A. Linde, and S. Nyman.** 1990. Healing of maxillary and mandibular bone defects using a membrane technique. An experimental study in monkeys. *Scand. J. Reconstr. Hand Surg.* **24**:13–19.

29. **Dahlin, C., A. Linde, J. Gottlow, and S. Nyman.** 1988. Healing of bone defects by guided tissue regeneration. *Plast. Reconstr. Surg.* **81**:672–676.

30. **Dahlin, C., L. Sennerby, U. Lekholm, A. Linde, and S. Nyman.** 1989. Generation of new bone around titanium implants using a membrane technique: an experimental study in rabbits. *Int. J. Oral Maxillofac. Implant.* **4**:19–25.

31. **Dreyer, W. P., and J. D. van Heerden.** 1986. The effect of citric acid on the healing of periodontal ligament-free, healthy roots, horizontally implanted against bone and gingival connective tissue. *J. Periodontal Res.* **21**:210–220.

32. **Ellegaard, B., T. Karring, R. Davies, and H. Löe.** 1974. New attachment after treatment of intrabony defects in monkeys. *J. Periodontol.* **45**:368–377.

33. **Ellegaard, B., T. Karring, M. Listgarten, and H. Löe.** 1973. New attachment after treatment of interradicular lesions. *J. Periodontol.* **44**:209–217.

34. **Ellegaard, B., T. Karring, and H. Löe.** 1975. The fate of vital and devitalized bone grafts in the healing of interradicular lesions. *J. Periodontal Res.* **10**:88–97.

35. **Gantes, B., M. Martin, S. Garrett, and J. Egelberg.** 1988. Treatment of periodontal furcation defects. II. Bone regeneration in mandibular class II defects. *J. Clin. Periodontol.* **15**:232–239.

36. **Gottlow, J., T. Karring, and S. Nyman.** 1990. Guided tissue regeneration following treatment of "recession type defects" in the monkey. *J. Periodontol.* **61**:680–685.

37. **Gottlow, J., S. Nyman, and T. Karring.** 1984. Healing following citric acid conditioning of roots implanted into bone and gingival connective tissue. *J. Periodontal Res.* **19**:214–220.

38. **Gottlow, J., S. Nyman, and T. Karring.** 1992. Maintenance of new attachment gained through guided tissue regeneration. *J. Clin. Periodontol.* **19**:315–317.

39. **Gottlow, J., S. Nyman, T. Karring, and J. Lindhe.** 1984. New attachment formation as the result of controlled tissue regeneration. *J. Clin. Periodontol.* **11**:494–503.

40. **Gottlow, J., S. Nyman, J. Lindhe, T. Karring, and J. Wennström.** 1986. New attachment formation in the human periodontium by guided tissue regeneration. Case reports. *J. Clin. Periodontol.* **13**:604–616.

41. **Handelsman, M., M. Davarpanah, and R. Celletti.** 1991. Guided tissue regeneration with and without citric acid treatment in vertical osseous defects. *Int. J. Periodontal Rest. Dent.* **11**:351–363.

42. **Hiatt, W. H., R. G. Schallhorn, and A. J. Aaronian.** 1978. The induction of new bone and cementum formation. IV. Microscopic examination of the periodontium following human bone and marrow allograft, autograft and nongraft periodontal regenerative procedures. *J. Periodontol.* **49**:495–512.

43. **Isidor, F., T. Karring, S. Nyman, and J. Lindhe.** 1986. The significance of coronal growth of the periodontal ligament tissue for new attachment formation. *J. Clin. Periodontol.* **13**:145–150.

44. **Karring, T., F. Isidor, S. Nyman, and J. Lindhe.** 1985. New attachment formation on teeth with a reduced but healthy periodontal ligament. *J. Clin. Periodontol.* **12:**51–60.

45. **Karring, T., S. Nyman, and J. Lindhe.** 1980. Healing following implantation of periodontitis affected roots into bone tissue. *J. Clin. Periodontol.* **7:**96–105.

46. **Karring, T., S. Nyman, J. Lindhe, and M. Sirirat.** 1984. Potentials for root resorption during periodontal wound healing. *J. Clin. Periodontol.* **11:**41–52.

47. **Kersten, B. G., A. D. H. Chamberlain, S. Khorsandi, U. M. E. Wikesjö, K. A. Selvig, and R. E. Nilvéus.** 1992. Healing of intrabony periodontal lesion following root conditioning with citric acid and wound closure including an expanded PTFE membrane. *J. Periodontol.* **63:**876–882.

48. **Klinge, B., R. Nilvéus, R. D. Kiger, and J. Egelberg.** 1981. Effect of flap placement and defect size on healing of experimental furcation defects. *J. Periodontal Res.* **16:**236–248.

49. **Lekovic, V., E. B. Kenney, F. A. Carranza, Jr., and V. Danilovic.** 1990. Treatment of class II furcation defects using porous hydroxyapatite in conjunction with a polytetrafluoroethylene membrane. *J. Periodontol.* **61:**575–578.

50. **Lekovic, V., E. B. Kenney, K. Kovacevic, and F. A. Carranza, Jr.** 1989. Evaluation of guided tissue regeneration in class II furcation defects. A clinical re-entry study. *J. Periodontol.* **60:**694–698.

51. **Lynch, S. E., G. R. Castilla, R. C. Williams, C. P. Kiritsy, H. T. Howell, M. S. Reddy, and H. N. Antoniades.** 1991. The effects of short-term application of a combination of platelet-derived and insulin-like growth factors on periodontal wound healing. *J. Periodontol.* **62:**458–467.

52. **Lynch, S. E., R. C. Williams, A. M. Polson, T. H. Howell, M. S. Reddy, U. E. Zappa, and H. N. Antoniades.** 1989. A combination of platelet-derived and insulin-like growth factors enhances periodontal regeneration. *J. Clin. Periodontol.* **16:**545–548.

53. **Magnusson, I., N. Claffey, G. Bogle, S. Garrett, and J. Egelberg.** 1985. Root resorption following periodontal flap procedures in monkeys. *J. Periodontal Res.* **20:**79–85.

54. **Magnusson, I., S. Nyman, T. Karring, and J. Egelberg.** 1985. Connective tissue attachment formation following exclusion of gingival connective tissue and epithelium during healing. *J. Periodontal Res.* **20:**201–208.

55. **Martin, M., B. Gantes, S. Garrett, and J. Egelberg.** 1988. Treatment of periodontal furcation defects. I. Review of the literature and description of a regenerative surgical technique. *J. Clin. Periodontol.* **15:**227–231.

56. **Moore, J. A., F. P. Ashkey, and C. A. Waterman.** 1987. The effect on healing of the application of citric acid during replaced flap surgery. *J. Clin. Periodontol.* **14:**130–135.

57. **Nilvéus, R.** 1978. Treatment of periodontal furcation pockets. Experimental studies in dogs. Ph.D. thesis. University of Lund, Malmö, Sweden.

58. **Nilvéus, R., O. Johansson, and J. Egelberg.** 1978. The effect of autogenous cancellous bone grafts on healing of experimental furcation defects in dogs. *J. Periodontal Res.* **13:**532–537.

59. **Nyman, S., J. Gottlow, T. Karring, and J. Lindhe.** 1982. The regenerative potential of the periodontal ligament. An experimental study in the monkey. *J. Clin. Periodontol.* **9:**257–265.

60. **Nyman, S., T. Karring, J. Lindhe, and S. Plantén.** 1980. Healing following implantation of periodontitis affected roots into gingival connective tissue. *J. Clin. Periodontol.* **7:**394–401.

61. **Nyman, S., N. P. Lang, D. Buser, and U. Brägger.** 1990. Bone regeneration adjacent to titanium dental implants using guided tissue regeneration: a report of two cases. *Int. J. Oral Maxillofac. Implant.* **5:**9–14.

62. **Nyman, S., J. Lindhe, and T. Karring.** 1981. Healing following surgical treatment and root demineralization in monkeys with periodontal disease. *J. Clin. Periodontol.* **8:**249–258.

63. **Nyman, S., J. Lindhe, T. Karring, and H. Rylander.** 1982. New attachment following surgical treatment of human periodontal disease. *J. Clin. Periodontol.* **9:**290–296.

64. **Patterson, R. L., C. K. Collings, and E. R. Zimmerman.** 1967. Autogenous implants in the alveolar process of the dog with induced periodontitis. *Periodontics* **5:**19–25.

65. **Pettersson, E. C., and I. Aukhil.** 1986. Citric acid conditioning of roots affects guided tissue regeneration in experimental periodontal wounds. *J. Periodontal Res.* **21:**543–552.

66. **Pini Prato, G., C. Tinti, G. Vincenzi, C. Magnani, P. Cortellini, and C. Clauser.** 1992. Guided tissue regeneration versus mucogingival surgery in the treatment of human buccal gingival recession. *J. Periodontol.* **63:**919–928.

67. **Polson, A. M., and M. P. Proye.** 1982. Effect of root surface alterations on periodontal healing. II. Citric acid treatment of the denuded root. *J. Clin. Periodontol.* **9:**441–454.

68. **Pontoriero, R., J. Lindhe, S. Nyman, T. Karring, E. Rosenberg, and F. Sanavi.** 1988. Guided tissue regeneration in degree II furcation-involved mandibular molars: a clinical study. *J. Clin. Periodontol.* **15:**247–254.

69. **Register, A. A.** 1973. Bone and cementum inducation by dentine, demineralized *in situ. J. Periodontol.* **44:**49–54.

70. **Renvert, S., and J. Egelberg.** 1981. Healing after treatment of periodontal intraosseous defects. II. Effect of citric acid conditioning of the root surface. *J. Clin. Periodontol.* **8:**459–473.

71. **Renvert, S., S. Garrett, R. G. Schallhorn, and J. Egelberg.** 1985. Osseous grafting and citric acid conditioning. *J. Clin. Periodontol.* **12:**441–455.

72. **Ririe, C. M., M. Crigger, and K. A. Selvig.** 1980. Healing of periodontal connective tissue following surgical wounding and application of citric acid in dogs. *J. Periodontal Res.* **15:**314–327.

73. **Rivault, A. F., P. D. Toto, S. Levy, and A. W. Gargiulo.** 1971. Autogenous bone grafts: osseous coagulum and osseous retrograde procedures in primates. *J. Periodontol.* **42:**787–796.

74. **Schallhorn, R., and P. McClain.** 1988. Combined osseous composite grafting, root conditioning, and guided tissue regeneration. *Int. J. Periodontal Rest. Dent.* **4:**9–31.

75. **Selvig, K. A., B. G. Kersten, D. H. Chamberlain, U. M. E. Wikesjo, and R. E. Nilvéus.** 1992. Regenerative surgery of intrabony periodontal defects using ePTFE barrier membranes: scanning electron microscopic evaluation of retrieved membranes versus clinical healing. *J. Periodontol.* **63:**974–978.

76. **Selvig, K. A., R. E. Nilvéus, L. Fitzmorris, B. Kersten, and S. Khorsandi.** 1990. Scanning electron microscopic observations of cell population and bacterial contamination of membranes used for guided periodontal tissue regeneration in humans. *J. Periodontol.* **61:**515–520.

77. **Selvig, K. A., C. M. Ririe, R. Nilvéus, and J. Egelberg.** 1981. Fine structure of new connective tissue attachment following acid treatment of experimental furcation pockets in dogs. *J. Periodontal Res.* **16:**123–129.

78. **Smith, B., R. Caffesse, C. Nasjleti, S. Kon, and W. Castelli.** 1987. Effects of citric acid and fibronectin and laminin application in treating periodontitis. *J. Clin. Periodontol.* **14:**396–402.

79. **Smith, B. A., W. E. Mason, E. C. Morrison, and R. G. Caffesse.** 1986. The effectiveness of citric acid as an adjunct to surgical reattachment procedures in humans. *J. Clin. Periodontol.* **13:**701–708.

80. **Smith, B. A., J. S. Smith, R. G. Caffesse, C. E. Nasjleti, D. E. Lopatin, and C. J. Kowalski.** 1987. Effect of citric acid and various concentrations of fibronectin on healing following periodontal flap surgery in dogs. *J. Periodontol.* **58:**667–673.

81. **Stahl, S. S., and S. J. Froum.** 1991. Healing of human suprabony lesions treated with guided tissue regeneration and coronally anchored flaps. Case reports. *J. Clin. Periodontol.* **18:**69–74.

82. **Stahl, S. S., S. J. Froum, and L. Kushner.** 1983. Healing responses of human intraosseous lesions following the use of debridement, grafting and citric acid root treatment. II. Clinical and histologic observations: one year postsurgery. *J. Periodontol.* **54:**325–338.

83. **Stahl, S., S. Froum, and D. Tarnow.** 1990. Human histologic responses to guided tissue regenerative techniques in intrabony lesions. Case reports on 9 sites. *J. Clin. Periodontol.* **17:**191–198.

84. **Tinti, C., G. Vincenzi, P. Cortellini, G. P. Pini Prato, and C. Clauser.** 1992. Guided tissue regeneration in the treatment of human facial recession. A 12-case report. *J. Periodontol.* **63:**554–560.

85. **Urist, M. R.** 1980. *Fundamental and Clinical Bone Physiology,* p. 348–353. J. B. Lippincott Co., Philadelphia.

86. **Wikesjö, U. M. E., N. Claffey, L. A. Christersson, L. C. Franzetti, R. J. Genco, V. P. Terranova, and J. Egelberg.** 1988. Repair of periodontal furcation defects in beagle dogs following reconstructive surgery including root surface demineralization with tetracycline hydrochloride and fibronectin application. *J. Clin. Periodontol.* **15:**73–80.

Molecular Pathogenesis of Periodontal Disease
Edited by Robert Genco et al.
© 1994 American Society for Microbiology, Washington, DC 20005

Chapter 32

Differential Regulation of Periodontal Ligament Cell Activities by Platelet-Derived Growth Factor, Insulin-Like Growth Factor I, and Epidermal Growth Factor

M.-I. Cho, N. Matsuda, P. R. Ramakrishnan, W.-L. Lin, and R. J. Genco

Periodontal regeneration is a complex phenomenon that requires repair of tooth supporting tissues such as periodontal ligament (PDL), root cementum, alveolar bone, and gingival epithelium and connective tissue. Repair of these tissues is achieved by migration, proliferation, and differentiation of PDL fibroblasts, cementoblasts, osteoblasts, and gingival epithelial cells and fibroblasts, respectively. More important, for successful periodontal regeneration, a series of orderly repair processes of these tissues are required. Migration and proliferation of PDL fibroblasts and synthesis of matrix components by the cells are needed for repair of the PDL in the early phases, while the differentiation of cementoblasts and osteoblasts is needed for the formation of new cementum and bone in the later phases of periodontal regeneration.

Growth factors are believed to be able to stimulate these cellular activities, and their clinical application has thus been proposed to promote periodontal regeneration (18, 27, 29, 32, 38, 40, 46). Indeed, some workers have already demonstrated enhanced periodontal regeneration in vivo (26, 27, 39). However, in spite of these in vivo findings and significant advances in the understanding of the chemical structures, functions, and mechanisms of action of various growth factors, their specific effects on PDL fibroblastic cells have not been clearly defined. In this chapter, we summarize our recent findings, with emphasis on the differential roles of platelet-derived growth factor (PDGF), insulin-like growth factor I (IGF-I), and epidermal growth factor (EGF) in directed migration, proliferation, and differ-

M.-I. Cho, P. R. Ramakrishnan, W.-L. Lin, and R. J. Genco • Department of Oral Biology and Periodontal Disease Research Center, State University of New York at Buffalo, Buffalo, New York 14214. *N. Matsuda* • Sunstar Inc., 3-1, Asahi-Machi, Takatsuki, Osaka, 569 Japan.

entiation of PDL fibroblastic cells, in order to elucidate the proper and effective clinical application of growth factors in promoting periodontal regeneration.

PDGF

Presence of α and β Subtypes of PDGF Receptors on PDL Fibroblastic Cells

PDGF is composed of two related but distinct polypeptide chains, A and B (3, 22, 23). Three dimeric isoforms of PDGF (AA, AB, and BB) with differences in their biological activities and potencies have been identified. The biological effects of PDGF are mediated via binding to specific cell surface receptors that differ in their abilities to bind the three isoforms. The α receptor binds all PDGF isoforms, while the β receptor binds PDGF-BB with high affinity and PDGF-AB with low affinity but does not bind PDGF-AA (4, 19, 43). To better understand the effects of PDGFs on PDL fibroblastic cells, the subtypes of PDGF receptors on these cells were identified by immunoprecipitation with antibodies specific to the α or β receptors (obtained from C.-H. Heldin, Ludwig Cancer Research Institute, Uppsala, Sweden). Antiserum to the β receptor precipitated two components with molecular masses of 148 and 172 kDa that represent the precursor and mature forms of the β receptor (Fig. 1). Antiserum to α receptor also precipitated the precursor and mature forms of the α receptor, which had molecular masses of 140 and 153 kDa, respectively (Fig. 1). These results demonstrate that rat PDL

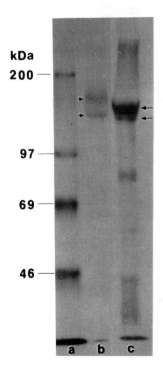

Figure 1. Immunoprecipitation of PDGF α and β receptors of PDL fibroblastic cells. Cells were labeled with [^{35}S]methionine (100 μCi/ml) in methionine-free Dulbecco modified Eagle medium containing 1% fetal bovine serum for 4 h. Cell lysates were passed through a Lens culinaris hemagglutinin affinity column (Sigma Chemical Co., St. Louis, Mo.) equilibrated with lysis buffer. The adsorbed fraction was eluted by buffer containing 10% α-D-mannoside. The receptors were then immunoprecipitated by incubating the adsorbed glycoproteins with 10 μl of rabbit antiserum to PDGF α receptor raised against a synthetic peptide containing the amino acid residues 1066 to 1084 of human PDGF α receptor (15) or with antiserum to PDGF β receptor raised against a synthetic peptide corresponding to the amino acid sequence 981 to 994 of murine PDGF β receptor (16). Both antisera were provided by C.-H. Heldin, Ludwig Cancer Research Institute, Uppsala, Sweden. The immunoprecipitates were run on sodium dodecyl sulfate–8% polyacrylamide gels, and the labeled proteins were detected by radioautography. The precursor (lower arrows) and mature (upper arrows) forms of the α and β receptors are shown in lanes b and c.

fibroblastic cells have both α and β types of PDGF receptors similar to those observed on fibroblasts derived from other connective tissues (15, 16).

Mitogenic Effect

PDGF is a potent mitogen for a variety of connective tissue cells (20, 37, 48). In an attempt to compare the mitogenic potencies of different isoforms of PDGFs, their mitogenic effects on PDL fibroblastic cells were assessed in vitro. All isoforms had a potent mitogenic effect. PDGF-BB was the most potent stimulator of mitogenesis, followed by PDGF-AA and -AB (Fig. 2). The combination of PDGF-AA and -BB had an additive mitogenic effect (Fig. 2). This result supports our finding that both α and β receptors are present on PDL fibroblastic cells and have the ability to transduce mitogenic signals. In addition, both recombinant human PDGF-BB and natural human PDGF are equally effective as mitogens in vitro, suggesting that recombinant PDGF-BB can be useful for clinical application to promote periodontal regeneration (29, 32, 40). Indeed, treatment of the periodontal lesion with a combination of PDGF-BB and IGF-I or dexamethasone (Dex) enhanced periodontal regeneration in animal studies (26, 27, 39).

Chemotactic Effect

PDGF is known as a potent chemoattractant for connective tissue cells (17, 37), and their chemotactic response to PDGF is mediated through the β receptor (44, 49). In our recent studies, however, PDL fibroblastic cells demonstrated a

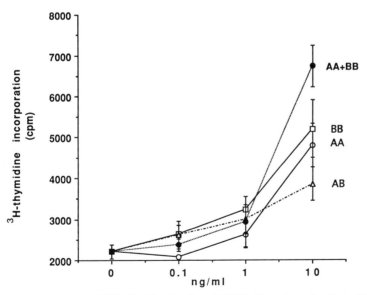

Figure 2. Mitogenic response of PDL fibroblastic cells to PDGFs. The mitogenic effect of PDGFs was assessed by measuring [³H]thymidine incorporation into proliferating rat PDL fibroblastic cells. The procedure of Matsuda et al. (29) was used.

marked dose-dependent chemotactic response to all forms of PDGFs (29), including PDGF-AA, which binds only to the α receptors. PDGF-BB is twice as potent as PDGF-AA at concentrations between 0.1 and 10 ng/ml (Fig. 3). These results indicate that both α and β receptors are involved in chemotaxis of PDL fibroblastic cells and that among the three isoforms of PDGF, PDGF-BB is the most potent chemoattractant for the cells.

Effect on Protein Synthesis

In the present study, PDGF-AB did not show any significant effect on total protein synthesis but exhibited a selective effect on collagen synthesis in vitro. Treatment of PDL fibroblastic cells with PDGF-AB (29) or transforming growth factor β increased collagen synthesis, as was observed in other connective tissue cells (8, 21, 37, 50). It is of interest to know whether other isoforms of PDGF have similar effects on synthesis of total protein and collagen.

IGF-I

Mitogenic Effect

The mitogenic effect of IGF-I on PDL fibroblastic cells was studied in vitro. IGF-I revealed a potent mitogenic effect between 10 and 100 ng/ml, which is comparable to the potency of PDGF-BB. However, IGF-I at concentrations below 10 ng/ml did not show any mitogenic effect (29).

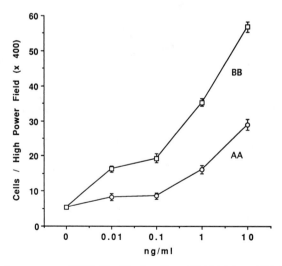

Figure 3. Chemotactic responses of PDL fibroblastic cells to PDGF-AA and PDGF-BB. Chemotactic effects of PDGF-AA and PDGF-BB on rat PDL fibroblastic cells were analyzed by the method of Matsuda et al. (29).

Chemotactic Effect

IGF-I also functions as a potent chemoattractant for PDL fibroblasts, but only at high concentrations. Its chemotactic effect on cells was dose dependent at concentrations between 0.1 and 100 ng/ml (29).

Effect on Cell Differentiation

IGF-I treatment has been known to increase new bone formation in vitro (6, 41) as well as in vivo (38, 45). This increase is due to the ability of IGF-I to stimulate proliferation of osteogenic cells, differentiation of these cells into osteoblasts, and synthesis of bone matrix components by osteoblasts (5, 6, 41, 47). The ability of IGF-I to induce the differentiation of PDL fibroblastic cells into cementoblasts and osteoblasts was studied by using the rat molar reimplantation and beagle dog models, respectively. Treatment of the root surface of the rat with IGF-I induced cementogenesis during periodontal regeneration. Cementoblasts appeared on the root surface at 6 days (Fig. 4a) and deposited the newly formed cementum at 8 days after reimplantation of the IGF-I treated molars (Fig. 4b). Similarly, the application of a mixture of type I collagen and IGF-I to the furcation defects of the second and fourth mandibular premolars of beagle dogs demonstrated increased new bone formation (Fig. 5b). This observation is in agreement with previous findings by Lynch et al. (26).

EGF

Unique Expression Pattern of EGF-R on PDL Fibroblasts

In our recent radioautographic studies on the localization of ^{125}I-EGF on PDL fibroblasts of the rat during their differentiation, we observed a unique expression pattern of EGF receptors (EGF-R) on the cells. A large number of EGF-R are present on the cell membranes of precursor cells of PDL fibroblasts (parafollicular cells) throughout their differentiation (14) as well as on mature PDL fibroblasts that have full synthetic activity in the functional PDL (12–14). Similarly, numerous EGF-R are expressed on undifferentiated preosteoblasts and prechondrocytes. However, the number of EGF-R on these cells falls dramatically as cell differentiation progresses, with fully differentiated osteoblasts and chondrocytes not expressing EGF-R (12, 13). These in vivo observations indicate that, unlike osteoblasts and chondrocytes, which lose their EGF-R during differentiation, mature PDL fibroblasts in the functional PDL continuously express a large number of EGF-R. We investigated the functional significance of the continued expression of EGF-R on PDL fibroblasts.

EGF-R on PDL Fibroblasts as a Phenotype Stabilizer

EGF has a variety of biological activities. Besides its well-known function as a potent mitogen for various cell types, EGF has nonmitogenic functions such as

408   Cho et al.

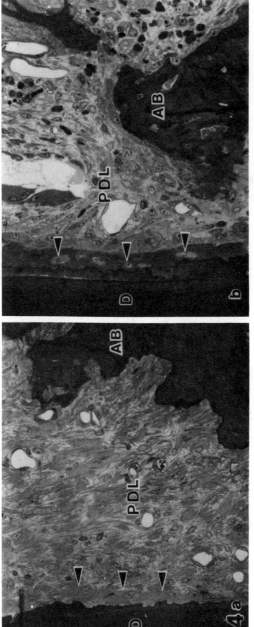

Figure 4. Effect of IGF-I on cementoblast differentiation. The effect of IGF-I on cementoblast differentiation was investigated with the rat molar reimplantation model. Sprague-Dawley rats (115 to 120 g body weight) were fed a powdered diet containing 0.4% β-aminopropionitrile for 5 days to achieve gentle tooth extraction by reducing the tensile strength of the PDL collagen. The first maxillary molars were extracted, washed in sterile water, and treated with bacterial collagenase to remove the PDL collagen fibers from the root surface. The mesial root surface was then demineralized by topical application of citric acid for 1 min and treated with IGF-I (2 to 3 μg). The molars were reimplanted into the sockets immediately after extraction of the corresponding molars of other animals. At 6 and 8 days after reimplantation, the animals were sacrificed, and the periodontal tissues around the first molars were demineralized and processed for morphological study (9). Note the appearance of cementoblasts (arrowheads) on the root surface at 6 days (a) and the newly formed cementum surrounding cementocytes (arrowheads) at 8 days after reimplantation (b). No cementum formation was observed on the root surface without IGF-I treatment. Magnification, ×1,200. AB, alveolar bone; PDL, periodontal ligament; D, dentin.

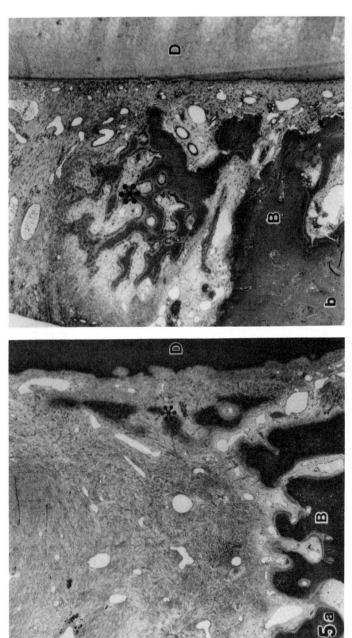

Figure 5. Effect of IGF-I on osteoblast differentiation. In order to study the effect of IGF-I on osteoblast differentiation, horizontal furcation periodontal defects (class III) were created around the second and fourth mandibular premolars of beagle dogs. The furcation area was then filled with either type I collagen purified from the dog skin (11) or collagen containing IGF-I (4 to 5 µg). Flaps were repositioned and sutured. The animals were sacrificed at 4 weeks after surgery. The tissues containing the lesions were demineralized and processed for morphological study (9). The site treated with a mixture of IGF-I and collagen demonstrates active new bone formation (*) (b) (magnification, ×170), whereas the one treated with collagen alone shows the presence of dense connective tissue in the furcation defect area and a small amount of new bone formation parallel to the root surface (*) (a) (magnification, ×150). B, bone; D, dentin.

inhibition of gastric secretion from the intestinal mucosa and promotion of healing in the corneal epithelium. EGF also accelerates some developmental processes, including palatal formation, eyelid opening, incisor eruption, and lung maturation (7). Interestingly, in spite of the expression of numerous EGF-R on PDL fibroblasts of the functional PDL, the fibroblasts demonstrate an extremely low labeling index ranging from 0.5 to 3% under normal physiological conditions (30, 35). Furthermore, EGF does not show significant mitogenic and chemotactic effects on PDL fibroblasts in vitro (29). Therefore, the need for the continued expression of numerous EGF-R on PDL fibroblasts cannot be suitably explained with the known physiological functions of EGF.

To determine the role of EGF-R on PDL fibroblasts, the expression of EGF-R on rat PDL fibroblastic cells was investigated in vitro in the absence or presence of Dex, EGF, or both. Dex was used because of its ability to induce osteogenic cell differentiation and elevate bone-specific marker proteins such as alkaline phosphatase (ALP), osteopontin, and osteocalcin (1, 24, 25, 31, 33). Treatment of PDL fibroblastic cells with Dex for 7 days elevated the ALP activity from 33.6 to 70.8 U/10^6 cells, whereas EGF or transforming growth factor α treatment reduced ALP activity to 19.4 U/10^6 cells. Treatment of the cells with a combination of Dex and EGF decreased the Dex-induced ALP activity to 41.8 U/10^6 cells at 7 days. In contrast, ROS cells, which are fully differentiated osteoblastic cells, do not express EGF-R and maintained a high ALP activity throughout culture, and the activity was not affected by EGF treatment (28). We used an ^{125}I-EGF binding assay, immunoprecipitation, and Northern (RNA) blot analysis (28) to investigate whether EGF-R expression is regulated by the treatment of EGF or Dex. Scatchard analysis of the binding assay showed that untreated PDL fibroblastic cells have both high- and low-affinity forms of EGF-R. Treatment of the cells with Dex did not change the dissociation constant for the receptors but reduced the total number of both forms of EGF-R on the cells equally by 37%. The data from both Northern blot and immunoprecipitation analyses demonstrated that the decreased ^{125}I-EGF-binding capacity observed in Dex-treated PDL fibroblastic cells resulted from the decreased synthesis of mRNA for EGF-R and the receptor protein.

On the basis of these results, we conclude that the up-regulation of EGF-R on PDL fibroblasts is associated with maintaining the cells in an undifferentiated state, while the down-regulation of EGF-R is related to the differentiation of cells into osteoblasts and/or cementoblasts (28).

Significance of EGF-R Expression on PDL Fibroblastic Cells

The alveolar bone undergoes continued remodeling both in physiological conditions and, more actively, under orthodontic treatment. The paravascular cells in the PDL are known to differentiate into osteoblasts (30, 34, 36). Also, cementum increases in thickness with age under normal physiological conditions (42). In pathological conditions following injury of the periodontium, deposition of a newly formed cementum layer was observed on dentin (2). In both cases, however, the formation of cementum was associated with the fibroblasts, not with cells with a cementoblastic phenotype. These observations suggested that PDL fibroblasts have the ability to differentiate into cementoblasts responsible for cementum formation (2, 10, 42), and thus, the PDL is likely a source of precursor cells for both osteoblasts and cementoblasts. The expression of numerous EGF-R on fibroblastic cells in the PDL, including PDL fibroblasts, paravascular cells, and preosteoblasts, suggests that these cells represent relatively undifferentiated cells and that the maintenance of their phenotype is under the control of the EGF-R/EGF system. In consequence, the EGF-R/EGF system on these cells may play an important role in maintaining the fibroblastic cell population in the PDL and in balancing formation of the alveolar bone and cementum by regulating their differentiation into osteoblasts and cementoblasts, respectively.

CONCLUSION

This chapter supports the general concept that the application of growth factors is useful for the promotion of periodontal regeneration. Also, information on the specific effects of growth factors on PDL fibroblastic cells allows us to select the proper growth factors for effective clinical application depending on the purpose. PDGF-BB, with the most potent mitogenic, chemotactic, and synthetic effects on PDL fibroblasts, may have the potential to achieve rapid repopulation of PDL fibroblastic cells and, consequently, rapid repair of the PDL. IGF-I may be useful for promoting cementogenesis and/or osteogenesis, while EGF may have clinical value in controlling ankylosis.

Acknowledgments. We thank C.-H. Heldin (Ludwig Cancer Research Institute, Uppsala, Sweden) for his generous supply of antibodies against PDGF α and β receptors.

This study was supported in part by a research grant from Sunstar Inc., Japan, and by U.S. Public Health Service grant DE 04898.

REFERENCES

1. **Aronow, M. A., L. C. Gerstenfeld, T. A. Owen, M. S. Tassinaria, G. S. Stein, and J. B. Lian.** 1990. Factors that promote progressive development of the osteoblast phenotype in cultured fetal calvaria cells. *J. Cell Physiol.* **143:**213–221.
2. **Beertsen, W., and V. Evert.** 1990. Formation of acellular root cementum in relation to dental and non-dental hard tissues in the rat. *J. Dent. Res.* **69:**1669–1673.
3. **Betsholtz, C., J. Johnson, C.-H. Heldin, B. Westermark, P. Lind, M. S. Urdea, R. Eddy, T. B. Shows, K. Philpott, A. L. Mellor, T. Z. Knott, and J. Scott.** 1986. cDNA sequence and chromosomal location of human platelet-derived growth factor A-chain and its expression in tumor cell lines. *Nature* (London) **320:**695–699.
4. **Bishayee, S., S. Majumdar, J. Khire, and M. Das.** 1989. Ligand induced dimerization of the platelet-derived growth factor receptor: monomer-dimer interconversion occurs independent of receptor phosphorylation. *J. Biol. Chem.* **264:**11699–11705.
5. **Canalis, E.** 1980. Effect of insulin-like growth factor-I on DNA and protein synthesis in cultured rat calvaria. *J. Clin. Invest.* **66:**709–719.
6. **Canalis, E., and J. B. Lian.** 1988. Effects of bone associated growth factors on DNA, collagen, and osteocalcin synthesis in cultured fetal rat calvaria. *Bone* **9:**243–246.
7. **Carpenter, G., and M. I. Wahl.** 1990. Epidermal growth factor. *Handbk. Exp. Pharmacol.* **95:**67–171.
8. **Centrella, M., T. L. McCarthy, and E. Canalis.** 1987. Transforming growth factor-β is a bifunctional

regulator of replication and collagen synthesis in osteoblast-enriched cell cultures from fetal rat bone. *J. Biol. Chem.* **262:**2869–2874.

9. **Cho, M. I., and P. R. Garant.** 1981. Role of microtubules in the organization of the Golgi complex and the secretion of collagen secretory granules by periodontal ligament fibroblasts. *Anat. Rec.* **199:**459–471.

10. **Cho, M. I., and P. R. Garant.** 1989. Radioautographic study of ³H-mannose utilization during cementoblast differentiation, formation of acellular cementum, and development of periodontal ligament principal fibers. *Anat. Rec.* **223:**209–222.

11. **Cho, M. I., Y. L. Lee, and P. R. Garant.** 1987. Immunocytochemical localization of extracellular matrix components. I. Collagen types I and III in healthy gingival connective tissue. *J. Periodontal Res.* **22:**313–319.

12. **Cho, M. I., Y. L. Lee, and P. R. Garant.** 1988. Radioautographic demonstration of receptors for epidermal growth factor in various cells of the oral cavity. *Anat. Rec.* **231:**14–24.

13. **Cho, M. I., Y. L. Lee, and P. R. Garant.** 1988. Periodontal ligament fibroblasts, preosteoblasts, and prechondrocytes express receptors for epidermal growth factor in vivo: a comparative radioautographic study. *J. Periodontal Res.* **23:**287–294.

14. **Cho, M. I., W.-L. Lin, and P. R. Garant.** 1991. Occurrence of epidermal growth factor-binding sites during differentiation of cementoblasts and periodontal ligament fibroblasts of the young rat: a light and electron microscopic radioautographic study. *Anat. Rec.* **231:**14–24.

15. **Claesson-Welsh, L., A. Eriksson, B. Westermark, and C.-H. Heldin.** 1989. cDNA cloning and expression of the human A-type platelet-derived growth factor (PDGF) receptor establishes structural similarity to the B-type PDGF receptor. *Proc. Natl. Acad. Sci. USA* **86:**4917–4921.

16. **Claesson-Welsh, L., A. Hammacher, B. Westermark, C.-H. Heldin, and M. Nister.** 1989. Identification and structural analysis of the A type receptor for platelet-derived growth factor. *J. Biol. Chem.* **264:**1742–1747.

17. **Deuel, T. F., R. S. Kawahara, T. A. Mustoe, and G. F. Pierce.** 1991. Growth factors and wound healing: platelet-derived growth factor as a model cytokine. *Annu. Rev. Med.* **42:**567–584.

18. **Graves, D. T., and D. L. Cochran.** 1990. Mesenchymal cell growth factors. *Crit. Rev. Oral Biol. Med.* **1:**17–36.

19. **Heldin, C.-H., A. Ernlund, C. Rorsman, and L. Ronnstrand.** 1989. Dimerization of β-type PDGF receptors occurs after ligand binding and is closely associated with receptor kinase activation. *J. Biol. Chem.* **264:**8905–8912.

20. **Heldin, C.-H., A. Wasteson, and B. Westermark.** 1985. Platelet-derived growth factor. *Mol. Cell Endocrinol.* **39:**169–187.

21. **Ignotz, R. A., and J. Massague.** 1986. Transforming growth factor-β stimulates the expression of fibronectin and collagen and their incorporation into the extracellular matrix. *J. Biol. Chem.* **26:**4337–4345.

22. **Johnson, A., C.-H. Heldin, A. Wasteson, B. Westermark, T. F. Deuel, J. S. Huang, P. H. Seeburg, A. Gray, A. Ullich, G. Scrace, P. Stroobant, and M. D. Waterfield.** 1984. The c-sis gene encodes a precursor of the B chain of platelet-derived growth factor. *EMBO J.* **3:**921–928.

23. **Josephs, S. F., C. Guo, L. Ratner, and F. Wong-Stall.** 1984. Human proto-oncogene nucleotide sequences corresponding to the transforming region of simian sarcoma virus. *Science* **223:**487–491.

24. **Kasugai, S., R. Todescan, T. Nagata, K.-L. Yao, W. T. Butler, and J. Sodek.** 1991. Expression of bone matrix proteins associated with mineralized tissue formation by adult rat bone marrow cells in vitro: inductive effects of dexamethasone on the osteoblastic phenotype. *J. Cell. Physiol.* **147:**11–120.

25. **Leboy, P. S., J. N. Beresford, C. Devlin, and M. E. Owen.** 1991. Dexamethasone induction of osteoblast mRNAs in rat marrow stromal cell culture. *J. Cell. Physiol.* **146:**370–378.

26. **Lynch, S. E., G. Ruiz de Castilla, R. C. Williams, C. P. Kiristy, T. H. Howell, M. S. Reddy, and H. N. Antoniades.** 1991. The effects of short term application of a combination of platelet-derived and insulin-like growth factors on periodontal wound healing. *J. Periodontol.* **62:**458–468.

27. **Lynch, S. E., R. C. Williams, A. M. Polson, T. H. Howell, M. S. Reddy, U. E. Zappa, and H. N. Antoniades.** 1989. A combination of platelet-derived growth factor and insulin-like growth factor enhances periodontal regeneration. *J. Clin. Periodontol.* **16:**545–548.

28. **Matsuda, N., N. M. Kumar, P. R. Ramakrishnan, W.-L. Lin, R. J. Genco, and M. I. Cho.** 1993. Evidence for up-regulation of epidermal growth factor receptors on rat periodontal ligament fibroblastic cells associated with stabilization of phenotype *in vitro. Arch. Oral Biol.* **38:**559–569.

29. **Matsuda, N., W. L. Lin, N. M. Kumar, M. I. Cho, and R. J. Genco.** 1992. Mitogenic, chemotactic, and synthetic responses of rat periodontal ligament fibroblastic cells to polypeptide growth factors in vitro. *J. Periodontol.* **63:**515–525.

30. **McCulloch, C. A. G., and A. H. Melcher.** 1983. Cell density and generation in the periodontal ligament of mice. *Am. J. Anat.* **167:**43–58.

31. **Nagata, T., C. G. Bellow, S. Kasugai, W. T. Butler, and J. Sodek.** 1991. Biosynthesis of bone sialoproteins [SPP-1 (secreted phosphoprotein-1, osteopontin), BSP (bone sialoprotein) and SPARC (osteonectin)] in association with mineralized tissue formation by fetal rat calvarial cells in culture. *Biochem. J.* **274:**513–520.

32. **Oates, T. W., C. A. Rouse, and D. L. Cochran.** 1993. Mitogenic effects of growth factors on human periodontal ligament cells in vitro. *J. Periodontol.* **64:**142–148.

33. **Owen, T. A., M. Aronow, V. Shalhoub, L. M. Barone, L. Wilming, M. S. Tassinari, M. B. Kennedy, S. Pickwinse, J. B. Lian, and G. S. Stein.** 1990. Progressive development of the rat osteoblast phenotype in vitro: reciprocal relationships in expression of genes associated with osteoblast proliferation and differentiation during formation of the bone extracellular matrix. *J. Cell. Physiol.* **143:**420–430.

34. **Roberts, W. E., and D. C. Chase.** 1981. Kinetics of cell proliferation and migration associated with orthodontically induced osteogenesis. *J. Dent. Res.* **60:**174–181.

35. **Roberts, W. E., and W. S. S. Jee.** 1974. Cell kinetics of orthodontically-stimulated and nonstimulated periodontal ligament in the rat. *Arch. Oral Biol.* **19:**17–21.

36. **Roberts, W. E., P. G. Mozsary, and E. Klingler.** 1982. Nuclear size as a cell-kinetic marker for osteoblast differentiation. *Am. J. Anat.* **165:**373–384.

37. **Ross, R., E. W. Raines, and D. F. Bowen-Pope.** 1986. The biology of platelet-derived growth factor. *Cell* **46:**155–169.

38. **Russel, S. M., and E. M. Spencer.** 1985. Local injections of human or rat growth hormone or purified human somatomedin C stimulate unilateral tibial epiphyseal growth in hypophysectomized rats *Endocrinology* **116:**2563–2565.

39. **Rutherford, R. B., C. E. Nierkrash, J. E. Kennedy, and M. F. Charette.** 1992. Platelet-derived and insulin-like growth factors stimulate regeneration of periodontal attachment in monkeys. *J. Periodontal Res.* **27:**285–290.

40. **Rutherford, R. B., M. D. TrailSmith, M. E. Ryan, and M. F. Charette.** 1992. Synergistic effects of dexamethasone on platelet-derived growth factor mitogenesis in vitro. *Arch. Oral Biol.* **37:**139–145.

41. **Scheven, B. A., and N. J. Hamilton.** 1991. Longitudinal bone growth in vitro: effects of insulin-like growth factor-I and growth hormone. *Acta Endocrinol.* **124:**602–607.

42. **Schroeder, H.** 1986. *The Periodontium,* p. 29–30. Springer Verlag, Berlin.

43. **Seifert, R. A., C. E. Hart, P. E. Phillips, J. W. Forstrom, R. Moss, M. J. Murray, and D. F. Bowen-Pope.** 1989. Two different subunits associate to create isoform-specific platelet-derived growth factors. *J. Biol. Chem.* **264:**8771–8778.

44. **Sieghbahn, A., A. Hammacher, B. Westermark, and C.-H. Heldin.** 1990. Differential effects of the various isoforms of platelet-derived growth factor on chemotaxis of fibroblasts, monocytes and granulocytes. *J. Clin. Invest.* **85:**916–920.

45. **Spencer, E. M., C. C. Liu, and G. A. Howard.** 1991. In vivo actions of insulin-like growth factor-I (IGF-I) on bone formation and resorption in rats. *Bone* **12:**21–26.

46. **Terranova, V. P., and U. M. E. Wikesjö.** 1987. Extracellular matrices and polypeptide growth factors as mediators of functions of cells of the periodontium. *J. Periodontol.* **58:**371–380.

47. **Thiebaud, D., K. W. Ng, D. M. Findlay, M. Harker, and T. J. Martin.** 1990. Insulin-like growth factor-I regulates mRNA levels of osteonectin and pro-α1 (I)-collagen in clonal preosteoblastic calvaria cells. *J. Bone Miner. Res.* **5:**761–767.

48. **Westermark, B.** 1990. The molecular and cellular biology of platelet-derived growth factor. *Acta Endocrinol.* **123:**131–142.

49. **Westermark, B., A. Siegbahn, C.-H. Heldin, and L. Claesson-Welsh.** 1990. β-type receptor for platelet-derived growth factor mediates a chemotactic response by means of ligand-induced activity of the receptor protein-tyrosine kinase. *Proc. Natl. Acad. Sci. USA* **87:**128–132.

50. **Wrana, J. E., M. Maeno, B. Hawrylyshyn, K. Yao, C. Domenicucci, and J. Sodek.** 1988. Differential effects of transforming growth factor-β on the synthesis of extracellular matrix proteins by normal fetal rat calvarial bone cell population. *J. Cell Biol.* **106:**915–924.

Molecular Pathogenesis of Periodontal Disease
Edited by Robert Genco et al.
© 1994 American Society for Microbiology, Washington, DC 20005

Chapter 33

Polypeptide Growth Factors: Molecular Mediators of Tissue Repair

Samuel E. Lynch and William V. Giannobile

Procedures used to treat periodontal diseases are currently deemed clinically successful if they slow the progression of the destructive process or aid in restoring a portion of one or more components of the periodontium. However, periodontal disease remains the leading cause of tooth loss in adults. Clearly, there is a need for new procedures or materials that promote periodontal wound healing and lead to restoration of the periodontal attachment apparatus. In this regard, a multitude of new procedures and materials are being tested for their abilities to promote periodontal wound healing, the ultimate goal being to develop a method to regenerate the entirety of the lost periodontium. Perhaps the most potent wound-healing agents currently being studied in dentistry and medicine are polypeptide growth factors.

The discovery of polypeptide growth factors over the past decade has led to major research in all areas of medicine on the effect of these agents on embryonic development and on wound healing. Growth factors appear to be nature's wound-healing hormones. They are potent regulators of the three key cellular events in tissue repair: mitogenesis (proliferation), migration, and metabolism. These events are also the key to regeneration of periodontal attachment apparatus structures. Preclinical results indicate that some of the polypeptide growth factors, in particular a combination of platelet-derived growth factor (PDGF) and insulin-like growth factor I (IGF-I), are potent stimulators of the proliferation and directed migration of periodontal ligament (PDL) and bone cells in vitro and of new PDL, cementum, and alveolar bone in vivo. This chapter summarizes recent developments utilizing polypeptide growth factors, including osteoinductive factors, for the promotion of tissue repair, with emphasis on their role in periodontal wound healing and regeneration of new attachment structures. Table 1 lists the best-characterized growth factors and their common abbreviations.

Samuel E. Lynch • Institute of Molecular Biology, Inc., One Innovation Drive, Worcester, Massachusetts 01605-4308.   ***William V. Giannobile*** • Harvard School of Dental Medicine, Boston, Massachusetts 02115.

Table 1. Growth factors used to promote healing

Name	Abbreviation
Platelet-derived growth factors	PDGF
Insulin-like growth factors I and II	IGF-I, IGF-II
Transforming growth factors β	TGF-β
Transforming growth factor α	TGF-α
Fibroblast growth factors (acidic and basic)	aFGF, bFGF
Epidermal growth factor	EGF
Nerve growth factor	NGF
Interleukin-1	IL-1
Bone morphogenetic proteins	BMP-1 to BMP-9

WOUND HEALING

Healing of wounds in both soft (e.g., gingival) and hard (e.g., alveolar bone) tissues involves a series of well-orchestrated, complex cell-cell and cell-macromolecule interactions. In the natural wound-healing process, multiple growth factors act in concert to form an intricate molecular language regulating the activity of cells within and adjacent to a wound (for reviews, see references 18 and 28). Normally, after acute injury involving subepithelial tissues, disruption of the wound vasculature leads to platelet aggregation. Activated platelets at the wound site release several important growth factors including PDGF, transforming growth factor β (TGF-β), an epidermal growth factor, (EGF)/TGF-α-like growth factor, and an endothelial cell growth factor (3). In addition, the plasma exudate provides an important source of IGFs (46). Cells immediately adjacent to the injured site also begin producing growth factors such as PDGF, IGF-I, TGF-α, and TGF-β within a few hours after injury (1, 2, 20, 51). Macrophages, which normally arrive at the wound site 2 to 4 days following injury and are responsible for cleaning away damaged tissue, provide another source of growth factors (45). Bone itself contains substantial amounts of several growth factors including TGF-β, PDGF, IGF, and fibroblast growth factor (FGF) in concentrations up to 20-fold greater than those in serum (21). Thus, several growth factors are specifically delivered to, produced at, or released from sites of acute injury. It is now thought that their presence provides the key stimulant to initiate the normal repair process.

The hypothesis currently being evaluated in periodontics is that periodontal disease, like other chronic nonhealing wounds (e.g., chronic skin ulcers), does not involve an acute injury with hemorrhage and activation of the typical acute-injury repair response that leads to the presence of relatively large quantities of growth factors at the wounded site. Rather, owing to the chronic nature of periodontal disease, the presence of growth factors at the periodontal lesion may be limited. Furthermore, although periodontal surgery induces an acute injury, the supply of growth factors released may be only sufficient in magnitude or quality to stimulate repair of the gingiva without stimulating to any great extent the cells within the remaining PDL and bone. The very limited activation of the PDL and bone cells following flap surgery has been documented (4). The addition or supplementation of appropriate growth factors at the defect site during flap surgery may thus promote proliferation of PDL and bone cells, stimulate their migration coronally, and increase their production of tissue components such as collagen and bone matrix, leading to increased regeneration of the periodontium.

Because periodontal therapy involves manipulations of both soft and hard tissues, it is important to understand the effects of growth factors on the healing of both tissue types.

GROWTH FACTORS AND SOFT TISSUE HEALING

It is now clear that the addition of several growth factors can accelerate many of the components of soft tissue healing, including reepithelialization, angiogenesis (new blood vessel formation), and provisional extracellular matrix and collagen synthesis. As shown in Table 2, the components that are most affected depend upon which growth factors are applied. Extensive wound-healing studies in animals have demonstrated that PDGF, EGF, TGF-α, TGF-β, and the FGFs can stimulate soft tissue wound repair (16, 19, 22, 32, 33, 38, 44, 49). The most potent are combinations of these factors such as PDGF plus IGF-I, PDGF plus TGF-α, and TGF-β plus IGF-I (32, 33).

Of the nonmineralized tissues of the periodontium, the PDL is the most problematic regarding regeneration. The PDL is composed, at least in part, of fibroblasts, blood vessels, nerves, and collagen. To achieve its full restoration following periodontal disease, its coronal reestablishment must be promoted. Thus, growth factors that promote PDL fibroblast proliferation and migration as well as collagen synthesis appear to be good candidates for enhancing new PDL formation (except that for a true new PDL to form, there must also be new cementum and new bone formation, a problem discussed below). The effects of growth factors on PDL-derived cells are summarized in Table 3. PDGF and TGF-β stimulate collagen synthesis by PDL-derived cells (36). Of the best-characterized growth factors, PDGF, IGF-I, and FGF are the most potent chemoattractants for fibroblasts derived

Table 2. Effects of growth factors on major soft tissue repair processes[a]

Growth factor	Reepithelialization	Angiogenesis	Provisional extracellular matrix	Collagen
PDGF	0/+[b]	+	+	+
TGF-β	+[c]/−	+ +	0/+[d]	+ + +
EGF or TGF-α	+ +	0/+	0	0
IGF-I	0	0	0	+
bFGF	+	+ + +	+ +	−
PDGF + IGF-I	+ +	+ +	+ +	+ + +
PDGF + TGF-α	+ +	+ +	+ +	+ +

[a] Adapted from reference 28. Symbols: +, marginally increased; + +, modestly increased; + + +, substantially increased; −, decreased; 0, no effect.
[b] Secondary to increased extracellular matrix and collagen production in impaired wound-healing methods.
[c] The effects of TGF-β1 on epithelialization appear to be dose and time dependent. TGF-β1 may increase reepithelialization secondary to increased granulation tissue production and epithelial cell migration, but its direct effect on epithelial proliferation is inhibitory.
[d] Primarily fibronectin.

Table 3. Effects of growth factors on PDL cells[a]

Growth factor	Mitogenesis	Migration	Metabolism[b]
PDGF	+ +	+ + +	+
IGF-I	+	+ + +	+
FGF	+ +	+ + +	−
TGF-β	−	0	+
EGF	+	0	0
PDGF plus IGF	+ + +	+ + +	+
TGF-β plus EGF	−	0	?

[a] Information is derived from Matsuda et al. (36), Terranova et al. (55), and Lynch et al. (30a). Symbols: −, inhibitory effect; 0, no effect; +, slight effect; + +, moderate effect; + + +, strong effect.
[b] As assessed by either total protein or collagen content.

from the PDL (36, 55). These growth factors are approximately 10,000 times more potent than fibronectin in inducing directed fibroblast migration. PDGF, IGF-I, or FGF can also stimulate proliferation of PDL-derived fibroblasts. No amount of fibronectin can stimulate the proliferation of fibroblasts to the extent achieved with nanogram amounts of these growth factors. PDGF can reduce the inhibitory effects of lipopolysaccharide on gingival fibroblast proliferation, thereby serving as another mechanism by which this growth factor exerts beneficial effects in wound healing in the periodontium (6). TGF-β and EGF do not, however, stimulate proliferation or migration of PDL-derived cells (36).

Greater potency can be achieved by combining growth factors. For example, PDGF and IGF-I in combination interact synergistically to promote greater proliferation of several types of mesenchymal cells and greater collagen deposition and more bone formation than can be achieved with either PDGF or IGF-I individually (29). Specifically, PDL fibroblasts proliferate more rapidly when exposed to the combination of PDGF and IGF-I than when exposed to these factors singularly (36). The PDGF–IGF-I combination stimulates periodontal regeneration and new attachment formation in dogs with naturally occurring disease and in non-human primates with ligature-induced disease, respectively (34, 35, 47), and promotes bone formation around endosseous dental implants in dogs (7, 31).

GROWTH FACTORS AND BONE HEALING

It has also now been clearly established that certain growth factors regulate bone formation and resorption. Alveolar bone formation occurs by the intramembranous route, as does bone formation in the calvariae. Numerous studies have been performed in vitro to determine the effects of growth factors on bone formation in calvariae (for reviews see references 11, 21, 30, and 37). The findings from these studies are summarized in Table 4 (11, 30). Acidic FGF (aFGF), basic FGF (bFGF), and PDGF increase a cell population of the osteoblast lineage capable of synthesizing collagen but do not directly increase collagen production by differentiated osteoblasts (5, 10). The effects of TGF-β appear to be highly dependent on the osteoblast cell type, dose applied, and local environment. TGF-β has been

Table 4. Effects of growth factors on bone formation and resorption[a]

Growth factor	Matrix synthesis		Matrix degradation or bone resorption	Osteoblast migration
	Direct effect	Secondary to mitogenesis		
aFGF	−	+	0	?
bFGF	−	+	0	?
PDGF	+/0	+	+	+
TGF-β	+/−[b]	+/[b]	+/−[b]	+
IGF-I	+	+	−	+
PDGF + IGF-I	+	+ +	−	?
TGF-β + IGF-I	+	+ +	?	?
PDGF + IGF-I + TGF-β	+	+ + +	?	?

[a] Adapted from reference 11. Symbols: +, increased; −, decreased; 0, no effect; ?, not determined.
[b] Dose and local environment dependent.

reported to stimulate or inhibit osteoblast proliferation (13, 52). TGF-β stimulates type I collagen, fibronectin, and osteonectin biosynthesis as well as bone matrix deposition (14, 42, 60). In addition, TGF-β decreases synthesis of metalloproteinases and plasminogen activator and increases synthesis of tissue inhibitor of metalloproteinase and plasminogen activator inhibitor, thus resulting in a decrease in connective tissue matrix degradation (17, 27, 40). TGF-β also appears to inhibit formation of osteoclast-like cells but may promote bone resorption via a prostaglandin-mediated mechanism (39, 54). Given TGF-β's diverse effects on bone matrix formation and resorption and its abundance in bone, it may act as a bone coupling factor, linking bone resorption and bone formation (8). The net result of repeated injections of TGF-β in animals is formation of cartilage, which, following cessation of injections, becomes bone via endochondral ossification (24, 25). In the same model system, injections of PDGF resulted in intramembranous bone formation, i.e., bone formation in the absence of cartilage (24).

The combination of PDGF and IGF-I increases bone matrix apposition in calvarial organ culture more than PDGF, IGF-I, or TGF-β individually (42). Likewise, the combination of TGF-β and IGF-I is more potent in stimulating matrix apposition than any individual growth factors (42). PDGF alone, similar to TGF-β, may promote calcium release via a prostaglandin-mediated mechanism; however, this effect is diminished when PDGF is combined with IGF-I (10, 15). Maximal proliferation of cultured adult human osteoblasts was seen with a combination of PDGF, IGF-I, TGF-β, and EGF (43).

Studies of the chemotactic effects of growth factors on osteoblasts would be particularly valuable to those hoping to use growth factors to achieve periodontal regeneration. The ability to enhance directed migration of osteoblasts is particularly important for periodontal regeneration, because bone is often present only at the apical portion of the periodontal lesion. Thus, for complete regeneration to occur, the bone-forming cells must migrate coronally, often several millimeters. Without potent chemotactic agents, it does not appear that such coronal migration of adult human osteoblasts will occur even in the presence of putative osteoconductive materials or barrier membranes. Recently, the effects of one growth factor (PDGF) on osteoblast influx or migration have been evaluated. It was demonstrated that bone cells migrate in a dose-dependent manner toward an increasing gradient of PDGF (56).

PERIODONTAL REGENERATION

The above information suggests advantages and disadvantages to the use of certain growth factors for periodontal regeneration. EGF and TGF-α, at least by themselves, do not appear likely to enhance periodontal regeneration, because both promote bone resorption, are weak promoters of connective tissue formation, and are strong promoters of epithelial migration and proliferation. aFGF and bFGF will likely promote the coronal migration of PDL fibroblasts. Since coronal migration of PDL cells is thought to be a prerequisite for new attachment, perhaps the FGFs could promote regeneration. FGFs may also have a role in regeneration because of their ability to stimulate neovascularization, a critical event in the process of wound healing (5). However, other studies indicate they may reduce collagen production by fibroblasts. Whether this reduction in collagen is significant enough to affect PDL and bone formation is not known. Both aFGF and bFGF are likely to promote proliferation of alveolar bone cells, which may indirectly lead to increased bone matrix formation, but no in vivo data for assessing the potency of FGF in this regard are currently available. TGF-β is the most potent individual growth factor for stimulating bone matrix apposition in culture and thus would likely be a potent stimulus for alveolar bone formation. However, the finding that TGF-β is not chemotactic for PDL cells raises some concerns, since the net result of rapid bone growth in the absence of coronal migration of PDL cells could be ankylosis. Precise titration of the amount of TGF-β administered may also be required to achieve cell stimulation and allow mineralization of the bone matrix to progress normally.

PDGF and IGF-I are chemotactic for PDL fibroblasts and thus will likely promote the coronal migration of PDL cells (36). In addition, in vitro data show that PDGF and IGF individually promote proliferation of osteoblasts and PDL-derived fibroblasts (10, 36) and that PDGF reduces the inhibitory effects of LPS on fibroblast proliferation (6). IGF-I primarily promotes collagen synthesis, and PDGF promotes noncollagen protein synthesis during bone matrix formation (10). Nearly all of these activities are increased even further when PDGF and IGF-I are combined (32, 33, 42, 53). The findings that (i) PDGF and IGF-I are chemotactic for fibroblasts derived from the PDL, (ii) PDGF and IGF-I interact synergistically to stimulate the proliferation of PDL-derived cells, (iii) the combination of PDGF and IGF-I promotes greater bone matrix formation than any individual growth factors, and (iv) PDGF and IGF-I interact synergistically to promote collagen deposition in vivo suggest that the PDGF–IGF-I combination would enhance regeneration of all the components of the periodontium.

Indeed, preclinical animal studies have shown that the PDGF–IGF-I combination has the capacity to stimulate new bone and cementum formation. When this growth factor combination was used to treat natural periodontal lesions in beagle dogs, substantial amounts of new bone and cementum were seen within 2 weeks following a single application (35). By 5 weeks after treatment, 40.6% defect fill was observed in treated sites compared to 6.9% defect fill in control sites (34). Normal maturation of the bone occurred within 3 to 6 months following treatment (30). A functionally oriented PDL connecting the new bone to the root surface was present. There was no increase in ankylosis. Independent studies with non-human primates have verified the substantial increase in new attachment that can be achieved with the use of the PDGF–IGF-I combination (48). A recent report evaluated the effects of TGF-β1, bFGF, and IGF-II on osseous regeneration in periodontal defects in dogs. The results demonstrated no enhancement in bone formation 4 weeks after a single application of these factors (50). Other purified growth factors and combinations remain to be tested in periodontal defects in vivo.

OSTEOINDUCTIVE FACTORS

Osteoinductive factors may have the potential to stimulate alveolar bone repair. Unlike the growth factors described above, the main action of osteoinductive factors is to cause undifferentiated pluripotential cells to differentiate into cartilage and bone-forming cells (for a review, see reference 58). In general, these factors do not stimulate proliferation and migration of osteoblasts. Bone morphogenetic proteins (BMP) are the classic example of osteoinductive factors. It now appears that there are at least nine BMP-like molecules (12). Other examples (Table 1) include osteogenin (BMP-3) (26) and osteoinductive protein (OP-1) (41). Osteoinductive proteins were first isolated from demineralized bone matrix. When partially purified BMP from demineralized bone matrix is implanted intraperitoneally into rats in a bone-derived collagenous matrix, it stimulates formation of a nodule of cartilage that undergoes remodeling into bone within the surrounding soft connective tissues (i.e., ectopic bone formation) (57). A deficiency of BMP-like proteins retards bone cell differentiation and may account for a failure of fractures to heal (58). Crude human BMP has been used to promote the healing of nonunion fracture sites (23). More recently, the use of recombinant proteins (e.g., BMP-2) has been shown to promote osseous union of segmental long-bone defects (61).

It may be reasonable to expect that BMP-like factors will stimulate bone formation if placed around periodontally involved teeth. A recent study using osteogenin found that it augmented new bone and cementum deposition when combined with demineralized freeze-dried bone (but not Collaplug) around sub-merged teeth (9). It did not significantly enhance new bone or cementum formation around nonsubmerged teeth compared to demineralized freeze-dried bone allo-grafts alone (9). Interestingly, ankylosis was observed in submerged defects grafted with demineralized bone plus osteogenin. This is not unexpected, since BMP-like molecules do not stimulate soft tissue or PDL formation and in fact act to cause bone formation within soft tissues. Within the PDL, this may lead to ankylosis. BMPs appear more likely to have a role in promoting bone formation around titanium dental implants (48).

Both growth factors and BMPs are found within bone matrix. It is possible, therefore, that demineralized freeze-dried bone contains small amounts of these proteins and that they are responsible for the bone-stimulating activity of demineralized freeze-dried bone allografts.

In summary, polypeptide growth factors possess greater abilities to control cell mitogenesis, migration, and metabolism than any other class of molecules currently in use in dentistry or medicine. Both in vitro cell and organ culture studies as well as in vivo wound healing studies in dogs and primates indicate that certain growth factors may be potent stimulators of periodontal wound healing. Their clinical applicability to periodontics awaits their large-scale production, purification, and formulation into delivery systems that control and maximize their actions. If these obstacles can be overcome, growth factors hold great promise for providing the periodontist with a means of stimulating periodontal regeneration.

Acknowledgments. This work was supported by grants from the NIH (DE08878), NIH Dentist-Scientist Award (DE00275-03), and the Institute of Molecular Biology, Inc.

REFERENCES

1. **Antoniades, H. N., T. Galanopoulos, J. Neville-Golden, C. P. Kiritsy, and S. E. Lynch.** 1991. Injury reduces *in vivo* expression of platelet-derived growth factor (PDGF) and PDGF receptor mRNAs in skin epithelial cells and PDGF and mRNA in connective tissue fibroblasts. *Proc. Natl. Acad. Sci. USA* **88**:565–569.

2. **Antoniades, H. N., T. Galanopoulos, J. Neville-Golden, C. P. Kiritsy, and S. E. Lynch.** 1993. Expression of growth factor and receptor mRNAs in skin epithelial cells following acute cutaneous injury. *Am. J. Pathol.* **142**:1099–1109.

3. **Assoian, R. K., G. R. Grotendorst, D. M. Muller, and M. B. Sporn.** 1984. Cellular transformation by coordinated action of three peptide growth factors from human platelets. *Nature* (London) **309**:804–806.

4. **Aukhil, I., and J. Iglhaut.** 1988. Periodontal ligament cell kinetics following experimental regenerative procedures. *J. Clin. Periodontol.* **15**:374–382.

5. **Baird, A., F. Esch, P. Mormede, L. N. Uenon, P. Bohlen, S. Y. Ying, W. B. Wehrenberg, and R. Guillemin.** 1986. Molecular characterization of fibroblast growth factor: distribution and biological activities in various tissues. *Recent Prog. Hormone Res.* **42**:143–200.

6. **Bartold, P. M., A. S. Narayanan, and R. C. Page.** 1992. Platelet-derived growth factor reduces the inhibitory effects of lipopolysaccharide on gingival fibroblast proliferation. *J. Periodontal Res.* **27**:499–505.

7. **Becker, W., S. E. Lynch, U. Lekholm, B. E. Becker, R. Caffesse, K. Donath, and R. Sanchez.** 1992. A comparison of ePTFE membranes alone or in combination with platelet derived growth factors and insulin-like growth factor-I or demineralized freeze-dried bone in promoting bone formation around immediate extraction socket implants. *J. Periodontol.* **63**:929–940.

8. **Bonewald, L. F., and G. R. Mundy.** 1990. Role of transforming growth factor-beta in bone remodeling. *Clin. Orthop.* **250**:261.

9. **Bowers, G., F. Felton, C. Middleton, P. Glynn, S. Sharp, J. Mellonig, R. Corio, J. Emerson, S. Park, J. Suzuki, S. Ma, E. Romberg, and A. H. Reddi.** 1991. Histologic comparison of regeneration in human intrabony defects when osteogenin is combined with demineralized freeze-dried bone allograft and with purified bovine collagen. *J. Periodontol.* **62**:690–702.

10. **Canalis, E., T. L. McCarthy, and M. Centrella.** 1988. Effects of platelet-derived growth factor on bone formation in vitro. *J. Cell. Physiol.* **140**:530.

11. **Canalis, E., T. L. McCarthy, and M. Centrella.** 1989. Growth factors and the skeletal system. *J. Endocrinol. Invest.* **12**:577–584.

12. **Celeste, A. J., J. A. Iannazzi, R. C. Taylor, R. M. Hewick, V. Rosen, E. A. Wang, and J. M.**

Wozney. 1990. Identification of new TGF-β family members present in bone-inductive protein purified from bovine bone. *Proc. Natl. Acad. Sci. USA* **87:**9843–9847.

13. **Centrella, M., T. L. McCarthy, and E. Canalis.** 1987. Transforming growth factor-beta is a bifunctional regulator or replication and collagen synthesis in osteoblast-enriched cell cultures from fetal rat bone. *J. Biol. Chem.* **262:**2869–2874.

14. **Centrella, M., T. L. McCarthy, and E. Canalis.** 1991. Transforming growth factor-beta and remodeling of bone. *J. Bone Joint Surg.* **73A:**1418–1428.

15. **Cochran, D. L., C. A. Rouse, S. E. Lynch, and D. T. Graves.** 1993. Effects of PDGF isoforms on calcium release from neonatal mouse calvaria. *Bone* **14:**53–58.

16. **Davidson, J., M. Klagsbrun, K. Hill, A. Buckley, R. Sullivan, P. Brewer, and S. Woodward.** 1985. Accelerated wound repair, cell proliferation, and collagen accumulation are produced by a cartilage-derived growth factor. *J. Cell Biol.* **100:**1219–1227.

17. **Edwards, D. R., G. Murphy, J. J. Reynolds, S. E. Whitham, A. J. P. Docherty. P. Angel, and J. K. Heath.** 1987. Transforming growth factor beta modulates expression of collagenase and metalloproteinase inhibitor. *EMBO J.* **6:**1899–1904.

18. **Graves, D. T., and D. L. Cochran.** 1990. Mesenchymal cell growth factors. *Crit. Rev. Oral Biol. Med.* **1:**17–36.

19. **Greenhalph, D. B., K. H. Sprugel, M. J. Murray, and R. Ross.** 1990. PDGF and FGF stimulate wound healing in the genetically diabetic mouse. *Am. J. Pathol.* **136:**1235–1246.

20. **Hansson, H. A., E. Jennische, and A. Skottner.** 1987. Regenerating endothelial cells express IGF-I immunoreactivity after arterial injury. *Cell Tissue Res.* **250:**499–505.

21. **Hauschka, P. V.** 1990. Growth factor effects in bone, p. 103–170. *In* B. K. Hall (ed.), *Bone.* Telford Press, Caldwell, N.J.

22. **Hebda, P. A.** 1988. Stimulatory effects of transforming growth factor-β and epidermal cell outgrowth from porcine skin explant cultures. *J. Invest. Dermatol.* **91:**440–445.

23. **Heckman, J. D., B. D. Boyan, T. B. Aufdemorte, and J. T. Abbott.** 1991. The use of bone morphogenetic protein in the treatment of non-union in a canine model. *J. Bone Joint Surg.* **73A:**750–764.

24. **Joyce, M. E., S. Jingushi, S. P. Scully, and M. E. Bolander.** 1990. Role of growth factors in fracture healing, p. 391–416. *In* M. Barbul (ed.), *Clinical and Experimental Approaches to Dermal and Epidermal Repair.* Wiley-Liss, New York.

25. **Joyce, M. E., A. B. Roberts, M. B. Sporn, and M. E. Bolander.** 1990. Transforming growth factor-β and the initiation of chondrogenesis and osteogenesis in the rat femur. *J. Cell Biol.* **110:**2195–2207.

26. **Katz, R. W., and A. H. Reddi.** 1988. Dissociative extraction and partial purification of osteogenin, a bone inductive protein, from rat tooth matrix by heparin affinity chromatography. *Biochem. Biophys. Res. Commun.* **157:**1253.

27. **Keska-Oja, J., R. Raghow, M. Sawdey, D. J. Loskutoff, A. E. Postlethwaite, A. H. Kang, and H. L. Moses.** 1988. Regulation of the mRNAs for type I plasminogen activator inhibitor, fibronectin, and type I procollagen by transforming growth factor-β. *J. Biol. Chem.* **263:**3111–3115.

28. **Kiritsy, C. P., and S. E. Lynch.** 1993. The role of growth factors in cutaneous wound healing: a review. *Crit. Rev. Oral Biol. Med.* **5:**21–52.

29. **Lynch, S. E.** 1991. Platelet-derived growth factor and insulin-like growth factor I: mediators of healing in soft tissue and bone wounds. *N. E. Soc. Periodont.* **13:**13–20.

30. **Lynch, S. E.** Growth factors: molecular modulators of repair and regeneration. *In* A. M. Polson (ed.), *Current Techniques and Concepts in Periodontal Therapy,* in press. Alan R. Liss, Inc., New York.

30a. **Lynch, S. E., et al.** Unpublished observations.

31. **Lynch, S. E., D. Buser, R. A. Hernandez, H. P. Weber, H. Stich, C. H. Fox, and R. C. Williams.** 1991. Effects of the platelet-derived growth factor/insulin-like growth factor combination on bone regeneration around titanium dental implants. Results of a pilot study in beagle dogs. *J. Periodontol.* **62:**710–716.

32. **Lynch, S. E., R. B. Colvin, and H. N. Antoniades.** 1989. Growth factors in wound healing: single and synergistic effects on partial thickness porcine skin wounds. *J. Clin. Invest.* **84:**640–646.

33. **Lynch, S. E., J. C. Nixon, R. B. Colvin, and H. N. Antoniades.** 1987. Role of platelet-derived

growth factor in wound healing: synergistic effects with other growth factors. *Proc. Natl. Acad. Sci. USA* **84**:7696–7700.

34. **Lynch, W. E., G. Ruiz de Castilla, R. C. Williams, C. P. Kiritsy, T. H. Howell, M. S. Reddy, and H. N. Antoniades.** 1991. The effects of short term application of a combination of platelet-derived and insulin-like growth factors on periodontal wound healing. *J. Periodontol.* **62**:458–467.

35. **Lynch, S. E., R. C. Williams, A. M. Polson, T. H. Howell, M. S. Reddy, U. E. Zappa, and H. N. Antoniades.** 1989. A combination of platelet-derived and insulin-like growth factors enhances periodontal regeneration. *J. Clin. Periodontol.* **16**:545–548.

36. **Matsuda, N., W.-L. Lin, M. Kumar, M. I. Cho, and R. J. Genco.** 1992. Mitogenic, chemotactic and synthetic responses of rat periodontal ligament cells to polypeptide growth factors in vitro. *J. Periodontol.* **63**:515–525.

37. **Mohan, S., and D. J. Baylink.** 1991. Bone growth factors. *Clin. Orthop. Relat. Res.* **263**:30–48.

38. **Mustoe, T. A., G. F. Pierce, A. Thomason, P. Gramates, M. B. Sporn, and T. F. Deuel.** 1987. Accelerated healing of incisional wounds in rats induced by transforming growth factor beta. *Science* **237**:1333–1336.

39. **Oreffo, R. O. C., L. Bonewald, S. M. Seyedin, D. Rosen, and G. R. Mundy.** 1989. Inhibition of osteoclastic bone resorption by growth factors. *J. Bone Miner. Res.* **4**:S154.

40. **Overall, C. M., J. L. Wrana, and J. Sodek.** 1989. Independent regulation of collagenase, 72-k Da progelatinase, and metalloproteinase inhibitor expression in human fibroblasts by transforming growth factor-β. *J. Biol. Chem.* **264**:1860–1869.

41. **Ozkaynak, E., D. C. Rueger, E. A. Drier, C. Corbett, R. J. Ridge, T. K. Sampath, and H. Oppermann.** 1990. OP-1 cDNA encodes osteogenic protein in the TGF-β family. *EMBO J.* **9**:2085–2093.

42. **Pfeilschifter, J., M. Oechsner, A. Naumann, R. G. K. Gronwald, H. W. Minne, R. Ziegler, T. K. Sampath, and H. Oppermann.** 1990. Stimulation of bone matrix apposition in vitro by local growth factors. A comparison between insulin-like growth factor-I platelet-derived growth factor and transforming growth factor-β. *Endocrinology* **127**:69–75.

43. **Piche, J. E., and D. T. Graves.** 1989. Study of the growth factor requirements of human bone-derived cells: a comparison with human fibroblasts. *Bone* **10**:131–138.

44. **Pierce, G. F., J. Vande Berg, R. Rudolph, J. Tarpley, and T. Mustoe.** 1991. Platelet-derived growth factor-BB and transforming growth factor beta selectively modulate glycosaminoglycans, collagen, and myofibroblasts in excisional wounds. *Am. J. Pathol.* **138**:629–646.

45. **Rappolee, D. A., D. Mark, M. J. Banda, and Z. Werb.** 1988. Wound macrophages express TGF-α and other growth factors in vivo: analysis of mRNA phenotyping. *Science* **241**:708–712.

46. **Rinderknecht, E., and R. E. Humbel.** 1976. Amino-terminal sequences of two polypeptides from human serum with nonsuppressible insulin-like and cell growth activities: evidence for structural homology with insulin B chain. *Proc. Natl. Acad. Sci. USA* **73**:4379.

47. **Rutherford, R. B., C. E. Nierkrash, J. E. Kennedy, and M. F. Charette.** 1992. Platelet-derived and insulin-like growth factors stimulate regeneration of periodontal attachment in monkeys. *J. Periodontal Res.* **27**:285–290.

48. **Rutherford, R. B., T. K. Sampath, D. C. Rueger, and T. D. Taylor.** 1992. Use of bovine osteogenic protein to promote rapid osseointegration of endosseous dental implants. *Int. J. Oral Maxillofac. Implants* **7**:297–301.

49. **Schultz, G. S., M. White, R. Mitchell, G. Brown, J. Lynch, D. R. Twardzik, and G. T. Todaro.** 1987. Epithelial wound healing enhanced by transforming growth factor-α and vaccinia growth factor. *Science* **235**:350–352.

50. **Sigurdsson, T. J., U. M. E. Wikesjo, G. C. Bogle, K. A. Selvig, and R. D. Finkelman.** 1992. Lack of regeneration of mineralized tissues in furcation defects in dogs following wound conditioning with IGF-II, bFGF and TGF-β1. *J. Periodontol.* **63**:1013.

51. **Sitaras, N. M., E. Sariban, P. Pantazis, B. Zetter, and H. N. Antoniades.** 1987. Human iliac artery endothelial cells express both genes encoding the chains of platelet derived growth factor (PDGF) and synthesize PDGF-like mitogens. *J. Cell Physiol.* **132**:376–380.

52. **Sporn, M. B., A. B. Roberts, L. M. Wakefield, and B. deCrombrugghe.** 1987. Some recent advances in the chemistry and biology of transforming growth factor-beta. *J. Cell Biol.* **105**:1039–1045.

53. **Stiles, C. D., G. T. Capone, C. D. Scher, H. N. Antoniades, J. J. VanWyk, and W. J. Pledger.**

1979. Dual control of cell growth by somatomedins and platelet-derived growth factors. *Proc. Natl. Acad. Sci. USA* **76:**1279–1283.

54. **Tashjian, A. H., Jr., E. F. Voelkel, M. Lazzaro, F. R. Singer, A. B. Roberts, R. Derynck, M. E. Winkler, and L. Levine.** 1985. Alpha and beta human transforming growth factors stimulate prostaglandin production and bone resorption in cultured mouse calvaria. *Proc. Natl. Acad. Sci. USA* **82:**4543–4548.

55. **Terranova, V. P., C. Odziemiec, K. S. Tweden, and D. P. Spadone.** 1989. Repopulation of dentin surfaces by periodontal ligament cells and endothelial cells: effects of basic fibroblast growth factor. *J. Periodontol.* **60:**293–301.

56. **Tsukamoto, T., T. Matsui, M. Fukase, and T. Fujita.** 1991. Platelet-derived growth factor B chain homodimer enhances chemotaxis and DNA synthesis in normal osteoblast-like cells (MC3T3-E1). *Biochem. Biophys. Res. Commun.* **175:**745–751.

57. **Urist, M. R.** 1965. Bone formation by autoinduction. *Science* **150:**893.

58. **Urist, M. R., R. J. DeLange, and G. A. M. Finerman.** 1983. Bone cell differentiation and growth factors. *Science* **220:**680.

59. **Wozney, J. M., V. Rosen, A. J. Celeste, L. M. Mitsock, M. J. Whitters, R. W. Kriz, R. M. Hewick, and E. A. Wang.** 1988. Novel regulations of bone formation: molecular clones and activities. *Science* **242:**1528.

60. **Wrana, J. L., M. Macho, B. Hawrylyshyn, K. L. Yao, C. Domenicucci, and J. Sodek.** 1988. Differential effects of transforming growth factor-β on the synthesis of extracellular matrix proteins by normal fetal rat calvarial bone cell populations. *J. Cell Biol.* **106:**915–924.

61. **Yasko, A. W., J. M. Lane, E. J. Fellinger, V. Rosen, J. M. Wozney, and E. A. Wang.** 1992. The healing of segmental bone defects, induced by recombinant human bone morphogenetic protein (rhBMP-2). *J. Bone Joint Surg.* **74A:**659–671.

Molecular Pathogenesis of Periodontal Disease
Edited by Robert Genco et al.
© 1994 American Society for Microbiology, Washington, DC 20005

Chapter 34

Role of Osteogenic (Bone Morphogenetic) Protein and Platelet-Derived Growth Factor in Periodontal Wound Healing

Bruce Rutherford, Marc Charette, and David Rueger

The periodontium is composed of distinct mineralized and nonmineralized tissues organized into a complex anatomical structure. This diversity of tissues and cells presents a challenging problem for regenerative therapy, as no single agent possesses the requisite spectrum of biological activities necessary to achieve restoration of this complex tissue architecture. Therefore, we have been testing the capacity of members of three classes of molecules, singly or in combination, as stimulators of new attachment formation. These are polypeptide growth factors, steroids, and the emerging class of morphogens as represented by the osteogenic proteins (OP) or bone morphogenetic proteins (BMPs).

BMPs form a subgroup of a larger family of structurally related proteins known as the transforming growth factor β (TGF-β) superfamily. Proteins of this family are implicated in diverse biological activities involving differentiation, tissue morphogenesis, regeneration, and repair. Most members of the BMP subgroup were originally identified by their capacity to induce bone in vivo in intra- or extraskeletal sites in mammals (41, 50). Relationships to other molecules identified from nonbony experimental systems have been made by comparing the protein and DNA sequences. This has led to the identification of the product of a gene from a mouse embryonic cDNA library (VgR-1) as a member of the BMP subgroup. No formal consensus for naming these molecules has yet emerged (Table 1). Therefore, we use here the names of specific members as given in the original publications, and we refer to the family as bone morphogenetic proteins.

We have tested the capacity of natural BMP as well as recombinant human osteogenic protein-1 (hOP-1) in assays of tissue repair and regeneration both in vivo (38) and in vitro (see below). Herein we review data pertaining to the devel-

Bruce Rutherford • Department of Cariology and General Dentistry, School of Dentistry, University of Michigan, 1101 North University, Ann Arbor, Michigan 48109-1078.   ***Marc Charette*** • Creative BioMolecules, 45 South Street, Hopkinton, Massachusetts 01748.   ***David Rueger*** • Creative BioMolecules, 35 South Street, Hopkinton, Massachusetts 01748.

Table 1. Alternative names for BMPs

Original name	Alternative names[a]
BMP-1[b]	
BMP-2a	BMP-2, DVR-2
Osteogenin	BMP-3, DVR-3
BMP-2b	BMP-4, DVR-4
BMP-5	DVR-5
mVgR-1	BMP-6, DVR-6
OP-1	BMP-7, DVR-7
OP-2	BMP-8, DVR-8

[a] DVR-2 through DVR-8 are recently proposed alternative nomenclature (19).

[b] A protein that copurified with other BMPs (53) that is now
known to be unrelated.

opment and use of BMPs as potential therapeutic agents for the promotion of repair or regeneration of oral tissues. In addition, we compare the BMPs as tissue morphogens to growth factors such as platelet-derived growth factor (PDGF) used in combination with dexamethasone (39). This chapter covers work published prior to April 1993. Other aspects of the BMPs are reviewed elsewhere in this volume.

THE TGF-β SUPERFAMILY

The TGF-β superfamily is widely represented in vertebrates and invertebrates. Examples among vertebrates are the Vg-1 gene product of *Xenopus laevis* (31), mullerian inhibiting substance (9), the inhibins and activins (15, 20, 49), and BMPs BMP-2, -3, -4, and -5 (10, 53), Vgr-1 (18), OP-1 (26), and OP-2 (28). More distantly related to these are growth and differentiation factors 1, 3, and 9 (24) and the TGF-βs themselves (21). Representing the invertebrates are the decapentaplegic protein and the 60A gene product of *Drosophila melanogaster* (46, 52). Recently, receptors for activin and TGF-β were cloned and identified as serine threonine kinases (14, 22, 23).

In their biological roles, these proteins act during embryogenesis and in adult animals. For example, the decapentaplegic protein is a determinant of dorsal-ventral axis formation and morphogenesis of the imaginal discs (29). VG-1 is localized to the vegetal hemisphere of *Xenopus* oocytes (51). Inhibins and activins control secretion of follicle-stimulating hormone (20, 49). Moreover, activins can induce the formation of mesoderm and anterior structures in *Xenopus* embryos. Mullerian inhibiting substance causes regression of the Mullerian duct during male development (9, 47). In adult animals, recombinant BMP-2 (50), BMP-4 (10), and OP-1 (42) induce de novo cartilage and bone formation in a subcutaneous implant assay in rats (44). In addition, recombinant OP-1 homodimers produced in mammalian cell culture have been effective in the repair of segmental bone defects in rabbit ulna (50) and induce reparative dentin formation in monkeys (40). The TGF-βs themselves are involved in a wide variety of differentiation and cell proliferation activities.

IDENTIFICATION OF OP-1

The history of the identification and purification of BMPs began in 1965, when Marshall Urist (48) demonstrated that the cellular events associated with embryonic bone development can be reproduced in heterotopic sites by implants of demineralized bone segments. This observation led to a search for the molecules in bone responsible for bone-forming activity. In the late 1960s and early 1970s, it was recognized that dentin also contained bone morphogenetic activity (4, 8).

Attempts to purify the bone morphogenetic substances took a significant step forward in 1981, when Sampath and Reddi demonstrated that demineralized bone powder could be inactivated by extraction with denaturants and that activity could be restored by reconstitution with the extract (44). This result further substantiated the existence of such bone morphogenetic molecules and provided an assay for their purification.

The bone morphogenetic activity from bovine bone has been purified, and the purified preparation is called BMP (41). The purified bovine BMP is composed of disulfide-linked dimers that migrate on sodium dodecyl sulfate gels as a diffuse band with an apparent molecular mass of about 30 to 35 kDa. Upon reduction, the dimers yield two subunits that migrate with molecular masses of 18 and 16 kDa. Both subunits are glycosylated. After chemical or enzymatic deglycosylation, the dimers migrate as a diffuse 27-kDa band that upon reduction yields two polypeptides that migrate at 16 and 14 kDa. The carbohydrate moiety does not appear to be essential for bone morphogenetic activity, since the deglycosylated proteins are capable of inducing bone formation in vivo. However, biological activity is lost when the protein is subjected to reducing conditions.

On the basis of the homology of tryptic peptide sequences (derived from the purified bovine BMP) with two known members of the TGF-β superfamily, a consensus gene was constructed for use as an oligonucleotide probe in screening human genomic libraries. This resulted in the isolation of a third product, termed osteogenic protein 1 (or OP-1), which was a new member of the TGF-β superfamily (26). The genomic clones were then used to isolate the corresponding cDNA clones that have been expressed in host cells. Subsequently, others isolated the OP-1 gene product, referring to it as BMP-7 (10).

Detailed sequence analysis of peptides generated by proteolytic digestion of purified bovine BMP showed that this preparation is composed of two gene products. The 18-kDa subunit is the product of the bovine equivalent of the OP-1 gene, and the 16-kDa subunit is the product of the bovine equivalent of the human BMP-2. Although examination of subfractioned samples of this bovine BMP preparation suggested that homodimers of these polypeptides were the predominant species, the existence of heterodimers could not be ruled out.

TISSUE LOCALIZATION AND BIOLOGIC ACTIVITY

Northern (RNA) blot analyses with probes for various family members indicate that mRNA for most of them is expressed in specific organs. OP-1 mRNA is expressed mainly in the kidneys and the bladder (27) but also in the brain, which is the sole site of growth and differentiation factor 1 expression. In contrast, BMP-3, BMP-4, BMP-5, and VgR-1 (BMP-6) are primarily expressed in the lungs. The lungs may participate in the growth regulation of bone and connective tissues in an endocrine manner, as proposed for the kidney (27); OP-2- and BMP-2-specific mRNAs were not detected in adult organs. OP-2 cDNA was found at low abundance in a hippocampus cDNA library and may be expressed at low levels in the brain. However, OP-2 mRNA was found at relatively high levels in 8-day-old mouse embryos, indicating a developmental role; in the adult animal, OP-2 and BMP-2 may be expressed in a more discrete location or primarily during tissue regeneration. While the timing of expression for BMP-5 and OP-1 seems to be directly related to growth, an inverse relationship is found for VgR-1 and BMP-4. The level of BMP-3 expression (in lungs) does not appear to change with the age of the animal (28).

Studies with recombinant hOP-1 have demonstrated that the homodimer of this polypeptide is capable of inducing new bone formation in vivo with a specific activity that is comparable to that of natural BMP (42). Homodimers of recombinant human BMP-2 have been reported to induce bone formation when they are implanted with rat collagen carrier in the rat subcutaneous assay model (50). In these studies, approximately 10 times more recombinant human BMP-2 was required to achieve the same level of bone-forming activity as that observed with the corresponding natural bovine bone-inductive protein preparations. Hammonds et al. (12) have recently reported that homodimers of recombinant human BMP-4, which is closely related to BMP-2, are also capable of inducing bone in vivo, whereas recombinant activin does not. TGF-β1, PDGF, epidermal growth factor, basic fibroblast growth factor, cartilage-derived growth factor, insulin, and bovine growth hormone have also been tested under the same assay conditions, and none of these growth factors induce bone formation (43).

COMPARISONS OF BMPs AND PDGF

Currently, the data permit initial, albeit tentative, comparisons between polypeptide growth factors as exemplified by PDGF and the BMPs that suggest different physiologic roles for each class of compounds. This chapter focuses on the function of these molecules in tissue repair and regeneration. For the purposes of this discussion, growth factors are defined as proteins originally identified as cell mitogens whose principle biologic function appears to involve regulation of cell proliferation and wound healing. Several reviews of the biochemistry and biology of PDGF and other growth factors have recently been published (1, 2, 6, 11, 34).

MOLECULAR STRUCTURE

There are two PDGF polypeptides (A and B) encoded by distinct genes. The mature molecule is a disulfide-linked dimer, and the three possible isoforms (AA,

AB, and BB) have been found in nature. BMP, like the members of the other larger TGF-β superfamily, are also disulfide-linked dimers with a conserved pattern of cysteine residues (designated the TGF-β domain). Reduction destroys biologic activity in both classes of molecules. OP-1 and BMP-2 were isolated from bovine bone as homodimers; however, it is not clear that BMP heterodimers exist in nature, as PDGF does.

All members of the TGF-β superfamily thus far characterized are initially produced as precursor proteins approximately four times larger than the mature form. The mature proteins are derived from the carboxy-terminal portion of the precursor protein and appear to be proteolytically processed by a trypsin-like protease. The pro domain of the precursor protein may be responsible for proper in vivo folding and transport of the mature region. In contrast, PDGF is found with molecular masses ranging from 28 to 35 kDa, a phenomenon thought to be due to proteolytic cleavage with no apparent effect on biologic activity (3).

The hOP-1 gene has been expressed in monkey kidney (BSC) and Chinese hamster ovary (CHO) cells, and the recombinant hOP-1 protein has been purified and characterized (42). The OP-1 gene product is secreted as a processed mature disulfide-linked homodimer with an apparent molecular mass of 30 to 38 kDa. Recombinant forms of all three isoforms of PDGF have been produced.

LOCALIZATION AND MODE OF ACTION

PDGF was originally inferred to be present in platelets (33) but subsequently has been shown to be produced by a variety of cells in addition to platelets (34). The BMPs, as described above, were originally identified in and purified from bone extracellular matrix. PDGF-like molecules have also been isolated from bone extracellular matrix (13). It is not clear, however, that any of the BMPs or PDGF are produced in bone in vivo, nor is the functional significance of these matrix-bound molecules established. PDGF does not induce ectopic bone formation in vivo.

PDGF is undetectable in plasma and is cleared from baboon blood with a half-life of less than 2 min (5). Hence, the mode of action of PDGF is most likely paracrine and/or autocrine rather than endocrine. Currently, there is no evidence of the circulation of the BMPs in plasma. However, the localization of BMP mRNAs in the kidneys, brain, and lung raises the possibility that these proteins have an endocrine function in addition to the putative paracrine mode of action implied by the ectopic bone formation assay (27, 28). Clarification of these questions awaits study with a complex set of probes and antibodies specific to each gene and protein.

BONE COLLAGEN AS A BIOCOMPATIBLE AND BIODEGRADABLE DELIVERY MATRIX

Optimum activity for the BMPs in vivo is achieved by combining them with a carrier molecule and implanting the combination as a solid mass (42, 50). Because PDGF does not circulate and has a rapid clearance rate, its presence in wounds is probably due to release at the site by platelets and other cells. Therefore, topical application of PDGF usually has relied on delivery with a carrier material. We (36) and others (17) have successfully used PDGF-BB combined with insulin-like growth factor I (IGF-I) in carboxymethyl cellulose to initiate periodontal regeneration in experimental animal models of adult periodontitis. PDGF and IGF-I in carboxymethyl cellulose are rapidly cleared (17). We found that this material is difficult to confine to the wound site, a potentially confounding factor in experimental trials. Therefore, in subsequent experiments (37), we opted to use the collagen carrier (described below) developed as a delivery vehicle for the BMPs (42). A useful carrier should (i) bind the active agents, (ii) accommodate each step of the cellular response during tissue formation, (iii) protect the active protein from nonspecific proteolysis, and (iv) be biocompatible and biodegradable. The collagen carrier may retard the release of the active agents relative to more soluble substances.

In most BMP research over the last decade, bone-derived collagen has been utilized as the delivery matrix. Published observations have shown that several key parameters of the collagen matrix, such as species specificity, surface charge, particle size, the presence of mineral, and the methodology for combining the collagen matrix with the BMP, can play critical roles in obtaining maximum biologic response (45). The species specificity attribute of bone-derived collagen was a special concern for human clinical trials, because limited availability, possible immunologic intolerance, and potential for transmission of disease made human bone-derived collagen matrix an undesirable carrier material (32, 45). However, a procedure was developed that resulted in preparations of bovine bone-derived collagen that showed reproducible activity in combination with recombinant OP-1 when tested in the well-characterized rat in vivo subcutaneous bone formation assay (35). The composition of the bovine bone-derived collagen matrix is predominantly insoluble type I collagen.

In addition to collagen, a variety of other biomaterials have been tested for use in supporting OP-1 biologic activities. These include various extracellular matrix components alone and in combination (collagen, fibronectin, glycosaminoglycans), calcium hydroxide, calcium phosphate, and poly acids alone and as copolymers (lactic, glycolic). Many of these materials, when formulated with OP-1, have shown some biologic activity. However, to date, none have produced results comparable to those obtained with bovine bone-derived collagen matrix (35).

DIFFERENTIAL ACTIVITIES OF OP-1 AND PDGF

We have conducted in vitro studies comparing the effects of OP-1 with those of the polypeptide growth factors PDGF and IGF-I and the synthetic glucocorticoid dexamethasone. These experiments utilized human fibroblasts cultured from biopsies of gingiva and the periodontal ligament remaining attached to fresh extracted roots (39). The cells were serially propagated in culture and used between passages 2 and 8. Cells were plated at various population densities in medium supplemented with 10% fetal calf serum. The cultures were then switched to medium supplemented with ITS-Plus (Collaborative Research, Cambridge, Mass.) to arrest proliferation and establish quiescence (39). The various combinations of growth factors and OP-1 were prepared in this medium at the concentrations described in the figure legends and applied once at the initiation of the experiment.

The data reveal substantial differences in the response of periodontal ligament fibroblasts to growth factors and OP-1. As we have previously described (39), the activities of PDGF-BB and IGF-I combined are additive as mitogens, and dexamethasone can substitute for in vitro IGF-I. PDGF and dexamethasone are synergistic, but dexamethasone is inactive alone. OP-1 at concentrations up to 200 ng/ml is not mitogenic for PLF (Fig. 1). In contrast, OP-1 at the same concentrations stimulated a significant increase in extractable alkaline phosphatase activity in these cells in a dose- and time-dependent manner (Fig. 2). PDGF-BB alone or in combination with either IGF-I or dexamethasone failed to alter periodontal ligament fibroblast alkaline phosphatase activity. Qualitatively similar data were obtained from cultures of gingival fibroblasts (not shown). Both PDGF and OP-1, on the other hand, stimulated proliferation of MC-3T3 cells, but only OP-1 induced significant increases in alkaline phosphatase activity (not shown). Others have shown that both PDGF (30) and OP-1 (41) are mitogenic for bone cell preparations.

We also directly compared the effects of OP-1 and a combination of PDGF-BB–dexamethasone in vivo. Single doses of OP-1 and the PDGF-dexamethasone combination were mixed with the collagen carrier described above and tested in

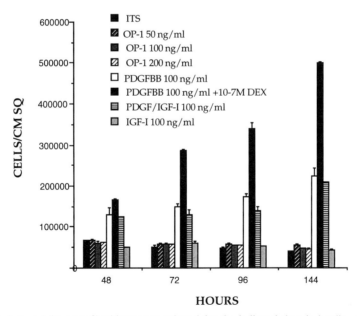

Figure 1. Periodontal ligament fibroblasts were cultured for the indicated time in insulin-, transferrin-, selenium-, and bovine serum albumin (ITS)-supplemented media with and without hOP-1, PDGF-BB, PDGF-BB–dexamethasone (DEX), PDGF-BB–IGF-I, or IGF-I alone at the indicated concentrations. Cells were counted using Coulter electronics. PDGF-BB and dexamethasone induced significantly more cells at each time point than any other treatment. PDGF-BB alone and in combination with IGF-I was significantly more effective than OP-1 or IGF-I alone (analysis of variation, Scheffe's F at $P < 0.01$).

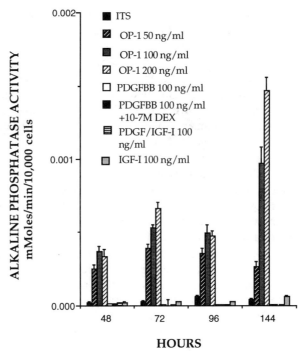

HOURS

Figure 2. Periodontal ligament fibroblasts were cultured for the indicated times in insulin-, transferrin-, selenium-, and bovine serum albumin (ITS)-supplemented media with and without hOP-1, PDGF-BB, PDGF-BB–dexamethasone (DEX), PDGF–IGF-I, or IGF-I alone at the indicated concentrations. The amount of extractable alkaline phosphatase activity was significantly increased in all OP-1-treated cultures compared with all other cultures and in the 200-ng/ml hOP-1-treated cultures compared with 50- or 100-ng/ml hOP-1-treated cultures at 144 h (analysis of variation, Scheffe's F at $P < 0.01$).

our ligature-induced monkey model of periodontitis (36). The PDGF-dexamethasone combination appeared to be more effective in regenerating the periodontium than OP-1 (data not shown); however, a range of doses of OP-1 should be tested in order to confirm this observation.

Natural bovine BMP rapidly induced bone formation in fresh extraction sockets in the presence or absence of dental implants (38). Others have demonstrated that growth factors may modulate bony remodeling adjacent to closely apposed implant surfaces placed in surgically prepared sites (16). In addition, Buser et al. (7) showed cementum with attached ligament fibers partially covering a titanium implant inserted in contact with a retained root tip. Implants inserted into bone but not in contact with the root osseointegrated without forming ligament-like structures. These interesting data demonstrate that a periodontal ligament is necessary for cementum and ligament production and support Melcher's hypothesis that tissues of the periodontium possess different tissue-forming potentials (25).

The data reviewed here may be interpreted as suggesting that growth factors as exemplified by PDGF and morphogens as exemplified by the BMPs play fundamentally different roles in the healing or regeneration of oral tissues, with the differences being dependent, at least to some extent, on the remaining tissues. In the fresh extraction sites, BMPs rapidly induce bone formation (38). In contrast, in regard to the formation of cementum and ligament, the periodontal ligament cells may be unresponsive to the BMPs, while osteoblasts (or precursors) are capable of responding to both growth factors and morphogens. PDGF, which is rapidly cleared from the wound, most likely triggers periodontal ligament cells to proliferate and/or migrate. Subsequently, an uncharacterized cascade of events leading to the formation of mineralized (cementum and bone) and unmineralized periodontal ligament occurs. Such events could involve the local production of tissue morphogens and/or growth factors that, acting alone or, more likely, in combination with other cell-modulating mechanisms such as integrin-mediated signals coming from extracellular matrix molecules, may lead to periodontal tissue regeneration.

Alternatively, all the tissues of the mature periodontium may possess cells capable of responding to morphogens and growth factors albeit to different degrees of sensitivity. Such sensitivities could be a consequence of cells expressing various numbers of the appropriate receptors for the cognate ligands. Both alternatives and variations thereof, important to the development of regenerative therapies, await further experimental testing.

Acknowledgments. This work was partially supported by grants from Creative BioMolecules and the National Institutes of Health. B. Rutherford has worked as consultant to Creative BioMolecules.

REFERENCES

1. **Aaronson, S. A.** 1991. Growth factors and cancer. *Science* **254:**1146–1152.
2. **Antoniades, H., and A. J. Owen.** 1982. Growth factors and regulation of cell growth. *Annu. Rev. Med.* **33:**445–461.
3. **Antoniades, H. N., and M. W. Hunkapillar.** 1983. Human platelet-derived growth factor (PDGF): amino terminal amino acid sequence. *Science* **220:**963–965.
4. **Bang, G., and M. R. Urist.** 1967. Bone induction in excavation chambers in matrix of decalcified dentin. *Arch. Surg.* **94:**781–789.
5. **Bowen-Pope, D. F., T. W. Malpass, D. M. Foster, and R. Ross.** 1984. Platelet-derived growth factor in vivo: levels, activity, and rate of clearance. *Blood* **64:**458–469.
6. **Bowen-Pope, D. F., A. Van Koppen, and G. Schatteman.** 1991. Is PDGF really important? Testing the hypotheses. *Trends Genet.* **7:**413–418.
7. **Buser, D., K. Warrer, and T. Karring.** 1990. Formation of a periodontal ligament around a titanium implant. *J. Periodontol.* **61:**597–601.
8. **Butler, W. T., A. Mikulski, M. R. Urist, G. Bridges, and S. Uyeno.** 1977. Noncollagenous proteins of a rat dentin matrix possessing bone morphogenetic activity. *J. Dent. Res.* **56:**228–232.
9. **Cate, R. L., R. J. Mattaliano, C. Hession, R. Tizard, R. A. Farber, J. M. Bertonis, G. Torres, B. P. Wallner, K. L. Ramachandran, R. C. Ragin, T. F. Manganaro, D. T. MacLaughlin, and P. K. Donohoe.** 1986. Isolation of the bovine and human genes for mullerian inhibiting substance and expression of the human gene in animal cells. *Cell* **45:**685–698.
10. **Celeste, A. J., J. A. Iannazzi, R. C. Taylor, R. M. Hewick, V. Rosen, and E. A. Wang.** 1990. Identification of transforming growth factor b family members present in bone-inductive protein purified from bovine bone. *Proc. Natl. Acad. Sci. USA* **87:**9843–9847.
11. **Graves, D. T., and D. L. Cochran.** 1990. Mesenchymal cell growth factors. *Crit. Rev. Oral Biol. Med.* **1:**17–36.
12. **Hammonds, R. G., Jr., R. Schwall, A. Dudley, L. Berkemeier, C. Lai, J. Lee, N. Cunningham, A. H. Reddi, W. I. Wood, and A. J. Mason.** 1991. Bone-inducing activity of mature BMP-2b produced from a hybrid BMP-2a/2b precursor. *J. Endocrinol.* **5:**149–155.

13. **Hauschka, P. V., T. L. Chen, and A. E. Mavrakos.** 1988. Polypeptide growth factors in bone matrix, p. 207–225. *In Cell and Molecular Biology of Vertebrate Hard Tissues.* Wiley, Chichester, England.

14. **Lin, H. Y., X.-F. Wang, E. Ng-Eaton, R. Weinberg, and H. F. Lodish.** 1992. Expression cloning of the TGF-β type II receptor, a functional transmembrane serine/threonine kinase. *Cell* **68:**775–785.

15. **Ling, N., S.-Y. Ying, N. Ueno, S. Shimaski, F. Esch, M. Hotta, and R. Guillemin.** 1986. Pituitary FSH is released by a heterodimer of the b-subunits from the two forms of inhibin. *Nature* (London):776–779.

16. **Lynch, S. E., D. Buser, R. A. Hernandez, H. P. Weber, H. Stich, C. H. Fox, and R. C. Williams.** 1991. Effects of the platelet-derived growth factor/insulin-like growth factor-I combination on bone regeneration and around titanium dental implants. Results of a pilot study in beagle dogs. *J. Periodontol.* **62:**710–716.

17. **Lynch, S. E., G. Ruiz de Castilla, R. C. Williams, C. P. Kiritsy, T. H. Howell, M. S. Reddy, and H. N. Antoniades.** 1991. The effects of short-term application of a combination of platelet-derived and insulin-like growth factors on periodontal wound healing. *J. Periodontol.* **62:**458–467.

18. **Lyons, K., J. L. Graycar, A. Lee, S. Hashmi, P. B. Lindquist, E. Y. Chen, B. L. M. Hogan, and R. Derynck.** 1989. Vgr-1, a mammalian gene related to Xenopus Vg-1, is a member of the transforming growth factor b gene superfamily. *Proc. Natl. Acad. Sci. USA* **86:**4554–4558.

19. **Lyons, K. M., C. M. Jones, and B. L. M. Hogan.** 1991. The DVR-gene family in embryonic development. *Trends Genet.* **7:**408–412.

20. **Mason, A. J., J. S. Hayflick, N. Ling, S. Shimaski, F. Esch, N. Ueno, S.-Y. Ying, R. Guillemin, H. Niall, and P. H. Seeburg.** 1985. Complementary DNA sequences of ovarian follicular fluid inhibin show precursor structure and homology with transforming growth factor-β. *Nature* (London) **318:**659–663.

21. **Massagué, J.** 1987. The TGF-β family of growth factor and differentiation factors. *Cell* **49:**437–438.

22. **Massagué, J.** 1992. Receptors for the TGF-beta family. *Cell* **69:**1067–1070.

23. **Mathews, L. S., and W. W. Vale.** 1991. Expression cloning of an activin receptor, a predicted transmembrane serine kinase. *Cell* **65:**973–982.

24. **McPherron, A., and S.-J. Lee.** 1993. GDF-3 and GDF-9: two new members of the transforming growth factor-b superfamily containing a novel pattern of cysteines. *J. Biol. Chem.* **268:**3444–3449.

25. **Melcher, A. H.** 1976. On the repair of periodontal tissues. *J. Periodontol.* **47:**256–260.

26. **Ozkaynak, E., D. C. Rueger, E. A. Drier, C. Corbett, R. J. Ridge, T. K. Sampath, and H. Oppermann.** 1990. OP-1 cDNA encodes an osteogenic protein in the TGF-beta family. *EMBO J.* **9:**2085–2093.

27. **Ozkaynak, E., P. N. J. Schnegelsberg, and H. Hoppermann.** 1991. Murine osteogenic protein (OP-1): high levels of mRNA in kidney. *Biochem. Biophys. Res. Commun.* **179:**116–123.

28. **Ozkaynak, E., P. N. J. Schnegelsberg, D. F. Jin, G. M. Clifford, F. D. Warren, E. A. Drier, and H. Oppermann.** 1992. Osteogenic protein-2. *J. Biol. Chem.* **267:**25220-25227.

29. **Padgett, R. W., R. D. St. Johnson, and W. M. Gelbart.** 1987. A transcript from a Drosophila pattern gene predicts a protein homologous to the transforming growth factor-b family. *Nature* (London) **325:**81–84.

30. **Piche, J. E., and D. T. Graves.** 1989. Study of the growth factor requirements of human bone-derived cells. *Bone* **10:**131–138.

31. **Rebagliati, M. R., D. L. Weeks, R. P. Harvey, and D. A. Melton.** 1985. Identification and cloning of localized maternal RNAs from xenopus eggs. *Cell* **42:**769–777.

32. **Reddi, H.** 1983. Extracellular bone matrix dependent local induction of cartilage and bone. *J. Rheum. Suppl.* **11:**67–69.

33. **Ross, R., J. A. Glomset, B. Kariya, and L. Harker.** 1974. A platelet-dependent serum factor that stimulates proliferation of arterial smooth muscle cells in vitro. *Proc. Natl. Acad. Sci. USA* **71:**1207–1210.

34. **Ross, R., E. W. Raines, and D. Bowen-Pope.** 1986. The biology of platelet-derived growth factor. *Cell* **46:**155–169.

35. **Rueger, D.** Unpublished observations.

36. **Rutherford, R. B., C. E. Niekrash, J. E. Kennedy, and M. F. Charette.** 1992. Platelet-derived

growth and insulin-like growth factors stimulate regeneration of periodontal attachment in monkeys. *J. Periodontal Res.* **27**:285–290.

37. **Rutherford, R. B., M. E. Ryan, J. E. Kennedy, M. M. Tucker, and M. F. Charette.** 1993. Platelet-derived growth factor and dexamethasone combined with a collagen matrix induce regeneration of the periodontium in monkeys. *J. Clin. Periodontol.* **20**:537–544.

38. **Rutherford, R. B., T. K. Sampath, D. C. Rueger, and T. D. Taylor.** 1992. Use of bovine osteogenic protein to promote rapid osseointegration of endosseous dental implants. *Int. J. Maxillofac. Implants* **7**:297–301.

39. **Rutherford, R. B., M. D. TrailSmith, M. E. Ryan, and M. F. Charette.** 1992. Synergistic effects of dexamethasone on platelet-derived growth factor mitogenesis in vitro. *Arch. Oral Biol.* **37**:139–145.

40. **Rutherford, R. B., J. Wahle, M. Tucker, D. Rueger, and M. Charette.** 1993. Recombinant human osteogenic protein-1 induces reparative dentin formation in monkeys. *Arch. Oral Biol.* **38**:571–576.

41. **Sampath, T. K., J. E. Coughlin, R. M. Whetstone, D. Banach, C. Corbett, R. J. Ridge, E. Ozkaynak, H. Oppermann, and D. C. Rueger.** 1990. Bovine osteogenic protein is composed of dimers of OP-1 and BMP-2A, two members of the transforming growth factor-beta superfamily. *J. Biol. Chem.* **265**:13198–13205.

42. **Sampath, T. K. J. C. Maliakal, P. V. Hauschka, W. K. Jones, H. Sasak, R. F. Tucker, K. H. White, J. E. Coughlin, M. M. Tucker, and R. H. Pang.** 1992. Recombinant human osteogenic protein-1 (hOP-1) induces new bone formation in vivo with a specific activity comparable with natural bovine osteogenic protein and stimulates osteoblast proliferation and differentiation in vitro. *J. Biol. Chem.* **267**:20352–20362.

43. **Sampath, T. K., N. Muthukumaran, and H. Reddi.** 1987. Isolation of osteogenin, an extracellular matrix-associated, bone inductive protein, by heparin affinity chromatography. *Proc. Natl. Acad. Sci. USA* **84**:7109–7113.

44. **Sampath, T. K., and H. Reddi.** 1981. Dissociative extraction and reconstitution of extracellular matrix components involved in local bone differentiation. *Proc. Natl. Acad. Sci. USA* **78**:7599–7603.

45. **Sampath, T. K., and A. H. Reddi.** 1983. Homology of bone-inductive proteins from human, monkey, bovine and rat extracellular matrix. *Proc. Natl. Acad. Sci. USA* **80**:6591–6595.

46. **Segal, D., and W. M. Gelbart.** 1985. Shortvein, a new component of the decapentaplegic gene complex in Drosophila melanogaster. *Genetics* **109**:119–143.

47. **Thomsen, G., T. Woolf, M. Whitman, S. Sokol, J. Vaughan, W. Vale, and D. A. Melton.** 1990. Activins are expressed early in *Xenopus* embryogenesis and can induce axial mesoderm and anterior structures. *Cell* **63**:485–493.

48. **Urist, M. R.** 1965. Bone: formation by autoinduction. *Science* **150**:893–899.

49. **Vale, W., J. Rivier, J. Vaughan, R. McClintock, A. Corrigan, W. Woo, D. Karr, and J. Spiess.** 1986. Purification and characterization of an FSH releasing protein from porcine ovarian follicular fluid. *Nature* (London) **321**:776–779.

50. **Wang, E. A., V. Rosen, J. S. D'Alessandro, M. Bauduy, P. Cordes, T. Harada, D. I. Israel, R. M. Hewick, K. M. Kerns, and P. LaPan.** 1990. Recombinant human bone morphogenetic protein induces bone formation. *Proc. Natl. Acad. Sci. USA* **87**:2220–2224.

51. **Weeks, D. L., and D. A. Melton.** 1987. A maternal mRNA localized to the vegetal hemisphere in Xenopus eggs codes for a growth factor related to TGF-β. *Cell* **51**:861–867.

52. **Wharton, K. A., G. H. Thomsen, and W. M. Gelbart.** 1991. Drosophila 60A gene, another transforming growth factor β family member, is closely related to human bone morphogenetic proteins. *Proc. Natl. Acad. Sci. USA* **88**:9214–9218.

53. **Wozney, J. M., V. Rosen, A. J. Celeste, L. M. Mitsock, M. J. Whitters, R. W. Kriz, R. M. Hewick, and E. A. Wang.** 1988. Novel regulators of bone formation: molecular clones and activities. *Science* **242**:1528–1534.

Molecular Pathogenesis of Periodontal Disease
Edited by Robert Genco et al.
© 1994 American Society for Microbiology, Washington, DC 20005

Chapter 35

Extracellular Matrix and Bone Morphogenetic Proteins: Molecular Approaches to Dentin and Periodontal Repair

A. H. Reddi

The aim of this chapter is to present a concise review of recent progress in the area of bone morphogenetic proteins (BMPs) with potential applications in dentin and periodontal repair. Repair and regeneration in general recapitulate embryonic development of tissues. A systematic study of the development of the tooth and associated periodontal structures will aid in the development of rational strategies for repairing dentin and periodontium. An alternative approach includes investigation of tissues with a well-known potential for repair, such as bone, and application of the principles gleaned to the tooth and periodontium. This approach is the theme of this brief discourse.

It is well known that during embryonic development and morphogenesis, reciprocal interactions between tissues result in induction of new phenotypes with attendant morphogenesis. In addition to primary induction in embryo, secondary inductions persist into postnatal life. A classic example is bone induction by demineralized bone matrix (22, 23, 28). Bone induction consists of a sequential developmental cascade with several steps: activation and chemotaxis of peripheral blood monocytes and mesenchymal stem cell progenitors, cell attachment via fibronectin, mitosis of stem cells, differentiation of bone, mineralization, remodeling, and hematopoietic marrow formation within the nascent ossicle (22).

THREE KEY STEPS IN THE REPAIR CASCADE

Chemotaxis, mitosis, and cell differentiation are the critical rate-limiting steps in development and repair of tissues. The cellular and molecular dissection of bone morphogenesis in response to demineralized bone matrix allowed the design of

A. H. Reddi • Laboratory of Musculoskeletal Cell Biology, Department of Orthopaedic Surgery, Johns Hopkins University School of Medicine, Baltimore, Maryland 21205.

experimental strategies in the quest for purification of the bioactive ingredient in the extracellular matrix of bone.

Polymorphonuclear leukocyte chemotaxis is transient (23); mesenchymal stem cell chemotaxis is lasting (23). The presence of several mitogenic growth factors can be inferred from the brisk proliferation of cells in the implant (22). The presence of differentiation factors in demineralized bone matrix is obvious as new bone formation is induced locally. It is well known that both bone and tooth morphogenesis are dependent on systemic factors (Table 1) and local factors (Table 2). The experimental model of implanting demineralized bone matrix in extraskeletal sites provides an opportunity to isolate the morphogenetic factors in the extracellular matrix of bone.

ISOLATION, CLONING, AND EXPRESSION OF BMPs

The demineralized extracellular matrix of diaphyseal bone is insoluble. Identification and isolation of bioactive BMPs from extracellular matrix in the solid state require prior solubilization. Chaotropic agents such as 4 M guanidine hydrochloride, 8 M urea, or 1% (wt/wt) sodium dodecyl sulfate dissociatively extract 3% of the matrix proteins. First, the bioassay of the extract alone or the insoluble residue revealed that neither was osteogenic. On the other hand, a reconstitution of the extract with residue restored the biological activity (26). This key experiment demonstrated that there is an optimal synergy between a soluble signal(s) in the extract and the insoluble substratum in initiating new bone differentiation in an extraskeletal site. The osteogenic protein was isolated by heparin-affinity chromatography and found to be identical to BMP-3 cloned by the incisive work of Wozney et al. (32). At present there are at least seven known members of the BMP family (Table 3). BMPs are highly conserved members of the transforming growth factor β superfamily (7, 18, 19, 32). BMPs are pleiotropic and influence a variety of functions including but not limited to initiation and promotion of cartilage and bone formation in vivo, maintenance of articular cartilage phenotype, chondrogenesis in chick limb bud cells, chemotaxis of mesenchymal cells and monocytes, mitosis of early progenitors, inhibition of proliferation in differentiated cells, and stimulation of dental pulp cells toward an odontoblast phenotype (17a).

Table 1. Systemic factors regulating bone and tooth morphogenesis

Growth hormone
Thyroxine
Insulin
Androgens
Estrogen
Progesterone
Calcitonin
Parathyroid hormone
Vitamins A, C, and D
Minerals, calcium, phosphate

Table 2. Local factors regulating bone morphogenesis

BMPs
Acidic fibroblast growth factor
Basic fibroblast growth factor
Colony-stimulating factors
Epidermal growth factor
Insulin-like growth factors I and II
Interferons
Interleukins
Platelet-derived growth factors
Transforming growth factors α and β
Tumor necrosis factors

BMPs appear to bind to type IV collagen (21) and to cell surface binding sites in an osteoblastic cell line MC3T3-E1 (20). Cross-linking experiments revealed two proteins at 200 and 70 kDa binding to recombinant BMP-4 (20). Binding affinity by Scatchard analysis demonstrated a high-affinity receptor with an apparent dissociation constant of 128 ± 40 pM. Further studies of the receptors for BMP are needed, as are studies of the regulation of BMP expression by other systemic and local factors (Tables 1 and 2). It is likely there is an autocrine stimulation of BMP transcription and translation by other members of the BMP superfamily.

DENTIN REPAIR BY BMPs

It is well known that the demineralized tooth matrix induces new bone formation in an extraskeletal site (2, 9–12, 27). It is not clear at present whether the bioactive constituent is a member of the BMP family or a related tooth protein with the same biological response of osteogenesis. BMP-2 and BMP-4 transcripts have been localized by in situ hybridization in developing teeth (13). It is likely that dentin matrix in teeth is a repository of putative dentin morphogenetic proteins with a role in reparative dentin formation. It is also known that dental pulp cells can be modulated to express dentinogenesis. Nakashima (15–17) made the important discovery that partially purified BMP induces reparative dentin in dogs. BMP apparently induces the differentiation of osteodentinocytes. It is note-

Table 3. Human BMPs[a]

BMP	Other names	Chromosome location	Bone induction
BMP-2	BMP-2a	20	+
BMP-3	Osteogenin	4	+
BMP-4	BMP-2b	14	+
BMP-5		6	+
BMP-6		6	+
Osteogenic protein-1	BMP-7	20	+
Osteogenic protein-2	BMP-8	?	?

[a] BMP-1 is not a member of the TGF-β superfamily

worthy that very recently, these observations have been extended to monkeys with the use of recombinant BMP-7 (OP-1) (25) in conjunction with a collagenous matrix (3, 26).

POTENTIAL ROLE OF BMPs IN PERIODONTAL REPAIR

The repair and ultimate regeneration of the periodontal attachment apparatus in periodontitis are challenging problems. The optimal restoration of bone, cementum, and periodontal ligament constitutes the complete functional repair of the periodontium (5). Although the BMPs induce copious amounts of new bone formation in the periodontium, one must avoid ankylosis of teeth (4, 29, 30).

The principle of guided tissue regeneration in periodontal therapy is based on deploying a mechanical barrier to epithelium while selectively permitting bone and periodontal ligament cells to breach the barrier (1, 4–6, 8, 31). The polytetrafluorethylene membrane blocks epithelial migration and promotes new cementum and connective tissue attachment (8). Conceptually, it will be invaluable to use a timed biodegradable barrier membrane to avoid a second operative procedure to remove the membrane. Such biodegradable membranes are currently under development, and it is likely that one can enrich these membranes with BMPs to inhibit epithelial ingrowth, initiate rapid new bone and cementum formation, and optimize periodontal ligament attachment in functional terms. The critical requirements for the geometry of the substratum and chemical composition of the collagenous substratum are well established for bone induction by BMPs (14, 24).

CHALLENGES AND OPPORTUNITIES FOR THE FUTURE

The rational and optimal use of BMPs with principles of guided tissue regeneration may usher us into the age of a totally synthetic biodegradable membrane on which recombinant human BMPs are immobilized. The goal of most tissue engineering scientists is to combine recent advances in biotechnology and biomaterials for the benefit of patients in dentistry and medicine. Thus, predictable periodontal regeneration and dentin repair may be a realistic goal as we enter the 21st century.

Acknowledgments. I thank Brenda Ludgood for excellent assistance in the preparation of this chapter.

I thank the National Institutes of Health for grant support.

REFERENCES

1. **Aukhil, I., E. Petterson, and G. Suggs.** 1986. Guided tissue regeneration. An experimental procedure in beagle dogs. *J. Periodontol.* **57:**727–734.
2. **Bang, G., and M. R. Urist.** 1967. Bone induction in excavation chambers in matrix of decalcified dentin. *Arch Surg.* **94:**781–789.
3. **Bimstein, E., and S. Shoshan.** 1984. Enhanced healing of tooth pulp wounds in the dog by enriched collagen solution as a capping agent. *Arch. Oral Biol.* **26:**97–101.

4. **Bowers, G. M., F. Felton, C. Middleton, D. Glynn, S. Sharp, J. Mellanig, R. Corio, J. Emerson, S. Park, J. Suzuki, S. Ma, E. Romberg, and A. H. Reddi.** 1991. Histologic comparison of regeneration in human intrabony defects when osteogenin is combined with demineralized freeze-dried bone allograft and with purified bovine collagen. *J. Periodontol.* **62:**690–702.

5. **Bowers, G. M., and A. H. Reddi.** 1991. Regenerating the periodontium in advanced periodontal disease. *J. Am. Dent. Assoc.* **122:**45–48.

6. **Caffesse, R. G., C. E. Nasjleti, G. B. Anderson, D. D. Lopatin, B. A. Smith, and E. C. Morrison.** 1991. Periodontal healing following guided tissue regeneration with citric acid and fibronectin application. *J. Periodontol.* **62:**21–29.

7. **Celeste, A. J., R. Taylor, N. Yamaji, J. Wang, J. Ross, and J. Wozney.** 1993. Molecular cloning of BMP-8: a protein present in bovine bone which is highly related to the BMP-5/6/7 subfamily of osteoinductive molecules. *J. Cell. Biochem.* **16F:**100.

8. **Gottlow, J., S. Nyman, J. Lindhe, T. Karring, and J. Wennstrom.** 1986. New attachment formation in the human periodontium by guided tissue regeneration. *J. Clin. Periodontol.* **13:**604–616.

9. **Huggins, C., S. Wiseman, and A. H. Reddi.** 1970. Transformation of fibroblasts by allogeneic and xenogeneic transplants of demineralized tooth and bone. *J. Exp. Med.* **132:**1250–1258.

10. **Inoue, T., D. A. Deporter, and A. H. Melcher.** 1986. Induction of chondrogenesis in muscle, skin, bone marrow and periodontal ligament by demineralized dentin and bone matrix in vivo and in vitro. *J. Dent. Res.* **65:**12–22.

11. **Katz, R. W., and A. H. Reddi.** 1988. Dissociative extraction and partial purification of osteogenin, a bone inductive protein from rat tooth matrix by heparin affinity chromatography. *Biochem. Biophys. Res. Commun.* **157:**1253–1257.

12. **Kawai, T., and M. Urist.** 1989. Bovine tooth-derived bone morphogenetic protein. *J. Dent. Res.* **68:**1069–1072.

13. **Lyons, K. M., R. W. Pelton, and B. L. M. Hogan.** 1990. Organogenesis and pattern formation in the mouse; RNA distribution patterns suggest a role for bone morphogenetic protein 2-A (BMP-2A). *Development* **109:**833–844.

14. **Ma, S., G. Chen, and A. H. Reddi.** 1990. Collaboration between collagenous matrix and osteogenin is required bone induction. *Ann. N.Y. Acad. Sci.* **580:**524–525.

15. **Nakashima, M.** 1989. Dentin induction by implants of autolyzed antigen extracted allogeneic dentin on amputated pulps of dogs. *Endodont. Dent. Traumatol.* **5:**279–286.

16. **Nakashima, M.** 1990. An ultrastructural study of the differentiation of mesenchymal cells in implants of allogeneic dentine matrix on the amputated dental pulp of the dog. *Arch. Oral Biol.* **35:**277–281.

17. **Nakashima, M.** 1990. The induction of reparative dentine in the amputated dental pulp of the dog by bone morphogenetic protein. *Arch. Oral Biol.* **35:**493–497.

17a. **Nakashima, M., and A. H. Reddi.** Unpublished observations.

18. **Ozkaynak, E., D. C. Rueger, E. A. Drier, C. Corbett, R. J. Ridge, T. K. Sampath, and H. Opperman.** 1992. Op-1 cDNA encodes an osteogenic protein in the TGF-β family. *EMBO J.* **9:**2085–2093.

19. **Ozkaynak, E., P. N. J. Schnegelsberg, D. F. Jin, G. M. Clifford, I. Warren, E. A. Drier, and H. Opperman.** 1992. Osteogenic protein-2. A new member of the TGF-β superfamily expressed early in embryogenesis. *J. Biol. Chem.* **267:**25220–25227.

20. **Paralkar, V. M., R. G. Hammonds, and A. H. Reddi.** 1991. Identification and characterization of cellular binding proteins (receptors) for recombinant human bone morphogenetic protein. *Proc. Natl. Acad. Sci. USA* **88:**3397–3401.

21. **Paralkar, V. M., A. K. N. Nandedkar, R. H. Pointer, H. K. Kleinman, and A. H. Reddi.** 1990. Interaction of osteogenin, a heparin binding bone morphogenetic protein with type IV collagen. *J. Biol. Chem.* **265:**17281–17284.

22. **Reddi, A. H.** 1981. Cell biology and biochemistry of endochondral bone development. *Collagen Relat. Res.* **1:**209–266.

23. **Reddi, A. H., and C. B. Huggins.** 1972. Biochemical sequences in the transformation of normal fibroblasts in adolescent rat. *Proc. Natl. Acad. Sci. USA* **69:**1601–1605.

24. **Ripamonti, U., S. Ma, and A. H. Reddi.** 1992. The critical role of geometry of porous hydroxyapatite delivery system in induction of bone by osteogenin, a bone morphogenetic protein. *Matrix* **12:**202–212.

25. **Rutherford, R. B., J. Wahle, M. Tucker, D. Rueger, and M. Charette.** 1993. Induction of reparative dentine formation in monkeys by recombinant human osteogenic protein-1. *Arch. Oral Biol.* **38:**571–576.
26. **Sampath, T. K., and A. H. Reddi.** 1981. Dissociative extraction and reconstitution of extracellular matrix components involved in local bone differentiation. *Proc. Natl. Acad. Sci. USA* **78:**7597–7603.
27. **Somerman, M. J., M. A. Nathanson, J. J. Sauk, and B. Manson.** 1987. Human dentin matrix induces cartilage formation *in vitro* by mesenchymal cells derived from embryonic muscle. *J. Dent. Res.* **66:**1551–1558.
28. **Urist, M. R.** 1965. Bone: formation by autoinduction. *Science* **150:**893–899.
29. **Urist, M. R., R. J. Delange, and G. A. Finerman.** 1983. Bone cell differentiation and growth factors. *Science* **220:**680–686.
30. **Wang, E. A., V. Rosen, J. S. D'Alessandro, M. Baudy, P. Cordes, T. Harada, D. F. Israel, R. M. Hewick, K. M. Kerns, P. Lapan, G. P. Luxemberg, D. McQuaid, I. K. Moutsatos, and J. M. Wozney.** 1990. Recombinant human bone morphogenetic protein induces bone formation. *Proc. Natl. Acad. Sci. USA* **87:**2220–2224.
31. **Warrer, K., and T. Karring.** 1992. Guided tissue regeneration combined with osseous grafting in suprabony periodontal lesions. *J. Clin. Periodontol.* **19:**373–380.
32. **Wozney, J. M., V. Rosen, A. J. Celeste, C. L. M. Mitsock, M. J. Whilters, R. W. Kyiz, R. M. Hewick, and E. A. Wang.** 1988. Molecular clones and activities. *Science* **242:**1528–1534.

Molecular Pathogenesis of Periodontal Disease
Edited by Robert Genco et al.
© 1994 American Society for Microbiology, Washington, DC 20005

Summary of Chapters 31 to 35

Sebastian G. Ciancio

Regeneration of the periodontium is a major goal of periodontology. For regeneration to occur, periodontal ligament (PDL)-, bone-, and cementum-forming cells must migrate coronally, proliferate, and synthesize new ligament, bone, and cementum matrix.

A variety of clinical treatment modalities are available for restoring the periodontal attachment apparatus, with the most predictable results available from guided tissue regeneration and bone grafting procedures. In reviewing this area in chapter 31, Gottlow explains that in guided tissue regeneration therapy, a resorbable or nonresorbable barrier is placed to prevent the soft tissue flap from making contact with the root during healing. At the same time, a space is formed between the barrier and the root that allows the natural regeneration of cementum, ligament, and bone.

Bone grafting materials or bone substitutes are frequently used in regenerative therapy. New attachment and bone fill of intrabony defects have been shown histologically in humans following the use of demineralized freeze-dried bone allografts. However, the actual influence of graft materials on the wound-healing process is still not clear.

Root surface demineralization with citric acid and tetracyclines has promoted new connective tissue attachment in animals. Studies in humans, however, have failed to detect clinically significant differences between acid treated and non-acid-treated sites.

Growth factors, extracellular matrix proteins, and other biochemical agents may enhance or replace current regenerative procedures and offer promise for future therapy.

At the cellular level, a series of ordered cellular activities are necessary for successful periodontal regeneration. Migration, proliferation of PDL fibroblastic cells, and synthesis of matrix components by the cells are required in the early phase of healing, while cell differentiation into cementoblasts and osteoblasts is needed for cementogenesis and osteogenesis in the later phase of healing. Several

Sebastian G. Ciancio • Department of Periodontology, State University of New York at Buffalo, 250 Squire Hall, Buffalo, New York 14214.

lines of investigation now indicate that certain growth factors such as platelet-derived growth factor (PDGF) and insulin-like growth factor I (IGF-I) are potent stimulators of cellular migration, proliferation, and matrix synthesis and thus may promote periodontal regeneration.

Data from in vitro studies by Cho and coworkers are given in chapter 32. They show that PDGF (particularly PDGF-BB) and IGF-I have potent mitogenic and chemotactic effects on PDL fibroblastic cells, and their combination further increases the mitogenic potency through a synergistic effect. In contrast, epidermal growth factor (EGF) and transforming growth factor β induce slight and inhibitory mitogenic effects, respectively. PDGF enhances collagen synthesis by the cells. Cho et al. also present data from studies in which maxillary molars with demineralized root surfaces were reimplanted into rats, revealing that treatment of these roots with IGF-I appears to enhance differentiation of PDL cells into cementoblasts and/or osteoblasts.

They also report that culture of PDL fibroblastic cells in the presence of dexamethasone increases alkaline phosphatase activity but decreases the number of EGF binding sites, synthesis of EGF-receptor protein, and expression of mRNA for EGF receptor. In contrast, EGF treatment of cells reverses the results. These observations suggest that PDGF-BB and IGF-I are potent stimulators of the proliferation and chemotaxis of PDL fibroblastic cells. In addition, IGF-I has the ability to induce differentiation of the cells into mineralized tissue-forming cells, while EGF stabilizes the PDL fibroblast phenotype by suppressing their differentiation.

In chapter 33, Lynch and Giannobile present results of studies of periodontal regeneration in dogs and primates in which, following treatment of lesions with a combination of PDGF and IGF-I, significant regeneration of alveolar bone occurred. They show that these growth factors could be used with implants to improve osseointegration in the dog and in the primate model.

Rutherford et al., in chapter 34, and Reddi, in chapter 35, present extensive data on bone morphogenetic proteins (BMPs), sometimes with contrasting views relative to BMP-7 and its role in osteogenesis. Rutherford et al. review the nomenclature of BMPs, including the renaming of VgR-1 as BMP-6, OP-1 as BMP-7, and OP-2 as BMP-8. BMP-7 gene activity has been detected in developing as well as adult mammalian tissues, including teeth, bone, brain, and kidneys, suggesting a role for this gene both during and after development. BMP-7 and BMP-2, which have been purified from adult bovine bone, are major components of the BMP activity of demineralized bone powder and are osteoinductive in vivo. Because of their putative roles in the development and their demonstrated capacities to induce ectopic bone formation in adult animals, the BMPs may be considered morphogens or differentiation factors representing a class of proteins distinct from growth factors. Growth factors such as PDGF, which stimulate cell proliferation, are found in adult bone and are associated with tissue repair but are not osteoinductive.

Reddi states that the type of bone formation is dependent on the microenvironment. Implantation of recombinant BMPs in muscle induces endochondral bone formation in vivo, while their implantation in the alveolar bone induces predominantly membranous bone formation. The multistep sequential development cascade consists of chemotaxis, mitosis, and differentiation of cartilage and bone. BMPs stimulate osteogenic and chondrogenic phenotypes. Osteogenin (BMP-3) and recombinant BMP-4 are equipotent in chemotaxis, limb bud chondrogenesis, cartilage maintenance, and in vivo bond induction. There are multiple isoforms of BMPs, raising the question of the biological relevance of this redundancy. However, the mode of action and the second messengers are not clear. BMPs appear to have cognate receptors, as demonstrated by iodinated BMP-2b (BMP-4). BMPs are members of the transforming growth factor β superfamily and include three distinct subfamilies: BMP-2, BMP-3, and BMP-7. Native BMP-3 and recombinant BMP-4 bind type IV collagen of the basement membrane. This novel connection may be the long-elusive mechanistic explanation for the requirement of angiogenesis and vascular invasion for bone morphogenesis. In his conclusion, Reddi states that BMPs may have a role in fracture repair, periodontal surgery, and reconstructive surgery.

Index